Lecture Notes in Applied and Computational Mechanics

Volume 73

For further volumes:
http://www.springer.com/series/4623

About this Series

This series aims to report new developments in applied and computational mechanics—quickly, informally and at a high level. This includes the fields of fluid, solid and structural mechanics, dynamics and control, and related disciplines. The applied methods can be of analytical, numerical and computational nature.

Jean-François Ganghoffer · Ivaïlo Mladenov
Editors

Similarity and Symmetry Methods

Applications in Elasticity and Mechanics of Materials

Springer

Editors
Jean-François Ganghoffer
LEMTA—Université de Lorraine
Vandoeuvre
France

Ivaïlo Mladenov
Institute of Biophysics and Biomedical Engineering
Bulgarian Academy of Sciences
Sofia
Bulgaria

ISSN 1613-7736 ISSN 1860-0816 (electronic)
ISBN 978-3-319-08295-0 ISBN 978-3-319-08296-7 (eBook)
DOI 10.1007/978-3-319-08296-7
Springer Cham Heidelberg New York Dordrecht London

Library of Congress Control Number: 2014944340

Printed on acid-free paper

Springer is part of Springer Science+Business Media (www.springer.com)

Preface

On June 06–09, 2013, the EUROMECH Workshop *Similarity, Symmetry and Group Theoretical Methods in Mechanics* took place in Varna, Bulgaria.

It brought together many scientists from European countries and USA, and focused on the current state of the art in the field of similarity methods in mechanics. The aim of this Workshop was to bring together researchers who apply similarity and symmetry analysis to theoretical and engineering problems in both solid and fluid mechanics, researchers who are developing significant extensions of these methods so that they can be applied more widely, and numerical analysts who develop and use such methods in numerical schemes.

The scientific program of the Workshop was built around main speakers who gave an overview of the field in the form of short lecture courses delivered by

Nail H. Ibragimov—Group Analysis as a Microscope of Mathematical Modeling,

George Bluman—Some Recent Developments in Finding Systematically Conservation Laws and Nonlocal Symmetries for Partial Differential Equations, and

Charles-Michel Marle—Symmetries of Hamiltonian Dynamical Systems, Momentum Maps and Reduction.

The two organizers are deeply grateful to EUROMECH for the provided support making possible the first in this new series of scientific meetings. This Springer volume contains lecture notes written by the principal speakers of the Workshop which are complemented by a few shorter contributions dealing with specific problems.

The Editors hope very much that this volume gives a modern overview of the similarity and symmetry methods and shows applications of this active field of research in mechanics and will serve as a reference in the years to come.

Nancy, April 2014 Jean-François Ganghoffer
Sofia Ivaïlo Mladenov

Contents

Some Recent Developments in Finding Systematically Conservation Laws and Nonlocal Symmetries for Partial Differential Equations

George Bluman and Zhengzheng Yang

Abstract This chapter presents recent developments in finding systematically conservation laws and nonlocal symmetries for partial differential equations. There is a review of local symmetries, including Lie's algorithm to find local symmetries in evolutionary form and their applications. The Direct Method for finding local conservation laws is reviewed and its relationship to and extension of Noether's theorem are discussed. Moreover, it is shown how symmetries, including discrete symmetries may yield additional conservation laws from known conservation laws. Systematic procedures are presented to seek nonlocally related PDE systems for a given PDE system with two independent variables. In particular, these procedures include the use of conservation laws, point symmetries, and subsystems (including subsystems arising after appropriate invertible transformations of variables) to obtain trees of equivalent nonlocally related PDE systems. In turn, it is shown how the calculation of point symmetries of such nonlocally related systems leads to the discovery of nonlocal symmetries for a given PDE system. The situation of systematically constructing useful nonlocally related systems in multidimensions is considered. Many illustrative examples are provided.

1 Introduction

This chapter is concerned with recent developments in finding conservation laws (CLs) and nonlocal symmetries for partial differential equations (PDEs). It focuses on recent research of the authors and some of the first author's collaborators, including Stephen Anco, Alexei Cheviakov, Temuer Chaolu, Jean-François Ganghoffer, Nataliya

G. Bluman (✉) · Z. Yang
Department of Mathematics, The University of British Columbia,
Vancouver V6T 1Z2, Canada
e-mail: bluman@math.ubc.ca

J.-F. Ganghoffer and I. Mladenov (eds.), *Similarity and Symmetry Methods*, Lecture Notes in Applied and Computational Mechanics 73, DOI: 10.1007/978-3-319-08296-7_1,

Ivanova, Sukeyuki Kumei, Ian Lisle, Alex Ma, Greg Reid, Vladimir Shtelen and Thomas Wolf. Much of the material in this chapter appears in more detail in [1, 2].

In the latter part of the 19th century, Sophus Lie initiated his studies on continuous groups of transformations (Lie groups of transformations) in order to put order to, and thereby extend systematically, the hodgepodge of heuristic techniques for solving ordinary differential equations (ODEs). In particular, Lie showed the following.

- The problem of finding a Lie group of point transformations leaving invariant a differential equation (*point symmetry* of a differential equation) is systematic and reduces to solving a related linear system of determining equations for the coefficients (infinitesimals) of its *infinitesimal generator.*
- A point symmetry of an ODE leads to reducing systematically the order of an ODE (irrespective of any imposed initial conditions).
- A point symmetry of a PDE leads to finding systematically special solutions called *invariant* (*similarity*) *solutions*.
- A point symmetry of a differential equation generates a one-parameter family of solutions from any known solution of the differential equation that is not an invariant solution.

However there were limitations to the applicability of Lie's work.

- There were a restricted number of applications for point symmetries, especially for PDE systems.
- Few differential equations have point symmetries.
- For PDE systems having point symmetries, the invariant solutions arising from point symmetries normally yield only a small submanifold of the solution manifold of the PDE system and hence few posed boundary value problems can be solved.
- There was the computational difficulty of finding point symmetries.

Since the end of the 19th century there have been significant extensions of Lie's work on symmetries of PDEs to extend its range of applicability.

- Further applications of point symmetries have been found to include linearizations, other mappings and solutions of boundary value problems. In particular, knowledge of the point symmetries of a nonlinear PDE system (contact symmetries in the case of a scalar PDE), allows one to determine whether the system can be mapped invertibly to a linear system and yields a procedure to find such a mapping when one exists [2–4]. Knowledge of the point symmetries of a linear PDE system with variable coefficients allows one to determine whether the system can be mapped invertibly to a linear system with constant coefficients and yields a procedure to find such a mapping when one exists [2, 3].
- Extensions of the spaces of symmetries of a given PDE system to include local symmetries (higher-order symmetries) as well as nonlocal symmetries [2, 5–8].
- Extension of the applications of symmetries to include variational symmetries that yield conservation laws for variational systems [2, 8].

- Extension of variational symmetries to more general multipliers and resulting conservation laws for essentially any given PDE system [2, 8–11].
- The discovery of further solutions that arise from the extension of Lie's method to the "nonclassical method" as well as other generalizations [2, 12, 13].
- The development of symbolic computation software to solve efficiently the (overdetermined) linear system of symmetry and/or multiplier determining equations as well as related calculations for solving the nonlinear systems of determining equations arising when one uses the nonclassical method [14–18].

1.1 What is a Symmetry of a PDE System and How to Find One?

A symmetry (discrete or continuous) of a PDE system is any transformation of its solution manifold into itself, i.e., a symmetry transforms (maps) any solution of a PDE system to a solution of the same system. In particular, continuous symmetries of a PDE system are continuous deformations of its solutions to solutions of the same PDE system. Hence continuous symmetries are defined topologically and not restricted to just point or local symmetries. Thus, in principle, any nontrivial PDE has symmetries. The problem is to find and use the symmetries of a given PDE system. Practically, to find symmetries of a given PDE system, one considers transformations, acting locally on the variables of some finite-dimensional space, which leave invariant the solution manifold of the PDE system and its differential consequences. However, these variables do not have to be restricted to just the independent and dependent variables of the given PDE system.

Higher-order symmetries (*local symmetries*) arise when the solutions of the linear determining equations for infinitesimals are allowed to depend on a finite number of derivatives of dependent variables of the PDE system.

- Infinitesimals for a point symmetry in evolutionary form allow at most linear dependence on first derivatives of dependent variables of a PDE system.
- Infinitesimals for a contact symmetry in evolutionary form (only exists for a scalar PDE) allow arbitrary dependence on at most first derivatives of the dependent variable of a scalar PDE.

In making the extension from point and contact symmetries to higher-order symmetries, it is essential to realize that the linear determining equations for local symmetries are the linearized system (*Fréchet derivative*) of the given PDE system that holds for all of its solutions. Globally, point and contact symmetries act on finite-dimensional spaces whereas higher-order symmetries act on infinite-dimensional spaces consisting of the dependent and independent variables of a given PDE system as well as all of their derivatives. Well-known integrable equations of mathematical physics such as the Korteweg-de-Vries equation have an infinite number of higher-order symmetries [19].

Another extension is to consider solutions of the determining equations where infinitesimals have an ad-hoc dependence on nonlocal variables such as integrals of the dependent variables [20–23]. For some PDEs, such *nonlocal symmetries* can be found formally through *recursion operators* that depend on inverse differentiation. Integrable equations such as the sine-Gordon and cubic Schrödinger equations have an infinite number of such nonlocal symmetries.

1.2 Conservation Laws

In her celebrated 1918 paper [5], Emmy Noether showed that if a DE system admits a variational principle, then any local transformation group leaving invariant the action integral for its Lagrangian density, i.e., a *variational symmetry*, yields a *local conservation law*. Conversely, any local CL of a variational DE system arises from a variational symmetry, and hence there is a direct correspondence between local CLs and variational symmetries (Noether's theorem).

However there are limitations in the use of Noether's theorem.

- Its application is restricted to variational systems. In particular, a given DE system, *as written*, is variational if and only if its linearized system is self-adjoint.
- One has the difficulty of finding local symmetries of the action integral. In general, not all local symmetries of a variational DE system are variational symmetries.
- The use of Noether's theorem to find local conservation laws is coordinate-dependent.

The *Direct Method* for finding CLs allows one to find local CLs more generally for a given DE system. A CL of a given DE system is a divergence expression that vanishes on all solutions of the DE system. Local CLs arise from scalar products formed by linear combinations of *local CL multipliers* (factors that are functions of independent and dependent variables and their derivatives) multiplying each DE in the system. This scalar product is annihilated by the Euler operators associated with each of its dependent variables without restricting these variables in the scalar product to solutions of the system of DEs, i.e., the dependent variables are replaced by arbitrary functions of the independent variables.

If a given DE system, *as written*, is variational, then local CL multipliers correspond to variational symmetries. In the variational situation, using the Direct Method, local CL multipliers satisfy a linear system of determining equations that includes the linearizing system of the given DE system augmented by additional determining equations that taken together correspond to the action integral being invariant under the associated variational symmetry.

More generally, in using the Direct Method for any given DE system, the local CL multipliers are the solutions of an easily found linear determining system that includes the adjoint system of the linearizing DE system [1, 2, 9–11].

For any set of local CL multipliers, usually one can directly find the fluxes and density of the corresponding local CL and, if this proves difficult, there is an integral formula that yields them without the need of a specific functional (Lagrangian) even in the case when the given DE system is variational [9–11].

One can compare the number of local symmetries and the number of local CLs of a given DE system. When a DE system is variational, i.e., its linearized system is self-adjoint, then local CLs arise from a subset of its local symmetries and the number of linearly independent local CLs cannot exceed the number of higher-order symmetries. In general, this will not be the case when a system is not variational. Here a given DE system can have more local conservation laws than local symmetries as well as vice versa.

For any given DE system, a transformation group (continuous or discrete) that leaves it invariant yields an explicit formula that maps a CL to a CL of the same system, whether or not the given system is variational. If the transformation group is a one-parameter Lie group of point (or contact) transformations, then in terms of a parameter expansion a given CL can map into more than one additional CL for the given DE system [2, 24].

1.3 Nonlocally Related Systems and Nonlocal Symmetries

Systematic procedures have been found to seek nonlocal symmetries of a given PDE system through applying Lie's algorithm to nonlocally related systems. In particular, to apply symmetry methods to PDE systems, one needs to work in some specific coordinate frame in order to perform calculations. A procedure to find symmetries that are nonlocal and yet are local in some related coordinate frame involves embedding a given PDE system in another PDE system obtained by adjoining nonlocal variables in such a way that the resulting nonlocally related PDE system is equivalent to the given system. Consequently, any local symmetry of the nonlocally related system yields a symmetry of the given system (The converse also holds). A local symmetry of the nonlocally related system, with the corresponding infinitesimals for the variables of the given PDE system having an essential dependence on nonlocal variables, yields a nonlocal symmetry of the given PDE system.

There are two known systematic ways to find such an embedding.

- Each local CL of a given PDE system yields a nonlocally related system. For each local CL, one can introduce a potential variable(s). Here the nonlocally related system is the given PDE system augmented by a corresponding potential system [2, 25–27].
- Each point symmetry of a given PDE system yields a nonlocally related system. Here, as a first step, the given PDE system naturally yields a locally related PDE system (intermediate system) arising from the canonical coordinates of the point symmetry. In turn, the intermediate system has a natural CL which yields a nonlocally related system (inverse potential system) for the given PDE system [28, 29]. The intermediate system plays the role of a potential system for the inverse potential system.

If a local symmetry of such a nonlocally related system has an essential dependence on nonlocal variables when projected to the given system, then it yields a nonlocal symmetry of the given PDE system. It turns out that many PDE systems have such systematically constructed nonlocal symmetries. Furthermore, one can find additional nonlocal symmetries of a given PDE system through seeking local symmetries of an equivalent subsystem of the given system or one of its constructed nonlocally related systems provided that such a subsystem is nonlocally related to the given PDE system.

There are many applications of nonlocally related systems.

- Invariant solutions of nonlocally related systems (arising from CLs or point symmetries) can yield further solutions of a given PDE system.
- Since a point symmetry-based or CL-based nonlocal symmetry is a local symmetry of a constructed nonlocally related system, it generates a one-parameter family of solutions from any known solution (that is not an invariant solution) of such a nonlocally related system. In turn, this yields a one-parameter family of solutions from any known solution of the given PDE system.
- Local CLs of such nonlocally related systems can yield nonlocal CLs of a given PDE system if their local CL multipliers have an essential dependence on nonlocal variables.

Still wider classes of nonlocally related systems can be constructed systematically for a given PDE system. One can further extend embeddings through the effective use of local CLs to systematically construct trees of nonlocally related but equivalent PDE systems. If a given PDE system has n local CLs, then each CL yields potentials and corresponding potential systems. From the n local CLs, one can directly construct up to $2^n - 1$ independent nonlocally related systems of PDEs by considering corresponding potential systems individually (n singlets), in pairs ($n(n-1)/2$ couplets), ..., taken all together (one n-plet). Any of these systems could lead to the discovery of new nonlocal symmetries and/or nonlocal CLs of the given PDE system or any of the other nonlocally related systems. Such nonlocal CLs could yield further nonlocally related systems, etc. Furthermore, subsystems of such nonlocally related systems could yield further nonlocally related systems. Correspondingly, a tree of nonlocally related, and equivalent, systems is constructed for a given PDE system [2, 30, 31].

The situation in the case of multidimensional PDE systems (i.e., with at least three independent variables) is especially interesting. Here one can show that nonlocal symmetries and nonlocal CLs arising from the CL-based approach cannot arise from potential systems unless they are augmented by gauge constraints [2, 32, 33].

There exist many applications of such systematically constructed nonlocally related systems that further extend the use of symmetry methods for PDE systems.

- Through such constructions, one can systematically relate Eulerian and Lagrangian coordinate descriptions of gas dynamics and nonlinear elasticity. In particular, for the Eulerian coordinate description, a subsystem of the potential system arising from conservation of mass, naturally yields the corresponding description in Lagrangian coordinates [2, 30, 31, 34, 35].
- For a given class of PDEs with constitutive functions, one finds trees of nonlocally related systems yielding symmetries and CLs with respect to various forms of its constitutive functions.

- One can systematically seek noninvertible mappings of nonlinear PDE systems to linear PDE systems. Consequently, further nonlinear PDE systems can be mapped into equivalent linear PDE systems beyond those obtained through invertible mappings [2, 27, 36].
- One can systematically extend the class of linear PDE systems with variable coefficients that can be mapped into equivalent linear PDE systems with constant coefficients through inclusion of noninvertible mappings [2, 37, 38].

The rest of this chapter is organized as follows. In Sect. 2, we review local symmetries, Lie's algorithm to find local symmetries in evolutionary form, applications of local symmetries and as examples consider the heat equation and the Kortweg-de Vries equation. In Sect. 3, we consider the construction of conservation laws, introduce the Direct Method and its relationship to Noether's theorem, and show how symmetries could yield additional CLs from known CLs. As examples, we consider nonlinear telegraph equations, the Korteweg-de Vries equation, the Klein-Gordon equation, and nonlinear wave equations. In Sect. 4, we present systematic procedures to seek nonlocally related systems and nonlocal symmetries of a given PDE system with two independent variables. We introduce conservation law and point symmetry based methods as well as the use of subsystems to obtain trees of equivalent nonlocally related PDE systems. As examples, we focus on nonlinear wave equations, nonlinear telegraph equations, planar gas dynamics equations, and nonlinear reaction diffusion equations. In Sect. 5, we consider the situation of nonlocality in multidimensions. We show that if one directly applies the CL-based method to a single CL, then it is necessary to append a gauge constraint relating potential variables of the resulting vector potential system when seeking nonlocal symmetries. Some open problems are discussed.

2 Local Symmetries

Lie's algorithm for seeking point symmetries can be extended to seek more general local symmetries admitted by PDE systems. In the extension of Lie's algorithm, one uses differential consequences of the given PDE system, i.e., invariance of a given PDE system is understood to include its differential consequences. Here it is important to consider the infinitesimal generators for point symmetries in their *evolutionary form* where the independent variables are themselves invariant and the action of a group of point transformations is strictly an action on the dependent variables of the PDE system, so that solutions are *directly mapped into other solutions* under the group action. This allows one to readily extend Lie's algorithm to seek *contact symmetries* (only existing for scalar PDEs) where now the components of infinitesimal generators for dependent variables can depend at most on the first derivatives of the dependent variable of a given scalar PDE (if this dependence is at most linear on the first derivatives, then a contact symmetry is a point symmetry).

A contact symmetry is equivalent to a point transformation acting on the space of the given independent variables, the dependent variable and its first derivatives and, through this, can be naturally extended to a point transformation acting on the space of the given independent variables, the dependent variable and its derivatives to any finite order greater than one.

Lie's algorithm can be still further extended by allowing the infinitesimal generators in evolutionary form to depend on derivatives of dependent variables to any finite order. This allows one to calculate symmetries that are called *higher-order symmetries*. In the scalar case, contact symmetries are first-order symmetries. Otherwise, higher-order symmetries are not equivalent to point transformations acting on a finite-dimensional space including the independent variables, the dependent variables and their derivatives to some finite order. Higher-order symmetries are local symmetries in the sense that the components of the dependent variables in their infinitesimal generators depend at most on a finite number of derivatives of a given PDE system's dependent variables so that their calculation only depends on the local behaviour of solutions of a given PDE system.

Local symmetries include point symmetries, contact symmetries and higher-order symmetries. Local symmetries are uniquely determined when infinitesimal generators are represented in evolutionary form.

Sophus Lie considered contact symmetries. Emmy Noether introduced the notion of higher-order symmetries in her celebrated paper on conservation laws [5]. The well-known infinite sequences of conservation laws of the Korteweg-de Vries (KdV) and sine-Gordon equations are directly related to admitted infinite sequences of local symmetries obtained through the use of recursion operators [19].

Consider a given scalar PDE of order k

$$R(x, t, u, \partial u, \ldots, \partial^k u) = 0 \tag{1}$$

with independent variables (x, t) and dependent variable $u(x, t)$; $\partial^j u$ denotes the jth order partial derivatives of $u(x, t)$ appearing in the PDE (1). In *evolutionary form*, the *local symmetries of order* p of a PDE (1), in terms of their infinitesimal generators

$$\eta(x, t, u, \partial u, \ldots, \partial^p u)\frac{\partial}{\partial u}$$

are the solutions $\eta(x, t, u, \partial u, \ldots, \partial^p u)$ of its linearized system (*Fréchet derivative*)

$$\left[\frac{\partial R}{\partial u}\eta + \frac{\partial R}{\partial u_x}\mathrm{D}_x\eta + \frac{\partial R}{\partial u_t}\mathrm{D}_t\eta + \frac{\partial^2 R}{\partial {u_x}^2}(\mathrm{D}_x)^2\eta + \cdots\right]_{\substack{R=0,\\ \mathrm{D}_x R=0,\\ \mathrm{D}_t R=0,\\ \vdots}} = 0$$

in terms of *total derivative operators*

$$\mathrm{D}_x = \frac{\partial}{\partial x} + u_x \frac{\partial}{\partial u} + u_{xx} \frac{\partial}{\partial u_x} + \cdots$$
$$\mathrm{D}_t = \frac{\partial}{\partial t} + u_t \frac{\partial}{\partial u} + u_{xt} \frac{\partial}{\partial u_x} + \cdots$$

and holding for all solutions $u = \theta(x, t)$ of the PDE (1) and its differential consequences.

A local symmetry of order p, $\eta(x, t, u, \partial u, \ldots, \partial^p u)\frac{\partial}{\partial u}$ (including its natural extension to action on derivatives), maps *any* solution $u = \theta(x, t)$ of PDE (1) (that is not an invariant solution of PDE (1)) into a one-parameter (ε) family of solutions of PDE (1) given by the expression

$$u = \left(\mathrm{e}^{\varepsilon\left(\eta \frac{\partial}{\partial u} + (\mathrm{D}_x \eta) \frac{\partial}{\partial u_x} + (\mathrm{D}_t \eta) \frac{\partial}{\partial u_t} + \cdots\right)} u \right)\Bigg|_{u=\theta(x,t)}$$

and is equivalent to the transformation

$$\begin{aligned} x^* &= x \\ t^* &= t \\ u^* &= \mathrm{e}^{\varepsilon\left(\eta \frac{\partial}{\partial u} + (\mathrm{D}_x \eta) \frac{\partial}{\partial u_x} + (\mathrm{D}_t \eta) \frac{\partial}{\partial u_t} + \cdots\right)} u \\ &= u + \varepsilon \eta(x, t, u, \partial u, \ldots, \partial^p u) + O(\varepsilon^2). \end{aligned}$$

If $p = 1$, then the first order symmetry is equivalent to the *contact symmetry*

$$\begin{aligned} x^* &= x + \varepsilon \frac{\partial \eta}{\partial u_x} + O(\varepsilon^2) \\ t^* &= t + \varepsilon \frac{\partial \eta}{\partial u_t} + O(\varepsilon^2) \\ u^* &= u + \varepsilon \left(u_x \frac{\partial \eta}{\partial u_x} + u_t \frac{\partial \eta}{\partial u_t} - \eta \right) + O(\varepsilon^2) \\ u_x^* &= u_x + \varepsilon \left(-u_x \frac{\partial \eta}{\partial u} - \frac{\partial \eta}{\partial x} \right) + O(\varepsilon^2) \\ u_t^* &= u_t + \varepsilon \left(-u_t \frac{\partial \eta}{\partial u} - \frac{\partial \eta}{\partial t} \right) + O(\varepsilon^2). \end{aligned}$$

If a first order symmetry has an infinitesimal of the form

$$\eta(x, t, u, \partial u) = \xi(x, t, u) u_x + \tau(x, t, u) u_t - \omega(x, t, u)$$

then it is equivalent to the *point symmetry*

$$x^* = x + \varepsilon\xi(x,t,u) + O(\varepsilon^2)$$
$$t^* = t + \varepsilon\tau(x,t,u) + O(\varepsilon^2)$$
$$u^* = u + \varepsilon\omega(x,t,u) + O(\varepsilon^2).$$

2.1 Example 1: The Heat Equation

The heat equation

$$R = u_t - u_{xx} = 0$$

has the point symmetries [12, 13]

$$\mathrm{X}_1 = u_x\frac{\partial}{\partial u}, \qquad \mathrm{X}_2 = u_t\frac{\partial}{\partial u}, \qquad \mathrm{X}_3 = (xu_x + 2tu_t)\frac{\partial}{\partial u}$$
$$\mathrm{X}_4 = (xtu_x + t^2u_t + [\tfrac{1}{4}x^2 + \tfrac{1}{2}t]u)\frac{\partial}{\partial u}$$
$$\mathrm{X}_5 = (tu_x + \tfrac{1}{2}xu)\frac{\partial}{\partial u}, \quad \mathrm{X}_6 = u\frac{\partial}{\partial u}.$$

2.2 Example 2: The Korteweg-de Vries Equation

The Korteweg-de Vries (KdV) equation

$$R = u_t + uu_x + u_{xxx} = 0$$

has an infinite sequence of higher-order symmetries given by

$$(\mathbf{R}^n)u_x, \qquad n = 0, 1, 2, \ldots$$

in terms of the recursion operator [19]

$$\mathbf{R} = (\mathrm{D}_x)^2 + \tfrac{2}{3}u + \tfrac{1}{3}u_x(\mathrm{D}_x)^{-1}.$$

Specifically, one obtains corresponding nonlocal symmetries

$$u_x\frac{\partial}{\partial u}, \quad (uu_x + u_{xxx})\frac{\partial}{\partial u}$$
$$(\tfrac{5}{6}u^2u_x + 4u_xu_{xx} + \tfrac{5}{3}uu_{xxx} + u_{xxxxx})\frac{\partial}{\partial u}, \ldots.$$

For a given PDE system, local symmetries can be used to determine

- specific invariant solutions.
- a one-parameter family of solutions from "any" known solution.
- whether it can be linearized by an invertible transformation and find the linearization when it exists [3, 4, 21].
- whether an inverse scattering transform exists.
- whether a given linear PDE with variable coefficients can be invertibly mapped into a linear PDE with constant coefficients and find such a mapping when it exists [39, 40].

3 Construction of Conservation Laws

In this section, we consider the problem of finding the *local conservation laws* for a given PDE system. In particular, we present the Direct Method for the construction of CLs. In the Direct Method one first derives the determining equations yielding the multipliers (*local CL multipliers*). Following this, one finds the fluxes and densities of corresponding local CLs. It is shown that a subset of the determining equations for local CL multipliers includes the adjoint equations of the determining equations yielding the local symmetries (in evolutionary form) of a given PDE system. The self-adjoint case is especially interesting since here the given PDE system is variational and thus the local CL multipliers are also local symmetries (the converse is false) of the given PDE system. A comparison is made with the classical Noether theorem. Further connections between symmetries and CLs are presented. In particular, it is shown how a symmetry of a PDE system maps a known CL to a CL of the same PDE system. In the case of a local symmetry it is shown that a parameter expansion could yield more than one new CL from a known CL.

3.1 Uses of Conservation Laws

Conservation laws can yield constants of motion for any posed boundary value problem for a given PDE system. For this reason, for global convergence of an approximation scheme, it is important to preserve CLs, at least those CLs considered to be of importance for a particular posed boundary value problem.

From knowledge of the local CL multipliers for a given nonlinear PDE system, one can determine whether it can be mapped invertibly to a linear PDE system and set up the equations to find such a mapping when one exists [2].

In Sect. 4, it will be shown how one can use local CLs to find nonlocally related systems for a given PDE system. In turn, invariant solutions arising from local symmetries of such a nonlocally related system could yield further solutions of the given PDE system beyond those obtained as invariant solutions arising from local symmetry reductions. Moreover, the computation of local CLs of a nonlocally related system could yield nonlocal CLs of a given PDE system and to noninvertible linearizations of nonlinear PDE systems.

3.2 Direct Method for Construction of Conservation Laws

Consider a given system $\mathbf{R}\{x; u\}$ of N PDEs of order k with n independent variables $x = (x^1, \ldots, x^n)$ and m dependent variables $u(x) = (u^1(x), \ldots, u^m(x))$

$$R^\sigma[u] = R^\sigma(x, u, \partial u, \ldots, \partial^k u) = 0, \quad \sigma = 1, \ldots, N. \tag{2}$$

A *local conservation law* of the PDE system (2) is an expression

$$\mathrm{D}_i \Phi^i[u] = \mathrm{D}_1 \Phi^1[u] + \cdots + \mathrm{D}_n \Phi^n[u] = 0 \tag{3}$$

holding for any solution of the PDE system (2). In (3), the operators $\mathrm{D}_i, i = 1, \ldots, n$ are total derivative operators.

Definition 1 A PDE system $\mathbf{R}\{x; u\}$ (2) is *totally non-degenerate* if (2) and its differential consequences have maximal rank and are locally solvable.

The proof of the following theorem appears in [11].

Theorem 1 *Suppose* $\mathbf{R}\{x; u\}$ *(2) is a totally non-degenerate PDE system. Then for every nontrivial local conservation law*

$$\mathrm{D}_i \Phi^i[u] = \mathrm{D}_i \Phi^i(x, u, \partial u, \ldots, \partial^r u) = 0$$

of (2), there exists a set of multipliers, called local conservation law multipliers,

$$\Lambda_\sigma[U] = \Lambda_\sigma(x, U, \partial U, \ldots, \partial^l U), \qquad \sigma = 1, \ldots, N$$

such that

$$\mathrm{D}_i \Phi^i[U] \equiv \Lambda_\sigma[U] R^\sigma[U]$$

holds for arbitrary $U(x)$.

Definition 2 The *Euler operator* with respect to U^j is the operator

$$\mathrm{E}_{U^j} = \frac{\partial}{\partial U^j} - \mathrm{D}_i \frac{\partial}{\partial U^j_i} + \cdots + (-1)^s \mathrm{D}_{i_1} \cdots \mathrm{D}_{i_s} \frac{\partial}{\partial U^j_{i_1 \cdots i_s}} + \cdots .$$

The proofs of the following two theorems follow from direct computations.

Theorem 2 *For* any divergence expression $\mathrm{D}_i \Phi^i[U]$, *one has*

$$\mathrm{E}_{U^j}(\mathrm{D}_i\Phi^i[U]) \equiv 0, \qquad j = 1, \ldots, m.$$

Theorem 3 *Let $F[U] = F(x, U, \partial U, \ldots, \partial^s U)$. Then*

$$\mathrm{E}_{U^j} F[U] \equiv 0, \qquad j = 1, \ldots, m$$

holds for arbitrary $U(x)$ if and only if

$$F[U] \equiv \mathrm{D}_i \Psi^i(x, U, \partial U, \ldots, \partial^{s-1} U)$$

for some set of functions $\{\Psi^i(x, U, \partial U, \ldots, \partial^{s-1} U)\}$.

The next theorem follows directly from Theorems 2 and 3.

Theorem 4 *A set of local multipliers $\{\Lambda_\sigma(x, U, \partial U, \ldots, \partial^l U)\}$ yields a divergence expression for PDE system (2) if and only if*

$$\mathrm{E}_{U^j}(\Lambda_\sigma(x, U, \partial U, \ldots, \partial^l U) R^\sigma(x, U, \partial U, \ldots, \partial^k U)) \equiv 0, \qquad j = 1, \ldots, m \tag{4}$$

holds for arbitrary $U(x)$.

3.2.1 Summary of Direct Method to Find Local CLs

The Direct Method to find local CLs for a given PDE system (2) can be summarized as follows. Further details can be found in [2, 10, 11].

1. Seek multipliers of the form $\Lambda_\sigma[U] = \Lambda_\sigma(x, U, \partial U, \ldots, \partial^l U)$ with derivatives $\partial^l U$ to some specified order l.
2. Obtain and solve the determining Eq. (4) to find the multipliers of local conservation laws.
3. For each set of multipliers, find the corresponding fluxes $\Phi^i[U] = \Phi^i(x, U, \partial U, \ldots, \partial^r U)$ satisfying the identity

$$\Lambda_\sigma[U] R^\sigma[U] \equiv \mathrm{D}_i \Phi^i[U]. \tag{5}$$

4. Consequently, one obtains the local CL

$$\mathrm{D}_i\Phi^i[u] = \mathrm{D}_i\Phi^i(x, u, \partial u, \ldots, \partial^r u) = 0$$

with fluxes $\Phi^i[u]$ holding for any solution of the PDE system (2).

The fluxes $\Phi^i[U] = \Phi^i(x, U, \partial U, \ldots, \partial^r U)$ in (5) can be found in the following ways:

- Directly manipulate the left-hand side of (5) to obtain the right-hand side divergence form.

- Treat the fluxes as unknowns in expression (5). Expand the right-hand side to set up a linear set of PDEs for the fluxes. Solve this linear set of PDEs.
- If one is unable to perform either of the first two ways successfully, then one can formally obtain the fluxes through use of an integral (homotopy) formula that appears in [11].

Example 1 Nonlinear Telegraph System

Consider the nonlinear telegraph system

$$\begin{aligned} R_1[u,v] &= v_t - (u^2+1)u_x - u = 0 \\ R_2[u,v] &= u_t - v_x = 0. \end{aligned} \tag{6}$$

We seek local CL multipliers of the form

$$\Lambda_1 = \xi[U,V] = \xi(x,t,U,V), \qquad \Lambda_2 = \varphi[U,V] = \varphi(x,t,U,V) \tag{7}$$

for the nonlinear telegraph system (6). In terms of the Euler operators

$$\mathrm{E}_U = \frac{\partial}{\partial U} - \mathrm{D}_x \frac{\partial}{\partial U_x} - \mathrm{D}_t \frac{\partial}{\partial U_t}, \quad \mathrm{E}_V = \frac{\partial}{\partial V} - \mathrm{D}_x \frac{\partial}{\partial V_x} - \mathrm{D}_t \frac{\partial}{\partial V_t}$$

the multipliers (7) yield a local CL of the nonlinear telegraph system (6) if and only if the determining equations

$$\begin{aligned} \mathrm{E}_U(\xi[U,V]R_1[U,V] + \varphi[U,V]R_2[U,V]) &\equiv 0 \\ \mathrm{E}_V(\xi[U,V]R_1[U,V] + \varphi[U,V]R_2[U,V]) &\equiv 0 \end{aligned} \tag{8}$$

hold for arbitrary differentiable functions $U(x,t)$, $V(x,t)$. It is straightforward to show that the Eq. (8) hold if and only if

$$\begin{aligned} \varphi_V - \xi_U &= 0 \\ \varphi_U - (U^2+1)\xi_V &= 0 \\ \varphi_x - \xi_t - U\xi_V &= 0 \\ (U^2+1)\xi_x - \varphi_t - U\xi_U - \xi &= 0. \end{aligned} \tag{9}$$

The five linearly independent solutions [41] of the linear determining system (9) are given by

$$\begin{aligned} &(\xi_1,\varphi_1) = (0,1), \qquad (\xi_2,\varphi_2) = (t, x - \tfrac{1}{2}t^2), \qquad (\xi_3,\varphi_3) = (1,-t) \\ &(\xi_4,\varphi_4) = (\mathrm{e}^{x+\frac{1}{2}U^2+V}, U\mathrm{e}^{x+\frac{1}{2}U^2+V}), \quad (\xi_5,\varphi_5) = (\mathrm{e}^{x+\frac{1}{2}U^2-V}, -U\mathrm{e}^{x+\frac{1}{2}U^2-V}). \end{aligned}$$

Correspondingly, through manipulation, one obtains the following five local conservation laws [41]

$$
\begin{aligned}
\mathrm{D}_t u + \mathrm{D}_x[-v] &= 0\\
\mathrm{D}_t[(x - \tfrac{1}{2}t^2)u + tv] + \mathrm{D}_x[(\tfrac{1}{2}t^2 - x)v - t(\tfrac{1}{3}u^3 + u)] &= 0\\
\mathrm{D}_t[v - tu] + \mathrm{D}_x[tv - (\tfrac{1}{3}u^3 + u)] &= 0\\
\mathrm{D}_t[\mathrm{e}^{x+\frac{1}{2}u^2+v}] + \mathrm{D}_x[-u\mathrm{e}^{x+\frac{1}{2}u^2+v}] &= 0\\
\mathrm{D}_t[\mathrm{e}^{x+\frac{1}{2}u^2-v}] + \mathrm{D}_x[u\mathrm{e}^{x+\frac{1}{2}u^2-v}] &= 0.
\end{aligned}
$$

Example 2 KdV Equation

As a second example, consider again the KdV equation [10]

$$R[u] = u_t + uu_x + u_{xxx} = 0. \tag{10}$$

It is convenient to also write (10) as

$$u_t = g[u] = -(uu_x + u_{xxx}). \tag{11}$$

Due to the evolutionary form of the KdV equation (10), it follows that all local CL multipliers are of the form $\Lambda[U] = \Lambda(t, x, U, \partial_x U, \dots, \partial_x^l U)$, $l = 1, 2, \dots$. Then $\mathrm{E}_U(\Lambda[U](U_t + UU_x + U_{xxx})) \equiv 0$ if and only if

$$
\begin{aligned}
&- \mathrm{D}_t\Lambda - U\mathrm{D}_x\Lambda - \mathrm{D}_x^3\Lambda + (U_t + UU_x + U_{xxx})\Lambda_U\\
&- \mathrm{D}_x((U_t + UU_x + U_{xxx})\Lambda_{\partial_x U}) + \cdots\\
&+ (-1)^l\mathrm{D}_x^l((U_t + UU_x + U_{xxx})\Lambda_{\partial_x^l U}) \equiv 0.
\end{aligned}
\tag{12}
$$

Note that the linear determining Eq. (12) is of the form

$$\alpha_1 + \alpha_2 U_t + \alpha_3\partial_x U_t + \cdots + \alpha_{l+2}\partial_x^l U_t \equiv 0 \tag{13}$$

where in Eq. (13) each coefficient α_i depends at most on t, x, U and x-derivatives of U. Since $U(x, t)$ is an arbitrary function in Eq. (13), it follows that each of the terms $U_t, \partial_x U_t, \dots, \partial_x^l U_t$ must be treated as independent variables in (13). Hence $\alpha_i = 0$, $i = 1, \dots, l + 2$. Thus Eq. (13) splits into an overdetermined linear system of $l + 2$ determining equations for the local multipliers $\Lambda(t, x, U, \partial_x U, \dots, \partial_x^l U)$, given by

$$\tilde{\mathrm{D}}_t\Lambda + U\mathrm{D}_x\Lambda + \mathrm{D}_x^3\Lambda = 0 \tag{14}$$

$$\sum_{k=1}^{l}(-\mathrm{D}_x)^k \Lambda_{\partial_x^k U} = 0 \tag{15}$$

$$(1-(-1)^q)\Lambda_{\partial_x^q U} + \sum_{k=q+1}^{l} \frac{k!}{q!(k-q)!}(-\mathrm{D}_x)^{k-q}\Lambda_{\partial_x^k U} = 0, \quad q = 1, \ldots, l-1 \tag{16}$$

$$(1-(-1)^l)\Lambda_{\partial_x^l U} = 0 \tag{17}$$

where $\tilde{\mathrm{D}}_t = \frac{\partial}{\partial t} + g[U]\frac{\partial}{\partial U} + (g[U])_x \frac{\partial}{\partial U_x} + \cdots$ is the total derivative operator restricted to the KdV equation, with $g[U] = -(UU_x + U_{xxx})$.

Now we seek local CL multipliers of the form $\Lambda[U] = \Lambda(x,t,U)$. Then the determining Eqs. (15)–(17) are satisfied and the determining Eq. (14) becomes

$$\begin{aligned}(\Lambda_t + U\Lambda_x + \Lambda_{xxx}) + 3\Lambda_{xxU}U_x + 3\Lambda_{xUU}U_x^2 \\ + \Lambda_{UUU}U_x^3 + 3\Lambda_{xU}U_{xx} + 3\Lambda_{UU}U_xU_{xx} = 0.\end{aligned} \tag{18}$$

Equation (18) holds for arbitrary values of x, t, U, U_x and U_{xx}. Hence Eq. (18) splits into six equations. Their solution yields the three local CL multipliers $\Lambda_1 = 1$, $\Lambda_2 = U$, $\Lambda_3 = tU - x$. In turn, after simple manipulations, these three multipliers yield the divergence expressions

$$\begin{aligned} U_t + UU_x + U_{xxx} &\equiv \mathrm{D}_t U + \mathrm{D}_x(\tfrac{1}{2}U^2 + U_{xx}) \\ U(U_t + UU_x + U_{xxx}) &\equiv \mathrm{D}_t(\tfrac{1}{2}U^2) + \mathrm{D}_x(\tfrac{1}{3}U^3 + UU_{xx} - \tfrac{1}{2}U_x^2) \\ (tU - x)(U_t + UU_x + U_{xxx}) &\equiv \mathrm{D}_t(\tfrac{1}{2}tU^2 - xU) \\ &\quad + \mathrm{D}_x(-\tfrac{1}{2}xU^2 + tUU_{xx} - \tfrac{1}{2}tU_x^2 - xU_{xx} + U_x). \end{aligned}$$

Thus the corresponding local conservation laws for the KdV Eq. (10) are given by

$$\begin{aligned} \mathrm{D}_t u + \mathrm{D}_x(\tfrac{1}{2}u^2 + u_{xx}) &= 0 \\ \mathrm{D}_t(\tfrac{1}{2}u^2) + \mathrm{D}_x(\tfrac{1}{3}u^3 + uu_{xx} - \tfrac{1}{2}u_x^2) &= 0 \\ \mathrm{D}_t(\tfrac{1}{2}tu^2 - xu) + \mathrm{D}_x(-\tfrac{1}{2}xu^2 + tuu_{xx} - \tfrac{1}{2}tu_x^2 - xu_{xx} + u_x) &= 0. \end{aligned}$$

One can show that there is only one additional local CL multiplier of the form $\Lambda[U] = \Lambda(x,t,U,U_x,U_{xx})$, given by

$$\Lambda_4 = U_{xx} + \tfrac{1}{2}U^2.$$

Moreover, one can show that in terms of the recursion operator

$$\mathbf{R}^*[U] = \mathrm{D}_x^2 + \tfrac{1}{3}U + \tfrac{1}{3}\mathrm{D}_x^{-1} \circ U \circ \mathrm{D}_x$$

the KdV equation has an infinite sequence of local CL multipliers given by

$$\Lambda_{2n} = (\mathbf{R}^*[U])^n U, \qquad n = 1, 2, \ldots .$$

General Expression Relating Local CL Multipliers and Solutions of Adjoint Equations.

Consider a given PDE system (2). Let $R^\sigma[U] = R^\sigma(x, U, \partial U, \ldots, \partial^k U)$, $\sigma = 1, \ldots, N$, where $U(x) = (U^1(x), \ldots, U^m(x))$ is *arbitrary* and $U(x) = u(x)$ solves the PDE system (2).

In terms of m *arbitrary* functions $V(x) = (V^1(x), \ldots, V^m(x))$, the *linearizing operator* $\mathrm{L}[U]$ associated with the PDE system (2) is given by

$$\mathrm{L}^\sigma_\rho[U]V^\rho \equiv \left[\frac{\partial R^\sigma[U]}{\partial U^\rho} + \frac{\partial R^\sigma[U]}{\partial U^\rho_i}\mathrm{D}_i + \cdots + \frac{\partial R^\sigma[U]}{\partial U^\rho_{i_1\ldots i_k}}\mathrm{D}_{i_1}\cdots\mathrm{D}_{i_k}\right]V^\rho,$$
$$\sigma = 1, \ldots, N$$

and, in terms of N *arbitrary* functions $W(x) = (W_1(x), \ldots, W_N(x))$, the *adjoint operator* $\mathrm{L}^*[U]$ associated with the PDE system (2) is given by

$$\mathrm{L}^{*\sigma}_\rho[U]W_\sigma \equiv \frac{\partial R^\sigma[U]}{\partial U^\rho}W_\sigma - \mathrm{D}_i\left(\frac{\partial R^\sigma[U]}{\partial U^\rho_i}W_\sigma\right) + \cdots$$
$$+ (-1)^k\mathrm{D}_{i_1}\cdots\mathrm{D}_{i_k}\left(\frac{\partial R^\sigma[U]}{\partial U^\rho_{i_1\cdots i_k}}W_\sigma\right), \qquad \rho = 1, \ldots, m.$$

In particular, $W_\sigma \mathrm{L}^\sigma_\rho[U]V^\rho - V^\rho \mathrm{L}^{*\sigma}_\rho[U]W_\sigma$ is a divergence expression.

Let

$$W_\sigma = \Lambda_\sigma[U] = \Lambda_\sigma(x, U, \partial U, \ldots, \partial^l U), \quad \sigma = 1, \ldots, N.$$

By direct calculation, in terms of Euler operators, one can show that

$$\mathrm{E}_{U^\rho}(\Lambda_\sigma[U]R^\sigma[U]) \equiv \mathrm{L}^{*\sigma}_\rho[U]\Lambda_\sigma[U] + F_\rho(R[U]) \tag{19}$$

with

$$F_\rho(R[U]) = \frac{\partial \Lambda_\sigma[U]}{\partial U^\rho}R^\sigma[U] - \mathrm{D}_i\left(\frac{\partial \Lambda_\sigma[U]}{\partial U^\rho_i}R^\sigma[U]\right) + \cdots$$
$$+ (-1)^l\mathrm{D}_{i_1}\cdots\mathrm{D}_{i_l}\left(\frac{\partial \Lambda_\sigma[U]}{\partial U^\rho_{i_1\cdots i_l}}R^\sigma[U]\right), \qquad \rho = 1, \ldots, m. \tag{20}$$

From (19), it follows that $\{\Lambda_\sigma(x, U, \partial U, \ldots, \partial^l U)\}_{\sigma=1}^N$ yields a set of local CL multipliers for the PDE system (2) if and only if the right-hand side of (19) vanishes for arbitrary $U(x)$. Moreover, since the expressions (20) vanish on any solution $U(x) = u(x)$ of $\mathbf{R}\{x; u\}$ (2), it follows that every set of local CL multipliers $\{\Lambda_\sigma(x, U, \partial U, \ldots, \partial^l U)\}_{\sigma=1}^N$ of the PDE system (2) must be a solution of its adjoint system of PDEs, which is the adjoint of its linearizing system of PDEs, when $U(x) = u(x)$ is a solution of $\mathbf{R}\{x; u\}$ (2), i.e.,

$$\mathrm{L}^{*\sigma}_{\ \rho}[u]\Lambda_\sigma[u] = 0, \qquad \rho = 1, \ldots, m. \tag{21}$$

The proof of the following theorem follows directly from expression (19).

Theorem 5 *Consider a given PDE system (2). A set of functions* $\{\Lambda_\sigma(x, U, \partial U, \ldots, \partial^l U)\}_{\sigma=1}^N$ *yields a set of local CL multipliers for PDE system (2) if and only if the identities*

$$\begin{aligned}\mathrm{L}^{*\sigma}_{\ \rho}[U]\Lambda_\sigma[U] + \frac{\partial \Lambda_\sigma[U]}{\partial U^\rho} R^\sigma[U] - \mathrm{D}_i\left(\frac{\partial \Lambda_\sigma[U]}{\partial U^\rho_i} R^\sigma[U]\right) + \cdots \\ + (-1)^l \mathrm{D}_{i_1} \cdots \mathrm{D}_{i_l}\left(\frac{\partial \Lambda_\sigma[U]}{\partial U^\rho_{i_1 \cdots i_l}} R^\sigma[U]\right) \equiv 0, \qquad \rho = 1, \ldots, m\end{aligned}$$

hold for m arbitrary functions $U(x) = (U^1(x), \ldots, U^m(x))$ *in terms of the components* $\{\mathrm{L}^{*\sigma}_{\ \rho}[U]\}$ *of the adjoint operator of the linearizing operator (Fréchet derivative) for the given PDE system (2).*

The derivation leading to Eq. (21) can be summarized in terms of the following theorem.

Theorem 6 *Consider a given PDE system (2). Suppose one has a set of local CL multipliers* $\{\Lambda_\sigma(x, U, \partial U, \ldots, \partial^l U)\}_{\sigma=1}^N$ *for the PDE system (2). Let* $\{\mathrm{L}^{*\sigma}_{\ \rho}[U]\}$ *be the components of the adjoint operator of the linearizing operator (Fréchet derivative) for the PDE system (2) and let* $U(x) = u(x) = (u^1(x), \ldots, u^m(x))$ *be any solution of the PDE system (2). Then* $\mathrm{L}^{*\sigma}_{\ \rho}[u]\Lambda_\sigma[u] = 0$.

The Situation When the Linearizing Operator is Self-adjoint

Definition 3 Let $\mathrm{L}[U]$, with its components $\mathrm{L}^\sigma_\rho[U]$, be the linearizing operator associated with a PDE system $\mathbf{R}\{x; u\}$ (2). The adjoint operator of $\mathrm{L}[U]$ is $\mathrm{L}^*[U]$, with components $\mathrm{L}^{*\sigma}_{\ \rho}[U]$. $\mathrm{L}[U]$ is a *self-adjoint* operator if and only if $\mathrm{L}[U] \equiv \mathrm{L}^*[U]$, i.e., $\mathrm{L}^\sigma_\rho[U] \equiv \mathrm{L}^{*\sigma}_{\ \rho}[U]$, $\sigma, \rho = 1, \ldots, m$.

One can show that a given PDE system, *as written*, has a variational formulation if and only if its associated linearizing operator is self-adjoint [8, 42, 43].

If the linearizing operator associated with a given PDE system is self-adjoint, then each set of local CL multipliers yields a local symmetry of the given PDE system. In particular, one has the following theorem.

Theorem 7 *Consider a given PDE system* $\mathbf{R}\{x; u\}$ *(2) with* $N = m$*, i.e., the number of dependent variables appearing in PDE system (2) is the same as the number of equations in PDE system (2). Suppose the associated linearizing operator* $\mathrm{L}[U]$ *for PDE system (2) is self-adjoint. Let* $\{\Lambda_\sigma(x, U, \partial U, \ldots, \partial^l U)\}_{\sigma=1}^m$ *be a set of local CL multipliers for (2). Let*

$$\eta^\sigma(x, u, \partial u, \ldots, \partial^l u) = \Lambda_\sigma(x, u, \partial u, \ldots, \partial^l u), \quad \sigma = 1, \ldots, m$$

where $U(x) = u(x)$ *is any solution of the PDE system (2). Then*

$$\eta^\sigma(x, u, \partial u, \ldots, \partial^l u)\frac{\partial}{\partial u^\sigma} \tag{22}$$

is a local symmetry of $\mathbf{R}\{x; u\}$.

Proof Since the hypothesis of Theorem 6 is satisfied with $\mathrm{L}[U] = \mathrm{L}^*[U]$, from the equations of this theorem it follows that in terms of the components of the associated linearizing operator $\mathrm{L}[U]$, one has

$$\mathrm{L}_\rho^\sigma[u]\Lambda_\sigma(x, u, \partial u, \ldots, \partial^l u) = 0, \qquad \rho = 1, \ldots, m \tag{23}$$

where $u = \theta(x)$ is any solution of the given PDE system (2). But the set of Eq. (23) is the set of determining equations for a local symmetry $\Lambda_\sigma(x, u, \partial u, \ldots, \partial^l u)\frac{\partial}{\partial u^\sigma}$ of PDE system (2). Hence (22) is a local symmetry of PDE system (2). □

The converse of Theorem 7 is false. In particular, suppose $\eta^\sigma(x, u, \partial u, \ldots, \partial^l u)\frac{\partial}{\partial u^\sigma}$ is a local symmetry of a PDE system $\mathbf{R}\{x; u\}$ (2) with a self-adjoint linearizing operator $\mathrm{L}[U]$. Let $\Lambda_\sigma(x, U, \partial U, \ldots, \partial^l U) = \eta^\sigma(x, U, \partial U, \ldots, \partial^l U)$, $\sigma = 1, \ldots, m$, where $U(x) = (U^1(x), \ldots, U^m(x))$ is arbitrary. Then it does not necessarily follow that $\{\Lambda_\sigma(x, U, \partial U, \ldots, \partial^l U)\}_{\sigma=1}^m$ is a set of local CL multipliers of $\mathbf{R}\{x; u\}$. This can be seen as follows: In the self-adjoint case, the set of local symmetry determining equations is a subset of the set of local multiplier determining equations. Here *each* local symmetry yields a set of local CL multipliers if and only *each* solution of the set of local symmetry determining equations also solves the remaining set of local multiplier determining equations.

To illustrate the situation, consider the following example of a nonlinear PDE whose linearizing operator is self-adjoint but the PDE has a point symmetry that does not yield a multiplier for a local CL

$$u_{tt} - u(uu_x)_x = 0. \tag{24}$$

It is easy to see that the PDE (24) has the scaling point symmetry $x \to \alpha x, u \to \alpha u$, corresponding to the infinitesimal generator

$$\mathrm{X} = (u - xu_x)\frac{\partial}{\partial u}. \tag{25}$$

The self-adjoint linearizing operator associated with PDE (24) is given by

$$\mathrm{L}[U] = \mathrm{D}_t^2 - U^2\mathrm{D}_x^2 - 2UU_x\mathrm{D}_x - 2UU_{xx} - U_x^2.$$

The determining equation for the local CL multipliers $\Lambda(t, x, U, U_t, U_x)$ of the PDE (24) is an identity holding for all values of the variables t, x, U, U_t, U_x, U_{tt}, U_{tx}, U_{xx}, U_{ttt}, U_{ttx}, U_{txx}, U_{xxx}, and splits into a system of two equations consisting of

$$\tilde{\mathrm{D}}_t^2\Lambda - U^2\mathrm{D}_x^2\Lambda - 2UU_x\mathrm{D}_x\Lambda - (2UU_{xx} + U_x^2)\Lambda = 0 \tag{26}$$

and

$$2\Lambda_U + \tilde{\mathrm{D}}_t\Lambda_{U_t} - \mathrm{D}_x\Lambda_{U_t} = 0 \tag{27}$$

in terms of the "restricted" total derivative operator $\tilde{\mathrm{D}}_t = \frac{\partial}{\partial t} + U_t\frac{\partial}{\partial U} + U_{tx}\frac{\partial}{\partial U_x} + g[U]\frac{\partial}{\partial U_t} + U_{txx}\frac{\partial}{\partial U_{xx}} + \mathrm{D}_t(g[U])\frac{\partial}{\partial U_{tt}}$ where $g[U] = U(UU_x)_x$.

Equation (26) is the determining equation for $\Lambda(t, x, u, u_t, u_x)\frac{\partial}{\partial u}$ to be a contact symmetry of the given PDE (24). If the contact symmetry satisfies the second determining Eq. (27) then it yields a local CL multiplier $\Lambda(t, x, U, U_t, U_x)$ of PDE (24). It is easy to check that the scaling symmetry (25) obviously satisfies the contact symmetry determining Eq. (26) but does not satisfy the second determining Eq. (27) when $u(x, t)$ is replaced by an arbitrary function $U(x, t)$. Hence the scaling symmetry (25) does not yield a local conservation law of PDE (24).

3.3 Noether's Theorem

In 1918, Emmy Noether presented her celebrated procedure (*Noether's theorem*) to find local CLs for a DE system that admits a variational principle.

When a given DE system admits a variational principle, then the extremals of the associated action functional yield the given DE system (the *Euler-Lagrange equations*). In this case, Noether showed that if a one-parameter local transformation leaves invariant the action functional (action integral), then one obtains the fluxes of a local CL through an explicit formula that involves the infinitesimals of the local transformation and the Lagrangian (Lagrangian density) of the action functional.

3.3.1 Euler-Lagrange Equations

Consider a functional $J[U]$ in terms of n independent variables $x = (x^1, \ldots, x^n)$ and m arbitrary functions $U = (U^1(x), \ldots, U^m(x))$ and their partial derivatives to order k, defined on a domain Ω

$$J[U] = \int_{\Omega} L[U]dx = \int_{\Omega} L(x, U, \partial U, \ldots, \partial^k U)dx. \tag{28}$$

In (28), the function $L[U] = L(x, U, \partial U, \ldots, \partial^k U)$ is called a *Lagrangian* and the functional $J[U]$ is called an *action integral.*

Consider an infinitesimal change $U(x) \to U(x) + \varepsilon v(x)$ where $v(x)$ is any function such that $v(x)$ and its derivatives to order $k-1$ vanish on the boundary $\partial\Omega$ of the domain Ω. The corresponding infinitesimal change (variation) in the Lagrangian $L[U]$ is given by

$$\begin{aligned}\delta L &= L(x, U + \varepsilon v, \partial U + \varepsilon\partial v, \ldots, \partial^k U + \varepsilon\partial^k v) - L(x, U, \partial U, \ldots, \partial^k U)\\ &= \varepsilon\left(\frac{\partial L[U]}{\partial U^i}v^i + \frac{\partial L[U]}{\partial U^i_j}v^i_j + \cdots + \frac{\partial L[U]}{\partial U^i_{j_1\cdots j_k}}v^i_{j_1\cdots j_k}\right) + O(\varepsilon^2).\end{aligned} \tag{29}$$

Let

$$\begin{aligned}W^l[U, v] &= v^i\left(\frac{\partial L[U]}{\partial U^i_l} + \cdots + (-1)^{k-1}\mathrm{D}_{j_1}\cdots\mathrm{D}_{j_{k-1}}\frac{\partial L[U]}{\partial U^i_{lj_1\cdots j_{k-1}}}\right)\\ &\quad + v^i_{j_1}\left(\frac{\partial L[U]}{\partial U^i_{j_1 l}} + \cdots + (-1)^{k-2}\mathrm{D}_{j_2}\cdots\mathrm{D}_{j_{k-1}}\frac{\partial L[U]}{\partial U^i_{j_1 l j_2\cdots j_{k-1}}}\right)\\ &\quad + \cdots + v^i_{j_1\cdots j_{k-1}}\frac{\partial L[U]}{\partial U^i_{j_1 j_2\cdots j_{k-1} l}}.\end{aligned} \tag{30}$$

After repeatedly using integration by parts, one can show that

$$\delta L = \varepsilon(v^i\mathrm{E}_{U^i}(L[U]) + \mathrm{D}_l W^l[U, v]) + O(\varepsilon^2) \tag{31}$$

where E_{U^i} is the Euler operator with respect to U^i. The corresponding variation in the action integral $J[U]$ is given by

$$\begin{aligned}\delta J &= J[U + \varepsilon v] - J[U] = \int_{\Omega} \delta L dx\\ &= \varepsilon\int_{\Omega}(v^i\mathrm{E}_{U^i}(L[U]) + \mathrm{D}_l W^l[U, v])dx + O(\varepsilon^2)\\ &= \varepsilon\Big(\int_{\Omega} v^i\mathrm{E}_{U^i}(L[U])dx + \int_{\partial\Omega} W^l[U, v]n^l d\sigma\Big) + O(\varepsilon^2).\end{aligned} \tag{32}$$

Hence if $U(x) = u(x)$ extremizes the action integral $J[U]$, then the $O(\varepsilon)$ term in δJ must vanish. Thus $\int_{\Omega} v^i\mathrm{E}_{u^i}(L[u])\mathrm{d}x = 0$ for an *arbitrary* function $v(x)$ defined on the domain Ω. Hence, if $U(x) = u(x)$ extremizes the action integral $J[U]$, then

$u(x)$ must satisfy the PDE system

$$\mathrm{E}_{u^i}(L[u]) = \frac{\partial L[u]}{\partial u^i} + \cdots + (-1)^k \mathrm{D}_{j_1} \cdots \mathrm{D}_{j_k} \frac{\partial L[u]}{\partial u^i_{j_1 \cdots j_k}} = 0, \quad i = 1, \ldots, m. \tag{33}$$

The Eq. (33) are called the *Euler-Lagrange equations* satisfied by an extremum $U(x) = u(x)$ of the action integral $J[U]$. Thus the following theorem has been proved.

Theorem 8 *If a smooth function $U(x) = u(x)$ is an extremum of an action integral (28), then $u(x)$ satisfies the Euler-Lagrange equations (33).*

3.3.2 Standard Formulation of Noether's Theorem

Definition 4 In the *standard formulation of Noether's theorem*, the action integral (28) is invariant under the one-parameter Lie group of point transformations

$$\begin{aligned} (x^*)^i &= x^i + \varepsilon \xi^i(x, U) + O(\varepsilon^2), \qquad i = 1, \ldots, n \\ (U^*)^\mu &= U^\mu + \varepsilon \eta^\mu(x, U) + O(\varepsilon^2), \qquad \mu = 1, \ldots, m \end{aligned} \tag{34}$$

with infinitesimal generator $\mathrm{X} = \xi^i(x, U)\frac{\partial}{\partial x^i} + \eta^\mu(x, U)\frac{\partial}{\partial U^\mu}$, if and only if $\int_{\Omega^*} L[U^*]dx^* = \int_\Omega L[U]dx$ where Ω^* is the image of Ω under the Lie group of point transformations (34).

The *Jacobian* of the one parameter Lie group of point transformations (34) is given by $\mathrm{J} = \det(\mathrm{D}_i(x^*)^j) = 1 + \varepsilon \mathrm{D}_i \xi^i(x, U) + O(\varepsilon^2)$. Then $\mathrm{d}x^* = \mathrm{J}\mathrm{d}x$. Moreover, $L[U^*] = \mathrm{e}^{\varepsilon \mathrm{X}} L[U]$ in terms of the infinitesimal generator X. Consequently, in the standard formulation of Noether's theorem, X is a point symmetry of $J[U]$ if and only if

$$0 = \int_\Omega (\mathrm{J}\mathrm{e}^{\varepsilon \mathrm{X}} - 1) L[U] dx = \varepsilon \int_\Omega (L[U]\mathrm{D}_i \xi^i(x, U) + \mathrm{X}^{(k)} L[U]) dx + O(\varepsilon^2) \tag{35}$$

holds for arbitrary $U(x)$ where $\mathrm{X}^{(k)}$ is the k-th extension (prolongation) of the infinitesimal generator X. Hence, if X is a point symmetry of $J[U]$, then the $O(\varepsilon)$ term in (35) must vanish. Thus $L[U]\mathrm{D}_i \xi^i(x, U) + \mathrm{X}^{(k)} L[U] \equiv 0$.

The one-parameter Lie group of point transformations (34) with infinitesimal generator X is equivalent to the one-parameter family of transformations in evolutionary form given by

$$\begin{aligned} (x^*)^i &= x^i, \qquad i = 1, \ldots, n \\ (U^*)^\mu &= U^\mu + \varepsilon[\eta^\mu(x, U) - U^\mu_i \xi^i(x, U)] + O(\varepsilon^2), \qquad \mu = 1, \ldots, m \end{aligned} \tag{36}$$

with k-th extended infinitesimal generator $\hat{\mathrm{X}}^{(k)} = \hat{\eta}^\mu[U]\frac{\partial}{\partial U^\mu} + \cdots$. Under transformation (36), $U(x) \rightarrow U(x) + \varepsilon v(x)$ has components $v^\mu(x) = \hat{\eta}^\mu[U] = \eta^\mu(x, U) - U_i^\mu \xi^i(x, U)$. Hence $\delta L = \varepsilon \hat{\mathrm{X}}^{(k)} L[U] + O(\varepsilon^2)$. Thus

$$\int_\Omega \delta L dx = \varepsilon \int_\Omega \hat{\mathrm{X}}^{(k)} L[U] dx + O(\varepsilon^2). \tag{37}$$

Consequently, after setting $v^\mu(x) = \hat{\eta}^\mu[U] = \eta^\mu(x, U) - U_i^\mu \xi^i(x, U)$, and comparing expressions (32) and (37), it follows that

$$\hat{\mathrm{X}}^{(k)} L[U] \equiv \hat{\eta}^\mu[U] \mathrm{E}_{U^\mu}(L[U]) + \mathrm{D}_l W^l[U, \hat{\eta}[U]]. \tag{38}$$

By direct calculation, one can show the following.

Lemma 1 *Let $F[U] = F(x, U, \partial U, \ldots, \partial^k U)$ be an arbitrary function of its arguments. Then, in terms of the extended infinitesimal generators $\mathrm{X}^{(k)}$ and $\hat{\mathrm{X}}^{(k)}$, one has the identity*

$$\mathrm{X}^{(k)} F[U] + F[U] \mathrm{D}_i \xi^i(x, U) \equiv \hat{\mathrm{X}}^{(k)} F[U] + \mathrm{D}_i(F[U] \xi^i(x, U)). \tag{39}$$

Theorem 9 Standard formulation of Noether's theorem. *Suppose a given PDE system is derivable from a variational principle, i.e., the given PDE system is a set of Euler-Lagrange equations (33) whose solutions $u(x)$ are extrema $U(x) = u(x)$ of an action integral $J[U]$ with Lagrangian $L[U]$. Suppose the one-parameter Lie group of point transformations (34) with infinitesimal generator X leaves invariant $J[U]$. Then*

1. *The identity*

$$\hat{\eta}^\mu[U] \mathrm{E}_{U^\mu}(L[U]) \equiv -\mathrm{D}_i(\xi^i(x, U) L[U] + W^i[U, \hat{\eta}[U]]) \tag{40}$$

   holds for arbitrary functions $U(x)$, i.e., $\{\hat{\eta}[U]\}_{\mu=1}^m$ is a set of local CL multipliers of the Euler-Lagrange system (33).
2. *The local conservation law*

$$\mathrm{D}_i(\xi^i(x, u) L[u] + W^i[u, \hat{\eta}[u]]) = 0 \tag{41}$$

   holds for any solution $u = \theta(x)$ of the Euler-Lagrange system (33).

Proof Let $F[U] = L[U]$ in the identity in Lemma 1. Then the identity

$$\hat{\mathrm{X}}^{(k)} L[U] + \mathrm{D}_i(L[U] \xi^i(x, U)) \equiv 0 \tag{42}$$

holds for arbitrary functions $U(x)$. Substitution for $\hat{\mathrm{X}}^{(k)} L[U]$ in (42) through (38) yields the identity (40). If $U(x) = u(x)$ solves the Euler-Lagrange system (33), then the left-hand-side of equation (40) vanishes. This yields the local conservation law (41). □

3.3.3 Extended Formulation of Noether's Theorem

One can extend the standard formulation of Noether's theorem to find additional local conservation laws arising from invariance under higher-order transformations through a generalization of Definition 4 for the invariance of an action integral $J[U]$. Here the action integral $J[U]$ is invariant under a one-parameter family of higher-order transformations if its integrand $L[U]$ is invariant to within a divergence.

Definition 5 Let $\hat{\mathrm{X}} = \hat{\eta}^\mu(x, U, \partial U, \ldots, \partial^s U)\frac{\partial}{\partial U^\mu}$ be the infinitesimal generator of a one-parameter family of local transformations (36) in evolutionary form with infinite extension $\hat{\mathrm{X}}^{(\infty)}$. Let $\hat{\eta}^\mu[U] = \hat{\eta}^\mu(x, U, \partial U, \ldots, \partial^s U)$. Here $\hat{\mathrm{X}}$ is a local symmetry of $J[U]$ if and only if the identity

$$\hat{\mathrm{X}}^{(\infty)} L[U] \equiv \mathrm{D}_i A^i[U] \tag{43}$$

holds for some set of functions $A^i[U] = A^i(x, U, \partial U, \ldots, \partial^r U), i = 1, \ldots, n$.

Theorem 10 Extended formulation of Noether's theorem. *Suppose a given PDE system is derivable from a variational principle, i.e., the given PDE system is a set of Euler-Lagrange equations (33) whose solutions $u(x)$ are extrema $U(x) = u(x)$ of an action integral $J[U]$ with Lagrangian $L[U]$. Suppose $\hat{\mathrm{X}} = \hat{\eta}^\mu[U]\frac{\partial}{\partial U^\mu}$ is a local symmetry of $J[U]$. Then*

1. *The identity*

$$\hat{\eta}^\mu[U]\mathrm{E}_{U^\mu}(L[U]) \equiv \mathrm{D}_i(A^i[U] - W^i[U, \hat{\eta}[U]]) \tag{44}$$

   holds for arbitrary functions $U(x)$, i.e., $\{\hat{\eta}^\mu[U]\}_{\mu=1}^m$ is a set of local CL multipliers for the Euler-Lagrange system (33).
2. *The local conservation law*

$$\mathrm{D}_i(W^i[u, \hat{\eta}[u]] - A^i[u]) = 0 \tag{45}$$

   holds for any solution $u = \theta(x)$ of the Euler-Lagrange system (33).

Proof For the one-parameter family of local transformations (36) with infinitesimal generator $\hat{\mathrm{X}} = \hat{\eta}^\mu[U]\frac{\partial}{U^\mu}$, it follows that the corresponding infinitesimal change $U(x) \to U(x) + \varepsilon v(x)$ has components $v^\mu(x) = \hat{\eta}^\mu[U]$. Consequently, $\delta L = \varepsilon\hat{\mathrm{X}}^{(\infty)} L[U] + O(\varepsilon^2)$. But $\delta L = \varepsilon(\hat{\eta}^\mu[U]\mathrm{E}_{U^\mu}(L[U]) + \mathrm{D}_i(W^i[U, \hat{\eta}[U]])) + O(\varepsilon^2)$. Hence it immediately follows that the identity

$$\hat{\mathrm{X}}^{(\infty)} L[U] \equiv \hat{\eta}^\mu[U]\mathrm{E}_{U^\mu}(L[U]) + \mathrm{D}_i(W^i[U, \hat{\eta}[U]]) \tag{46}$$

holds for arbitrary functions $U(x)$. Since $\hat{\mathrm{X}} = \hat{\eta}^{\mu}[U]\frac{\partial}{U^{\mu}}$ is a local symmetry of $J[U]$, it follows that Eq. (43) holds. Substitution for $\hat{\mathrm{X}}^{(\infty)}L[U]$ in (46) through (43) yields the identity (44). If $U(x) = u(x)$ solves the Euler-Lagrange system (33), then the left-hand-side of Eq. (44) vanishes. This yields the local conservation law (45). □

The following theorem shows that any local conservation law obtained through the standard formulation of Noether's theorem can be obtained through the extended formulation of Noether's theorem.

Theorem 11 *If a local conservation law is obtained through the standard formulation of Noether's theorem, then this local conservation law can be obtained through its extended formulation.*

Proof Suppose the one-parameter Lie group of point transformations (34) with infinitesimal generator X yields a local CL of a given PDE system derivable from a variational principle with Euler-Lagrange system (33). Then the identity (42) holds. Consequently,

$$\hat{\mathrm{X}}^{(k)}L[U] = \hat{\mathrm{X}}^{(\infty)}L[U] = \mathrm{D}_i A^i[U] \tag{47}$$

where $A^i[U] = -\mathrm{D}_i(L[U]\xi^i(x, U))$. But Eq. (47) is just the condition for X to be a local symmetry of $J[U]$. Consequently, one obtains the same local conservation law from the extended formulation of Noether's theorem. □

3.3.4 Limitations of Noether's Theorem

There are several limitations in using Noether's theorem to find the local conservation laws of a given PDE system.

1. There is the difficulty of finding variational symmetries. To find the variational symmetries of a given DE system arising from a variational principle, first one determines the local symmetries $\mathrm{X} = \eta^{\sigma}[u]\frac{\partial}{\partial u^{\sigma}}$ of the Euler-Lagrange equations (33). Then for each local symmetry, one checks if X leaves invariant the Lagrangian $L[U]$ to within a divergence. Note that since all local conservation laws, obtainable by Noether's theorem, arise from local CL multipliers, one can simply use the Direct Method to check whether a local symmetry is a variational symmetry.
2. A given system of DEs is not variational as written. A given system of differential equations, as written, is variational if and only if its linearized system (Fréchet derivative) is self-adjoint. Consequently, it is necessary, but far from sufficient, that a given system of DEs, *as written*, must be of even order, have the same number of equations in the system as its number of dependent variables and be non-dissipative to directly admit a variational principle.
3. Artifices can make a given system of DEs variational that is not variational, as written. Such artifices include

- The use of multipliers. As an example, the PDE

$$u_{tt} + H'(u_x)u_{xx} + H(u_x) = 0 \tag{48}$$

as written, does not admit a variational principle since its linearized equation $\varsigma_{tt} + H'(u_x)\varsigma_{xx} + (H''(u_x) + H'(u_x))\varsigma_x = 0$ is not self-adjoint. However, the equivalent PDE $\mathrm{e}^x[u_{tt} + H'(u_x)u_{xx} + H(u_x)] = 0$, obtained after multiplying PDE (48) by e^x, is self-adjoint!
- The use of a contact transformation. As an example, the ODE

$$y'' + 2y' + y = 0 \tag{49}$$

as written, obviously does not admit a variational principle. But the point transformation $x \to X = x$, $y \to Y = y\mathrm{e}^x$, maps the ODE (49) to the variational ODE $Y'' = 0$. However, it is well-known that every second order ODE, written in solved form, can be mapped into $Y'' = 0$ by some contact transformation but there is no finite algorithm to find such a transformation.
- The use of a differential substitution. As an example, the KdV equation (11), as written, obviously does not admit a variational principle since it is of odd order. But the well-known differential substitution $u = v_x$ yields the equivalent transformed KdV equation $v_{xxxx} + v_x v_{xx} + v_{xt} = 0$, that is the Euler-Lagrange equation for an extremum $V(x,t) = v(x,t)$ of the action integral with Lagrangian $L[V] = \frac{1}{2}(V_{xx})^2 - \frac{1}{6}(V_x)^3 - \frac{1}{2}V_x V_t$.

4. Noether's theorem is coordinate-dependent. The use of Noether's theorem to obtain a local conservation law is coordinate-dependent since the action of a contact transformation can transform a DE having a variational principle to one that does not have one. *On the other hand it is well-known that local conservation laws are coordinate-independent in the sense that a contact transformation maps a local CL of a given DE into a local CL of the transformed DE.*
5. The artifice of a Lagrangian itself for finding the local CLs of a given DE system. One should be able to expect to directly find the local conservation laws of a given DE system without the need to find a related action integral whether or not the given DE system is variational.

3.4 Further Comments on the Direct Method to Find Local Conservation Laws vis-á-vis Noether's Theorem

The Direct Method to find local CLs addresses limitations of Noether's theorem as follows.

1. In principle, the Direct Method can be used to find local conservation laws for *any* DE system, no matter how it is written, whereas the direct application of Noether's theorem requires the linearized system of a given DE system to be self-adjoint.

Essentially, the Direct Method finds all local CLs of a given DE system. Note that Noether's theorem can only be used to find local CLs. As seen in Theorems 9 and 10, Noether's theorem is also a multiplier method.

2. In the Direct Method, no functional is required unlike the situation for Noether's theorem. Local CLs are constructed directly. In the Direct Method, local CL multipliers correspond to symmetries of a given DE system if and only if its linearization operator is self-adjoint.

Example 1 Klein-Gordon Equation

As an example to compare the use of Noether's theorem and the Direct Method to find local CLs, consider the Klein-Gordon equation

$$u_{tx} - u^n = 0, \qquad n \neq 0, 1. \tag{50}$$

The PDE (50) has the scaling point symmetry

$$x^* = \alpha^{1-n}x, \qquad t^* = t, \qquad u^* = \alpha u \tag{51}$$

with the corresponding infinitesimal generator $\mathrm{X} = (u - (1-n)xu_x)\frac{\partial}{\partial u}$. One can show that the Klein-Gordon equation (50) is variational with action functional $J[U] = \int L[U]dtdx$; $L[U] = -\frac{1}{2}U_tU_x + \frac{1}{n+1}U^{n+1}$. We now show that the point symmetry (51) of the PDE (50) does not yield a local CL of this PDE from the presented three points of view.

1. Standard formulation of Noether's theorem. Let $x^* = \alpha^{1-n}x, t^* = t, U^* = \alpha U$. Then $J[U^*] = J[\alpha U] = \int L[U^*]dt^*dx^* = \alpha^{1-n}\int L[\alpha U]dtdx$. But $L[\alpha U] = \alpha^{1+n}L[U]$. Hence $J[U^*] = \alpha^2 J[U] \neq J[U]$ for any value of $\alpha \neq 1$. Thus the point symmetry (51) of the Klein-Gordon equation (50) yields no local CL.
2. Extended formulation of Noether's theorem. Here, by direct calculation, one can show that the extended infinitesimal generator $\mathrm{X}^{(\infty)}$ of the infinitesimal generator X of the point symmetry (51) yields

$$\begin{aligned}\mathrm{X}^{(\infty)}L[U] = U^n(U - xU_x(1-n)) - \tfrac{1}{2}(U_x(U_t - xU_{xt}(1-n)) \\ + U_t(U_x - xU_{xx}(1-n))).\end{aligned} \tag{52}$$

The right-hand side of the expression (52) does not yield a divergence. The best way to show this is through applying the Euler operator with respect to U to the right-hand side of (52). In particular, $\mathrm{E}_U(\mathrm{X}^{(\infty)}L[U]) \equiv 2(U_{xt} + U^n) \neq 0$. Hence the extended formulation of Noether's theorem yields no local CL.
3. Application of the Direct Method. Here $\mathrm{E}_U[(U - xU_x(1-n))(U_{tx} - U)] \neq 0$ for an arbitrary function $U(x,t)$. Hence the point symmetry (51) of the Klein-Gordon equation (50) yields no local CL multiplier and thus no local CL.

Example 2 Nonlinear Wave Equation

Now we use the nonlinear wave equation

$$u_{tt} - (c^2(u)u_x)_x = 0 \tag{53}$$

as an example to show how the Direct Method finds the fluxes for a local CL from a known local CL multiplier. In particular, one can show that $\Lambda[U] = xt$ is a local CL multiplier for the PDE (53). Then

$$xt(U_{tt} - (c^2(U)U_x)_x) = \mathrm{D}_t(T[U]) + \mathrm{D}_x(X[U]) \tag{54}$$

for some functions $T[U] = T(x, t, U, U_x, U_t)$, $X[U] = X(x, t, U, U_x, U_t)$. Consequently, the Eq. (54) becomes

$$\begin{aligned} xt(U_{tt} - 2c(U)c'(U)U_x^2 - c^2(U)U_{xx}) = {} & T_t + T_U U_t + T_{U_t} U_{tt} + T_{U_x} U_{tx} \\ & + X_x + X_U U_x + X_{U_t} U_{tx} + X_{U_x} U_{xx}. \end{aligned} \tag{55}$$

Equating to zero the coefficients of U_{xx}, U_{tt}, U_{tx}, U_x^2, U_t, U_x, and the rest of the terms in Eq. (55) straightforwardly yields the fluxes $T[U] = xtU_t - xU$, $X[U] = -xtc^2(U)U_x + t\int c^2(U)dU$.

3.5 Use of Symmetries to Seek Further Conservation Laws from a Known Conservation Law

It is now shown how any symmetry (discrete or continuous) of a given PDE system $\mathbf{R}\{x; u\}$ (2) maps any CL of (2) into a CL of (2). Usually, no additional CL of (2) is obtained.

A symmetry of a PDE system induces a symmetry that leaves invariant the linear determining system for its local CL multipliers. Hence it follows that if one determines the action of a symmetry on a set of local CL multipliers $\{\Lambda_\sigma[U]\}$ for a known local CL of $\mathbf{R}\{x; u\}$ to obtain another set of local CL multipliers $\{\hat{\Lambda}_\sigma[U]\}$, then a priori one can determine whether an additional local CL is obtained for $\mathbf{R}\{x; u\}$.

In particular, suppose the invertible point transformation

$$x = x(\tilde{x}, \tilde{u}), \qquad u = u(\tilde{x}, \tilde{u}) \tag{56}$$

with its inverse transformation given by $\tilde{x} = \tilde{x}(x, u)$, $\tilde{u} = \tilde{u}(x, u)$, is a symmetry of a PDE system (2). Then corresponding to each PDE in (2), with solutions $u(x)$ replaced by arbitrary functions $U(x)$, and $\tilde{u}(x)$ replaced by $\tilde{U}(x)$, one has

$$R^\alpha[U] = A^\alpha_\beta[\tilde{U}]R^\beta[\tilde{U}] \tag{57}$$

holding for some set of functions $\{A^{\alpha}_{\beta}[U]\}$. Consequently, by direct calculation, one can prove the following theorem. For details, see [2, 24].

Theorem 12 *Under a point transformation (56), with $u(x)$ replaced by $U(x)$ and $\tilde{u}(x)$ replaced by $\tilde{U}(x)$, in terms of any given set of functions $\{\Phi^i[U]\}$, there exists a corresponding set of functions $\{\Psi^i[\tilde{U}]\}$ such that*

$$\mathrm{J}[\tilde{U}]\mathrm{D}_i\Phi^i[U] = \tilde{\mathrm{D}}_i\Psi^i[\tilde{U}] \tag{58}$$

where the Jacobian determinant

$$\mathrm{J}[\tilde{U}] = \frac{\mathrm{D}(x^1,\ldots,x^n)}{\mathrm{D}(\tilde{x}^1,\ldots,\tilde{x}^n)} = \begin{vmatrix} \tilde{\mathrm{D}}_1 x^1 & \cdots & \tilde{\mathrm{D}}_1 x^n \\ & \cdots & \\ \vdots \;\; \vdots & \vdots & \vdots \\ \tilde{\mathrm{D}}_n x^1 & & \tilde{\mathrm{D}}_n x^n \end{vmatrix} \tag{59}$$

and

$$\Psi^{i_1}[\tilde{U}] = \pm \begin{vmatrix} \Phi^1[U] & \Phi^2[U] & \cdots & \Phi^n[U] \\ \tilde{\mathrm{D}}_{i_2} x^1 & & \cdots & \tilde{\mathrm{D}}_{i_2} x^n \\ \vdots & \vdots & \vdots & \vdots \\ \tilde{\mathrm{D}}_{i_n} x^1 & & \cdots & \tilde{\mathrm{D}}_{i_n} x^n \end{vmatrix}. \tag{60}$$

By direct calculation, one can prove the following theorem with details appearing in [24].

Theorem 13 *Suppose the point transformation (56) is a symmetry of* $\mathbf{R}\{x; u\}$ *(2) and $\{\Lambda_\sigma[U]\}$ is a set of local CL multipliers for* $\mathbf{R}\{x; u\}$ *with fluxes $\{\Phi^i[U]\}$. Then*

$$\hat{\Lambda}_\beta[\tilde{U}]R^\beta[\tilde{U}] = \tilde{\mathrm{D}}_i\Psi^i[\tilde{U}] \tag{61}$$

where

$$\hat{\Lambda}_\beta[\tilde{U}] = \mathrm{J}[\tilde{U}]A^{\alpha}_{\beta}[\tilde{U}]\Lambda_\alpha[U], \qquad \beta = 1,\ldots,N \tag{62}$$

with the components of the derivatives in $\{\Lambda_\alpha[U]\}$ expressed in terms of the prolongation of the point transformation (56). In Eq. (61), the functions $\Psi^i[\tilde{U}]$ are yielded by determinant (60). In Eq. (62), *the functions $A^{\alpha}_{\beta}[\tilde{U}]$ are obtained through Eq. (57), and the Jacobian* $\mathrm{J}[\tilde{U}]$ *is yielded by the determinant (59).*

After replacing $\tilde{x}^i$ by x^i, $\tilde{U}^\alpha$ by U^α, etc., in Eq. (62), one obtains the following corollary.

Corollary 1 *If $\{\Lambda_\alpha[U]\}$ is a set of local CL multipliers for the PDE system* $\mathbf{R}\{x; u\}$ *(2) that has the symmetry (56), then $\{\hat{\Lambda}_\beta[U]\}$ yields a set of local CL multipliers for* $\mathbf{R}\{x; u\}$ *where $\{\hat{\Lambda}_\beta[U]\}$ is given by (62) after replacing $\tilde{x}^i$ by x^i, $\tilde{U}^\sigma$ by U^σ, $\tilde{U}^\sigma_i$ by U^σ_i, etc. The set of local CL multipliers $\{\hat{\Lambda}_\beta[U]\}$ yields a new local CL of PDE* *system (2) if and only if this set is nontrivial on all solutions $U = u(x)$ of PDE system (2), i.e., $\hat{\Lambda}_\beta[u] \not\equiv c\Lambda_\beta[u]$, $\beta = 1, \ldots, N$, for some constant c.*

Now suppose the symmetry (56) is a one-parameter Lie group of point transformations

$$x = x(\tilde{x}, \tilde{U}; \varepsilon) = \mathrm{e}^{\varepsilon\tilde{\mathrm{X}}}\tilde{x}, \qquad U = U(\tilde{x}, \tilde{U}; \varepsilon) = \mathrm{e}^{\varepsilon\tilde{\mathrm{X}}}\tilde{U} \tag{63}$$

in terms of its infinitesimal generator (and extensions) $\tilde{\mathrm{X}} = \xi^j(\tilde{x}, \tilde{U})\frac{\partial}{\partial \tilde{x}^j} + \eta^\sigma(\tilde{x}, \tilde{U})\frac{\partial}{\partial \tilde{U}^\sigma}$.

If Eq. (61) holds, then from Eq. (58) and the Lie group properties of (63), it follows that

$$\mathrm{J}[U; \varepsilon]\mathrm{e}^{\varepsilon\mathrm{X}}(\Lambda_\sigma[U]R^\sigma[U]) = \mathrm{D}_i\Psi^i[U; \varepsilon] \tag{64}$$

in terms of the infinitesimal generator (and its extensions) $\mathrm{X} = \xi^j(x, U)\frac{\partial}{\partial x^j} + \eta^\sigma(x, U)\frac{\partial}{\partial U^\sigma}$. Then, after expanding both sides of Eq. (64) in terms of power series in ε, one obtains an expression of the form

$$\sum_p \varepsilon^p \hat{\Lambda}_\sigma[U; p]R^\sigma[U] = \sum_p \varepsilon^p \mathrm{D}_i\left(\frac{1}{p!}\frac{\mathrm{d}^p}{\mathrm{d}\varepsilon^p}\Psi^i[U; \varepsilon]\right)\Big|_{\varepsilon=0}. \tag{65}$$

Corresponding to the sequence of sets of local CL multipliers $\{\hat{\Lambda}_\sigma[U; p]\}$, $p = 1, 2, \ldots$, arising in expression (65), one obtains a sequence of local CLs

$$\mathrm{D}_i\left(\frac{\mathrm{d}^p}{\mathrm{d}\varepsilon^p}\Psi^i[u; \varepsilon]\right)\Big|_{\varepsilon=0} = 0, \qquad p = 1, 2, \ldots$$

for PDE system (2) from its known local CL $\mathrm{D}_i\Phi^i[u] = 0$.

Example 1 A Nonlinear Telegraph System

Consider the nonlinear telegraph PDE system

$$\begin{aligned} v_t + (1 - 2\mathrm{e}^{2u})u_x - \mathrm{e}^u &= 0 \\ v_x - u_t &= 0. \end{aligned} \tag{66}$$

The PDE system (66) has the set of local CL multipliers

$$\begin{aligned} \Lambda_1 &= \xi = \mathrm{e}^{-\frac{1}{2}(U+t/\sqrt{2})}\sin(\tfrac{1}{2}(V + (x + 2\mathrm{e}^U)/\sqrt{2})) \\ \Lambda_2 &= \varphi = -\mathrm{e}^{-\frac{1}{2}(U+t/\sqrt{2})}(\sqrt{2}\mathrm{e}^U\sin(\tfrac{1}{2}(V + (x + 2\mathrm{e}^U)/\sqrt{2})) \\ &\quad + \cos(\tfrac{1}{2}(V + (x + 2\mathrm{e}^U)/\sqrt{2}))) \end{aligned}$$

and corresponding fluxes

$$T = -2e^{-\frac{1}{2}(u+t/\sqrt{2})}\cos(\tfrac{1}{2}(v + (x + 2e^u)/\sqrt{2}))$$
$$X = 2e^{-\frac{1}{2}(u+t/\sqrt{2})}(\sqrt{2}e^u\cos(\tfrac{1}{2}(v + (x + 2e^u)/\sqrt{2})) - \sin(\tfrac{1}{2}(v + (x + 2e^u)/\sqrt{2}))).$$

The nonlinear telegraph PDE system (66) obviously has the discrete reflection symmetry $(t, x, u, v) = (-\tilde{t}, \tilde{x}, \tilde{u}, -\tilde{v})$ and the translational point symmetry $(t, x, u, v) = (\tilde{t}, \tilde{x}, \tilde{u}, \tilde{v} + \varepsilon)$. One can show that for the above local CL of PDE system (66), these symmetries yield three additional local CLs as follows.

1. Reflection symmetry applied to the above local CL.
2. Translation symmetry applied to the above local CL.
3. Reflection symmetry applied again to the local CL found in (2).

For further details, see [41].

Example 2 Another Nonlinear Telegraph System

Consider another nonlinear telegraph PDE system given by

$$v_t - (\operatorname{sech}^2 u)u_x + \tanh u = 0, \qquad v_x - u_t = 0. \tag{67}$$

The PDE system (67) has the set of local CL multipliers

$$\Lambda_1 = \xi = e^x(2x + t^2 - V^2 - 2\log(\cosh U)), \qquad \Lambda_2 = \varphi = 2e^x(V\tanh U - t)$$

and corresponding fluxes

$$T = e^x(2tu - \tfrac{1}{3}v^3 + v(t^2 + 2x - 2\log(\cosh u)))$$
$$X = e^x((v^2 - t^2 - 2x + 2(1 + \log(\cosh u)))\tanh u - 2(vt + u)).$$

The nonlinear telegraph PDE system (67) has the point symmetries with infinitesimal generators given respectively by

$$\mathrm{X}_1 = \frac{\partial}{\partial t}, \qquad \mathrm{X}_2 = v\frac{\partial}{\partial t} + \tanh u\frac{\partial}{\partial x} + \frac{\partial}{\partial u} + t\frac{\partial}{\partial v}.$$

One can show that for the above local CL of PDE system (67), these two point symmetries yield three additional local CLs as follows.

1. The $O(\varepsilon)$, $O(\varepsilon^2)$ terms that result from applying the translation symmetry X_1 to the above local CL yield two additional local CLs.

2. The action of the second point symmetry X_2 on the additional $O(\varepsilon)$ local CL, obtained in (1), yields a third additional CL.

For further details, see [41].

4 Nonlocally Related Systems and Nonlocal Symmetries

Often a given PDE system has no local symmetry or no local conservation law. Even if a given PDE system has a local symmetry, it may not be useful for the problem at hand. The aim is to extend existing methods for finding local symmetries and local CLs to PDE systems that are nonlocally related and equivalent to a given PDE system in order to seek nonlocal symmetries and nonlocal CLs for a given PDE system. Two systematic and natural ways will be presented to find such nonlocally related systems for a given PDE system. In particular, it will be shown that for any PDE system, each local CL as well as each point symmetry systematically yields a nonlocally related system. Further systematic extensions for seeking additional nonlocally related systems will also be presented.

4.1 Conservation Law-based Method to Obtain Nonlocally Related Systems and Nonlocal Symmetries: Subsystems

Initially, we focus on the situation of a scalar PDE with two independent variables. As will be seen, no extra complication arises for a PDE system with two independent variables. But the situation for a PDE system with three or more independent variables is more complicated as will be seen in Sect. 5.

For a local conservation law

$$\mathrm{D}_t T(x,t,u,\partial u,\ldots,\partial^r u) + \mathrm{D}_x X(x,t,u,\partial u,\ldots,\partial^r u) = 0 \tag{68}$$

of a given scalar PDE

$$R[u] = R(x,t,u,\partial u,\ldots,\partial^k u) = 0 \tag{69}$$

one can form an equivalent augmented *potential system* P given by

$$\begin{aligned} \frac{\partial v}{\partial t} &= X(x,t,u,\partial u,\ldots,\partial^r u) \\ \frac{\partial v}{\partial x} &= -T(x,t,u,\partial u,\ldots,\partial^r u) \\ R(x&,t,u,\partial u,\ldots,\partial^k u) = 0. \end{aligned} \tag{70}$$

If $(u(x,t), v(x,t))$ solves the potential system P, then $u(x,t)$ solves the given scalar PDE (69). Conversely, if $u(x,t)$ solves the given scalar PDE (69), then there exists a solution $(u(x,t), v(x,t))$ of the potential system P since the integrability condition $v_{xt} = v_{tx}$ is satisfied due to the existence of the local CL (68). But the equivalence relationship is *nonlocal* and *non-invertible* since for any solution $u(x,t)$ of the given scalar PDE (69), if $(u(x,t), v(x,t))$ solves the potential system P, then so does $(u(x,t), v(x,t) + C)$ for any constant C.

Consequently, any symmetry (CL) of the given scalar PDE (69) yields a symmetry (CL) of the equivalent potential system P. Conversely, any symmetry (CL) of the potential system P yields a symmetry (CL) of the given scalar PDE (69).

Now suppose the equivalent potential system P has a point symmetry given by an infinitesimal generator

$$\xi(x,t,u,v)\frac{\partial}{\partial x} + \tau(x,t,u,v)\frac{\partial}{\partial t} + \omega(x,t,u,v)\frac{\partial}{\partial u} + \varphi(x,t,u,v)\frac{\partial}{\partial v}. \tag{71}$$

The point symmetry (71) of the potential system P yields a *nonlocal symmetry* of the given scalar PDE (69) if and only if its infinitesimal components satisfy the relationship

$$(\xi_v)^2 + (\tau_v)^2 + (\omega_v)^2 \not\equiv 0. \tag{72}$$

Hence, through a local CL of the PDE (69), a nonlocal symmetry of (69) can be obtained from a point symmetry (71) of the nonlocally related potential system P given by the PDE system (70) if the components of the point symmetry (71) satisfy the inequality (72).

The converse is also true. In particular, suppose a scalar PDE (69) has a point symmetry given by the infinitesimal generator

$$\alpha(x,t,u)\frac{\partial}{\partial x} + \beta(x,t,u)\frac{\partial}{\partial t} + \gamma(x,t,u)\frac{\partial}{\partial u}. \tag{73}$$

The point symmetry (73) of the PDE (69) yields a *nonlocal symmetry* of the potential system P if and only if the potential system P has no corresponding point symmetry of the form $\alpha(x,t,u)\frac{\partial}{\partial x} + \beta(x,t,u)\frac{\partial}{\partial t} + \gamma(x,t,u)\frac{\partial}{\partial u} + \delta(x,t,u,v)\frac{\partial}{\partial v}$ for some function $\delta(x,t,u,v)$.

Next, we show how to obtain further nonlocally related systems for a given PDE system.

4.1.1 Use of n Local CLs to Obtain up to $2^n - 1$ Nonlocally Related Systems

Suppose there are n local CL multipliers $\{\Lambda_i(x,t,U,\partial U,\ldots,\partial^q U)\}_{i=1}^n$ yielding n independent local CLs of a given scalar PDE. Let v^i be the potential variable arising from the local CL multiplier $\Lambda_i[U]$. Then one obtains n *singlet* potential systems P^i, $i = 1,\ldots,n$. Moreover, one can consider potential systems in *couplets*

$\{P^i, P^j\}_{i,j=1}^n$ with two potential variables; in *triplets* $\{P^i, P^j, P^k\}_{i,j,k=1}^n$ with three potential variables; . . . ; in an *n-plet* $\{P^1, \ldots, P^n\}$ with n potential variables. Consequently from n local CLs of a given scalar PDE, one obtains $2^n - 1$ *distinct potential systems*!

Moreover, starting from *any* one of these $2^n - 1$ potential systems, one can continue the process. In particular, if one of these potential systems has N "local" CLs, in principle one could obtain up to $2^N - 1$ further distinct potential systems. However, not all local CLs of these $2^n - 1$ potential systems yield additional potential systems. *In particular, one can show that if a set of local CL multipliers depends only on independent variables (x, t) then no additional potential system is obtained.* See [2, 30, 31] for further details.

Any potential system could yield additional nonlocal symmetries or additional nonlocal CLs for any other potential system or the "given" PDE. Furthermore, one of the constructed potential systems could be a "given" PDE system. A more direct way of seeing this will be presented in Sect. 5 through the symmetry-based method for obtaining nonlocally related systems.

4.1.2 Nonlocally Related Subsystems

Definition 6 Suppose one has a given PDE system $\mathbf{S}\{x, t; u^1, \ldots, u^M\}$ with the indicated M dependent variables. A *subsystem* excluding a dependent variable, say u^M, is *nonlocally related* to the given system $\mathbf{S}\{x, t; u^1, \ldots, u^M\}$ if u^M cannot be directly expressed from the equations of $\mathbf{S}\{x, t; u^1, \ldots, u^M\}$ in terms of x, t, the remaining dependent variables $u^1, \ldots, u^{M-1}$, and their derivatives.

Subsystems for consideration can arise following an interchange of one or more of the dependent and independent variables of a given system $\mathbf{S}\{x, t; u^1, \ldots, u^M\}$. Consequently, for a given PDE system, one obtains a tree of nonlocally related (but equivalent) PDE systems arising from local conservation laws and subsystems. *Each PDE system in such a tree is equivalent in the sense that the solution set for any system in the tree can be found from the solution set for any other PDE system in the tree through a connection formula.* Due to the equivalence of the solution sets and the nonlocal relationship between PDE systems in a tree, it follows that any coordinate-independent method of analysis (quantitative, analytical, numerical, perturbation, etc.) when applied to some PDE system in a tree may yield simpler computations and/or results that cannot be obtained when the method of analysis is directly applied to any particular PDE system in a tree. In particular, it is important to note that a "given" system could be any system in such a tree!!

Example 1 Nonlinear Wave Equation

Suppose a given PDE $\mathbf{U}\{x, t; u\}$ is the nonlinear wave equation

$$u_{tt} = (c^2(u)u_x)_x. \tag{74}$$

Directly, one obtains the singlet potential system (local CL multiplier is 1) **UV**$\{x, t; u, v\}$ given by

$$v_x - u_t = 0, \qquad v_t - c^2(u)u_x = 0. \tag{75}$$

Through the invertible point transformation (hodograph transformation) $x = x(u, v)$, $t = t(u, v)$, the potential system **UV**$\{x, t; u, v\}$ becomes the invertibly equivalent PDE system **XT**$\{u, v; x, t\}$ given by

$$x_v - t_u = 0, \qquad x_u - c^2(u)t_v = 0. \tag{76}$$

One can show that there are only three additional local CL multipliers of the form $\Lambda(x, t, U) = xt, x, t$ for the nonlinear wave equation **U**$\{x, t; u\}$ (74) for an *arbitrary* wave speed $c(u)$. This yields three additional singlet potential systems **UA**$\{x, t; u, a\}$, **UB**$\{x, t; u, b\}$, and **UW**$\{x, t; u, w\}$, respectively given by the PDE systems

$$a_x - x[tu_t - u] = 0, \qquad a_t - t[xc^2(u)u_x - \int c^2(u)du] = 0 \tag{77}$$

$$b_x - xu_t = 0, \qquad b_t - [xc^2(u)u_x - \int c^2(u)du] = 0 \tag{78}$$

and

$$w_x - [tu_t - u] = 0, \qquad w_t - tc^2(u)u_x = 0. \tag{79}$$

Nonlocally related subsystems **T**$\{u, v; t\}$ and **X**$\{u, v; x\}$ arise from **UV**$\{x, t; u, v\}$ through **XT**$\{u, v; x, t\}$ after one respectively deletes the dependent variables x and t from **XT**$\{u, v; x, t\}$

$$t_{vv} - c^{-2}(u)t_{uu} = 0 \tag{80}$$

and

$$x_{vv} - (c^{-2}(u)x_u)_u = 0. \tag{81}$$

One can show that the symmetry classifications of the PDEs (80) and (81) are "equivalent" [25]. Hence we concentrate on **T**$\{u, v; t\}$. Since the PDE **T**$\{u, v; t\}$ (80) is linear and self-adjoint, it follows that any solution of **T**$\{u, v; t\}$ yields a local CL multiplier for **T**$\{u, v; t\}$. Four of these local CL multipliers, for an arbitrary wave speed $c(u)$, are given by $\Lambda(u, v, T) = c^2(u), uc^2(u), vc^2(u), uvc^2(u)$. These yield three additional singlet potential systems **TP**$\{u, v; t, p\}$, **TQ**$\{u, v; t, q\}$, **TR**$\{u, v; t, r\}$, respectively given by

$$p_v - (ut_u - t) = 0, \qquad p_u - uc^2(u)t_v = 0 \tag{82}$$

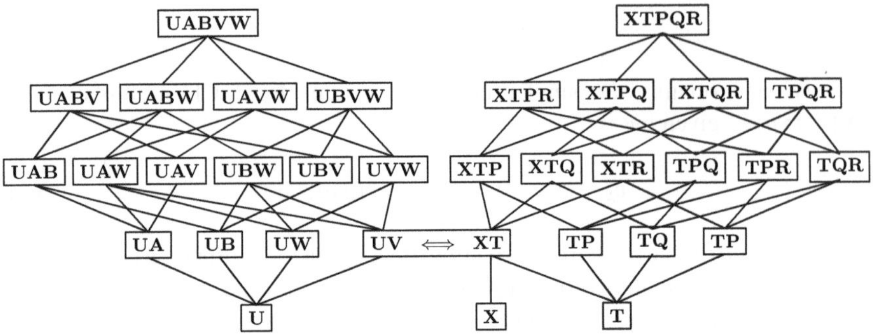

Fig. 1 A tree of nonlocally related systems for the nonlinear wave equation (74) for arbitrary wave speed $c(u)$

$$q_v - vt_u = 0, \qquad q_u + c^2(u)(t - vt_v) = 0 \tag{83}$$

and

$$r_v - v(ut_u - t) = 0, \qquad r_u - uc^2(u)(vt_v - t) = 0. \tag{84}$$

Consequently, one obtains the following (*far from exhaustive*) tree (Fig. 1) of nonlocally related systems for the nonlinear wave equation **U**$\{x, t; u\}$ (74), holding for an arbitrary wave speed $c(u)$.

The point symmetry classification for the nonlinear wave equation **U**$\{x, t; u\}$ (74) is given in [44]. The point symmetry classifications for the potential system **XT**$\{u, v; x, t\}$ (76) (of course, it is exactly the same as that for the potential system **UV**$\{x, t; u, v\}$ (75)) and the subsystem **T**$\{u, v; t\}$ (80) is given in [25]. A partial point symmetry classification for the potential system **TP**$\{u, v; t, p\}$ (82) can be adapted from results presented in [45]. The complete point symmetry classifications for the potential systems **UA**$\{x, t; u, a\}$ (77), **UB**$\{x, t; u, b\}$ (78), **UW**$\{x, t; u, w\}$ (79), **TP**$\{u, v; t, p\}$ (82), and **TQ**$\{u, v; t, q\}$ (83) are given in [46]. Many nonlocal symmetries of the nonlinear wave equation are found from each of these nonlocally related systems in terms of specific forms of the nonlinear wave speed $c(u)$. In particular, the following additional nonlocal symmetries of the nonlinear wave equation **U**$\{x, t; u\}$ (74) have been found.

For the potential system **UB**$\{x, t; u, b\}$ (78), setting $F(u) = \int c^2(u)\, du$, one finds that if $F(u)$ satisfies the ODE

$$\frac{F''(u)}{F'(u)^2} = \frac{4F(u) + 2C_1}{(F(u) + C_2)^2 + C_3}$$

in terms of arbitrary constants C_1, C_2, C_3, then the potential system **UB**$\{x, t; u, b\}$ (78) has the point symmetry

$$\mathrm{X} = (F(u)+C_1)x\frac{\partial}{\partial x}+b\frac{\partial}{\partial t}+\frac{(F(u)+C_2)^2+C_3}{F'(u)}\frac{\partial}{\partial u}+(2C_2b-(C_2{}^2+C_3)t)\frac{\partial}{\partial b}$$

that is a nonlocal symmetry of the nonlinear wave equation $\mathbf{U}\{x, t; u\}$ (74).

For the potential system $\mathbf{UW}\{x, t; u, w\}$ (79), if the wave speed $c(u)$ satisfies the ODE

$$\frac{c'(u)}{c(u)} = -\frac{2u + C_1}{u^2 + C_2}$$

in terms of arbitrary constants C_1, C_2, then it has the point symmetry

$$\mathrm{X} = w\frac{\partial}{\partial x} + (u + C_1)t\frac{\partial}{\partial t} + (u^2 + C_2)\frac{\partial}{\partial u} - C_2x\frac{\partial}{\partial w}$$

that is a nonlocal symmetry of the nonlinear wave equation $\mathbf{U}\{x, t; u\}$ (74).

The potential system $\mathbf{TP}\{u, v; t, p\}$ (82), for $c(u) = u^{-2}\mathrm{e}^{1/u}$, has the point symmetries

$$\mathrm{X}_1 = (pu - 2tv(u + 1))\frac{\partial}{\partial t} - 2u^2v\frac{\partial}{\partial u} + (u^2 + \mathrm{e}^{2/u})\frac{\partial}{\partial v} + tu^{-1}\mathrm{e}^{2/u}\frac{\partial}{\partial p}$$

$$\mathrm{X}_2 = t(u + 1)\frac{\partial}{\partial t} + u^2\frac{\partial}{\partial u} - v\frac{\partial}{\partial v}$$

that are both nonlocal symmetries of the nonlinear wave equation $\mathbf{U}\{x, t; u\}$ (74).

For the potential system $\mathbf{TR}\{u, v; t, r\}$ (84), new nonlocal symmetries are found for $\mathbf{U}\{x, t; u\}$ (74) from the point symmetries of $\mathbf{TR}\{u, v; t, r\}$ when $c(u) = u^{-4/3}$.

For details and a table of listed nonlocal symmetries derived from the above tree of nonlocally related systems for the nonlinear wave equation $\mathbf{U}\{x, t; u\}$ (74), see [46].

Example 2 Nonlinear Telegraph Equation

Suppose a given PDE $\mathbf{U}\{x, t; u\}$ is the nonlinear telegraph (NLT) equation

$$u_{tt} - (F(u)u_x)_x - (G(u))_x = 0. \tag{85}$$

Case (a) For *arbitrary* $F(u)$, $G(u)$, one obtains two singlet potential systems $\mathbf{UV}_1\{x, t; u, v_1\}$ and $\mathbf{UV}_2\{x, t; u, v_2\}$ respectively given by the PDE systems

$$v_{1x} - u_t = 0, \qquad v_{1t} - (F(u)u_x + G(u)) = 0 \tag{86}$$

and

$$v_{2x} - (tu_t - u) = 0, \qquad v_{2t} - t(F(u)u_x + G(u)) = 0. \tag{87}$$

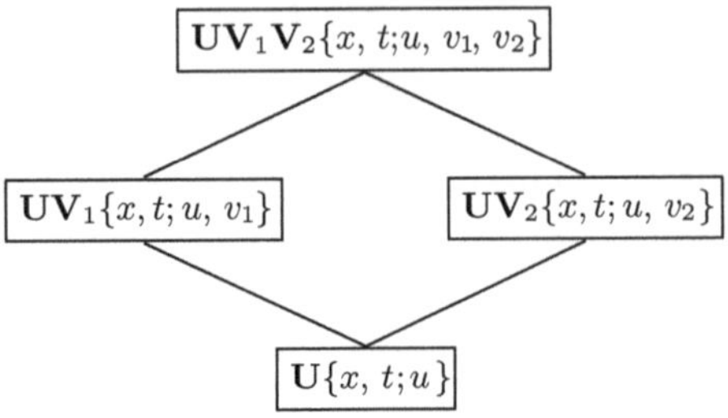

Fig. 2 Tree of nonlocally related PDE systems for the NLT equation (85) for arbitrary $F(u)$, $G(u)$

Case (b) For *arbitrary* $G(u)$, $F(u) = G'(u)$, one obtains two additional singlet potential systems $\mathbf{UB}_3\{x, t; u, b_3\}$ and $\mathbf{UB}_4\{x, t; u, b_4\}$ respectively given by the PDE systems

$$b_{3x} - \mathrm{e}^x u_t = 0, \qquad b_{3t} - \mathrm{e}^x F(u) u_x = 0 \tag{88}$$

and

$$b_{4x} - \mathrm{e}^x (t u_t - u) = 0, \qquad b_{4t} - t\mathrm{e}^x F(u) u_x = 0. \tag{89}$$

Case (c) For *arbitrary* $F(u)$, $G(u) = u$, in addition to the singlet potential systems $\mathbf{UV}_1\{x, t; u, v_1\}$ (86) and $\mathbf{UV}_2\{x, t; u, v_2\}$ (87), one again obtains two further singlet potential systems $\mathbf{UC}_3\{x, t; u, c_3\}$ and $\mathbf{UC}_4\{x, t; u, c_4\}$ respectively given by the PDE systems

$$\begin{aligned} c_{3x} - ((x - \tfrac{1}{2}t^2)u_t + tu) &= 0 \\ c_{3t} - (x - \tfrac{1}{2}t^2)(F(u)u_x + u) + \int F(u)du &= 0 \end{aligned} \tag{90}$$

and

$$\begin{aligned} c_{4x} + (\tfrac{1}{6}t^3 - tx)u_t + (x - \tfrac{1}{2}t^2)u &= 0 \\ c_{4t} + (\tfrac{1}{6}t^3 - tx)(F(u)u_x + u) + t\int F(u)du &= 0. \end{aligned} \tag{91}$$

The corresponding trees of nonlocally related systems for the NLT equation are illustrated in Figs. 2 and 3.

In the cases where $F(u)$ and $G(u)$ are power law functions, see [47] for tabulations of nonlocal symmetries and nonlocal conservation laws for the NLT equation $\mathbf{U}\{x, t; u\}$ (85), arising for many of the above listed nonlocally related systems.

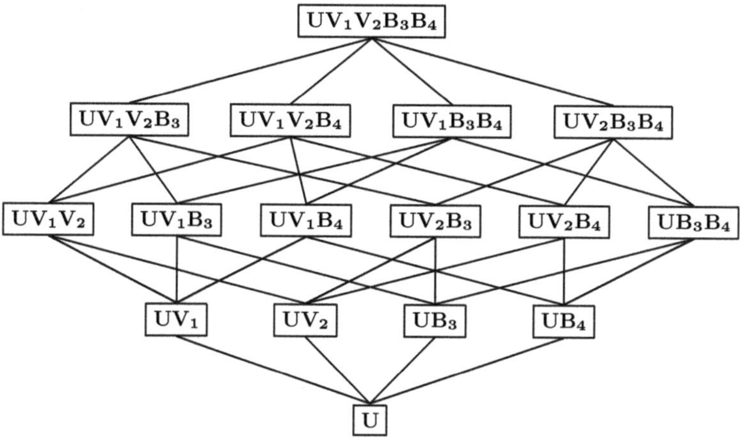

Fig. 3 Tree of nonlocally related PDE systems for the NLT equation (85) for arbitrary $G(u)$, $F(u) = G'(u)$

4.1.3 Conservation Law and Symmetry Classification Problems for the NLT Equation U$\{x, t; u\}$ and its Potential System UV$_1\{x, t; u, v_1\}$

Now we consider symmetry and conservation law classification problems for the NLT equation **U**$\{x, t; u\}$ (85) and its potential system **UV**$_1\{x, t; u, v_1\}$ (86). For specific $(F(u), G(u))$ pairs, the CL classification problem for **UV**$_1\{x, t; u, v_1\}$ yields additional CLs and hence further potential systems for consideration [41].

Nonlocal Symmetries of **U**$\{x, t; u\}$ *Arising from Point Symmetries of* **UV**$_1\{x, t; u, v_1\}$.

The potential system **UV**$_1\{x, t; u, v_1\}$ has a point symmetry corresponding to the infinitesimal generator

$$X = \xi(x, t, u, v_1)\frac{\partial}{\partial x} + \tau(x, t, u, v_1)\frac{\partial}{\partial t} + \eta(x, t, u, v_1)\frac{\partial}{\partial u} + \varphi(x, t, u, v_1)\frac{\partial}{\partial v_1} \quad (92)$$

if and only if the coefficients of (92) satisfy the determining equations

$$\begin{aligned}
\xi_{v_1} - \tau_u &= 0 \\
\eta_u - \varphi_{v_1} + \xi_x - \tau_t &= 0 \\
G(u)[\eta_{v_1} + \tau_x] + \eta_t - \varphi_x &= 0 \\
\xi_u - F(u)\tau_{v_1} &= 0 \\
\varphi_u - G(u)\tau_u - F(u)\eta_{v_1} &= 0 \\
G(u)\xi_{v_1} + \xi_t - F(u)\tau_x &= 0 \\
F(u)[\varphi_{v_1} - \tau_t + \xi_x - \eta_u - 2G(u)\tau_{v_1}] - F'(u)\eta &= 0 \\
G(u)[\varphi_{v_1} - \tau_t - G(u)\tau_{v_1}] - F(u)\eta_x - G'(u)\eta + \varphi_t &= 0
\end{aligned} \tag{93}$$

for *arbitrary* values of x, t, u, v_1.

The solution of the determining Eq. (93) appears in [48] and the resulting nonlocal symmetries for the NLT equation $\mathbf{U}\{x, t; u\}$ (85) are summarized by the following theorem.

Theorem 14 *A point symmetry of the potential system* $\mathbf{UV}_1\{x, t; u, v_1\}$ *(86) yields a nonlocal symmetry of the NLT equation* $\mathbf{U}\{x, t; u\}$ *(85) if and only if the pair of constitutive functions* $(F(u), G(u))$ *satisfies the first order ODE system*

$$\begin{aligned}
(c_3u + c_4)F'(u) - 2(c_1 - c_2 - G(u))F(u) &= 0 \\
(c_3u + c_4)G'(u) + G^2(u) - (c_1 - 2c_2 + c_3)G(u) - c_5 &= 0
\end{aligned} \tag{94}$$

in terms of arbitrary constants $c_1, \ldots, c_5$. *For any pair* $(F(u), G(u))$ *satisfying (94), the potential system* $\mathbf{UV}_1\{x, t; u, v_1\}$ *(86) has the point symmetry (92) with*

$$\begin{aligned}
\xi &= c_1x + \int F(u)du \\
\tau &= c_2t + v_1 \\
\eta &= c_3u + c_4 \\
\varphi &= c_5t + (c_1 - c_2 + c_3)v_1
\end{aligned}$$

which is a (nonlocal) potential symmetry of the scalar NLT equation $\mathbf{U}\{x, t; u\}$ *(85).*

Modulo translations and scalings in u and G and scalings in F (involving 5/7 parameters), one obtains six distinct classes for $(F(u), G(u))$ for which the scalar NLT equation $\mathbf{U}\{x, t; u\}$ (85) has a potential symmetry. These classes are summarized in Table 1.

Point Symmetry Classification of the Scalar NLT Equation $\mathbf{U}\{x, t; u\}$ (85)

The NLT equation $\mathbf{U}\{x, t; u\}$ (85) has a point symmetry corresponding to the infinitesimal generator $\mathrm{X} = \xi(x, t, u)\frac{\partial}{\partial x} + \tau(x, t, u)\frac{\partial}{\partial t} + \eta(x, t, u)\frac{\partial}{\partial u}$ if and only if the determining equations

Table 1 Classification table for potential symmetries of the NLT equation (85)

Relationship	$G(u)$	$F(u)$
$F(u)=\frac{u^\beta}{\alpha}G'(u)$	$\frac{u^{2\alpha}-1}{u^{2\alpha}+1}$	$\frac{4u^{2\alpha+\beta-1}}{(u^{2\alpha}+1)^2}$
	$\frac{u^{2\alpha}+1}{u^{2\alpha}-1}$	$-\frac{4u^{2\alpha+\beta-1}}{(u^{2\alpha}-1)^2}$
$F(u)=\frac{u^\beta}{\alpha}G'(u)$	$\tan(\alpha\ln u)$	$u^{\beta-1}\sec^2(\alpha\ln u)$
$F(u)=u^\beta G'(u)$	$(\ln u)^{-1}$	$-u^{\beta-1}(\ln u)^{-2}$
$F(u)=e^{2\beta u}G'(u)$	$\tan u$	$e^{2\beta u}\sec^2 u$
$F(u)=e^{2\beta u}G'(u)$	$\tanh u$	$e^{2\beta u}\mathrm{sech}^2 u$
	$\coth u$	$-e^{2\beta u}\mathrm{csch}^2 u$
$F(u)=e^{2\beta u}G'(u)$	u^{-1}	$-u^{-2}e^{2\beta u}$

Table 2 Classes of $(F(u), G(u))$ yielding additional point symmetries of the scalar NLT equation $\mathbf{U}\{x,t;u\}$ (85)

$G(u)$	$F(u)$	Admitted additional point symmetries
e^u	$e^{(\alpha+1)u}$	$2\alpha x\frac{\partial}{\partial x}+(\alpha-1)t\frac{\partial}{\partial t}+2\frac{\partial}{\partial u}$
$u^{\alpha+\beta+1}$	u^α	$2\beta x\frac{\partial}{\partial x}+(\alpha+2\beta)t\frac{\partial}{\partial t}-2u\frac{\partial}{\partial u}$
u^{-1}	u^{-2}	$t\frac{\partial}{\partial t}+u\frac{\partial}{\partial u}$, $e^x\frac{\partial}{\partial x}-ue^x\frac{\partial}{\partial u}$
$\ln u$	u^α	$2(\alpha+1)x\frac{\partial}{\partial x}+(\alpha+2)t\frac{\partial}{\partial t}+2u\frac{\partial}{\partial u}$
u	$e^{\alpha u}$	$2\alpha x\frac{\partial}{\partial x}+\alpha t\frac{\partial}{\partial t}+2\frac{\partial}{\partial u}$
u^{-3}	u^{-4}	$2t\frac{\partial}{\partial t}+u\frac{\partial}{\partial u}$, $t^2\frac{\partial}{\partial t}+tu\frac{\partial}{\partial u}$

$$\xi_u=\tau_x=\tau_u=\eta_{uu}=\xi_t=0$$
$$2F(u)[-\tau_t+\xi_x]-F'(u)\eta=0$$
$$\eta_{tt}-F(u)\eta_{xx}-G'(u)\eta_x=0$$
$$2\eta_{tu}-\tau_{tt}=0$$
$$F(u)[2\eta_{xu}-\xi_{xx}]+\xi_{tt}+2F'(u)\eta_x-G'(u)[\xi_x-2\tau_t]+G''(u)\eta=0$$

are satisfied for *arbitrary* values of x, t, and u.

For arbitrary $(F(u), G(u))$, the scalar NLT equation $\mathbf{U}\{x,t;u\}$ (85) is only invariant under translations in x and t. The classification of its point symmetries for specific forms of $(F(u), G(u))$, modulo scalings and translations in u, is presented in Table 2.

The following theorem holds. See [48] for details.

Theorem 15 *A point symmetry of the scalar NLT equation* $\mathbf{U}\{x,t;u\}$ *(85) yields a point symmetry of the NLT potential system* $\mathbf{UV}_1\{x,t;u,v_1\}$ *(86) for all cases except when* $(F(u),G(u))=(u^{-4},u^{-3})$. *In this case, its admitted point symmetry* $t^2\frac{\partial}{\partial t}+tu\frac{\partial}{\partial u}$ *yields a nonlocal symmetry of the NLT potential system* $\mathbf{UV}_1\{x,t;u,v_1\}$ *(86).*

Local Conservation Laws of the Potential System $\mathbf{UV}_1\{x,t;u,v_1\}$.

$\{\Lambda_1(x,t,U,V),\ \Lambda_2(x,t,U,V)\}$ is a set of local CL multipliers for the NLT potential system $\mathbf{UV}_1\{x,t;u,v_1\}$ (86) if and only if the equations

$$
\begin{aligned}
&E_U\left(\Lambda_1\left(V_x-U_t\right)+\Lambda_2\left(V_t-\left(F(U)U_x+G(U)\right)\right)\right)\equiv 0\\
&E_V\left(\Lambda_1\left(V_x-U_t\right)+\Lambda_2\left(V_t-\left(F(U)U_x+G(U)\right)\right)\right)\equiv 0
\end{aligned}
\tag{95}
$$

hold for *arbitrary* differentiable functions $(U(x,t), (V(x,t))$. Equations (95) yield the system of determining equations

$$
\begin{aligned}
\frac{\partial\Lambda_2}{\partial V}-\frac{\partial\Lambda_1}{\partial U}&=0\\
\frac{\partial\Lambda_2}{\partial U}-F(U)\frac{\partial\Lambda_1}{\partial V}&=0\\
\frac{\partial\Lambda_2}{\partial x}-\frac{\partial\Lambda_1}{\partial t}-G(U)\frac{\partial\Lambda_1}{\partial V}&=0\\
F(U)\frac{\partial\Lambda_1}{\partial x}-\frac{\partial\Lambda_2}{\partial t}-\frac{\partial}{\partial U}[G(U)\Lambda_1]&=0.
\end{aligned}
\tag{96}
$$

One can show that for any solution of (96), the fluxes for the corresponding local CLs of the potential NLT system $\mathbf{UV}_1\{x,t;u,v_1\}$ (86) are given by

$$
\begin{aligned}
X(x,t,u,v_1)=&-\int_a^u\Lambda_1(x,t,s,b)ds-\int_b^{v_1}\Lambda_2(x,t,u,s)ds\\
&-G(a)\int^x\Lambda_1(s,t,a,b)ds
\end{aligned}
$$

$$
T(x,t,u,v_1)=\int_a^u\Lambda_2(x,t,s,b)ds+\int_b^{v_1}\Lambda_1(x,t,u,s)ds.
$$

One can show [41] that the solution of the determining system (96) reduces to the study of the system of two functions given by

$$
\begin{aligned}
d(U)&=G'^2F'''-3G'G''F''+[3G''^2-G'G''']F'\\
h(U)&=G'^2G^{(4)}-4G'G''G'''+3G''^3.
\end{aligned}
$$

Three cases arise

$$
\begin{aligned}
&d(U)\not\equiv 0, h(U)\equiv 0\\
&d(U)\not\equiv 0, h(U)\not\equiv 0\\
&d(U)=h(U)\equiv 0.
\end{aligned}
$$

The results are summarized as follows.

When $d(U)\not\equiv 0$, $h(U)\equiv 0$, the resulting local CL multipliers for the potential system $\mathbf{UV}_1\{x,t;u,v_1\}$ are indicated in Table 3.

Table 3 $d(U) \not\equiv 0, h(U) \equiv 0$

$F(U)$	$G(U)$	Local CL multipliers
Arbitrary	U	$(\Lambda_1, \Lambda_2) = (t, x - \frac{1}{2}t^2)$, $(\Lambda_1, \Lambda_2) = (1, -t)$
Arbitrary	$1/U$	$(\Lambda_1, \Lambda_2) = (U, V)$, $(\Lambda_1, \Lambda_2) = (UV, \frac{1}{2}V^2 + x + \int^U sF(s)ds)$

Table 4 $d(U) \not\equiv 0, h(U) \not\equiv 0$

Relationship	Local CL multipliers
$\gamma F - G' = \frac{\alpha}{\gamma}(G+\beta)^2$	$(\Lambda_1, \Lambda_2) = (\varphi_1, \varphi_2) = e^{\gamma x + \frac{\alpha}{\gamma}\int^U (G(s)+\beta)ds} e^{\sqrt{\alpha}(\beta t + V)}(1, \frac{\sqrt{\alpha}}{\gamma}(G(U)+\beta))$, $(\Lambda_1, \Lambda_2) = (\varphi_1, -\varphi_2) = (x, -t, U, -V)$
$\gamma F - G' = \frac{\alpha}{\gamma}$	$(\Lambda_1, \Lambda_2) = (\psi_1, \psi_2) = e^{\gamma x + \sqrt{\alpha} t}(1, \frac{\sqrt{\alpha}}{\gamma})$, $(\Lambda_1, \Lambda_2) = (\psi_1, -\psi_2)(x, -t)$
$\gamma F = G'$	$(\Lambda_1, \Lambda_2) = e^{\gamma x}(t, \frac{1}{\gamma})$, $(\Lambda_1, \Lambda_2) = e^{\gamma x}(V, \frac{1}{\gamma}G(U))$, $(\Lambda_1, \Lambda_2) = e^{\gamma x}(1, 0)$

When $d(U) \not\equiv 0$, $h(U) \not\equiv 0$, the resulting local CL multipliers for the potential system $\mathbf{UV}_1\{x, t; u, v_1\}$ are indicated in Table 4.

When $d(U) = h(U) \equiv 0$, using symmetry analysis (substitution and invariance of the ODE under a solvable three-parameter Lie group of point transformations), the ODE $h(U) = 0$ can be solved in terms of elementary functions (for $G(U)$). Then note that $F(U) = G(U) + \text{const}$ is a particular solution of the resulting linear ODE $d(U) = 0$. In turn, this leads to its general solution. Consequently, for $F(U) = \beta_1 G^2(U) + \beta_2 G(U) + \beta_3$, ${\beta_2}^2 \neq 4\beta_1\beta_3$, there are four highly nontrivial CLs when $G(U) = U, 1/U, e^U, \tanh U, \tan U$. In the case of a "perfect square" ${\beta_2}^2 = 4\beta_1\beta_3$, there are only two local CLs. For details, see [41].

The NLT potential system $\mathbf{UV}_1\{x, t; u, v_1\}$ (86) is not variational. In the case of a variational system, each set of local CL multipliers of the system must correspond to a local symmetry of the system written in evolutionary form. Hence, in the variational situation, for any pair of constitutive functions $(F(u), G(u))$, the number of sets of local CL multipliers is at most equal to the number of local symmetries. Note that for the PDE system $\mathbf{UV}_1\{x, t; u, v_1\}$ (86), for many pairs of constitutive functions $(F(u), G(u))$, the number of sets of local CL multipliers (which of course do not correspond to local symmetries) exceeds the number of local symmetries.

Example 3 Planar Gas Dynamics Equations

Suppose the given PDE system is the planar gas dynamics (PGD) equations. In the *Eulerian* description, the corresponding Euler PGD system $\mathbf{E}\{x, t; v, p, \rho\}$ is given by

$$\begin{aligned} \rho_t + (\rho v)_x &= 0 \\ \rho(v_t + v v_x) + p_x &= 0 \\ \rho(p_t + v p_x) + B(p, \rho^{-1}) v_x &= 0 \end{aligned} \tag{97}$$

where $v(x, t)$ is the velocity of the gas, $p(x, t)$ is the pressure, and $\rho(x, t)$ is the mass density of the gas. In the Eulerian system $\mathbf{E}\{x, t; v, p, \rho\}$ (97), in terms of the *entropy density* $S(p, \rho)$, the constitutive function $B(p, \rho^{-1})$ is given by $B(p, \rho^{-1}) = -\rho^2 S_\rho / S_p$.

In the *Lagrangian* description, in terms of Lagrange mass coordinates $s = t$, $y = \int_{x_0}^{x} \rho(\xi) d\xi$, the corresponding Lagrange PGD system $\mathbf{L}\{y, s; v, p, q\}$ is given by

$$\begin{aligned} q_s - v_y &= 0 \\ v_s + p_y &= 0 \\ p_s + B(p, q) v_y &= 0 \end{aligned} \tag{98}$$

with $q = 1/\rho$.

It is now shown that the potential system framework, based on using local CLs, yields a direct connection between the Euler system (97) and the Lagrange system (98). As well, as a consequence, other equivalent descriptions are derived. The Euler system $\mathbf{E}\{x, t; v, p, \rho\}$ (97) is used as the given PDE system. The first equation of the Euler system is written as a local CL, corresponding to conservation of mass. Through this equation, a potential variable $r(x, t)$ is introduced and leads to the Euler potential system $\mathbf{G}\{x, t; v, p, \rho, r\}$ given by

$$\begin{aligned} r_x - \rho &= 0 \\ r_t + \rho v &= 0 \\ \rho(v_t + v v_x) + p_x &= 0 \\ \rho(p_t + v p_x) + B(p, \rho^{-1}) v_x &= 0. \end{aligned} \tag{99}$$

Now consider an interchange of dependent and independent variables in $\mathbf{G}\{x, t; v, p, \rho, r\}$ with $r = y$, $t = s$ as independent variables and x, v, p, $q = 1/\rho$ as dependent variables to obtain the system $\mathbf{G}_0\{y, s; x, v, p, q\}$, invertibly equivalent to $\mathbf{G}\{x, t; v, p, \rho, r\}$ (99), given by

$$\begin{aligned} x_y - q &= 0 \\ x_s - v &= 0 \\ v_s + p_y &= 0 \\ p_s + B(p, q) v_y &= 0. \end{aligned} \tag{100}$$

A nonlocally related subsystem of $\mathbf{G}_0\{y, s; x, v, p, q\}$ (100) is obtained by excluding its dependent variable x through the integrability condition $x_{ys} = x_{sy}$. The resulting subsystem is the Lagrange system $\mathbf{L}\{y, s; v, p, q\}$ (98)!

A second CL of the Euler system $\mathbf{E}\{x, t; v, p, \rho\}$ (97) is obtained from its set of local CL multipliers $(\Lambda_1, \Lambda_2, \Lambda_3) = (V, 1, 0)$. This yields a second potential variable w. The resulting couplet system $\mathbf{W}\{x, t; v, p, \rho, r, w\}$ that includes the potential variables r and w is given by the PDE system

$$\begin{aligned} r_x - \rho &= 0 \\ r_t + \rho v &= 0 \\ w_x + r_t &= 0 \\ w_t + p + v w_x &= 0 \\ \rho(p_t + v p_x) + B(p, \rho^{-1}) v_x &= 0. \end{aligned} \tag{101}$$

The third equation of the couplet system $\mathbf{W}\{x, t; v, p, \rho, r, w\}$ (101), which is a local CL as written, yields a third potential variable z to yield an additional potential system $\mathbf{Z}\{x, t; v, p, \rho, r, w, z\}$ given by

$$\begin{aligned} r_x - \rho &= 0 \\ r_t + \rho v &= 0 \\ z_t - w &= 0 \\ z_x + r &= 0 \\ w_t + p + v w_x &= 0 \\ \rho(p_t + v p_x) + B(p, \rho^{-1}) v_x &= 0. \end{aligned} \tag{102}$$

The Lagrange system $\mathbf{L}\{y, s; v, p, q\}$ (98) has a nonlocally related subsystem obtained by excluding its dependent variable v through the integrability condition $v_{ys} = v_{sy}$. The resulting subsystem $\underline{\mathbf{L}}\{y, s; p, q\}$ is given by

$$q_{ss} + p_{yy} = 0, \qquad p_s + B(p, q) q_s = 0. \tag{103}$$

The resulting tree of nonlocally related systems, including two additional subsystems, is illustrated in Fig. 4.

Now treating the Lagrange system $\mathbf{L}\{y, s; v, p, q\}$ (98) as a given PDE system, from its three sets of local CL multipliers given by $(1, 0, 0)$, $(0, 1, 0)$, and $(y, s, 0)$, one can obtain the three singlet potential systems $\mathbf{LW}_1\{y, s; v, p, q, w_1\} = \mathbf{G}_0\{y, s; x, v, p, q\}$ (100), $\mathbf{LW}_2\{y, s; v, p, q, w_2\}$ and $\mathbf{LW}_3\{y, s; v, p, q, w_3\}$ respectively given by

$$\begin{aligned} w_{1y} - q &= 0 \\ w_{1s} - v &= 0 \\ v_s + p_y &= 0 \\ p_s + B(p, q) v_y &= 0 \end{aligned} \tag{104}$$

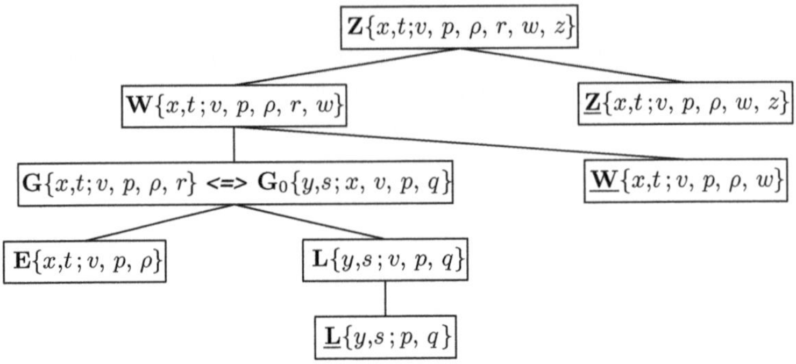

Fig. 4 Tree of nonlocally related PDE systems for PGD equations $\mathbf{E}\{x, t; v, p, \rho\}$ (97)

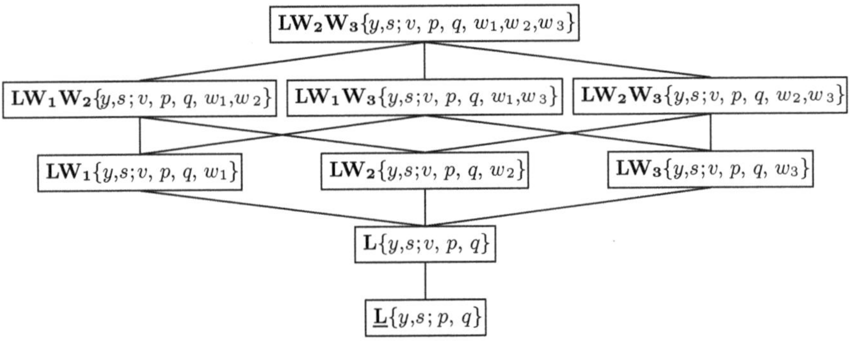

Fig. 5 Extension of tree of nonlocally related PDE systems for the Lagrange PGD system $\mathbf{L}\{y, s; v, p, q\}$ (98)

$$\begin{aligned} q_s - v_y &= 0 \\ w_{2y} - v &= 0 \\ w_{2s} + p &= 0 \\ p_s + B(p, q)v_y &= 0 \end{aligned} \tag{105}$$

and

$$\begin{aligned} w_{3y} - sv - yq &= 0 \\ w_{3s} + sp - yv &= 0 \\ v_s + p_y &= 0 \\ p_s + B(p, q)v_y &= 0. \end{aligned} \tag{106}$$

The extension of the tree illustrated in Fig. 4 is exhibited in Fig. 5.

Additional local CLs arise for the Lagrange system $\mathbf{L}\{y, s; v, p, q\}$ (98) when one considers sets of local CL multipliers of the form $\{\Lambda_i(y, s, V, P, Q)\}$, $i = 1$,

2, 3. After solving the corresponding determining equations, one can show that the resulting sets of local CL multipliers are given by

$$\Lambda_1 = \alpha y - \beta P + B(P, Q)\mu_3 + \delta$$
$$\Lambda_2 = \alpha s + \beta V + \nu$$
$$\Lambda_3 = \Lambda_3(y, P, Q)$$

where $\alpha, \beta, \nu, \delta$ are arbitrary constants and $\Lambda_3(y, P, Q)$ is any solution of the PDE

$$\frac{\partial \Lambda_3}{\partial Q} - \frac{\partial}{\partial P}(B(P, Q)\Lambda_3) + \beta = 0.$$

The additional local CLs that arise (for an *arbitrary* constitutive function $B(p, q)$) for the Lagrange system $\mathbf{L}\{y, s; v, p, q\}$ (98) include

- Conservation of energy $\frac{\partial}{\partial s}(\frac{1}{2}v^2 + K(p, q)) + \frac{\partial}{\partial y}(pv) = 0$ where $K(p, q)$ is any solution of the PDE $K_q - B(p, q)K_p + p = 0$.
- Conservation of entropy $\frac{\partial}{\partial s}S(p, q) = 0$ where $S(p, q)$ is any solution of the PDE $S_q - B(p, q)S_p = 0$.

In the case of a Lagrange PGD system $\mathbf{L}\{y, s; v, p, q\}$ (98), with a generalized polytropic equation of state given by

$$B(p, q) = \frac{M(p)}{q}, \qquad M''(p) \neq 0 \tag{107}$$

one can show that for local CL multipliers restricted to dependence on the independent variables (y, s), still only the three exhibited singlet potential systems (104)–(106) arise. For a generalized polytropic equation of state (107), the local symmetries arising for $\mathbf{L}\{y, s; v, p, q\}$ (98) and its resulting singlet, doublet and triplet potential systems that arise from the potential systems (104)–(106), as well as its subsystem $\underline{\mathbf{L}}\{y, s; p, q\}$ (103), are exhibited in [30].

The following remarks are noted.

- The exhibited extended trees of nonlocally related PDE systems hold for an *arbitrary* constitutive function $B(p, q)$.
- Either the Euler system $\mathbf{E}\{x, t; v, p, \rho\}$ (97) or the Lagrange system $\mathbf{L}\{y, s; v, p, q\}$ (98) can play the role of the given system in the tree.
- In a beautiful paper [49], a complete group classification with respect to the constitutive function $B(p, q)$ is given separately for the Euler and Lagrange systems but the connections between the systems are heuristic.
- To systematically construct nonlocal symmetries of the Euler and Lagrange systems, one needs to do the group classification problem for *all* PDE systems in an extended tree as well as consider other possible extended trees for specific constitutive functions followed by appropriate point symmetry analyses.

- For a Chaplygin gas given by $B(p,q) = -p/q$, one can show that the Lagrange subsystem $\underline{\mathbf{L}}\{y,s;\ p,q\}$ (103) has the point symmetry (which could not be exhibited in [49] due to its heuristic approach) $\mathbf{X} = -y^2\frac{\partial}{\partial y} - py\frac{\partial}{\partial p} + 3yq\frac{\partial}{\partial q}$ that in turn yields a nonlocal symmetry for both the Euler and Lagrange systems.
- Further extended trees arise for the PGD equations for specific constitutive functions:

  – $B(p, 1/\rho) = \rho(1+\mathrm{e}^p)$: Here the Euler potential system $\mathbf{G}\{x,t;\ v,p,\rho,r\}$ (99) has the family of local CLs given by $\mathrm{D}_t\left(\frac{f(r)\mathrm{e}^p}{1+\mathrm{e}^p}\right) + \mathrm{D}_x\left(\frac{f(r)v\mathrm{e}^p}{1+\mathrm{e}^p}\right) = 0$, for arbitrary $f(r)$. Such a local CL can be used to replace the fourth equation of $\mathbf{G}\{x,t;\ v,p,\rho,r\}$ (99) through introduction of a potential variable c and yields the corresponding potential system

$$
\begin{aligned}
r_x - \rho &= 0\\
r_t + \rho v &= 0\\
r_x(v_t + vv_x) + p_x &= 0\\
c_x + \mathrm{e}^p f(r)/(1+\mathrm{e}^p) &= 0\\
c_t - v\mathrm{e}^p f(r)/(1+\mathrm{e}^p) &= 0.
\end{aligned}
$$

  – For a Chaplygin gas given by $B(p, 1/\rho) = -p\rho$, the Euler potential system $\mathbf{G}\{x,t;\ v,p,\rho,r\}$ (99) has the family of local CLs given by $\mathrm{D}_t\left(\frac{f(r)}{p}\right) + \mathrm{D}_x\left(\frac{f(r)v}{p}\right) = 0$, for arbitrary $f(r)$. Such a local CL yields the corresponding potential system

$$
\begin{aligned}
r_x - \rho &= 0\\
r_t + \rho v &= 0\\
r_x(v_t + vv_x) + p_x &= 0\\
d_x + f(r)/p &= 0\\
d_t - vf(r)/p &= 0.
\end{aligned}
\tag{108}
$$

Here one can show that additional nonlocal symmetries arise for the Chaplygin gas Euler system $\mathbf{E}\{x,t;\ v,p,\rho\}$ (97) through the calculation of point symmetries for the potential system (108) only when $f(r) = r$, $f(r) = \text{const}$. For $f(r) = r$, the Chaplygin gas potential system (108) has the point symmetries $\mathbf{X}_{\mathbf{D}_1} = \left(-\frac{t^3}{6} + \mathrm{d}t\right)\frac{\partial}{\partial x} + \left(d - \frac{t^2}{2}\right)\frac{\partial}{\partial v} + rt\frac{\partial}{\partial p} - \frac{rt\rho}{p}\frac{\partial}{\partial p}$ and $\mathbf{X}_{\mathbf{D}_2} = \left(-\frac{t^2}{2} + d\right)\frac{\partial}{\partial x} + -t\frac{\partial}{\partial v} + r\frac{\partial}{\partial p} - \frac{r\rho}{p}\frac{\partial}{\partial p}$. The symmetry $\mathbf{X}_{\mathbf{D}_1}$ is a nonlocal symmetry for both the Euler and Lagrange systems and consequently was not able to be exhibited in [49]. On the other hand, the symmetry $\mathbf{X}_{\mathbf{D}_2}$ is a nonlocal symmetry for the Euler system but a local symmetry for the Lagrange system.

4.2 Symmetry-based Method to Obtain Nonlocally Related Systems and Nonlocal Symmetries

It is now shown that any point symmetry of a given PDE system systematically yields an equivalent nonlocally related PDE system. To illustrate the situation, consider as an example the nonlinear reaction diffusion equation

$$u_t - u_{xx} = Q(u). \tag{109}$$

One can show that for any nonlinear reaction term $Q(u)$, the PDE (109) has no local conservation laws. Hence the CL-based method yields no nonlocally related systems for the PDE (109). On the other hand, note that the PDE (109) is invariant under translations in x and t.

Consider the invariance of PDE (109) under translations in x. After an interchange of the variables x and u, the PDE (109) becomes the invertibly equivalent PDE

$$x_t = \frac{x_{uu} - Q(u)x_u^3}{x_u^2}. \tag{110}$$

Accordingly, we introduce two auxiliary dependent variables $v = x_u$, $w = x_t$, and consider the *intermediate PDE system*

$$v = x_u, \qquad w = x_t, \qquad w = \frac{v_u - Q(u)v^3}{v^2}. \tag{111}$$

By its construction, the intermediate PDE system (111) is locally related to the given scalar PDE (109). Now consider the subsystem (*inverse potential system*) of the intermediate system (111) that is obtained by excluding x through the integrability condition $x_{ut} = x_{tu}$, namely

$$v_t = w_u, \qquad w = \frac{v_u - Q(u)v^3}{v^2}. \tag{112}$$

The intermediate system (111) (and hence the given PDE (109)) is nonlocally related to the inverse potential system (112). This follows from the intermediate system (111) being the potential system of the PDE system (112) with the potential variable x arising from the first equation in the inverse potential system (112), which is a local CL as written. Moreover, excluding w from the inverse potential system (112), one obtains the scalar PDE

$$v_t = \left(\frac{v_u - Q(u)v^3}{v^2}\right)_u \tag{113}$$

which is clearly nonlocally related to the given PDE (109) since the PDE (109) has no local CLs.

Hence through the example of the nonlinear reaction diffusion equation (109), one essentially sees that any point symmetry of a given PDE system naturally yields a nonlocally related system. This will be seen more explicitly as follows.

4.2.1 Construction of a Nonlocally Related System from a Point Symmetry

Consider a given PDE system

$$R^{\sigma}(x,t,u,\partial u,\ldots,\partial^k u)=0, \qquad \sigma=1,\ldots,N \tag{114}$$

where $u=(u^1(x,t),\ldots,u^m(x,t))$. Suppose the PDE system (114) has a point symmetry

$$\mathrm{X}=\xi(x,t,u)\frac{\partial}{\partial x}+\tau(x,t,u)\frac{\partial}{\partial t}+\eta^i(x,t,u)\frac{\partial}{\partial u^i}. \tag{115}$$

Let $X(x,t,u)$, $T(x,t,u)$, $U^1(x,t,u)$, …, $U^m(x,t,u)$ be corresponding canonical coordinates so that the point symmetry X of the PDE system transforms to $\mathrm{Y}=\frac{\partial}{\partial U^1}$, i.e., the PDE system (114) transforms invertibly to a PDE system invariant under translations in U^1 given by

$$\hat{R}^{\sigma}(X,T,\hat{U},\partial U,\ldots,\partial^k U)=0, \qquad \sigma=1,\ldots,N \tag{116}$$

with $\hat{U}=(U^2,\ldots,U^m)$, $U=(U^1,\ldots,U^m)$.

Now consider the *intermediate PDE system*, obtained after introducing two auxiliary dependent variables $\alpha=U^1_T$, $\beta=U^1_X$

$$\begin{aligned}
&\alpha=U^1_T\\
&\beta=U^1_X\\
&\tilde{R}^{\sigma}(X,T,\hat{U},\alpha,\beta,\partial\hat{U},\ldots,\partial^{k-1}\alpha,\partial^{k-1}\beta,\partial^k\hat{U})=0, \qquad \sigma=1,\ldots,N
\end{aligned} \tag{117}$$

where $\tilde{R}^{\sigma}(X,T,\hat{U},\alpha,\beta,\partial\hat{U},\ldots,\partial^{k-1}\alpha,\partial^{k-1}\beta,\partial^k\hat{U})$ is obtained from $\hat{R}^{\sigma}(X,T,\hat{U},\partial U,\ldots,\partial^k U)$ after making the appropriate substitutions. By construction, the intermediate system (117) is locally equivalent to the given PDE system (114). Excluding the dependent variable U^1 from the intermediate system (117), one obtains the equivalent *inverse potential system*

$$\begin{aligned}
&\alpha_X=\beta_T\\
&\tilde{R}^{\sigma}(X,T,\hat{U},\alpha,\beta,\partial\hat{U},\ldots,\partial^{k-1}\alpha,\partial^{k-1}\beta,\partial^k\hat{U})=0, \quad \sigma=1,\ldots,N.
\end{aligned} \tag{118}$$

The inverse potential system (118) is nonlocally related to the given PDE system (114) since the intermediate system (117) is the potential system for the inverse potential system (118) with its dependent variable U^1 playing the role of the potential variable arising from the displayed CL of the inverse potential system (118).

Consequently, the following theorem has been proved.

Theorem 16 *Any point symmetry of a PDE system (114) yields an equivalent nonlocally related PDE system (inverse potential system) given by the PDE system (118).*

This theorem can be extended to the situation of three or more independent variables. Here the resulting inverse potential system has curl-type CLs.

4.2.2 The Special Situation When the Given PDE is an Evolutionary Scalar PDE

When a given PDE system (114) is an evolutionary scalar PDE, then another related PDE system naturally arises. The situation is summarized by the following theorem whose proof is immediately obvious.

Theorem 17 *Suppose a given PDE is an evolutionary scalar PDE invariant under a point symmetry. Without loss of generality, here the given PDE can be taken to be of the form*

$$u_t = F(x, t, u_1, \ldots, u_k) \tag{119}$$

with $u_i = \frac{\partial^i u}{\partial x^i}$. *Let* $\beta = u_x$. *Then the scalar PDE*

$$\beta_t = \mathrm{D}_x F(x, t, \beta, \ldots, \beta_{k-1}) \tag{120}$$

is a locally related subsystem of the corresponding inverse potential system resulting from the invariance of the PDE (119) under translations in u.

Example Nonlinear Wave Equation

As an example, consider again the nonlinear wave equation (74) and its nonlocally related potential system (75). The invariance of the potential system (75) under translations in t and v shows that the PDE system (75) is invariant under the point symmetry with the infinitesimal generator

$$\mathrm{X} = \frac{\partial}{\partial v} - \frac{\partial}{\partial t}. \tag{121}$$

Corresponding canonical coordinates are represented by the point transformation

$$X = x, \qquad T = u, \qquad U = t + v, \qquad V = v \tag{122}$$

with the potential system (75) invariant under translations in V. The point transformation (122) maps the potential system (75) into the invertibly related PDE system

$$
\begin{aligned}
V_X U_T - V_T U_X - 1 &= 0 \\
V_T + c^2(T)(U_X - V_X) &= 0
\end{aligned}
\tag{123}
$$

which is invariant under translations in U and V.

From the invariance of the PDE system (123) under translations in V, one accordingly introduces auxiliary dependent variables $\alpha(X, T), \beta(X, T)$ to obtain the locally related intermediate system

$$
\begin{aligned}
\alpha &= V_T \\
\beta &= V_X \\
\beta U_T - \alpha U_X - 1 &= 0 \\
\alpha + c^2(T)(U_X - \beta) &= 0.
\end{aligned}
\tag{124}
$$

Excluding V from the intermediate system (124), one obtains the inverse potential system

$$
\begin{aligned}
\beta_T &= \alpha_X \\
\beta U_T - \alpha U_X - 1 &= 0 \\
\alpha + c^2(T)(U_X - \beta) &= 0.
\end{aligned}
\tag{125}
$$

It is straightforward to exclude the dependent variables α and β from the last two equations of the inverse potential system (125) to obtain its locally related scalar PDE

$$
\begin{aligned}
U_{TT} + c^4(T)U_{XX} + c^2(T)[2U_{TX}U_T U_X - U_{XX}U_T^2 \\
-U_{TT}U_X^2 - 2U_{TX}] + 2c(T)c'(T)[U_X^2 U_T - U_X] = 0.
\end{aligned}
\tag{126}
$$

In [29], it is shown that the scalar PDE (126) is nonlocally related to the scalar nonlinear wave equation (74) through comparison of the symmetry classifications of these two PDEs.

When $c(u) = u^{-2}$, one can show [28, 29] that the PDE (126) has the point symmetry $U^2\frac{\partial}{\partial U} + TU\frac{\partial}{\partial T} - \frac{U}{T^3}\frac{\partial}{\partial X}$ that yields a previously unknown nonlocal symmetry of both the nonlinear wave equation (74) and the potential system (75).

Further details and examples of the symmetry-based method to obtain nonlocally related systems and nonlocal symmetries are presented in [28, 29].

5 Nonlocality in Multidimensions

In the multidimensional situation ($n \geq 3$ independent variables), a local conservation law for a given PDE system yields $\frac{1}{2}n(n-1)$ potential variables. It will be shown that a local symmetry of the resulting potential system *always* corresponds to a local symmetry of the given PDE system (As we have seen, this is not the situation for $n = 2$ independent variables).

In the conservation law-based approach, to obtain nonlocal symmetries of a given PDE system it is necessary to augment the potential system by a *gauge constraint.*

5.1 Divergence-type CLs and Corresponding Potential Systems

Consider a PDE system with N PDEs of order k with $n \geq 3$ independent variables $x = (x^1, \dots, x^n)$ and m dependent variables $u(x) = (u^1(x), \dots, u^m(x))$

$$R^\sigma[u] = R^\sigma(x, u, \partial u, \dots, \partial^k u) = 0, \qquad \sigma = 1, \dots, N. \tag{127}$$

Suppose the PDE system (127) has a divergence-type CL given by

$$\text{div}\ \Phi[u] = \mathrm{D}_i \Phi^i[u] \equiv \mathrm{D}_i \Phi^i(x, u, \partial u, \dots, \partial^r u) = 0. \tag{128}$$

From Poincaré's lemma, the local CL (128) yields $\frac{1}{2}n(n-1)$ potential variables $v^{jk}(x) = -v^{kj}(x)$. This leads to a set of n potential equations

$$\Phi^i[u] \equiv \mathrm{D}_j v^{ij}, \qquad i = 1, \dots, n \tag{129}$$

equivalent to the local CL (128). The corresponding *potential system* is the union of the given PDE system (127) and the set of potential equations (129). This potential system is nonlocally related and equivalent to the given PDE system (127). In turn the potential system has the *gauge freedom* invariance given by the transformation

$$v^{ij} \to \mathrm{D}_k w^{ijk} \tag{130}$$

where the functions $w^{ijk}(x)$ are $\frac{1}{6}n(n-1)(n-2)$ *arbitrary* functions that are the components of a totally antisymmetric tensor, i.e., the constructed potential system has an infinite number of point symmetries (gauge symmetries) through the transformation (130) in terms of the infinitesimal generator

$$\mathrm{X}_{\text{gauge}} = \mathrm{D}_k w^{ijk}(x) \frac{\partial}{\partial v^{ij}}. \tag{131}$$

As it stands, the potential system is *underdetermined* due to the gauge freedom (130).

Now assume that the given PDE system (127) is *determined* in the sense that it does not have symmetries that involve *arbitrary functions* of *all* independent variables $x = (x^1, \dots, x^n)$. In particular, suppose the potential system has a local symmetry

$$\mathrm{X} = \eta^\mu(x, u, \partial u, \dots, \partial^P u, v, \partial v, \dots \partial^Q v) \frac{\partial}{\partial u^\mu} + \zeta^{\alpha\beta}[u, v] \frac{\partial}{\partial v^{\alpha\beta}}. \tag{132}$$

Then the potential system has local symmetries given by the commutator $[\mathrm{X}_{\text{gauge}}, \mathrm{X}]$ that project to the symmetries

$$\left(\alpha^{ij}\frac{\partial\eta^\mu}{\partial v^{ij}} + (\mathrm{D}_{i_1}\alpha^{ij})\frac{\partial\eta^\mu}{\partial v^{ij}_{i_1}} + \cdots + (\mathrm{D}_{i_1}\cdots\mathrm{D}_{i_Q}\alpha^{ij})\frac{\partial\eta^\mu}{\partial v^{ij}_{i_1\cdots i_Q}}\right)\frac{\partial}{\partial u^\mu} \tag{133}$$

of the PDE system (127) with $\alpha^{ij}(x) = \mathrm{D}_k w^{ijk}(x)$, and $v^{ij}_{i_1\cdots i_R} = \mathrm{D}_{i_1}\cdots\mathrm{D}_{i_R}\alpha^{ij}$ denoting derivatives of v^{ij}. In the infinitesimal generator (133), $\alpha^{ij}(x)$ and each of its derivatives are arbitrary functions of $x = (x^1, \ldots, x^n)$. Since the given PDE system (127) is a *determined* system, it follows that the symmetry (133) is a symmetry of the given PDE system (127) if and only if $\frac{\partial\eta^\mu}{\partial v^{ij}} = \frac{\partial\eta^\mu}{\partial v^{ij}_{i_1}} = \cdots = \frac{\partial\eta^\mu}{\partial v^{ij}_{i_1\cdots i_Q}} \equiv 0$. Thus each local symmetry of the *underdetermined* potential system, arising from a divergence-type conservation law, yields only a local symmetry of the given *determined* PDE system (127).

Hence if a potential system arising from a divergence-type conservation law of a given PDE system (127) is to be used to seek a nonlocal symmetry of the PDE system (127) from a point symmetry of the potential system, *it is necessary to augment* the potential system with auxiliary constraint equations (*gauge constraints*) to obtain a *determined potential system.*

Definition 7 A *gauge constraint* has the property that the augmented potential system is equivalent to the given PDE system (127), i.e., every solution of the augmented potential system yields a solution of the given PDE system (127) and, conversely, every solution of the given PDE system (127) yields a solution of the augmented potential system.

Some examples of gauges (relating potential variables) include

- divergence (Coulomb) gauge
- spatial gauge
- Poincaré gauge
- Lorentz gauge (a form of divergence gauge)
- Cronstrom gauge (a form of Poincaré gauge).

For details on these gauges, see [32].

Example Wave Equation

As an example, consider the wave equation

$$u_{tt} - u_{xx} - u_{yy} = 0 \tag{134}$$

which is already a divergence-type CL. Correspondingly, one has the vector potential $v = (v^0, v^1, v^2)$ and the underdetermined potential system given by

$$
\begin{aligned}
u_t &= v^2_x - v^1_y \\
-u_x &= v^0_y - v^2_t \\
-u_y &= v^1_t - v^0_x.
\end{aligned} \tag{135}
$$

Now consider the equivalent augmented constrained system obtained by appending the Lorentz gauge

$$
v^0_t - v^1_x - v^2_y = 0 \tag{136}
$$

to the underdetermined potential system (135) to obtain the determined potential system

$$
\begin{aligned}
u_t &= v^2_x - v^1_y \\
-u_x &= v^0_y - v^2_t \\
-u_y &= v^1_t - v^0_x \\
0 &= v^0_t - v^1_x - v^2_y.
\end{aligned} \tag{137}
$$

One can show [32] that the determined potential system (137) has six point symmetries that yield nonlocal symmetries as well as nonlocal CLs of the wave equation (134). One such point symmetry is given by the infinitesimal generator

$$
\begin{aligned}
\mathrm{X} = {} & (yv^1 - xv^2 - tu)\frac{\partial}{\partial u} - (2tv^0 + xv^1 + yv^2)\frac{\partial}{\partial v^0} \\
& - (xv^0 + 2tv^1 - yu)\frac{\partial}{\partial v^1} - (yv^0 + 2tv^2 + xu)\frac{\partial}{\partial v^2}.
\end{aligned}
$$

The other listed gauges yield no nonlocal symmetries from point symmetries of the corresponding determined potential systems.

5.2 *Systematic Procedures to Seek Nonlocal Symmetries in Multidimensions*

In the multidimensional situation ($n \geq 3$ independent variables), four systematic procedures (some with known examples) are presented to search for nonlocal symmetries of a given PDE system through seeking local symmetries of an equivalent nonlocally related PDE system.

- Potential systems arising from divergence-type conservation laws (of degree r; $1 < r \leq n-1$) augmented with gauge constraints to yield a determined potential system.
- Determined potential systems arising from curl-type conservation laws (of degree 1).
- Determined nonlocally related systems arising from admitted point symmetries. Here, each point symmetry of a given PDE system systematically yields a deter-

mined inverse potential system connected to an intermediate system through a curl-type conservation law of degree 1 [2, 50, 51].
- Determined nonlocally related subsystems.

In the case of three independent variables ($n = 3$), two types of local CLs arise.

- Degree 2 CLs (divergence-type CLs).
- Degree 1 CLs (curl-type CLs).

Potential systems arising from lower degree CLs ($r < n - 1$) essentially correspond to particular gauge constraints for underdetermined potential systems arising from divergence-type CLs.

Examples illustrating the types of nonlocal symmetries that can arise as described above appear in [50, 51].

5.3 Some Open Problems in Multidimensions

There are many open problems in seeking systematically nonlocal symmetries for multidimensional PDE systems. These include the following.

- Find examples of *nonlinear* PDE systems for which nonlocal symmetries arise as local symmetries of a potential system following from divergence-type CLs appended with gauge constraints.
- Find efficient procedures to obtain "useful" gauge constraints (eg, yielding nonlocal symmetries/nonlocal CLs) for potential systems arising from divergence-type CLs (as well as for underdetermined potential systems arising from lower-degree CLs). Can one rule out specific families of gauges for particular classes of potential systems?
- Find further examples of lower-degree CLs for PDE systems of physical importance. CLs of degree one (curl-type) are of particular interest since corresponding potential systems are determined. Examples to-date suggest that lower-degree CLs are rare and only arise when a given PDE system has a special geometrical structure. Of course, divergence-type CLs are common!
- Find examples of PDE systems of physical interest admitting point symmetries that in turn yield nonlocal symmetries of the systems.
- Find useful subsystems and useful means of obtaining subsystems (including in the two-dimensional case). Progress has been made in this direction [28, 29].
- Extend the work on obtaining nonlocally related systems to multidimensions for continuum mechanics systems such as gas dynamics equations and equations of dynamical nonlinear elasticity. A start on this has been made in [52].

References

1. Bluman, G.W., Anco, S.C.: Symmetry and Integration Methods for Differential Equations. Springer, New York (2002)
2. Bluman, G.W., Cheviakov, A.F., Anco, S.C.: Applications of Symmetry Methods to Partial Differential Equations. Springer, New York (2010)
3. Kumei, S., Bluman, G.W.: When nonlinear differential equations are equivalent to linear differential equations. SIAM J. Appl. Math. **42**, 1157–1173 (1982)
4. Bluman, G.W., Kumei, S.: Symmetry-based algorithms to relate partial differential equations. I. Local symmetries. EJAM **1**, 189–216 (1990)
5. Noether, E.: Invariante Variationsprobleme. Nachr. König. Gesell. Wissen. Göttingen, Math.-Phys., 235–257 (1918)
6. Anderson, R.L., Kumei, S., Wulfman, C.E.: Generalization of the concept of invariance of differential equations. Phys. Rev. Lett. **28**, 988–991 (1972)
7. Anderson, R.L., Ibragimov, N.H.: Lie-Bäcklund Transformations in Applications. SIAM, Philadelphia (1979)
8. Olver, P.J.: Applications of Lie Groups to Differential Equations. Springer, New York (1986)
9. Anco, S.C., Bluman, G.W.: Direct construction of conservation laws from field equations. Phys. Rev. Lett. **78**, 2869–2873 (1997)
10. Anco, S.C., Bluman, G.W.: Direct construction method for conservation laws of partial differential equations. Part I: examples of conservation law classifications. EJAM **13**, 545–566 (2002)
11. Anco, S.C., Bluman, G.W.: Direct construction method for conservation laws of partial differential equations. Part II: general treatment. EJAM **13**, 567–585 (2002)
12. Bluman, G.W.: Construction of solutions to partial differential equations by the use of transformation groups. Ph.D. Thesis, California Institute of Technology, Pasadena, CA (1967)
13. Bluman, G.W., Cole, J.D.: General similarity solution of the heat equation. J. Math. Mech. **18**, 1025–1042 (1969)
14. Hereman, W.: Review of symbolic software for lie symmetry analysis. Math. Comput. Model. **25**, 115–132
15. Wolf, T.: Investigating differential equations with CRACK, LiePDE, applsymm and ConLaw. In: Grabmeier, J., Kaltofen, E., Weispfenning. (eds.) Handbook of Computer Algebra, Foundations, Applications, Systems, vol. 37, pp. 465–468. Springer, New York (2002)
16. Wolf, T.: A comparison of four approaches to the calculation of conservation laws. EJAM **13**, 129–152 (2002)
17. Cheviakov, A.F.: GeM software package for computation of symmetries and conservation laws of differential equations. Comput. Phys. Commun. **176**, 48–61 (2007)
18. Cheviakov, A.F.: "GeM": a maple module for symmetry and conservation law computation for PDEs/ODEs. http://www.math.usask.ca/cheviakov/gem/ (2013)
19. Olver, P.J.: Evolution equations possessing infinitely many symmetries. J. Math. Phys. **18**, 1212–1215 (1977)
20. Konopolchenko, B.G., Mokhnachev, V.G.: On the group theoretical analysis of differential equations. J. Phys. **A13**, 3113–3124 (1980)
21. Kumei, S.: A group analysis of nonlinear differential equations. Ph.D. Thesis, University of British Columbia, Vancouver, BC (1981)
22. Kapcov, O.V.: Extension of the symmetry of evolution equations. Sov. Math. Dokl. **25**, 173–176 (1982)
23. Pukhnachev, V.V.: Equivalence transformations and hidden symmetry of evolution equations. Sov. Math. Dokl. **35**, 555–558 (1987)
24. Bluman, G.W., Temuerchaolu, Anco, S.C.: New conservation laws obtained directly from symmetry action on a known conservation law. JMAA **322**, 233–250 (2006)

25. Bluman, G.W., Kumei, S.: On invariance properties of the wave equation. J. Math. Phys. **28**, 307–318 (1987)
26. Bluman, G.W., Kumei, S., Reid, G.J.: New classes of symmetries of partial differential equations. J. Math. Phys. **29**, 806–811; Erratum. J. Math. Phys. **29**, 2320 (1988)
27. Bluman, G.W., Kumei, S.: Symmetries and Differential Equations. Springer, New York (1989)
28. Yang, Z.: Nonlocally related partial differential equation systems, the nonclassical method and applications. Ph.D Thesis, University of British Columbia, Vancouver, BC (2013)
29. Bluman, G.W., Yang, Z.: A symmetry-based method for constructing nonlocally related PDE systems. J. Math. Phys. **54**, 093504 (2013)
30. Bluman, G.W., Cheviakov, A.F.: Framework for potential systems and nonlocal symmetries: algorithmic approach. J. Math. Phys. **46**, 123506 (2005)
31. Bluman, G.W., Cheviakov, A.F., Ivanova, N.M.: Framework for nonlocally related partial differential equations systems and nonlocal symmetries: extension, simplification, and examples. J. Math. Phys. **47**, 113505 (2006)
32. Anco, S.C., Bluman, G.W.: Nonlocal symmetries and nonlocal conservation laws of Maxwell's equations. J. Math. Phys. **38**, 3508–3532 (1997)
33. Anco, S.C., The, D.: Symmetries, conservation laws, and cohomology of Maxwell's equations using potentials. Acta Appl. Math. **89**, 1–52 (2005)
34. Bluman, G.W., Cheviakov, A.F., Ganghoffer, J.-F.: Nonlocally related PDE systems for one-dimensional nonlinear elastodynamics. J. Eng. Math. **62**, 203–221 (2008)
35. Bluman, G.W., Cheviakov, A.F., Ganghoffer, J.-F.: On the nonlocal symmetries, group invariant solutions and conservation laws of the equations of nonlinear dynamical compressible elasticity. In: Proceedings of IUTAM Symposium on Progress in the Theory and Numerics of Configurational Mechanics, pp. 107–120. Springer, New York (2009)
36. Bluman, G.W., Kumei, S.: Symmetry-based algorithms to relate partial differential equations. II. Linearization by nonlocal symmetries. EJAM **1**, 217–223 (1990)
37. Bluman, G.W., Shtelen, V.M.: New classes of Schroedinger equations equivalent to the free particle equation through non-local transformations. J. Phys. **A29**, 4473–4480 (1996)
38. Bluman, G.W., Shtelen, V.M.: Nonlocal transformations of Kolmogorov equations into the backward heat equation. JMAA **291**, 419–437 (2004)
39. Bluman, G.W.: On the transformation of diffusion processes into the Wiener process. SIAM J. Appl. Math. **39**, 238–247 (1980)
40. Bluman, G.W.: On mapping linear partial differential equations to constant coefficient equations. SIAM J. Appl. Math. **43**, 1259–1273 (1983)
41. Bluman, G.W.: Temuerchaolu: Conservation laws for nonlinear telegraph equations. JMAA **310**, 459–476 (2005)
42. Volterra, V.: Leçons sur les Functions de Lignes. Gauthier-Villars, Paris (1913)
43. Vainberg, M.M.: Variational Methods for the Study of Nonlinear Operators. Holden-Day, San Francisco (1964)
44. Ames, W.F., Lohner, R.J., Adams, E.: Group properties of $u_{tt} = [f(u)u_x]_x$. Int. J. Nonlinear Mech. **16**, 439–447 (1981)
45. Ma, A.: Extended group analysis of the wave equation. M.Sc. Thesis, University of British Columbia, Vancouver, BC (1991)
46. Bluman, G.W., Cheviakov, A.F.: Nonlocally related systems, linearization and nonlocal symmetries for the nonlinear wave equation. JMAA **333**, 93–111 (2007)
47. Bluman, G.W.: Temuerchaolu: Comparing symmetries and conservation laws of nonlinear telegraph equations. J. Math. Phys. **46**, 073513 (2005)
48. Bluman, G.W.: Temuerchaolu, Sahadevan, R.: Local and nonlocal symmetries for nonlinear telegraph equations. J. Math. Phys. **46**, 023505 (2005)
49. Akhatov, I.S., Gazizov, R.K., Ibragimov, N.H.: Nonlocal symmetries. Heuristic approach. J. Sov. Math. **55**, 1401–1450 (1991)

50. Cheviakov, A.F., Bluman, G.W.: Multidimensional partial differential equation systems: generating new systems via conservation laws, potentials, gauges, subsystems. J. Math. Phys. **51**, 103521 (2010)
51. Cheviakov, A.F., Bluman, G.W.: Multidimensional partial differential equation systems: nonlocal symmetries, nonlocal conservation laws, exact solutions. J. Math. Phys. **51**, 103522 (2010)
52. Bluman, G.W., Ganghoffer, J.-F.: Connecting Euler and Lagrange systems as nonlocally related systems of dynamical nonlinear elasticity. Arch. Mech. **63**, 363–382 (2011)

Construction of Conservation Laws Using Symmetries

Nail H. Ibragimov

Abstract The concept of *nonlinear self-adjointness* of differential equations, introduced by the author in 2010, is discussed in detail. All linear equations and systems are nonlinearly self-adjoint. Moreover, the class of nonlinearly self-adjoint equations includes all nonlinear equations and systems having at least one *local* conservation law. It follows, in particular, that the *integrable systems* possessing infinite set of *Lie-Bäcklund* symmetries (*higher-order tangent transformations*) are nonlinearly self-adjoint. An explicit formula for conserved vectors associated with symmetries is provided for all nonlinearly self-adjoint differential equations and systems. The number of equations contained in the systems under consideration can be different from the number of dependent variables. A utilization of conservation laws for constructing exact solutions is discussed and illustrated by computing non-invariant solutions of the Chaplygin equations in gas dynamics.

1 Nonlinear Self-Adjointness

The concept of self-adjointness of nonlinear equations was introduced [1, 2] for constructing conservation laws associated with symmetries of differential equations. To extend the possibilities of the new method for constructing conservation laws the notion of quasi self-adjointness was suggested in [3]. I introduce here the general concept of *nonlinear self-adjointness*. It embraces the previous notions of self-adjoint

N. H. Ibragimov (✉)
Laboratory "Group Analysis of Mathematical Models in Natural and Engineering Sciences",
Ufa State Aviation Technical University, Ufa 450 000, Russia
e-mail: nailhib@gmail.com

N. H. Ibragimov
Research Centre ALGA: Advances in Lie Group Analysis,
Department of Mathematics and Science, Blekinge Institute of Technology,
SE-371 79 Karlskrona, Sweden

J.-F. Ganghoffer and I. Mladenov (eds.), *Similarity and Symmetry Methods*, Lecture Notes in Applied and Computational Mechanics 73, DOI: 10.1007/978-3-319-08296-7_2,

and quasi self-adjoint equations and includes the linear self-adjointness as a particular case. But the set of nonlinearly self-adjoint equations is essentially wider and contains, in particular, all *linear equations* and *nonlinear equations and systems having at least one local conservation law*, including the so-called integrable systems. The construction of conservation laws demonstrates a practical significance of the nonlinear self-adjointness. Namely, *conservation laws can be associated with symmetries for all nonlinearly self-adjoint differential equations and systems.*

1.1 Preliminaries

1.1.1 Notation

We will use the following notation. The independent variables are denoted by

$$x = (x^1, \ldots, x^n).$$

The dependent variables are

$$u = (u^1, \ldots, u^m).$$

They are used together with their first-order partial derivatives $u_{(1)}$

$$u_{(1)} = \{u_i^\alpha\}, \qquad u_i^\alpha = D_i(u^\alpha)$$

and higher-order derivatives $u_{(2)}, \ldots, u_{(s)}, \ldots,$ where

$$\begin{aligned} u_{(2)} &= \{u_{ij}^\alpha\}, & u_{ij}^\alpha &= D_i D_j(u^\alpha), \ldots \\ u_{(s)} &= \{u_{i_1 \cdots i_s}^\alpha\}, & u_{i_1 \cdots i_s}^\alpha &= D_{i_1} \cdots D_{i_s}(u^\alpha). \end{aligned}$$

Here D_i is the total differentiation with respect to x^i

$$D_i = \frac{\partial}{\partial x^i} + u_i^\alpha \frac{\partial}{\partial u^\alpha} + u_{ij}^\alpha \frac{\partial}{\partial u_j^\alpha} + \cdots. \tag{1}$$

A locally analytic function $f(x, u, u_{(1)}, \ldots, u_{(k)})$ of any finite number of the variables $x, u, u_{(1)}, u_{(2)}, \ldots$ is called a *differential function*. The set of all differential functions is denoted by $\mathcal{A}$. For more details see [4, Chap. 8].

1.1.2 Linear Self-Adjointness

Recall that the adjoint operator F^* to a linear operator F in a Hilbert space H with a scalar product (u, v) is defined by

$$(Fu, v) = (u, F^*v), \qquad u, v \in H. \tag{2}$$

Let us consider, for the sake of simplicity, the case of one dependent variable u and denote by H the Hilbert space of real valued functions $u(x)$ such that $u^2(x)$ is integrable. The scalar product is given by

$$(u, v) = \int_{\mathbb{R}^n} u(x)v(x)\mathrm{d}x.$$

Let F be a linear differential operator in H. Its action on the dependent variable u is denoted by $F[u]$. The Definition (2) of the adjoint operator F^* to F

$$(F[u], v) = (u, F^*[v])$$

can be written, using the divergence theorem, in the simple form

$$vF[u] - uF^*[v] = D_i(p^i) \tag{3}$$

where v is a new dependent variable, and p^i are any functions of x, u, v, $u_{(1)}$, $v_{(1)}, \ldots$ It is manifest from Eq. (3) that the operators F and F^* are mutually adjoint

$$\left(F^*\right)^* = F. \tag{4}$$

In other words, the adjointness of linear operators is a *symmetric relation.*

The linear operator F is said to be self-adjoint if $F^* = F$. In this case we say that the equation $F[u] = 0$ is self-adjoint. Thus, the self-adjointness of a linear equation $F[u] = 0$ can be expressed by the equation

$$F^*[v]\Big|_{v=u} = F[u]. \tag{5}$$

1.1.3 Adjoint Equations to Nonlinear Differential Equations

Let us consider a system of m differential equations (linear or nonlinear)

$$F_\alpha\big(x, u, u_{(1)}, \ldots, u_{(s)}\big) = 0, \qquad \alpha = 1, \ldots, m \tag{6}$$

with m dependent variables $u = (u^1, \ldots, u^m)$. Equation (6) involve the partial derivatives $u_{(1)}, \ldots, u_{(s)}$ up to order s.

Definition 1 The *adjoint equation* to equation (6) are given by

$$F^*_\alpha\big(x, u, v, u_{(1)}, v_{(1)}, \ldots, u_{(s)}, v_{(s)}\big) = 0, \qquad \alpha = 1, \ldots, m \tag{7}$$

with

$$F^*_\alpha\big(x, u, v, u_{(1)}, v_{(1)}, \ldots, u_{(s)}, v_{(s)}\big) = \frac{\delta \mathcal{L}}{\delta u^\alpha} \tag{8}$$

where $\mathcal{L}$ is the *formal Lagrangian* for Eq. (6) defined by [1]

$$\mathcal{L} = v^\beta F_\beta \equiv \sum_{\beta=1}^{m} v^\beta F_\beta. \tag{9}$$

Here $v = (v^1, \ldots, v^m)$ are new dependent variables, $v_{(1)}, \ldots, v_{(s)}$ are their derivatives, e.g., $v_{(1)} = \{v_i^\alpha\}$, $v_i^\alpha = D_i(v^\alpha)$. We use $\delta/\delta u^\alpha$ for the Euler-Lagrange operator

$$\frac{\delta}{\delta u^\alpha} = \frac{\partial}{\partial u^\alpha} + \sum_{s=1}^{\infty}(-1)^s D_{i_1} \cdots D_{i_s} \frac{\partial}{\partial u^\alpha_{i_1 \cdots i_s}}, \qquad \alpha = 1, \ldots, m$$

so that

$$\frac{\delta(v^\beta F_\beta)}{\delta u^\alpha} = \frac{\partial(v^\beta F_\beta)}{\partial u^\alpha} - D_i\left(\frac{\partial(v^\beta F_\beta)}{\partial u_i^\alpha}\right) + D_i D_k\left(\frac{\partial(v^\beta F_\beta)}{\partial u_{ik}^\alpha}\right) - \cdots.$$

The total differentiation (1) is extended to the new dependent variables

$$D_i = \frac{\partial}{\partial x^i} + u_i^\alpha \frac{\partial}{\partial u^\alpha} + v_i^\alpha \frac{\partial}{\partial v^\alpha} + u_{ij}^\alpha \frac{\partial}{\partial u_j^\alpha} + v_{ij}^\alpha \frac{\partial}{\partial v_j^\alpha} + \cdots. \tag{10}$$

The adjointness of *nonlinear* equations is not a symmetric relation. In other words, nonlinear equations, unlike the linear ones, do not obey the condition (4) of mutual adjointness. Instead, the following equation holds

$$\big(F^*\big)^* = \hat{F} \tag{11}$$

where $\hat{F}$ is the *linear approximation* to F defined as follows. We use the temporary notation $F[u]$ for the left-hand side of Eq. (6) and consider $F[u + w]$ by letting $w \ll 1$. Then neglecting the nonlinear terms in w we define $\hat{F}$ by the equation

[1] See [2]. An approach in terms of variational principles is developed in [5].

$$F[u+w] \approx F[u] + \hat{F}[w]. \tag{12}$$

For linear equations we have $\hat{F} = F$, and hence Eq. (11) is identical with Eq. (4). Let us illustrate Eq. (11) by the equation

$$F \equiv u_{xy} - \sin u = 0. \tag{13}$$

Equation (8) yields

$$F^* \equiv \frac{\delta}{\delta u}[v(u_{xy} - \sin u)] = v_{xy} - v\cos u \tag{14}$$

and

$$\left(F^*\right)^* \equiv \frac{\delta}{\delta v}[w(v_{xy} - v\cos u)] = w_{xy} - w\cos u. \tag{15}$$

Let us find $\hat{F}$ by using Eq. (12). Since $\sin w \approx w, \ \cos w \approx 1$ when $w \ll 1$, we have

$$\begin{aligned} F[u+w] \equiv & \ (u+w)_{xy} - \sin(u+w) \\ &= u_{xy} + w_{xy} - \sin u \cos w - \sin w \cos u \\ &\approx u_{xy} - \sin u + w_{xy} - w\cos u \\ &= F[u] + w_{xy} - w\cos u. \end{aligned}$$

Hence, by (12) and(15), we have

$$\hat{F}[w] = w_{xy} - w\cos u = \left(F^*\right)^* \tag{16}$$

in accordance with Eq. (11).

1.1.4 The Case of One Dependent Variable

Let us consider the differential equation

$$F\big(x, u, u_{(1)}, \ldots, u_{(s)}\big) = 0 \tag{17}$$

with one dependent variable u and any number of independent variables. In this case Definition 1 of the adjoint equation is written

$$F^*\big(x, u, v, u_{(1)}, v_{(1)}, \ldots, u_{(s)}, v_{(s)}\big) = 0 \tag{18}$$

where

$$F^*\big(x, u, v, u_{(1)}, v_{(1)}, \ldots, u_{(s)}, v_{(s)}\big) = \frac{\delta(vF)}{\delta u}. \tag{19}$$

1.1.5 Construction of Adjoint Equations to Linear Equations

The following statement has been formulated in [1, 2].

Proposition 1 *In the case of linear differential equations and systems, the adjoint equations determined by Eq.* (8) *and by Eq.* (3) *coincide.*

Proof The proof is based on the statement (see Proposition 8 in Sect. 2.1.2) that a function $Q(u, v)$ is a divergence, i.e., $Q = D_i(h^i)$, if and only if

$$\frac{\delta Q}{\delta u^\alpha} = 0, \qquad \frac{\delta Q}{\delta v^\alpha} = 0, \qquad \alpha = 1, \ldots, m. \tag{20}$$

Let the adjoint operator F^* be constructed according to Eq. (3). Let us consider the case of many dependent variables and write Eq. (3) as follows

$$v^\beta F_\beta[u] = u^\beta F^*_\beta[v] + D_i(p^i). \tag{21}$$

Applying to (21) the variational differentiations and using Eq. (20) we obtain

$$\frac{\delta(v^\beta F_\beta[u])}{\delta u^\alpha} = \delta^\beta_\alpha F^*_\beta[v] \equiv F^*_\alpha[v].$$

Hence, (8) coincides with $F^*_\alpha[v]$ given by (3).

Conversely, let $F^*[v]$ be given by (8), i.e.,

$$F^*_\beta[v] = \frac{\delta(v^\gamma F_\gamma[u])}{\delta u^\beta} \cdot$$

Consider the expression Q defined by

$$Q = v^\beta F_\beta[u] - u^\beta F^*_\beta[v] \equiv v^\beta F_\beta[u] - u^\beta \frac{\delta(v^\gamma F_\gamma[u])}{\delta u^\beta} \cdot$$

Applying to the first expression for Q the variational differentiations $\delta/\delta u^\alpha$ we obtain

$$\frac{\delta Q}{\delta u^\alpha} = \frac{\delta(v^\beta F_\beta[u])}{\delta u^\alpha} - \delta^\beta_\alpha F^*_\beta[v] \equiv F^*_\alpha[v] - \delta^\beta_\alpha F^*_\beta[v] = 0.$$

Applying $\delta/\delta v^\alpha$ to the second expression for Q we obtain

$$\frac{\delta Q}{\delta v^\alpha} = \delta^\beta_\alpha F_\beta[u] - \frac{\delta}{\delta v^\alpha}\left[u^\beta \frac{\delta(v^\gamma F_\gamma[u])}{\delta u^\beta}\right] \equiv F_\alpha[u] - \frac{\delta}{\delta v^\alpha}\left[u^\beta \frac{\delta(v^\gamma F_\gamma[u])}{\delta u^\beta}\right] \cdot$$

The reckoning shows that

$$\frac{\delta}{\delta v^\alpha}\left[u^\beta \frac{\delta(v^\gamma F_\gamma[u])}{\delta u^\beta}\right] = F_\alpha[u]. \tag{22}$$

Thus Q solves Eq. (20) and hence Eq. (21) is satisfied. This completes the proof.

Remark 1 Let us discuss the proof of Eq. (22) in the case of a second-order linear operator for one dependent variable

$$F[u] = a^{ij}(x)u_{ij} + b^i(x)u_i + c(x)u.$$

Then we have

$$u\frac{\delta(vF[u])}{\delta u} = u\left[cv - vD_i(b^i) + vD_iD_j(a^{ij}) - b^i v_i + 2v_iD_j(a^{ij}) + a^{ij}v_{ij}\right].$$

Whence, after simple calculations we obtain

$$\begin{aligned}\frac{\delta}{\delta v}\left[u\frac{\delta(vF[u])}{\delta u}\right] &= \left[cu + b^i u_i + a^{ij}u_{ij}\right]\\ &\quad + \left\{D_iD_j(a^{ij}u) - D_i(a^{ij}u_j) - D_i[uD_j(a^{ij})]\right\}\end{aligned}$$

and, noting that the expression in the braces vanishes, arrive at Eq. (22).

Let us illustrate Proposition 1 by the following simple example.

Example 1 Consider the heat equation

$$F[u] \equiv u_t - u_{xx} = 0 \tag{23}$$

and construct the adjoint operator to the linear operator

$$F = D_t - D_x^2 \tag{24}$$

by using Eq. (3). Noting that due to

$$\begin{aligned} vu_t &= D_t(uv) - uv_t\\ vu_{xx} &= D_x(vu_x) - v_xu_x = D_x(vu_x - uv_x) + uv_{xx}\end{aligned}$$

we have

$$vF[u] \equiv v(u_t - u_{xx}) = u(-v_t - v_{xx}) + D_t(uv) + D_x(uv_x - vu_x).$$

Hence

$$vF[u] - u(-v_t - v_{xx}) = D_t(uv) + D_x(uv_x - vu_x).$$

Therefore, denoting $t = x^1,\ x = x^2$, we obtain Eq. (3) with $F^*[v] = -v_t - v_{xx}$ and $p^1 = uv,\ p^2 = uv_x - vu_x$. Thus, the adjoint operator to the linear operator (24) is

$$F^* = -D_t - D_x^2 \tag{25}$$

and the adjoint equation to the heat Eq. (23) is written $-v_t - v_{xx} = 0$, or

$$v_t + v_{xx} = 0. \tag{26}$$

The derivation of the adjoint Eq. (26) and the adjoint operator (25) by the definition (19) is much simpler. Indeed, we have

$$F^* = \frac{\delta(vu_t - vu_{xx})}{\delta u} = -D_t(v) - D_x^2(v) = -(v_t + v_{xx}).$$

1.1.6 Self-Adjointness and Quasi Self-Adjointness

Recall that a linear differential operator F is called a *self-adjoint operator* if it is identical with its adjoint operator, $F = F^*$. Then the equation $F[u] = 0$ is also said to be self-adjoint. Thus, the self-adjointness of a *linear differential equation* $F[u] = 0$ means that the adjoint equation $F^*[v] = 0$ coincides with $F[u] = 0$ upon the substitution $v = u$. This property has been extended to nonlinear equations in [2]. It will be called here the *strict self-adjointness* and defined as follows.

Definition 2 We say that the differential equation (17) is strictly self-adjoint if the adjoint Eq. (18) becomes equivalent to the original Eq. (17) upon the substitution

$$v = u. \tag{27}$$

It means that the equation

$$F^*\big(x, u, u, \ldots, u_{(s)}, u_{(s)}\big) = \lambda\, F\big(x, u, \ldots, u_{(s)}\big) \tag{28}$$

holds with a certain (in general, variable) coefficient λ.

Example 2 The Korteweg-de Vries (KdV) equation

$$u_t = u_{xxx} + uu_x$$

is strictly self-adjoint [1]. Indeed, its adjoint Eq. (18) has the form

$$v_t = v_{xxx} + uv_x$$

and coincides with the KdV equation upon the substitution (27).

In the case of linear equations the strict self-adjointness is identical with the usual self-adjointness of linear equations.

Example 3 Consider the linear equation

$$u_{tt} + a(x)u_{xx} + b(x)u_x + c(x)u = 0. \tag{29}$$

According to Eqs. (18) and (19), the adjoint equation to equation (29) is

$$\frac{\delta}{\delta u}\{v[u_{tt} + a(x)u_{xx} + b(x)u_x + c(x)u]\} \equiv D_t^2(v) + D_x^2(av) - D_x(bv) + cv = 0.$$

Upon substituting $v = u$ and performing the differentiations it becomes

$$u_{tt} + au_{xx} + (2a' - b)u_x + (a'' - b' + c)u = 0. \tag{30}$$

According to Definition 2, Eq. (29) is strictly self-adjoint if Eq. (30) coincides with Eq. (29). This is possible if

$$b(x) = a'(x). \tag{31}$$

Definition 2 is too restrictive. Moreover, it is inconvenient in the case of systems with several dependent variables $u = (u^1, \ldots, u^m)$ because in this case Eq. (27) is not uniquely determined as it is clear from the following example.

Example 4 Let us consider the system of two equations

$$u_y^1 + u^2u_x^2 - u_t^2 = 0, \qquad u_y^2 - u_x^1 = 0 \tag{32}$$

with two dependent variables, $u = (u^1, u^2)$, and three independent variables t, x, y. Using the formal Lagrangian (9)

$$\mathcal{L} = v^1(u_y^1 + u^2u_x^2 - u_t^2) + v^2(u_y^2 - u_x^1)$$

and Eq. (8) we write the adjoint Eq. (7), changing their sign, in the form

$$v_y^2 + u^2v_x^1 - v_t^1 = 0, \qquad v_y^1 - v_x^2 = 0. \tag{33}$$

If we use here the substitution (27), $v = u$ with $v = (v^1, v^2)$, i.e., let

$$v^1 = u^1, \qquad v^2 = u^2$$

then the adjoint system (33) becomes

$$u_y^2 + u^2u_x^1 - u_t^1 = 0, \qquad u_y^1 - u_x^2 = 0$$

which is not connected with the system (32) by the equivalence relation (28). But if we set

$$v^1 = u^2, \qquad v^2 = u^1$$

the adjoint system (33) coincides with the original system (32).

The concept of quasi self-adjointness generalizes Definition 2 and is more convenient for dealing with systems (6). This concept was formulated in [3] as follows. The system (6) is *quasi self-adjoint* if the adjoint system (7) becomes equivalent to the original system (6) upon a substitution

$$v = \varphi(u) \tag{34}$$

such that its derivative does not vanish in a certain domain of u

$$\varphi'(u) \neq 0, \qquad \text{where} \qquad \varphi'(u) = \left\| \frac{\partial \varphi^\alpha(u)}{\partial u^\beta} \right\|. \tag{35}$$

Remark 2 The substitution (34) defines a mapping

$$v^\alpha = \varphi^\alpha(u), \qquad \alpha = 1, \ldots, m$$

from the m-dimensional space of variables $u = (u^1, \ldots, u^m)$ into the m-dimensional space of variables $v = (v^1, \ldots, v^m)$. It is assumed that this mapping is continuously differentiable. The condition (35) guarantees that it is invertible, and hence Eqs. (7) and (6) are equivalent. The equivalence means that the following equations hold with certain coefficients λ_α^β

$$F_\alpha^*\big(x, u, \varphi, \ldots, u_{(s)}, \varphi_{(s)}\big) = \lambda_\alpha^\beta\, F_\beta\big(x, u, \ldots, u_{(s)}\big), \quad \alpha = 1, \ldots, m \tag{36}$$

where

$$\varphi = \{\varphi^\alpha(u)\}, \quad \varphi_{(\sigma)} = \{D_{i_1} \cdots D_{i_\sigma}\big(\varphi^\alpha(u)\big)\}, \quad \sigma = 1, \ldots, s. \tag{37}$$

It can be shown that the matrix $\|\lambda_\alpha^\beta\|$ is invertible due to the condition (35).

Example 5 The quasi self-adjointness of nonlinear wave equations of the form

$$u_{tt} - u_{xx} = f(t, x, u, u_t, u_x)$$

is investigated in [6]. The results of the paper [6] show that, e.g., the equation

$$u_{tt} - u_{xx} + u_t^2 - u_x^2 = 0 \tag{38}$$

is quasi self-adjoint and that in this case the substitution (34) has the form

$$v = \mathrm{e}^u. \tag{39}$$

Indeed, the adjoint equation to equation (38) is written

$$v_{tt} - v_{xx} - 2vu_{tt} - 2u_t v_t + 2vu_{xx} + 2u_x v_x = 0. \tag{40}$$

After the substitution (39) the left-hand side of Eq. (40) takes the form(36)

$$v_{tt} - v_{xx} - 2vu_{tt} - 2u_t v_t + 2vu_{xx} + 2u_x v_x = -\mathrm{e}^u[u_{tt} - u_{xx} + u_t^2 - u_x^2]. \tag{41}$$

It is manifest from Eq. (41) that v given by (39) solves the adjoint Eq. (40) if one replaces u by any solution of Eq. (38).

In constructing conservation laws one can relax the condition (35). Therefore I generalize the previous definition of quasi self-adjointness as follows.

Definition 3 The system (6) is said to be quasi self-adjoint if the adjoint Eq. (7) are satisfied for all solutions u of the original system (6) upon a substitution

$$v^\alpha = \varphi^\alpha(u), \qquad \alpha = 1, \ldots, m \tag{42}$$

such that

$$\varphi(u) \neq 0. \tag{43}$$

In other words, the Eq. (36) hold after the substitution (42), where not all $\varphi^\alpha(u)$ vanish simultaneously.

Remark 3 The condition (43), unlike (35), does not guarantee the equivalence of Eqs. (7) and (6) because the matrix $\|\lambda_\alpha^\beta\|$ may be singular.

Example 6 It is well known that the linear heat Eq. (23) is not self-adjoint (not strictly self-adjoint in the sense of Definition 2). It is clear from Eqs. (23) and (26). Let us test Eq. (23) for quasi self-adjointness. Letting $v = \varphi(u)$, we obtain

$$v_t = \varphi' u_t, \qquad v_x = \varphi' u_x, \qquad v_{xx} = \varphi' u_{xx} + \varphi'' u_x^2$$

and the condition (36) is written

$$\varphi'(u)[u_t + u_{xx}] + \varphi''(u)u_x^2 = \lambda[u_t - u_{xx}].$$

Whence, comparing the coefficients of u_t in both sides, we obtain $\lambda = \varphi'(u)$. Then the above equation becomes

$$\varphi'(u)[u_t + u_{xx}] + \varphi''(u)u_x^2 = \varphi'(u)[u_t - u_{xx}].$$

This equation yields that $\varphi'(u) = 0$. Hence, Eq. (23) is quasi self-adjoint with the substitution $v = C$, where C is any non-vanishing constant. This substitution does not satisfy the condition (35).

Example 7 Let us consider the Fornberg-Whitham equation [7]

$$u_t - u_{txx} - uu_{xxx} - 3u_x u_{xx} + uu_x + u_x = 0. \tag{44}$$

Equations (18) and (19) give the following adjoint equation

$$F^* \equiv -v_t + v_{txx} + uv_{xxx} - uv_x - v_x = 0. \tag{45}$$

It is manifest from the Eqs. (44) and (45) that the Fornberg-Whitham equation is not strictly self-adjoint. Let us test it for quasi self-adjointness. Inserting in (45) the substitution $v = \varphi(u)$ and its derivatives

$$v_t = \varphi' u_t, \quad v_x = \varphi' u_x, \quad v_{xx} = \varphi' u_{xx} + \varphi'' u_x^2, \quad v_{tx} = \varphi' u_{tx} + \varphi'' u_t u_x, \ldots$$

then writing the condition (36) and comparing the coefficients for $u_t,\ u_{tx},\ u_{xx}, \ldots$ one can verify that $\varphi'(u) = 0$. Hence, Eq. (44) is quasi self-adjoint but does not satisfy the condition (35).

1.2 Strict Self-Adjointness Via Multipliers

It is commonly known that numerous linear equations used in practice, e.g., linear evolution equations, are not self-adjoint in the classical meaning of the self-adjointness. Likewise, useful nonlinear equations such as the nonlinear heat equation, the Burgers equation, etc. are not strictly self-adjoint. We will see here that these and many other equations can be rewritten in a strictly self-adjoint equivalent form by using multipliers. The general discussion of this approach will be given in Sect. 1.3.7.

1.2.1 Motivating Examples

Example 8 Let us consider the following second-order nonlinear equation

$$u_{xx} + f(u)u_x - u_t = 0. \tag{46}$$

Its adjoint Eq. (18) is written

$$v_{xx} - f(u)v_x + v_t = 0. \tag{47}$$

It is manifest that the substitution $v = u$ does not map Eq. (47) into Eq. (46). Hence Eq. (46) is not strictly self-adjoint.

Let us clarify if Eq. (46) can be written in an equivalent form

$$\mu(u)[u_{xx} + f(u)u_x - u_t] = 0 \tag{48}$$

with a certain multiplier $\mu(u) \neq 0$ so that Eq. (48) is strictly self-adjoint. The formal Lagrangian for Eq. (48) is

$$\mathcal{L} = v\mu(u)[u_{xx} + f(u)u_x - u_t].$$

We have

$$\begin{aligned}\frac{\delta \mathcal{L}}{\delta u} &= D_x^2[\mu(u)v] - D_x[\mu(u)f(u)v] + D_t[\mu(u)v] \\ &\quad + \mu'(u)v[u_{xx} + f(u)u_x - u_t] + \mu(u)f'(u)vu_x\end{aligned}$$

whence, upon performing the differentiations

$$\frac{\delta \mathcal{L}}{\delta u} = \mu v_{xx} + 2\mu' v u_{xx} + 2\mu' u_x v_x + \mu'' v u_x^2 - \mu f v_x + \mu v_t.$$

The strict self-adjointness requires that

$$\left.\frac{\delta \mathcal{L}}{\delta u}\right|_{v=u} = \lambda[u_{xx} + f(u)u_x - u_t].$$

This provides the following equation for the unknown multiplier $\mu(u)$

$$(\mu + 2u\mu')u_{xx} + (2\mu' + u\mu'')u_x^2 - \mu f u_x + \mu u_t = \lambda[u_{xx} + f(u)u_x - u_t]. \tag{49}$$

Since the right side of Eq. (49) does not contain u_x^2 we should have $2\mu' + u\mu'' = 0$, whence $\mu = C_1 u^{-1} + C_2$. Furthermore, comparing the coefficients of u_t in both sides of Eq. (49) we obtain $\lambda = -\mu$. Now Eq. (49) takes the form

$$(C_2 - C_1 u^{-1})u_{xx} - (C_1 u^{-1} + C_2) f u_x = -(C_1 u^{-1} + C_2)[u_{xx} + f(u)u_x]$$

and yields $C_2 = 0$. Thus, $\mu = C_1 u^{-1}$. We can let $C_1 = -1$ and formulate the result.

Proposition 2 *Equation* (46) *becomes strictly self-adjoint if we rewrite it in the form*

$$\frac{1}{u}[u_t - u_{xx} - f(u)u_x] = 0. \tag{50}$$

Example 9 One can verify that the n-th order nonlinear evolution equation

$$\frac{\partial u}{\partial t} - f(u)\frac{\partial^n u}{\partial x^n} = 0, \qquad f(u) \neq 0 \tag{51}$$

with one spatial variable x is not strictly self-adjoint. The following statement shows that it becomes strictly self-adjoint after using an appropriate multiplier.

Proposition 3 *Equation* (51) *becomes strictly self-adjoint upon rewriting it in the following equivalent form*

$$\frac{1}{uf(u)}\left[\frac{\partial u}{\partial t} - f(u)\frac{\partial^n u}{\partial x^n}\right] = 0. \tag{52}$$

Proof Multiplying Eq. (51) by $\mu(u)$ and taking the formal Lagrangian

$$\mathcal{L} = v\mu(u)[u_t - f(u)u_n]$$

where $u_n = D_x^n(u)$, we have

$$\frac{\delta \mathcal{L}}{\delta u} = -D_t[\mu(u)v] - D_x^n[\mu(u)f(u)v] + v\mu'(u)u_t - v[\mu(u)f(u)]'u_n.$$

Noting that $-D_t[\mu(u)v] + v\mu'(u)u_t = -\mu(u)v_t$ and letting $v = u$ we obtain

$$\left.\frac{\delta \mathcal{L}}{\delta u}\right|_{v=u} = -\mu(u)u_t - D_x^n[\mu(u)f(u)u] - [\mu(u)f(u)]'uu_n.$$

If we take $\mu(u) = [uf(u)]^{-1}$, then $\mu(u)f(u)u = 1,\ \mu(u)f(u) = u^{-1}$, and hence

$$\left.\frac{\delta \mathcal{L}}{\delta u}\right|_{v=u} = -\frac{1}{uf(u)}[u_t - f(u)u_n].$$

Thus, Eq. (52) satisfies the strict self-adjointness condition (28) with $\lambda = -1$. □

1.2.2 Linear Heat Equation

Taking in (50) $f(u) = 0$, we rewrite the classical linear heat equation $u_t = u_{xx}$ in the following strictly self-adjoint form

$$\frac{1}{u}[u_t - u_{xx}] = 0. \tag{53}$$

This result can be extended to the heat equation

$$u_t - \Delta u = 0 \tag{54}$$

where Δu is the Laplacian with n variables $x = (x^1, \ldots, x^n)$. Namely, the strictly self-adjoint form of Eq. (54) is

$$\frac{1}{u}\left[u_t - \Delta u\right] = 0. \tag{55}$$

Indeed, the formal Lagrangian (9) for Eq. (55) has the form

$$\mathcal{L} = \frac{v}{u}\left[u_t - \Delta u\right].$$

Substituting it in (19) we obtain

$$F^* = -D_t\left(\frac{v}{u}\right) - \Delta\left(\frac{v}{u}\right) - \frac{v}{u^2}\left[u_t - \Delta u\right].$$

Upon letting $v = u$ it becomes

$$F^* = -\frac{1}{u}\left[u_t - \Delta u\right].$$

Hence, Eq. (55) satisfies the condition (28) with $\lambda = -1$.

1.2.3 Nonlinear Heat Equation

Consider the nonlinear heat equation $u_t - D_x\left(k(u)u_x\right) = 0$, or

$$u_t - k(u)u_{xx} - k'(u)u_x^2 = 0. \tag{56}$$

Its adjoint equation has the form

$$v_t + k(u)v_{xx} = 0.$$

Therefore it is obvious that (56) does not satisfy Definition 2. But it becomes strictly self-adjoint if we rewrite it in the form

$$\frac{1}{u}\left[u_t - k(u)u_{xx} - k'(u)u_x^2\right] = 0. \tag{57}$$

Indeed, the formal Lagrangian (9) for Eq. (57) is written

$$\mathcal{L} = \frac{v}{u}\left[u_t - k(u)u_{xx} - k'(u)u_x^2\right].$$

Substituting it in (19) we obtain

$$F^* = -D_t\left(\frac{v}{u}\right) - D_x^2\left(\frac{v}{u}\,k(u)\right) + 2D_x\left(\frac{v}{u}\,k'(u)u_x\right)$$
$$-\frac{v}{u}\,k'(u)u_{xx} - \frac{v}{u}\,k''(u)u_x^2 - \frac{v}{u^2}\left[u_t - k(u)u_{xx} - k'(u)u_x^2\right].$$

Letting here $v = u$ we have

$$F^* = -\frac{1}{u}\left[u_t - k(u)u_{xx} - k'(u)u_x^2\right].$$

Hence, Eq. (55) satisfies the strict self-adjointness condition (28) with $\lambda = -1$.

1.2.4 The Burgers Equation

Taking in (50) $f(u) = u$ we obtain the strictly self-adjoint form

$$\frac{1}{u}\left[u_t - u_{xx}\right] - u_x = 0 \tag{58}$$

of the Burgers equation $u_t = u_{xx} + uu_x$.

1.2.5 Heat Conduction in Solid Hydrogen

According to [8], the heat conduction in solid crystalline molecular hydrogen at low pressures is governed by the nonlinear equation (up-to positive constant coefficient)

$$u_t = u^2 \Delta u. \tag{59}$$

It is derived from the Fourier equation

$$\rho\, c_* \frac{\partial T}{\partial t} = \nabla \cdot (k\, \nabla T)$$

using the empirical information that the density ρ at low pressures has a constant value, whereas the specific heat c_* and the thermal conductivity k have the estimations

$$c_* \cong T^3, \qquad k \cong T^3\left(1 + T^4\right)^{-2}.$$

It is also shown in [8] that the one-dimensional Eq. (59)

$$u_t = u^2 u_{xx} \tag{60}$$

is related to the linear heat equation by a non-point transformation (Eq. (5) in [8]). A similar relation was found in [9] for another representation of Eq. (60). The non-point transformation of Eq. (60) to the linear heat equation

$$w_s = w_{\xi\xi} \tag{61}$$

is written in [10] as the differential substitution

$$t = s, \qquad x = w, \qquad u = w_\xi. \tag{62}$$

It is also demonstrated in [10], Sect. 20, that Eq. (60) is the unique equation with nontrivial Lie-Bäcklund symmetries among the equations of the form

$$u_t = f(u) + h(u, u_x), \qquad f'(u) \neq 0.$$

The connection between Eq. (60) and the heat equation is treated in [11] as a reciprocal transformation [11]. It is shown in [12] that this connection, together with its extensions, allows to solve certain moving boundary problems in nonlinear heat conduction.

Our Example 9 from Sect. 1.2.1 reveals one more remarkable property of Eq. (60). Namely, taking $n = 2$ and $f(u) = u^2$ in Eq. (52) we see that Eq. (60) *becomes strictly self-adjoint if we rewrite it in the form*

$$\frac{u_t}{u^3} = \frac{u_{xx}}{u}. \tag{63}$$

1.2.6 Harry Dym Equation

Taking in Example 9 from Sect. 1.2.1 $n = 3$ and $f(u) = u^3$ we see that the Harry Dym equation

$$u_t - u^3 u_{xxx} = 0 \tag{64}$$

becomes strictly self-adjoint upon rewriting it in the form

$$\frac{u_t}{u^4} - \frac{u_{xxx}}{u} = 0.$$

1.2.7 Kompaneets Equation

The equations considered in Sects. 1.2.1–1.2.6 are quasi self-adjoint. For example, for Eq. (51) we have

$$F^* = -v_t - D_x^n(f(u)v) - vf'(u)u_n$$

whence making the substitution

$$v = \frac{1}{f(u)}$$

we obtain

$$F^* = \frac{f'}{f^2}\,u_t - \frac{f'}{f}\,u_n = \frac{f'}{f^2}[u_t - f(u)\,u_n].$$

Hence, Eq. (51) is quasi self-adjoint.

Example 10 The Kompaneets equation

$$u_t = \frac{1}{x^2}\,D_x\left[x^4(u_x + u + u^2)\right] \tag{65}$$

provides an example of an equation that is not quasi self-adjoint. Indeed, Eq. (65) has the formal Lagrangian

$$\mathcal{L} = v[-u_t + x^2u_{xx} + (x^2 + 4x + 2x^2u)u_x + 4x(u + u^2)].$$

The calculation yields the following adjoint equation to (65)

$$\frac{\delta\mathcal{L}}{\delta u} \equiv v_t + x^2v_{xx} - x^2(1 + 2u)v_x + 2(x + 2xu - 1)v = 0. \tag{66}$$

Letting $v = \varphi(u)$ one obtains

$$\begin{aligned}\frac{\delta\mathcal{L}}{\delta u}\bigg|_{v=\varphi(u)} &= \varphi'(u)u_t + x^2u_{xx} - x^2(1 + 2u)u_x]\\ &\quad + \varphi''(u)x^2u_x^2 + 2(x + 2xu - 1)\varphi(u).\end{aligned}$$

Writing the quasi self-adjointness condition (36) in the form

$$\frac{\delta\mathcal{L}}{\delta u}\bigg|_{v=\varphi(u)} = \lambda[-u_t + x^2u_{xx} + (x^2 + 4x + 2x^2u)u_x + 4x(u + u^2)]$$

and comparing the coefficients for u_t in both sides one obtains $\lambda = -\varphi'(u)$, so that the quasi self-adjointness condition takes the form

$$\varphi'(u)[u_t + x^2u_{xx} - x^2(1+2u)u_x] + \varphi''(u)x^2u_x^2 + 2(x + 2xu - 1)\varphi(u)$$
$$= \varphi'(u)[u_t - x^2u_{xx} - (x^2 + 4x + 2x^2u)u_x - 4x(u + u^2)].$$

Comparing the coefficients for u_{xx} in both sides we obtain $\varphi'(u) = 0$. Then the above equation becomes $(x + 2xu - 1)\varphi(u) = 0$ and yields $\varphi(u) = 0$. Hence the Kompaneets equation is not quasi self-adjoint because the condition (43) is not satisfied.

But we can rewrite Eq. (65) in the strictly self-adjoint form by using a more general multiplier than above, namely, the multiplier

$$\mu = \frac{x^2}{u}\cdot \tag{67}$$

Indeed, upon multiplying by this μ Eq. (65) is written

$$\frac{x^2}{u}u_t = \frac{1}{u}D_x\left[x^4(u_x + u + u^2)\right].$$

Its formal Lagrangian

$$\mathcal{L} = \frac{v}{u}\left\{-x^2u_t + D_x\left[x^4(u_x + u + u^2)\right]\right\}$$

satisfies the strict self-adjointness condition (28) with $\lambda = -1$

$$\left.\frac{\delta\mathcal{L}}{\delta u}\right|_{v=u} = -\frac{1}{u}\left\{-x^2u_t + D_x\left[x^4(u_x + u + u^2)\right]\right\}.$$

Remark 4 Note that $v = x^2$ solves Eq. (66) for any u. The connection of this solution with the multiplier (67) is discussed in Sect. 1.3.7. See also Sect. 1.4.

1.3 General Concept of Nonlinear Self-Adjointness

Motivated by the examples discussed in Sects. 1.1 and 1.2 as well as other similar examples, I have suggested in [13] the general concept of *nonlinear self-adjointness* of systems consisting of any number of equations with m dependent variables. This concept encapsulates Definition 2 of strict self-adjointness and Definition 3 of quasi self-adjointness. The new concept has two different features. They are expressed below by two different but equivalent definitions.

1.3.1 Two Definitions and Their Equivalence

Definition 4 The system of $\overline{m}$ differential equations (compare with Eq. 6)

$$F_{\bar{\alpha}}\big(x, u, u_{(1)}, \ldots, u_{(s)}\big) = 0, \qquad \bar{\alpha} = 1, \ldots, \overline{m} \tag{68}$$

with m dependent variables $u = (u^1, \ldots, u^m)$ is said to be *nonlinearly self-adjoint* if the *adjoint equations*

$$F^*_{\alpha}\big(x, u, v, u_{(1)}, v_{(1)}, \ldots, u_{(s)}, v_{(s)}\big) \equiv \frac{\delta(v^{\bar{\beta}} F_{\bar{\beta}})}{\delta u^{\alpha}} = 0, \quad \alpha = 1, \ldots, m \tag{69}$$

are satisfied for all solutions u of the original system (68) upon a substitution

$$v^{\bar{\alpha}} = \varphi^{\bar{\alpha}}(x, u), \qquad \bar{\alpha} = 1, \ldots, \overline{m} \tag{70}$$

such that

$$\varphi(x, u) \neq 0. \tag{71}$$

In other words, the following equations hold

$$F^*_{\alpha}\big(x, u, \varphi(x, u), \ldots, u_{(s)}, \varphi_{(s)}\big) = \lambda^{\bar{\beta}}_{\alpha} F_{\bar{\beta}}\big(x, u, \ldots, u_{(s)}\big), \quad \alpha = 1, \ldots, m \tag{72}$$

where $\lambda^{\bar{\beta}}_{\alpha}$ are undetermined coefficients, and $\varphi_{(\sigma)}$ are derivatives of (70)

$$\varphi_{(\sigma)} = \{D_{i_1} \cdots D_{i_\sigma}\big(\varphi^{\bar{\alpha}}(x, u)\big)\}, \qquad \sigma = 1, \ldots, s.$$

Here v and φ are the $\overline{m}$-dimensional vectors

$$v = (v^1, \ldots, v^{\overline{m}}), \qquad \varphi = (\varphi^1, \ldots, \varphi^{\overline{m}})$$

and Eq. (71) means that not all components $\varphi^{\bar{\alpha}}(x, u)$ of φ vanish simultaneously.

Remark 5 If the system (68) is over-determined, i.e., $\overline{m} > m$, then the adjoint system (69) is sub-definite since it contains $m < \overline{m}$ equations for $\overline{m}$ new dependent variables v. Vise versa, if $\overline{m} < m$, then the system (68) is sub-definite and the adjoint system (69) is over-determined.

Remark 6 The adjoint system (69), upon substituting there any solution $u(x)$ of Eq. (68), becomes a linear homogeneous system for the new dependent variables $v^{\bar{\alpha}}$. The essence of Eq. (72) is that for the self-adjoint system (68) there exist functions (70) that provide a non-trivial (not identically zero) solution to the adjoint system (69) for all solutions of the original system (68). This property can be taken as the following alternative definition of the nonlinear self-adjointness.

Definition 5 The system (68) is nonlinearly self-adjoint if there exist functions $v^{\bar{\alpha}}$ given by (70) that solve the adjoint system (69) for all solutions $u(x)$ of Eq. (68) and satisfy the condition (71).

Proposition 4 *The above two definitions are equivalent.*

Proof Let the system (68) be nonlinearly self-adjoint by Definition 4. Then, according to Remark 6, the system (68) satisfies the condition of Definition 5.

Conversely, let the system (68) be nonlinearly self-adjoint by Definition 5. Namely, let the functions $v^{\bar{\alpha}}$ given by (70) and satisfying the condition (71) solve the adjoint system (69) for *all* solutions $u(x)$ of Eq. (68). This is possible if and only if Eq. (72) hold. Then the system (68) is nonlinearly self-adjoint by Definition 4. □

Example 11 It has been mentioned in Example 2 that the KdV equation

$$u_t = u_{xxx} + uu_x \tag{73}$$

is strictly self-adjoint. In terms of Definition 5 it means that $v = u$ solves the adjoint equation

$$v_t = v_{xxx} + uv_x \tag{74}$$

for all solutions of the KdV Eq. (73). One can verify that the general substitution of the form (70), $v = \varphi(t, x, u)$, satisfying Eq. (72) is given by

$$v = A_1 + A_2 u + A_3(x + tu) \tag{75}$$

where A_1, A_2, A_3 are arbitrary constants. One can also check that v given by Eq. (75) solves the adjoint Eq. (74) for all solutions u of the KdV equation. The solution $v = x + tu$ is an invariant of the Galilean transformation of the KdV equation and appears in different approaches (see [10, Sect. 22.5] and [14]). Thus, the KdV equation is nonlinearly self-adjoint with the substitution (75).

Proposition 5 *Any linear equation is nonlinearly self-adjoint.*

Proof This property is the direct consequence of Definition 5 because the adjoint equation $F^*[v] = 0$ to a linear equation $F[u] = 0$ does not involve the variable u. □

1.3.2 Remark on Differential Substitutions

One can further extend the concept of self-adjointness by replacing the *point-wise* substitution (70) with *differential* substitutions of the form

$$v^{\bar{\alpha}} = \varphi^{\bar{\alpha}}(x, u, u_{(1)}, \ldots, u_{(r)}), \qquad \bar{\alpha} = 1, \ldots, \overline{m}. \tag{76}$$

Then Eq. (72) will be written, e.g., in the case $r = 1$, as follows

$$F_\alpha^*\big(x, u, \varphi, \ldots, u_{(s)}, \varphi_{(s)}\big) = \lambda_\alpha^{\bar\beta} F_{\bar\beta} + \lambda_\alpha^{j\bar\beta} D_j(F_{\bar\beta}). \tag{77}$$

Example 12 The reckoning shows that the equation

$$u_{xy} = \sin u \tag{78}$$

is not self-adjoint via a point-wise substitution $v = \varphi(x, y, u)$, but it is self-adjoint in the sense of Definition 4 with the following differential substitution

$$v = \varphi(x, y, u_x, u_y) \equiv A_1[xu_x - yu_y] + A_2u_x + A_3u_y \tag{79}$$

where $A_1,\ A_2,\ A_3$ are arbitrary constants. The adjoint equation to equation (78) is

$$v_{xy} - v\cos u = 0$$

and the self-adjointness condition (77) with the function φ given by (79) is satisfied in the form

$$\begin{aligned}\varphi_{xy} - \varphi\cos u = {} & (A_1x + A_2)D_x(u_{xy} - \sin u)\\ & + (A_3 - A_1y)D_y(u_{xy} - \sin u).\end{aligned} \tag{80}$$

1.3.3 Nonlinear Heat Equation

One-Dimensional Case

Let us apply the new viewpoint to the nonlinear heat Eq. (56), $u_t = (k(u)u_x)_x$, discussed in Sect. 1.2.3. We will take it in the expanded form

$$u_t - k(u)u_{xx} - k'(u)u_x^2 = 0, \qquad k(u) \neq 0. \tag{81}$$

The adjoint Eqs. (18) to (81) is

$$v_t + k(u)v_{xx} = 0. \tag{82}$$

We take the substitution (70) written together with the necessary derivatives

$$\begin{aligned} v &= \varphi(t, x, u)\\ v_t &= \varphi_u u_t + \varphi_t, \qquad v_x = \varphi_u u_x + \varphi_x\\ v_{xx} &= \varphi_u u_{xx} + \varphi_{uu}u_x^2 + 2\varphi_{xu}u_x + \varphi_{xx}\end{aligned} \tag{83}$$

and arrive at the following self-adjointness condition (72)

$$\varphi_u u_t + \varphi_t + k(u)[\varphi_u u_{xx} + \varphi_{uu} u_x^2 + 2\varphi_{xu} u_x + \varphi_{xx}]$$
$$= \lambda[u_t - k(u)u_{xx} - k'(u)u_x^2]. \qquad (84)$$

The comparison of the coefficients of u_t in both sides of Eq. (84) yields $\lambda = \varphi_u$. Then, comparing the terms with u_{xx} we see that $\varphi_u = 0$. Hence Eq. (84) reduces to

$$\varphi_t + k(u)\varphi_{xx} = 0 \qquad (85)$$

and yields $\varphi_t = 0, \ \varphi_{xx} = 0$, whence $\varphi = C_1 x + C_2$, where $C_1, C_2 = \text{const}$. We have demonstrated that Eq. (81) is nonlinearly self-adjoint by Definition 4 and that the substitution (70) has the form

$$v = C_1\, x + C_2. \qquad (86)$$

The same result can be easily obtained by using Definition 5. We look for the solution of the adjoint Eq. (82) in the form $v = \varphi(t, x)$. Then Eq. (82) has the form (85). Since it should be satisfied for all solutions u of Eq. (81), we obtain $\varphi_t = 0, \ \varphi_{xx} = 0$, and hence Eq. (86).

Multi-dimensional Case

The similar analysis can be applied to the nonlinear heat equation with several variables $x = (x^1, \dots, x^n)$

$$u_t = \nabla \cdot (k(u)\nabla u) \qquad (87)$$

or

$$u_t - k(u)\, \Delta u - k'(u)|\nabla u|^2 = 0. \qquad (88)$$

The reckoning shows that the adjoint Eqs. (18) to (88) is written

$$v_t + k(u)\, \Delta v = 0. \qquad (89)$$

It is easy to verify the nonlinear elf-adjointness by Definition 5. Namely, searching for the solution of the adjoint Eq. (89) in the form $v = \varphi(t, x^1, \dots, x^n)$, one obtains

$$\varphi_t + k(u)\, \Delta\varphi = 0$$

whence

$$\varphi_t = 0, \qquad \Delta\varphi = 0.$$

We conclude that Eq. (88) is self-adjoint and that the substitution (70) is given by

$$v = \varphi(x^1, \dots, x^n) \qquad (90)$$

where $\varphi(x^1, \dots, x^n)$ is any solution of the n-dimensional Laplace equation $\Delta\varphi = 0$.

1.3.4 Anisotropic Nonlinear Diffusion Equation

Two-Dimensional Case

Consider the diffusion equation

$$u_t = (f(u)u_x)_x + (g(u)u_y)_y \tag{91}$$

in an anisotropic two-dimensional medium (see [15–17], vol. 1, Sect. 10.8) with arbitrary functions $f(u)$ and $g(u)$. The adjoint equation is

$$v_t + f(u)v_{xx} + g(u)v_{yy} = 0. \tag{92}$$

Using Definition 5 we obtain the following equations for nonlinear self-adjointness of Eq. (91)

$$\varphi_t = 0, \qquad \varphi_{xx} = 0, \qquad \varphi_{yy} = 0. \tag{93}$$

Integrating equations (93) we obtain the following substitution (70)

$$v = C_1\, xy + C_2\, x + C_3\, y + C_4. \tag{94}$$

Three-Dimensional Case

The three-dimensional anisotropic nonlinear diffusion equation has the following form (see [15–17], vol. 1, Sect. 10.9)

$$u_t = (f(u)u_x)_x + (g(u)u_y)_y + (h(u)u_z)_z\,. \tag{95}$$

Its adjoint equation is

$$v_t + f(u)v_{xx} + g(u)v_{yy} + h(u)v_{zz} = 0. \tag{96}$$

Equation (95) is nonlinearly self-adjoint. In this case the substitution (94) is replaced by

$$v = C_1\, xyz + C_2\, xy + C_3\, xz + C_4\, yz + C_5\, x + C_6\, y + C_7\, z + C_8. \tag{97}$$

1.3.5 Nonlinear Wave Equations

One-Dimensional Case

Consider the following one-dimensional nonlinear wave equation

$$u_{tt} = (k(u)u_x)_x, \qquad k(u) \neq 0 \tag{98}$$

or in the expanded form

$$u_{tt} - k(u)u_{xx} - k'(u)u_x^2 = 0. \tag{99}$$

The adjoint Eqs. (18) to (98) is written

$$v_{tt} - k(u)v_{xx} = 0. \tag{100}$$

Proceeding as in One-Dimensional Case or applying Definition 5 to Eqs. (99) and (100) by letting $v = \varphi(t, x)$, we obtain the following equations that guarantee the nonlinear self-adjointness of Eq. (98)

$$\varphi_{tt} = 0, \qquad \varphi_{xx} = 0. \tag{101}$$

Integrating equation (101) we obtain the following substitution

$$v = C_1\, tx + C_2\, t + C_3\, x + C_4. \tag{102}$$

Multi-Dimensional Case

The multi-dimensional version of Eq. (98) with $x = (x^1, \ldots, x^\nu)$ is written

$$u_{tt} = \nabla \cdot (k(u)\nabla u) \tag{103}$$

or

$$u_{tt} - k(u)\,\Delta u - k'(u)|\nabla u|^2 = 0. \tag{104}$$

The adjoint equation is

$$v_{tt} - k(u)\,\Delta v = 0. \tag{105}$$

Using Definition 5 and searching the solution of the adjoint Eq. (105) in the form $v = \varphi(t, x^1, \ldots, x^\nu)$, we obtain the equations

$$\varphi_{tt} = 0, \qquad \Delta\varphi = 0.$$

Solving them we arrive at the following substitution (70)

$$v = a(x)t + b(x) \tag{106}$$

where $a(x)$ and $b(x)$ solve the ν-dimensional Laplace equation

$$\Delta a(x^1, \dots, x^\nu) = 0, \qquad \Delta b(x^1, \dots, x^\nu) = 0.$$

Hence Eq. (103) is nonlinearly self-adjoint.

Nonlinear Vibration of Membranes

Vibrations of a uniform membrane whose tension varies during deformations are described by the following Lagrangian

$$L = \frac{1}{2}\left[u_t^2 - k(u)\left(u_x^2 + u_y^2\right)\right], \qquad k'(u) \neq 0. \tag{107}$$

The corresponding Euler-Lagrange equation

$$\frac{\partial L}{\partial u} - D_t\left(\frac{\partial L}{\partial u_t}\right) - D_x\left(\frac{\partial L}{\partial u_x}\right) - D_y\left(\frac{\partial L}{\partial u_y}\right) = 0$$

provides the nonlinear wave equation

$$u_{tt} = k(u)\,(u_{xx} + u_{yy}) + \frac{1}{2}\,k'(u)(u_x^2 + u_y^2). \tag{108}$$

Note that Eq. (108) differs from the two-dimensional nonlinear wave Eq. (104) by the coefficient $1/2$. Let us find out if this difference affects self-adjointness.

By applying (69) to the formal Lagrangian of Eq. (108) we obtain

$$F^* = v_{tt} - k(u)\,(v_{xx} + v_{yy}) - k'(u)(u_x v_x + u_y v_y + v u_{xx} + v u_{yy})$$
$$- \frac{v}{2}\,k''(u)(u_x^2 + u_y^2).$$

We take the substitution (70) together with the necessary derivatives (see Eq. 83)

$$\begin{aligned}
v &= \varphi(t, x, y, u), & v_t &= \varphi_u u_t + \varphi_t \\
v_x &= \varphi_u u_x + \varphi_x, & v_y &= \varphi_u u_y + \varphi_y \\
v_{xx} &= \varphi_u u_{xx} + \varphi_{uu} u_x^2 + 2\varphi_{xu} u_x + \varphi_{xx} \\
v_{yy} &= \varphi_u u_{yy} + \varphi_{uu} u_y^2 + 2\varphi_{yu} u_y + \varphi_{yy} \\
v_{tt} &= \varphi_u u_{tt} + \varphi_{uu} u_t^2 + 2\varphi_{tu} u_t + \varphi_{tt}
\end{aligned} \tag{109}$$

and substitute the expressions (109) in the self-adjointness condition (72)

$$F^*\Big|_{v=\varphi} = \lambda[u_{tt} - k(u)\,(u_{xx} + u_{yy}) - \frac{1}{2}\,k'(u)(u_x^2 + u_y^2)].$$

Comparing the coefficients of u_{tt} we obtain $\lambda = \varphi_u$. Then we compare the coefficients of u_{xx} and obtain $\varphi\,k'(u) = 0$. This equation yields $\varphi = 0$ because $k'(u) \neq 0$. Thus, the condition (71) is not satisfied for the point-wise substitution (70). Further investigation of Eq. (108) for the nonlinear self-adjointness requires differential substitutions.

1.3.6 Anisotropic Nonlinear Wave Equation

Two-Dimensional Case

The two-dimensional anisotropic nonlinear wave equation is (see [15–17, vol. 1, Sect. 12.6])

$$u_{tt} = (f(u)u_x)_x + (g(u)u_y)_y\,. \tag{110}$$

Its adjoint equation has the form

$$v_{tt} - f(u)v_{xx} - g(u)v_{yy} = 0. \tag{111}$$

Proceeding as in Sect. 1.3.4 we obtain the following equations that guarantee the self-adjointness of Eq. (110)

$$\varphi_{tt} = 0, \qquad \varphi_{xx} = 0, \qquad \varphi_{yy} = 0. \tag{112}$$

Integrating equation (112) we obtain the following substitution (70)

$$v = C_1\,txy + C_2\,tx + C_3\,ty + C_4\,xy + C_5\,t + C_6\,x + C_7\,y + C_8. \tag{113}$$

Remark 7 I provide here detailed calculations in integrating Eq. (112). The general solution to the linear second-order equation $\varphi_{tt} = 0$ is given by

$$\varphi = A(x, y)t + B(x, y) \tag{114}$$

with arbitrary functions $A(x, y)$ and $B(x, y)$. Substituting this expression for φ in the second and third Eq. (112) and splitting with respect to t we obtain the following equations for $A(x, y)$ and $B(x, y)$

$$A_{xx} = 0, \qquad A_{yy} = 0, \qquad B_{xx} = 0, \qquad B_{yy} = 0.$$

Substituting the general solution

$$A = a_1(y)x + a_2(y)$$

of the equation $A_{xx} = 0$ in the equation $A_{yy} = 0$ and splitting with respect to x, we obtain $a_1'' = 0,\ a_2'' = 0$, whence

$$a_1 = c_{11}y + c_{12}, \qquad a_2 = c_{21}y + c_{22}$$

where $c_{11}, \ldots, c_{22}$ are arbitrary constants. Substituting these in the above expression for A we obtain

$$A = c_{11}xy + c_{12}x + c_{21}y + c_{22}.$$

Proceeding likewise with the equations for $B(x, y)$, we have

$$B = d_{11}xy + d_{12}x + d_{21}y + d_{22}$$

with arbitrary constant coefficients $d_{11}, \ldots, d_{22}$. Finally, we substitute the resulting A and B in the expression (114) for φ and, changing the notation, arrive at (113).

Three-Dimensional Case

The three-dimensional anisotropic nonlinear wave equation

$$u_{tt} = (f(u)u_x)_x + (g(u)u_y)_y + (h(u)u_z)_z \tag{115}$$

has the following adjoint equation

$$v_{tt} - f(u)v_{xx} - g(u)v_{yy} - h(u)v_{zz} = 0. \tag{116}$$

In this case Eq. (112) are replaced by

$$\varphi_{tt} = 0, \qquad \varphi_{xx} = 0, \qquad \varphi_{yy}, \qquad \varphi_{zz} = 0$$

and yield the following substitution (70)

$$\begin{aligned} v = {} & C_1txyz + C_2txy + C_3txz + C_4tyz + C_5tx + C_6ty + C_7tz \\ & + C_8xy + C_9xz + C_{10}yz + C_{11}t + C_{12}x + C_{13}y + C_{14}z + C_{15}. \end{aligned} \tag{117}$$

1.3.7 From Nonlinear to Strict Self-Adjointness

The approach of this section is not used for constructing conservation laws. But it may be useful for other applications of the nonlinear self-adjointness.

Theorem 1 *The differential equation* (17)

$$F\big(x, u, u_{(1)}, \ldots, u_{(s)}\big) = 0 \tag{118}$$

is nonlinearly self-adjoint (Definition 4) *if and only if it becomes strictly self-adjoint* (Definition 2) *upon rewriting in the equivalent form*

$$\mu(x, u) F\big(x, u, u_{(1)}, \ldots, u_{(s)}\big) = 0, \qquad \mu(x, u) \neq 0 \tag{119}$$

with an appropriate multiplier $\mu(x, u)$.

Proof We will write the condition (72) for nonlinear self-adjointness of Eq. (118) in the form

$$\left.\frac{\delta(vF)}{\delta u}\right|_{v=\varphi(x,u)} = \lambda(x, u) F\big(x, u, u_{(1)}, \ldots, u_{(s)}\big). \tag{120}$$

Furthermore, invoking that the Eqs. (119) and (118) are equivalent, we will write the condition (28) for strict self-adjointness of Eq. (119) in the form

$$\left.\frac{\delta(w\mu F)}{\delta u}\right|_{w=u} = \tilde{\lambda}(x, u) F\big(x, u, u_{(1)}, \ldots, u_{(s)}\big). \tag{121}$$

Since w is a dependent variable and $\mu = \mu(x, u)$ is a certain function of x, u, the variational derivative in the left-hand side of (121) can be written as follows

$$\begin{aligned}\frac{\delta(w\mu F)}{\delta u} &= w\frac{\partial \mu}{\partial u} F + \mu w \frac{\partial F}{\partial u} - D_i\left(\mu w \frac{\partial F}{\partial u_i}\right) + D_i D_j \left(\mu w \frac{\partial F}{\partial u_{ij}}\right) - \cdots \\ &= w\frac{\partial \mu}{\partial u} F + \frac{\delta(vF)}{\delta u}\end{aligned}$$

where v is the new dependent variable instead of w defined by the formula

$$v = \mu(x, u) w. \tag{122}$$

Now the left side of Eq. (121) is written

$$\left.\frac{\delta(w\mu F)}{\delta u}\right|_{w=u} = u\frac{\partial \mu}{\partial u} F + \left.\frac{\delta(vF)}{\delta u}\right|_{v=u\mu(x,u)}. \tag{123}$$

Let us assume that Eq. (118) is nonlinearly self-adjoint. Then Eq. (120) holds with a certain given function $\varphi(x, u)$. Therefore, we take the multiplier

$$\mu(x, u) = \frac{\varphi(x, u)}{u} \tag{124}$$

and reduce Eq. (123) to the following form

$$\frac{\delta(w\mu F)}{\delta u}\bigg|_{w=u} = \left(\lambda + \frac{\partial\varphi}{\partial u} - \frac{\varphi}{u}\right) F.$$

This proves that Eq. (121) holds with

$$\tilde{\lambda} = \frac{\partial\varphi}{\partial u} - \frac{\varphi}{u} + \lambda.$$

Hence, Eq. (119) with the multiplier μ given by (124) is strictly self-adjoint.

Let us assume now that Eq. (119) with a certain multiplier $\mu(x,u)$ is strictly self-adjoint. Then Eq. (121) holds. Therefore, if we take the function φ defined by (see 124)

$$\varphi(x,u) = u\mu(x,u) \tag{125}$$

Equation (123) yields

$$\frac{\delta(vF)}{\delta u}\bigg|_{v=\varphi(x,u)} = \left(\tilde{\lambda} - u\frac{\partial\mu}{\partial u}\right) F.$$

It follows that Eq. (120) holds with

$$\lambda = \tilde{\lambda} - u\frac{\partial\mu}{\partial u}\cdot$$

We conclude that Eq. (118) is nonlinearly self-adjoint, thus completing the proof. □

Example 13 The multiplier (67) used in Example 10 and the function $\varphi = x^2$ that provides a solution of the adjoint Eq. (66) to the Kompaneets equation are related by Eq. (125).

Example 14 Let us consider the one-dimensional nonlinear wave Eq. (99),

$$u_{tt} - k(u)u_{xx} - k'(u)u_x^2 = 0.$$

If we substitute in (124) the function φ given by the right-hand side of (102) we will obtain the multiplier that maps Eq. (99) into the strictly self-adjoint equivalent form. For example, taking (102) with $C_1 = C_3 = C_4 = 0,\ C_2 = 1$ we obtain the multiplier

$$\mu = \frac{t}{u}\cdot$$

The corresponding equivalent equation to equation (99) has the formal Lagrangian

$$\mathcal{L} = \frac{tv}{u}[u_{tt} - k(u)u_{xx} - k'(u)u_x^2].$$

We have

$$\frac{\delta \mathcal{L}}{\delta u} = D_t^2\left(\frac{tv}{u}\right) - \frac{tv}{u^2}\, u_{tt} - D_x^2\left(\frac{tv}{u}\, k(u)\right) - \frac{tv}{u}\, k'(u) u_{xx} + \frac{tv}{u^2}\, k(u) u_{xx}$$

$$+ 2D_x\left(\frac{tv}{u}\, k'(u) u_x\right) - \frac{tv}{u}\, k''(u) u_x^2 + \frac{tv}{u^2}\, k'(u) u_x^2.$$

Letting here $v = u$ we see that the strict self-adjointness condition is satisfied in the following form

$$\left.\frac{\delta \mathcal{L}}{\delta u}\right|_{v=u} = -\frac{t}{u}[u_{tt} - k(u) u_{xx} - k'(u) u_x^2].$$

1.4 Generalized Kompaneets Equation

1.4.1 Introduction

The equation

$$\frac{\partial n}{\partial t} = \frac{1}{x^2} \frac{\partial}{\partial x}\left[x^4\left(\frac{\partial n}{\partial x} + n + n^2\right)\right] \tag{126}$$

known as the Kompaneets equation or the *photon diffusion equation*, was derived independently by Kompaneets[2] [18] and Weymann [19]. They take as a starting point the kinetic equations for the distribution function of a photon gas[3] and arrive, at certain idealized conditions, at Eq. (126). This equation provides a mathematical model for describing the time development of the energy spectrum of a low energy homogeneous photon gas interacting with a rarefied electron gas via the Compton scattering. Here n is the density of the photon gas (photon number density), t is time and x is connected with the photon frequency ν by the formula

$$x = \frac{h\nu}{kT_e} \tag{127}$$

where h is Planck's constant and kT_e is the *electron temperature* with the standard notation k for Boltzmann's constant. According to this notation, $h\nu$ has the meaning of the *photon energy*. The nonrelativistic approximation is used, i.e., it is assumed that the electron temperatures satisfy the condition $kT_e \ll mc^2$, where m is the

---

[2] He mentions in his paper that the work has been done in 1950 and published in *Report # 336* of the Institute of Chemical Physics of the USSR Academy of Sciences.

[3] Weymann uses Dreicer's kinetic equation [20] for a photon gas interacting with a plasma which is slightly different from the equation used by Kompaneets.

electron mass and c is the light velocity. The term *low energy photon gas* means that $h\nu \ll mc^2$.

The question arises if the idealized conditions assumed in deriving Eq. (126) may be satisfied in the real world. For discussions of theoretical and observational evidences for such possibility in astrophysical environments, for example in intergalactic gas, see e.g., [21, 22] and the references therein. See also the recent publication [23].

1.4.2 Discussion of Self-Adjointness of the Kompaneets Equation

For unifying the notation, the dependent variable n in Eq. (126) will be denoted by u and Eq. (126) will be written further in the form

$$u_t = \frac{1}{x^2} D_x\big[x^4(u_x + u + u^2)\big]. \tag{128}$$

Writing it in the expanded form

$$u_t = x^2 u_{xx} + (x^2 + 4x + 2x^2 u)u_x + 4x(u + u^2) \tag{129}$$

we have the following formal Lagrangian for Eq. (128)

$$\mathcal{L} = v[-u_t + x^2 u_{xx} + (x^2 + 4x + 2x^2 u)u_x + 4x(u + u^2)].$$

Working out the variational derivative of this formal Lagrangian

$$\frac{\delta \mathcal{L}}{\delta u} = D_t(v) + D_x^2(x^2 v) - D_x[(x^2 + 4x + 2x^2 u)v] + 2x^2 v u_x + 4x(1 + 2u)v$$

we obtain the adjoint equation to equation (128)

$$\frac{\delta \mathcal{L}}{\delta u} \equiv v_t + x^2 v_{xx} - x^2(1 + 2u)v_x + 2(x + 2xu - 1)v = 0. \tag{130}$$

If $v = \varphi(u)$, then

$$v_t = \varphi'(u)u_t, \qquad v_x = \varphi'(u)u_x, \qquad v_{xx} = \varphi'(u)u_{xx} + \varphi''(u)u_x^2.$$

It follows that the quasi self-adjointness condition (36)

$$\left.\frac{\delta \mathcal{L}}{\delta u}\right|_{v=\varphi(u)} = \lambda[-u_t + x^2 u_{xx} + (x^2 + 4x + 2x^2 u)u_x + 4x(u + u^2)]$$

is not satisfied.

Let us check if this condition is satisfied in the more general form (72)

$$\frac{\delta \mathcal{L}}{\delta u}\bigg|_{v=\varphi(t,x,u)} = \lambda[-u_t + x^2 u_{xx} + (x^2 + 4x + 2x^2 u)u_x + 4x(u + u^2)]. \quad (131)$$

In this case

$$\begin{aligned} v_t &= D_t[\varphi(t,x,u)] = \varphi_u u_t + \varphi_t \\ v_x &= D_x[\varphi(t,x,u)] = \varphi_u u_x + \varphi_x \\ v_{xx} &= D_x(v_x) = \varphi_u u_{xx} + \varphi_{uu} u_x^2 + 2\varphi_{xu} u_x + \varphi_{xx}. \end{aligned} \quad (132)$$

Inserting (132) in the expression for the variational derivative given by (130) and singling out in Eq. (131) the terms containing u_t and u_{xx}, we obtain the following equation

$$\varphi_u[u_t + x^2 u_{xx}] = \lambda[-u_t + x^2 u_{xx}].$$

Since this equation should be satisfied identically in u_t and u_{xx}, it yields $\lambda = \varphi_u = 0$. Hence $\varphi = \varphi(t,x)$ and Eq. (131) becomes

$$\varphi_t + x^2 \varphi_{xx} - x^2(1 + 2u)\varphi_x + 2(x + 2xu - 1)\varphi = 0. \quad (133)$$

This equation should be satisfied identically in t, x and u. Therefore we nullify the coefficient for u and obtain

$$x\varphi_x - 2\varphi = 0$$

whence

$$\varphi(t,x) = c(t)x^2.$$

Substitution in Eq. (133) yields $c'(t) = 0$. Hence, $v = \varphi(t,x) = Cx^2$ with arbitrary constant C. Since $\lambda = 0$ in (131) and the adjoint Eq. (130) is linear and homogeneous in v, one can let $C = 1$. Thus, we have demonstrated the following statement.

Proposition 6 *The adjoint Eq.* (130) *has the solution*

$$v = x^2 \quad (134)$$

for any solution u of Eq. (128). *In another words,* the Kompaneets Eq. (128) is nonlinearly self-adjoint *with the substitution* (70) *given by* (134).

Remark 8 The substitution (134) does not depend on u. The question arises on existence of a substitution $v = \varphi(t,x,u)$ involving u if we rewrite Eq. (128) in an equivalent form

$$\alpha(t,x,u)[-u_t + x^2 u_{xx} + (x^2 + 4x + 2x^2 u)u_x + 4x(u + u^2)] = 0 \qquad (129')$$

with an appropriate multiplier $\alpha \neq 0$. This question is investigated in next section for a more general model.

1.4.3 The Generalized Model

In the original derivation of Eq. (126) the following more general equation appears accidentally (see [18], Eqs. (9), (10) and their discussion)

$$\frac{\partial n}{\partial t} = \frac{1}{g(x)} \frac{\partial}{\partial x}\left[g^2(x)\left(\frac{\partial n}{\partial x} + f(n)\right)\right] \qquad (135)$$

with undetermined functions $f(u)$ and $g(x)$. Then, using a physical reasoning, Kompaneets takes $f(u) = n(1+n)$ and $g(x) = x^2$. This choice restricts the symmetry properties of the model significantly. Namely, Eq. (126) has only the time-translational symmetry with the generator

$$X = \frac{\partial}{\partial t} \cdot \qquad (136)$$

The symmetry (136) provides only one invariant solution, namely the stationary solution $n = n(x)$ defined by the Riccati equation

$$\frac{dn}{dx} + n^2 + n = \frac{C}{x^4} \cdot$$

The generalized model (135) can be used for extensions of symmetry properties via the methods of *preliminary group classification* [24, 25]. In this way, exact solutions known for particular approximations to the Kompaneets equation can be obtained. This may also lead to new approximations of the solutions by taking into account various inevitable perturbations of the idealized situation assumed in the Kompaneets model (126).

So, we will take with minor changes in notation the generalized model (135)

$$u_t = \frac{1}{h(x)} D_x\{h^2(x)[u_x + f(u)]\}, \qquad h'(x) \neq 0. \qquad (137)$$

It is written in the expanded form as follows

$$u_t = h(x)\big(u_{xx} + f'(u)u_x\big) + 2h'(x)\big(u_x + f(u)\big). \qquad (138)$$

We will write Eq. (138) in the equivalent form similar to (129′)

$$\alpha(t,x,u)\big[-u_t + h(x)\big(u_{xx} + f'(u)u_x\big) + 2h'(x)\big(u_x + f(u)\big)\big] = 0 \qquad (139)$$

where $\alpha \neq 0$. This provides the following formal Lagrangian

$$\mathcal{L} = v\,\alpha(t,x,u)\big[-u_t + h(x)\big(u_{xx} + f'(u)u_x\big) + 2h'(x)\big(u_x + f(u)\big)\big] \tag{140}$$

where v is a new dependent variable. For this Lagrangian, we have

$$\begin{aligned}\frac{\delta \mathcal{L}}{\delta u} &= D_t(v\alpha) + D_x^2[h(x)v\alpha] - D_x[h(x)f'(u)v\alpha + 2h'(x)v\alpha]\\ &\quad + h(x)f''(u)\,v\,\alpha\,u_x + 2h'(x)f'(u)v\alpha\\ &\quad + v\alpha_u\big[-u_t + h(x)\big(u_{xx} + f'(u)u_x\big) + 2h'(x)\big(u_x + f(u)\big)\big].\end{aligned}$$

The reckoning shows that

$$\begin{aligned}\frac{\delta \mathcal{L}}{\delta u} &= D_t(v\alpha) + hD_x^2(v\alpha) - hf'D_x(v\alpha) + (h'f' - h'')v\alpha\\ &\quad + v\alpha_u\big[-u_t + (u_{xx} + f'u_x)h + 2(u_x + f)h'\big].\end{aligned} \tag{141}$$

Now we write the condition for the self-adjointness of Eq. (138) in the form

$$\left.\frac{\delta \mathcal{L}}{\delta u}\right|_{v=\varphi(t,x,u)} = \lambda\big[-u_t + (u_{xx} + f'u_x)h + 2(u_x + f)h'\big] \tag{142}$$

with an undetermined coefficient λ. Substituting (141) in (142) we have

$$\begin{aligned}&D_t(\varphi\alpha) + hD_x^2(\varphi\alpha) - hf'D_x(\varphi\alpha) + (h'f' - h'')\varphi\alpha\\ &\quad + \varphi\alpha_u\big[-u_t + (u_{xx} + f'u_x)h + 2(u_x + f)h'\big]\\ &= \lambda\big[-u_t + (u_{xx} + f'u_x)h + 2(u_x + f)h'\big].\end{aligned} \tag{143}$$

Here $\varphi = \varphi(t,x,u),\ \alpha = \alpha(t,x,u)$ and consequently (see (132))

$$\begin{aligned}D_t(\varphi\alpha) &= (\varphi\alpha)_u\,u_t + (\varphi\alpha)_t\\ D_x(\varphi\alpha) &= (\varphi\alpha)_u\,u_x + (\varphi\alpha)_x\\ D_x^2(\varphi\alpha) &= (\varphi\alpha)_u\,u_{xx} + (\varphi\alpha)_{uu}\,u_x^2 + 2(\varphi\alpha)_{xu}\,u_x + (\varphi\alpha)_{xx}.\end{aligned} \tag{144}$$

We substitute (141) in Eq. (143), equate the coefficients for u_t in both sides of the resulting equation and obtain $(\varphi\alpha)_u - \varphi\alpha_u = -\lambda$. Hence

$$\lambda = -\alpha\varphi_u.$$

Using this expression for λ and equating the coefficients for hu_{xx} in both sides of Eq. (143) we get $(\varphi\alpha)_u + \varphi\alpha_u = -\alpha\varphi_u$. It follows that $(\varphi\alpha)_u = 0$ and hence

$$\alpha\varphi = k(t,x).$$

Now Eq. (143) becomes

$$k_t + h(x)k_{xx} - h''(x)k + f'(u)[h'(x)k - h(x)k_x] = 0.$$

If $f''(u) \neq 0$, the above equation splits into two equations

$$h'(x)k - h(x)k_x = 0, \qquad k_t + h(x)k_{xx} - h''(x)k.$$

The first of these equations yields $k(t, x) = c(t)h(x)$, and then the second equation shows that $c'(t) = 0$. Hence, $k = C\,h(x)$ with $C = \text{const}$. Letting $C = 1$, we have

$$\alpha\varphi = h(x). \tag{145}$$

Equation (145) can be satisfied by taking, e.g.,

$$\alpha = \frac{h(x)}{u}, \qquad \varphi = u. \tag{146}$$

Thus, we have proved the following statement.

Proposition 7 *Equation* (137) *written in the equivalent form*

$$\frac{h(x)}{u}\, u_t = \frac{1}{u}\, D_x\big\{h^2(x)[u_x + f(u)]\big\} \tag{147}$$

is strictly self-adjoint. In another words, the adjoint equation to equation (147) *coincides with* (147) *upon the substitution*

$$v = u. \tag{148}$$

In particular, let us verify by direct calculations that the original Eq. (128) becomes strictly self-adjoint if we rewrite it in the equivalent form

$$\frac{x^2}{u}\, u_t = \frac{1}{u}\, D_x\big[x^4(u_x + u + u^2)\big]. \tag{149}$$

Equation (149) reads

$$-\frac{x^2}{u}\, u_t + \frac{x^4}{u}\, u_{xx} + \Big[(x^4 + 4x^3)\frac{1}{u} + 2x^4\Big] u_x + 4x^3(1 + u) = 0 \tag{150}$$

and has the formal Lagrangian

$$\mathcal{L} = -x^2\frac{v}{u}\, u_t + x^4\frac{v}{u}\, u_{xx} + \Big[(x^4 + 4x^3)\frac{v}{u} + 2x^4 v\Big] u_x + 4x^3(v + uv).$$

Accordingly, the adjoint equation to equation (150) is written

$$D_t\left(x^2\frac{v}{u}\right)+D_x^2\left(x^4\frac{v}{u}\right)-D_x\left[(x^4+4x^3)\frac{v}{u}+2x^4v\right]$$
$$+x^2\frac{v}{u^2}u_t-x^4\frac{v}{u^2}u_{xx}-(x^4+4x^3)\frac{v}{u^2}u_x+4x^3v=0.$$

Letting here $v=u$ one has $v/u=1$ and after simple calculations arrives at Eq. (150).

1.5 Quasi Self-Adjoint Reaction-Diffusion Models

Let us consider the one-dimensional reaction-diffusion model described by the following system (see e.g., [26])

$$\frac{\partial u}{\partial t}=f(u,v)+A\frac{\partial^2 u}{\partial x^2}+\frac{\partial}{\partial x}\left(\phi(u,v)\frac{\partial v}{\partial x}\right)$$
$$\frac{\partial v}{\partial t}=g(u,v)+B\frac{\partial^2 v}{\partial x^2}+\frac{\partial}{\partial x}\left(\psi(u,v)\frac{\partial u}{\partial x}\right). \tag{151}$$

It is convenient to write Eq. (151) in the form

$$D_t(u)=AD_x^2(u)+D_x\left[\phi(u,v)D_x(v)\right]+f(u,v)$$
$$D_t(v)=BD_x^2(v)+D_x\left[\psi(u,u)D_x(u)\right]+g(u,v). \tag{152}$$

The total differentiations have the form

$$D_t=\frac{\partial}{\partial t}+u_t\frac{\partial}{\partial u}+v_t\frac{\partial}{\partial v}+u_{tt}\frac{\partial}{\partial u_t}+u_{tx}\frac{\partial}{\partial u_x}+v_{tt}\frac{\partial}{\partial v_t}+v_{tx}\frac{\partial}{\partial v_x}+\cdots$$
$$D_x=\frac{\partial}{\partial x}+u_x\frac{\partial}{\partial u}+v_x\frac{\partial}{\partial v}+u_{tx}\frac{\partial}{\partial u_t}+u_{xx}\frac{\partial}{\partial u_x}+v_{tx}\frac{\partial}{\partial v_t}+v_{xx}\frac{\partial}{\partial v_x}+\cdots \tag{153}$$

and Eq. (152) are written

$$u_t=Au_{xx}+\phi v_{xx}+\left[\phi_u u_x+\phi_v v_x\right]v_x+f$$
$$v_t=Bv_{xx}+\psi u_{xx}+\left[\psi_u u_x+\psi_v v_x\right]u_x+g. \tag{154}$$

The formal Lagrangian for the system (154) is

$$\mathcal{L}=z(Au_{xx}-u_t+\phi v_{xx}+\phi_u u_x v_x+\phi_v v_x^2+f)$$
$$+w(Bv_{xx}-v_t+\psi u_{xx}+\psi_u u_x^2+\psi_v u_x v_x+g) \tag{155}$$

where z and w are new dependent variables. Equation (8) are written

$$F_1^* = \frac{\delta \mathcal{L}}{\delta u} = D_x^2\left(\frac{\partial \mathcal{L}}{\partial u_{xx}}\right) - D_t\left(\frac{\partial \mathcal{L}}{\partial u_t}\right) - D_x\left(\frac{\partial \mathcal{L}}{\partial u_x}\right) + \frac{\partial \mathcal{L}}{\partial u}$$

$$F_2^* = \frac{\delta \mathcal{L}}{\delta v} = D_x^2\left(\frac{\partial \mathcal{L}}{\partial v_{xx}}\right) - D_t\left(\frac{\partial \mathcal{L}}{\partial v_t}\right) - D_x\left(\frac{\partial \mathcal{L}}{\partial v_x}\right) + \frac{\partial \mathcal{L}}{\partial v}\cdot$$

Substituting here the expression (155) for $\mathcal{L}$ we obtain after simple calculations the following adjoint Eq. (69) to the system (154)

$$Az_{xx} + z_t + \psi_v v_x w_x - \phi_u v_x z_x + \psi w_{xx} + zf_u + wg_u = 0 \tag{156}$$

$$Bw_{xx} + w_t + \phi_u u_x z_x - \psi_v u_x w_x + \phi z_{xx} + zf_v + wg_v = 0. \tag{157}$$

Let us investigate the system (154) for quasi self-adjointness (Definition 3). We write the left-hand sides of Eqs. (156) and (157) as linear combinations of the left-hand sides of Eq. (154)

$$\begin{aligned} &Az_{xx} + z_t + \psi_v v_x w_x - \phi_u v_x z_x + \psi w_{xx} + zf_u + wg_u \\ &= (Au_{xx} - u_t + \phi v_{xx} + \phi_u u_x v_x + \phi_v v_x^2 + f)P \\ &\quad + (Bv_{xx} - v_t + \psi u_{xx} + \psi_u u_x^2 + \psi_v u_x v_x + g)Q \end{aligned} \tag{158}$$

$$\begin{aligned} &Bw_{xx} + w_t + \phi_u u_x z_x - \psi_v u_x w_x + \phi z_{xx} + zf_v + wg_v \\ &= (Au_{xx} - u_t + \phi v_{xx} + \phi_u u_x v_x + \phi_v v_x^2 + f)M \\ &\quad + (Bv_{xx} - v_t + \psi u_{xx} + \psi_u u_x^2 + \psi_v u_x v_x + g)N \end{aligned} \tag{159}$$

where P, Q, M and N are undetermined coefficients. We write the substitution (42) in the form

$$z = Z(u, v), \qquad w = W(u, v) \tag{160}$$

and insert in the left-hand sides of Eqs. (158) and (159) these expressions for z, w together with their derivatives

$$\begin{aligned} z_t &= Z_u u_t + Z_v v_t, \qquad z_x = Z_u u_x + Z_v v_x \\ z_{xx} &= Z_u u_{xx} + Z_v v_{xx} + Z_{uu} u_x^2 + 2Z_{uv} u_x v_x + Z_{vv} v_x^2 \\ w_t &= W_u u_t + W_v v_t, \quad w_x = W_u u_x + W_v v_x \\ w_{xx} &= W_u u_{xx} + W_v v_{xx} + W_{uu} u_x^2 + 2W_{uv} u_x v_x + W_{vv} v_x^2. \end{aligned}$$

Equating the coefficients in front of u_t and v_t in both sides of Eqs. (158) and (159) we obtain

$$P = -Z_u, \qquad Q = -Z_v, \qquad N = -W_v, \qquad M = -W_u. \tag{161}$$

Now we calculate the coefficients for u_{xx} and v_{xx}, take into account Eq. (161) and arrive at the following system of equations

$$\begin{aligned} 2AZ_u + \psi Z_v + \psi W_u &= 0, \qquad & (A+B)Z_v + \phi Z_u + \psi W_v &= 0 \\ 2BW_v + \phi Z_v + \psi W_u &= 0, \qquad & (A+B)W_u + \phi Z_u + \psi W_v &= 0. \end{aligned} \tag{162}$$

Equation (162) provide a linear homogeneous algebraic equations for the quantities

$$Z_u, \quad Z_v, \quad W_u, \quad W_u$$

with the matrix

$$\begin{pmatrix} 2A & \psi & \psi & 0 \\ \phi & A+B & 0 & \psi \\ 0 & \phi & \phi & 2B \\ \phi & 0 & A+B & \psi \end{pmatrix}.$$

This matrix has an inverse because its determinant is equal to

$$4(A+B)^2(\phi\psi - AB)$$

which does not vanish in the case of arbitrary A, B, ϕ and ψ. Hence, Eq. (162) yield

$$Z_u = Z_v = W_u = W_u = 0. \tag{163}$$

It follows that $Z(u, v) = C_1$, $W(u, v) = C_2$. Thus, the substitution (42) has the form

$$z = C_1, \qquad w = C_2 \tag{164}$$

with arbitrary constants C_1, C_2. Then Eqs. (158) and (159) become

$$(C_1 f + C_2 g)_u = 0, \qquad (C_1 f + C_2 g)_v = 0$$

and yield

$$\tilde{f} + \tilde{g} = C,$$

where $\tilde{f} = C_1 f$, $\tilde{g} = C_2 g$, and $C = \text{const}$. Since $\tilde{f}$ and $\tilde{g}$, along with f and g, are arbitrary functions, we can omit the "tilde" and write

$$f + g = C. \tag{165}$$

Equation (165) provides the necessary and sufficient condition for the quasi self-adjointness of the system (151). Thus, we have proved the following statement.

Theorem 2 *The system* (151) *is quasi self-adjoint if and only if it has the form*

$$\frac{\partial u}{\partial t} = f(u,v) + A\frac{\partial^2 u}{\partial x^2} + \frac{\partial}{\partial x}\left(\phi(u,v)\frac{\partial v}{\partial x}\right)$$
$$\frac{\partial v}{\partial t} = C - f(u,v) + B\frac{\partial^2 v}{\partial x^2} + \frac{\partial}{\partial x}\left(\psi(u,v)\frac{\partial u}{\partial x}\right) \tag{166}$$

where $\phi(u,v),\ \psi(u,v),\ f(u,v)$ *are arbitrary functions and* A, B, C *are arbitrary constants. The substitution* (42) *is given by* (164).

Remark 9 If we replace (160) by the general substitution (70), i.e., take

$$z = Z(t,x,u,v), \qquad w = W(t,x,u,v) \tag{167}$$

then Eq. (164) will be replaced by

$$z = Z(t,x), \qquad w = W(t,x) \tag{168}$$

with functions $Z(t,x),\ W(t,x)$ satisfying the following equations

$$(\psi_v W - \phi_u Z)_x = 0 \tag{169}$$

$$AZ_{xx} + Z_t + \psi W_{xx} + (fZ + gW)_u = 0$$
$$BW_{xx} + W_t + \phi Z_{xx} + (fZ + gW)_v = 0. \tag{170}$$

1.6 A Model of an Irrigation System

Let us consider the second-order nonlinear partial differential equation

$$C(\psi)\psi_t = [K(\psi)\psi_x]_x + \left[K(\psi)\,(\psi_z - 1)\right]_z - S(\psi). \tag{171}$$

It serves as a mathematical model for investigating certain irrigation systems (see [15–17], vol. 2, Sect. 9.8 and references therein). The dependent variable ψ denotes the soil moisture pressure head, $C(\psi)$ is the specific water capacity, $K(\psi)$ is the unsaturated hydraulic conductivity, $S(\psi)$ is a source term. The independent variables are the time t, the horizontal axis x and the vertical axis z which is taken to be positive downward.

The adjoint equation (69) to equation (171) has the form

$$C(\psi)v_t + K(\psi)\left[v_{xx} + v_{zz}\right] + K'(\psi)v_z - S'(\psi)v = 0. \tag{172}$$

It follows from (172) that Eq. (171) is not nonlinearly self-adjoint if $C(\psi)$, $K(\psi)$ and $S(\psi)$ are arbitrary functions. Indeed, using Definition 5 of the nonlinear self-adjointness and nullifying in (172) the term with $S'(\psi)$ we obtain $v = 0$. Hence, the condition (71) of the nonlinear self-adjointness is not satisfied.

However, Eq. (171) can be nonlinearly self-adjoint if there are certain relations between the functions $C(\psi)$, $K(\psi)$ and $S(\psi)$. For example, let us suppose that the specific water capacity $C(\psi)$ and the hydraulic conductivity $K(\psi)$ are arbitrary, but the source term $S(\psi)$ is related with $C(\psi)$ by the equation

$$S'(\psi) = aC(\psi), \qquad a = \text{const.} \tag{173}$$

Then Eq. (172) becomes $C(\psi)[v_t - av] + K(\psi)\left[v_{xx} + v_{zz}\right] + K'(\psi)v_z = 0$ and yields

$$v_z = 0, \qquad v_{xx} = 0, \qquad v_t - av = 0. \tag{174}$$

We solve the first two Eq. (174) and obtain

$$v = p(t)x + q(t).$$

We substitute this in the third Eq. (174)

$$[p'(t) - ap(t)]x + q'(t) - aq(t) = 0$$

split it with respect to x and obtain $p'(t) - ap(t) = 0, \quad q'(t) - aq(t) = 0$ whence

$$p(t) = b\mathrm{e}^{at}, \qquad q(t) = l\mathrm{e}^{at}, \qquad b, l = \text{const.}$$

Thus, Eq. (171) satisfying the condition Eq. (173) is nonlinearly self-adjoint, and the substitution (70) has the form

$$v = (bx + l)\mathrm{e}^{at}. \tag{175}$$

One can obtain various nonlinearly self-adjoint equations (171) by considering other relations between $C(\psi)$, $K(\psi)$ and $S(\psi)$ different from (173).

1.7 Krichever–Novikov Equation

Let us consider the Krichever-Novikov equation [27] written in the form

$$F \equiv u_t - u_{xxx} + \frac{3}{2}\frac{u_{xx}^2}{u_x} - \frac{P(u)}{u_x} = 0 \tag{176}$$

where $P(u)$ a polynomial of degree four with distinct roots. The nonlinear self-adjointness of this equation has been investigated recently in [28]. Namely, it is shown that Eq. (176) satisfies the nonlinear self-adjointness condition in the form

$$F^*\big|_{v=\varphi} = \mu_0 F + \mu_1 D_x(F) + \mu_2 D_x^2(F) + \mu_3 D_x^3(F) + \mu_4 D_x^4(F) \tag{177}$$

with

$$\mu_0 = 2\frac{u_{xx}}{u_x^4}\,P'(u) - \frac{1}{u_x^2}\,P''(u)$$

$$\mu_1 = 2\frac{u_{xxxx}}{u_x^3} - 12\frac{u_{xxx}u_{xx}}{u_x^4} + 12\frac{u_{xx}^3}{u_x^5} + \frac{2}{u_x^3}\,P'(u) - 8\frac{u_{xx}}{u_x^5}\,P(u)$$

$$\mu_2 = 4\frac{u_{xxx}}{u_x^3} - 9\frac{u_{xx}^2}{u_x^4} + \frac{2}{u_x^4}\,P(u), \qquad \mu_3 = 4\frac{u_{xx}}{u_x^3}, \qquad \mu_4 = -\frac{1}{u_x^2}$$

where $v = \varphi$ is the differential substitution (76) with the following φ

$$\varphi = \frac{u_{xxxx}}{u_x^2} + 3\,\frac{u_{xx}^3}{u_x^4} - 4\,\frac{u_{xx}u_{xxx}}{u_x^3} - 2\,\frac{P(u)u_{xx}}{u_x^4} + \frac{P'(u)}{u_x^2}\cdot \tag{178}$$

2 Construction of Conservation Laws Using Symmetries

The well-known Noether's theorem [29] states that if a variational integral is invariant with respect to a one-parameter group of transformations then a certain formula (see further Theorem 3) provides a conservation law for the corresponding Euler-Lagrange equation. Thus, according to Noether's theorem, the invariance of the variational integral is a *sufficient condition* for existence of the conservation law. It has been proved in [30] that the *necessary and sufficient condition* for existence of this conservation law is the invariance of the extremal values (the values on the *solutions of the Euler-Lagrange equations*) of the variational integral. The result extends to multi-parameter symmetry groups.

The new approach based on the concept of nonlinear self-adjointness allows to extend a connection between symmetries and conservation laws significantly.

2.1 *Discussion of the Operator Identity*

2.1.1 Operator Identity and Alternative Proof of Noether's Theorem

Let us discuss some consequences of the operator identity[4]

$$X + D_i(\xi^i) = W^\alpha \frac{\delta}{\delta u^\alpha} + D_i \mathrm{N}^i. \tag{179}$$

Here

$$X = \xi^i \frac{\partial}{\partial x^i} + \eta^\alpha \frac{\partial}{\partial u^\alpha} + \zeta_i^\alpha \frac{\partial}{\partial u_i^\alpha} + \zeta_{i_1 i_2}^\alpha \frac{\partial}{\partial u_{i_1 i_2}^\alpha} + \cdots \tag{180}$$

$$W^\alpha = \eta^\alpha - \xi^j u_j^\alpha, \qquad \alpha = 1, \ldots, m \tag{181}$$

$$\frac{\delta}{\delta u^\alpha} = \frac{\partial}{\partial u^\alpha} + \sum_{s=1}^{\infty} (-1)^s D_{i_1} \cdots D_{i_s} \frac{\partial}{\partial u_{i_1 \cdots i_s}^\alpha}, \qquad \alpha = 1, \ldots, m \tag{182}$$

and

$$\mathrm{N}^i = \xi^i + W^\alpha \frac{\delta}{\delta u_i^\alpha} + \sum_{s=1}^{\infty} D_{i_1} \cdots D_{i_s} (W^\alpha) \frac{\delta}{\delta u_{i i_1 \cdots i_s}^\alpha}, \qquad i = 1, \ldots, n \tag{183}$$

where the Euler-Lagrange operators with respect to derivatives of u^α are obtained from (182) by replacing u^α by the corresponding derivatives, e.g.,

$$\frac{\delta}{\delta u_i^\alpha} = \frac{\partial}{\partial u_i^\alpha} + \sum_{s=1}^{\infty} (-1)^s D_{j_1} \cdots D_{j_s} \frac{\partial}{\partial u_{i j_1 \cdots j_s}^\alpha}. \tag{184}$$

The coefficients ξ^i, η^α in (180) are arbitrary *differential functions* (see Sect. 1.1.1) and the other coefficients are determined by the prolongation formulae

$$\zeta_i^\alpha = D_i(W^\alpha) + \xi^j u_{ij}^\alpha, \qquad \zeta_{i_1 i_2}^\alpha = D_{i_1} D_{i_2}(W^\alpha) + \xi^j u_{j i_1 i_2}^\alpha, \ldots . \tag{185}$$

The derivation of Eq. (179) is essentially based on Eq. (185).

Remark 10 If we write the operator (180) in the equivalent canonical form[5]

---

[4] The operator identity (179) was called in [31] the *Noether identity* and used for simplifying the proof of Noether's theorem. A simple proof of the identity (179) can be found in [32] where Eq. (179) is written as Eq. (19).

[5] One can verify that if one uses the canonical form of the operator X, then the operator identity (179) becomes identical with Eqs. (3), (6) in Noether's paper [29] except for the notation. Noether

$$X = W^\alpha \frac{\partial}{\partial u^\alpha} + \zeta_i^\alpha \frac{\partial}{\partial u_i^\alpha} + \zeta_{i_1 i_2}^\alpha \frac{\partial}{\partial u_{i_1 i_2}^\alpha} + \cdots \tag{186}$$

then the prolongation formulae (185) become simpler

$$\zeta_i^\alpha = D_i(W^\alpha), \qquad \zeta_{i_1 i_2}^\alpha = D_{i_1} D_{i_2}(W^\alpha), \ldots . \tag{187}$$

The original proof of Noether's theorem [29] is based on the calculus of variations. An alternative proof of this theorem has been given in [31] (see also [4, 10]). Let us outline the latter proof based on the identity (179).

We consider the Euler-Lagrange equations

$$\frac{\delta \mathcal{L}}{\delta u^\alpha} = 0, \qquad \alpha = 1, \ldots, m. \tag{188}$$

If we assume that the operator (180) is admitted by Eq. (188) and that the variational integral

$$\int \mathcal{L}(x, u, u_{(1)}, \ldots) \mathrm{d}x$$

is invariant under the transformations of the group with the generator X then the following equation holds

$$X(\mathcal{L}) + D_i(\xi^i)\mathcal{L} = 0. \tag{189}$$

Therefore, if we act on $\mathcal{L}$ by both sides of the identity (179)

$$X(\mathcal{L}) + D_i(\xi^i)\mathcal{L} = W^\alpha \frac{\delta \mathcal{L}}{\delta u^\alpha} + D_i[\mathsf{N}^i(\mathcal{L})]$$

and take into account Eqs. (188) and (189), we see that the vector with the components

$$C^i = \mathsf{N}^i(\mathcal{L}), \qquad i = 1, \ldots, n \tag{190}$$

satisfies the conservation equation

$$D_i(C^i)\Big|_{(188)} = 0. \tag{191}$$

For practical applications, when we deal with law order Lagrangians $\mathcal{L}$, it is convenient to restrict the operator (183) on the derivatives involved in $\mathcal{L}$ and write the expressions (190) in the expanded form

(Footnote 5 continued)
comments that in the case of the first-order Lagrangians her Eq. (3) is identical with the *central equation of Lagrange* (Eqs. (4) and (5) in [29]).

$$
\begin{aligned}
C^i = \xi^i \mathcal{L} + W^\alpha \left[\frac{\partial \mathcal{L}}{\partial u_i^\alpha} - D_j\left(\frac{\partial \mathcal{L}}{\partial u_{ij}^\alpha}\right) + D_j D_k\left(\frac{\partial \mathcal{L}}{\partial u_{ijk}^\alpha}\right) - \cdots\right] \\
+ D_j\left(W^\alpha\right)\left[\frac{\partial \mathcal{L}}{\partial u_{ij}^\alpha} - D_k\left(\frac{\partial \mathcal{L}}{\partial u_{ijk}^\alpha}\right) + \cdots\right] + D_j D_k\left(W^\alpha\right)\left[\frac{\partial \mathcal{L}}{\partial u_{ijk}^\alpha} - \cdots\right].
\end{aligned} \tag{192}
$$

Thus, Noether's theorem can be formulated as follows.

Theorem 3 *If the operator* (180) *is admitted by Eq.* (188) *and satisfies the condition* (189) *of the invariance of the variational integral, then the vector* (192) *constructed by Eq.* (192) *satisfies the conservation law* (191).

Remark 11 The identity (179) is valid also in the case when the coefficients ξ^i, η^α of the operator X involve not only the local variables $x, u, u_{(1)}, u_{(2)}, \ldots$ but also nonlocal variables (see Sect. 2.5.6). Accordingly, the formula (192) associates conserved vectors with nonlocal symmetries as well.

Remark 12 If the invariance condition (189) is replaced by the divergence condition

$$X(\mathcal{L}) + D_i(\xi^i)\mathcal{L} = D_i(B^i)$$

then the identity (179) leads to the conservation law (191) where the conserved vector (190) is replaced with

$$C^i = \mathsf{N}^i(\mathcal{L}) - B^i, \qquad i = 1, \ldots, n. \tag{193}$$

2.1.2 Test for Total derivative and for Divergence

I recall here the well-known necessary and sufficient condition for a differential function to be divergence, or total derivative in the case of one independent variable.

One can easily derive from the Definition (1) of the total differentiation D_i the following lemmas (see also [4], Sect. 8.4.1).

Lemma 1 *The following infinite series of equations hold*

$$
\begin{aligned}
\frac{\partial}{\partial u^\alpha} D_i &= D_i \frac{\partial}{\partial u^\alpha} \\
D_j \frac{\partial}{\partial u_j^\alpha} D_i &= D_i \frac{\partial}{\partial u^\alpha} + D_i D_j \frac{\partial}{\partial u_j^\alpha} \\
D_j D_k \frac{\partial}{\partial u_{jk}^\alpha} D_i &= D_i D_k \frac{\partial}{\partial u_k^\alpha} + D_i D_j D_k \frac{\partial}{\partial u_{jk}^\alpha} \\
&\cdots\cdots\cdots\cdots\cdots\cdots
\end{aligned}
$$

Lemma 2 *The following operator identity holds for every i and α*

$$\frac{\delta}{\delta u^\alpha} D_i = 0.$$

Proof Using Lemma 1 and manipulating with summation indices we obtain

$$\frac{\delta}{\delta u^\alpha} D_i = \left(\frac{\partial}{\partial u^\alpha} - D_j \frac{\partial}{\partial u_j^\alpha} + D_j D_k \frac{\partial}{\partial u_{jk}^\alpha} - D_j D_k D_l \frac{\partial}{\partial u_{jkl}^\alpha} + \cdots\right) D_i$$
$$= \frac{\partial}{\partial u^\alpha} D_i - D_i \frac{\partial}{\partial u^\alpha} - D_i D_j \frac{\partial}{\partial u_j^\alpha} + D_i D_k \frac{\partial}{\partial u_k^\alpha} + D_i D_j D_k \frac{\partial}{\partial u_{jk}^\alpha}$$
$$- D_i D_k D_l \frac{\partial}{\partial u_{kl}^\alpha} - \cdots = 0.$$

□

Proposition 8 *A differential function* $f(x, u, u_{(1)}, \ldots, u_{(s)}) \in \mathcal{A}$ *is divergence*

$$f = D_i(h^i), \qquad h^i(x, u, \ldots, u_{(s-1)}) \in \mathcal{A} \tag{194}$$

if and only if the following equations hold identically in $x, u, u_{(1)}, \ldots$

$$\frac{\delta f}{\delta u^\alpha} = 0, \qquad \alpha = 1, \ldots, m. \tag{195}$$

The statement that the relation (194) implies (195) follows immediately from Lemma 2. For the proof of the inverse statement that (195) implies (194), see [33], Chap. 4, Sect. 3.5, and [32]. See also [4], Sect. 8.4.1.

We will use Proposition 8 also in the particular case of one independent variable x and one dependent variable $u = y$. Then it is formulated as follows.

Proposition 9 *A differential function* $f(x, y, y', \ldots, y^{(s)}) \in \mathcal{A}$ *is the total derivative*

$$f = D_x(g), \qquad g(x, y, y', \ldots, y^{(s-1)}) \in \mathcal{A} \tag{196}$$

if and only if the following equation holds identically in $x, y, y', \ldots$

$$\frac{\delta f}{\delta y} = 0. \tag{197}$$

Here $\delta f / \delta y$ is the Euler-Lagrange operator (184)

$$\frac{\delta}{\delta y} = \frac{\partial}{\partial y} - D_x \frac{\partial}{\partial y'} + D_x^2 \frac{\partial}{\partial y''} - D_x^3 \frac{\partial}{\partial y'''} + \cdots . \tag{198}$$

2.1.3 Adjoint Equation to Linear ODE

Let us consider an arbitrary sth-order linear ordinary differential operator

$$L[y] = a_0 y^{(s)} + a_1 y^{(s-1)} + \cdots + a_{s-2} y'' + a_{s-1} y' + a_s y \tag{199}$$

where $a_i = a_i(x)$. We know from Sect. 1.1.5 that the adjoint operator to (199) can be calculated by using Eq. (8). I give here the independent proof based on the operator identity (179).

Proposition 10 *The adjoint operator to* (199) *can be calculated by the formula*

$$L^*[z] = \frac{\delta(zL[y])}{\delta y} \cdot \tag{200}$$

Proof Let

$$X = w\frac{\partial}{\partial y} + w'\frac{\partial}{\partial y'} + w''\frac{\partial}{\partial y''} + \cdots \tag{201}$$

be the operator (186) with one independent variable x and one dependent variable $u = y$, where the prolongation formulae (187) are written using the notation

$$w' = D_x(w), \qquad w'' = D_x^2(w), \; \ldots \; . \tag{202}$$

In this notation the operator (183) is written

$$\mathsf{N} = w\,\frac{\delta}{\delta y'} + w'\,\frac{\delta}{\delta y''} + w''\,\frac{\delta}{\delta y'''} + \cdots \, .$$

Having in mind its application to the differential function $L[y]$ given by (199) we consider the following restricted form of N

$$\mathsf{N} = w\,\frac{\delta}{\delta y'} + w'\,\frac{\delta}{\delta y''} + \cdots + w^{(s-1)}\,\frac{\delta}{\delta y^{(s)}} \cdot \tag{203}$$

The identity (179) has the form

$$X = w\frac{\delta}{\delta y} + D_x\mathsf{N} \, . \tag{204}$$

We act by both sides of this identity on $zL[y]$, where z is a new dependent variable:

$$X(zL[y]) = w\frac{\delta(zL[y])}{\delta y} + D_x\mathsf{N}(zL[y]) \, . \tag{205}$$

Since the operator (201) does not act on the variables x and z, we have

$$X(zL[y]) = zX(L[y]). \tag{206}$$

Furthermore we note that

$$X(L[y]) = L[w]. \tag{207}$$

Inserting (206) and (207) in Eq. (205) we obtain

$$zL[w] - w\frac{\delta(zL[y])}{\delta y} = D_x(\Psi) \tag{208}$$

where Ψ is a quadratic form $\Psi = \Psi[w, z]$ defined by

$$\Psi = \mathsf{N}(zL[y]). \tag{209}$$

After replacing w with y Eq. (208) coincides with Eq. (3) for the adjoint operator

$$zL[y] - yL^*[z] = D_x(\psi) \tag{210}$$

where $L^*[z]$ is given by the formula (200) and $\psi = \psi[y, z]$ is defined by

$$\psi[y, z] = \Psi[w, z]\big|_{w=y} \equiv \mathsf{N}(zL[y])\big|_{w=y}. \tag{211}$$

□

Remark 13 Let us find the explicit formula for ψ in Eq. (210) We write the operator N given by Eq. (203) in the expanded form

$$\begin{aligned}\mathsf{N} = w &\left[\frac{\partial}{\partial y'} - D_x\frac{\partial}{\partial y''} + \cdots + (-D_x)^{s-1}\frac{\partial}{\partial y^{(s)}}\right]\\ &+ w'\left[\frac{\partial}{\partial y''} - D_x\frac{\partial}{\partial y'''} + \cdots + (-D_x)^{s-2}\frac{\partial}{\partial y^{(s)}}\right]\\ &+ w^{(s-2)}\left[\frac{\partial}{\partial y^{(s-1)}} - D_x\frac{\partial}{\partial y^{(s)}}\right] + w^{(s-1)}\,\frac{\delta}{\delta y^{(s)}}\end{aligned}$$

act on $zL[y]$ written in the form

$$zL[y] = a_s yz + a_{s-1}y'z + a_{s-2}y''z + \cdots + a_1 y^{(s-1)}z + a_0 y^{(s)}z$$

and obtain Ψ. We replace w with y in $\Psi = \Psi[w, z]$ and $\psi = \psi[y, z]$

$$\begin{aligned}\psi[y, z] = y&\left[a_{s-1}\,z - (a_{s-2}\,z)' + \cdots + (-1)^{s-1}(a_0\,z)^{(s-1)}\right]\\ &+ y'\left[a_{s-2}\,z - (a_{s-3}\,z)' + \cdots + (-1)^{s-2}(a_0\,z)^{(s-2)}\right]\\ &+ y^{(s-2)}\left[a_1\,z - (a_0\,z)'\right] + y^{(s-1)}\,a_0\,z.\end{aligned} \tag{212}$$

The expression (212) is obtained in the classical literature using integration by parts (see e.g., [34, Chap. 5, Sect. 4, Eq. (31′)]).

2.1.4 Conservation Laws and Integrating Factors for Linear ODEs

Consider an sth-order homogeneous linear ordinary differential equation

$$L[y] = 0 \tag{213}$$

where $L[y]$ is the operator defined by Eq. (199). If $L[y]$ is a total derivative

$$L[y] = D_x\left(\psi(x, y, y', \ldots, y^{(s-1)})\right) \tag{214}$$

the Eq. (213) can be written as a conservation law

$$D_x\left(\psi(x, y, y', \ldots, y^{(s-1)})\right) = 0$$

whence upon integration one obtains a linear equation of order $s - 1$

$$\psi(x, y, y', \ldots, y^{(s-1)}) = C_1. \tag{215}$$

We can also reduce the order of the non-homogeneous equation

$$L[y] = f(x) \tag{216}$$

by rewriting it in the conservation form

$$D_x\left[\psi(x, y, y', \ldots, y^{(s-1)}) - \int f(x)\mathrm{d}x\right] = 0. \tag{217}$$

Integrating it once we obtain the non-homogeneous linear equation of order $s - 1$

$$\psi(x, y, y', \ldots, y^{(s-1)}) = C_1 + \int f(x)\mathrm{d}x.$$

Example 15 Consider the second-order equation

$$y'' + y' \sin x + y \cos x = 0.$$

We have

$$y'' + y' \sin x + y \cos x = D_x(y' + y \sin x).$$

Therefore the second-order equation in question reduces to the first-order equation

$$y' + y \sin x = C_1.$$

Integrating the latter equation we obtain the general solution

$$y = \left[C_2 + C_1 \int \mathrm{e}^{-\cos x}\, \mathrm{d}x \right] \mathrm{e}^{\cos x}$$

to our second-order equation. Dealing likewise with the non-homogeneous equation

$$y'' + y' \sin x + y \cos x = 2x$$

we obtain its general solution

$$y = \left[C_2 + \int \left(C_1 + x^2 \right) \mathrm{e}^{-\cos x}\, \mathrm{d}x \right] \mathrm{e}^{\cos x}.$$

If $L[y]$ in Eq. (213) is not a total derivative, one can find an appropriate factor $\phi(x) \neq 0$, called an *integrating factor*, such that $\phi(x)L[y]$ becomes a total derivative

$$\phi(x)L[y] = D_x \left(\psi(x, y, y', \ldots, y^{(s-1)}) \right). \tag{218}$$

A connection between integrating factors and the adjoint equations for linear equations is well known in the classical literature (see e.g., [34], Chap. 5, Sect. 4). Proposition 9 gives a simple way to establish this connection and prove the following statement.

Proposition 11 *A function* $\phi(x)$ *is an integrating factor for Eq.* (213) *if and only if*

$$z = \phi(x), \qquad \phi(x) \neq 0 \tag{219}$$

is a solution of the adjoint equation [6] *to Eq.* (213)

$$L^*[z] = 0. \tag{220}$$

Knowledge of a solution (219) *to the adjoint equation* (220) *allows to reduce the order of Eq.* (213) *by integrating Eq.* (218)

$$\psi(x, y, y', \ldots, y^{(s-1)}) = C_1. \tag{221}$$

Here C_1 *is an arbitrary constants and* ψ *defined according to Eqs.* (209) and (210), *i.e.,*

$$\psi = \mathsf{N}(zL[y])\big|_{w=y}. \tag{222}$$

[6] This statement is applicable to nonlinear ODEs as well, see [35].

Proof If (219) is a solution of the adjoint Eq. (220), we substitute it in Eq. (210) and arrive at Eq. (218). Hence $\phi(x)$ is an integrating factor for Eq. (213). Conversely, if $\phi(x)$ is an integrating factor for Eq. (213), then Eq. (218) is satisfied. Now Proposition 9 yields

$$\frac{\delta(\phi(x)L[y])}{\delta y} = 0.$$

Hence (219) is a solution of the adjoint Eq. (220). Finally, Eq. (222) follows from (210). □

Example 16 Let us apply the above approach to the first-order equation

$$y' + P(x)y = Q(x). \tag{223}$$

Here $L[y] = y' + P(x)y$. The adjoint Eq. (220) is written

$$z' - P(x)z = 0.$$

Solving it we obtain the integrating factor

$$z = \mathrm{e}^{\int P(x)\mathrm{d}x}.$$

Therefore we rewrite Eq. (223) in the equivalent form

$$\left[y' + P(x)y\right]\mathrm{e}^{\int P(x)\mathrm{d}x} = Q(x)\mathrm{e}^{\int P(x)\mathrm{d}x} \tag{224}$$

and compute the function Ψ given by Eq. (209)

$$\Psi = \mathrm{N}(zL[y]) = w\frac{\partial}{\partial y'}[z(y' + P(x)y)] = wz = w\mathrm{e}^{\int P(x)\mathrm{d}x}.$$

Equation (222) yields

$$\psi = y\mathrm{e}^{\int P(x)\mathrm{d}x}. \tag{225}$$

Now we can take (224) instead of Eq. (216) and write it in the form (217) with ψ given by (225). Then we obtain

$$D_x\left[y\mathrm{e}^{\int P(x)\mathrm{d}x} - \int Q(x)\mathrm{e}^{\int P(x)\mathrm{d}x}\mathrm{d}x\right] = 0$$

whence

$$y\mathrm{e}^{\int P(x)\mathrm{d}x} = C_1 + \int Q(x)\mathrm{e}^{\int P(x)\mathrm{d}x}\mathrm{d}x.$$

Solving the latter equation for y we obtain the general solution of Eq. (223)

$$y=\left[C_1+\int Q(x)\mathrm{e}^{\int P(x)\mathrm{d}x}\mathrm{d}x\right]\mathrm{e}^{-\int P(x)\mathrm{d}x}. \tag{226}$$

Example 17 Let us consider the second-order homogeneous equation

$$y''+\frac{\sin x}{x^2}y'+\left(\frac{\cos x}{x^2}-\frac{\sin x}{x^3}\right)y=0. \tag{227}$$

Its left-hand side does not satisfy the total derivative condition (197) because

$$\frac{\delta}{\delta y}\left[y''+\frac{\sin x}{x^2}y'+\left(\frac{\cos x}{x^2}-\frac{\sin x}{x^3}\right)y\right]=\frac{\sin x}{x^2}\cdot$$

Therefore we will apply Proposition 11. The adjoint equation to equation (227) is written

$$z''-\frac{\sin x}{x^2}z'+\frac{\sin x}{x^3}z=0.$$

We take its obvious solution $z=x$, substitute it in Eq. (209) and using (211) find

$$\Psi=\mathrm{N}\left[xy''+\frac{\sin x}{x}y'+\left(\frac{\cos x}{x}-\frac{\sin x}{x^2}\right)y\right]=\frac{\sin x}{x}w-w+xw'.$$

Therefore Eq. (221) is written

$$xy'+\left(\frac{\sin x}{x}-1\right)y=C_1.$$

Integrating this first-order linear equation we obtain the general solution of Eq. (227)

$$y=\left(C_2+C_1\int\frac{1}{x^2}\mathrm{e}^{\int\frac{\sin x}{x^2}\mathrm{d}x}\mathrm{d}x\right)x\mathrm{e}^{-\int\frac{\sin x}{x^2}\mathrm{d}x}. \tag{228}$$

2.1.5 Application of the Operator Identity to Linear PDEs

Using the operator identity (179) one can easily extend the Eqs. (210) and (211) for linear ODEs to linear partial differential equations and systems. Let us consider the second-order linear operator

$$L[u]=a^{ij}(x)u_{ij}+b^i(x)u_i+c(x)u \tag{229}$$

considered in Sect. 1.1.5, Remark 1. The adjoint operator is

$$L^*[v] \equiv \frac{\delta(vF[u])}{\delta u} = D_i D_j(a^{ij} v) - D_i(b^i v) + cv. \tag{230}$$

Let us take the operator identity (179)

$$X = W\frac{\delta}{\delta u} + D_i \mathsf{N}^i \tag{231}$$

where X is the operator (186) with one dependent variable u

$$X = W\frac{\partial}{\partial u} + W_i\frac{\partial}{\partial u_i} + W_{ij}\frac{\partial}{\partial u_{ij}}$$

and N^i are the operators (183)

$$\mathsf{N}^i = W\,\frac{\delta}{\delta u_i} + W_j\,\frac{\delta}{\delta u_{ij}} = W\left[\frac{\partial}{\partial u_i} - D_j\frac{\partial}{\partial u_{ij}}\right] + W_j\,\frac{\partial}{\partial u_{ij}}\cdot$$

We use above the notation $W_i = D_i(W),\ W_{ij} = D_i D_j(W)$. Now we proceed as in Sect. 2.1.3. Namely, we act on $vL[u]$ by both sides of the identity (231)

$$X(vL[u]) = W\frac{\delta(vL[u])}{\delta u} + D_i\mathsf{N}^i(vL[u])$$

take into account that X does not act on the variables $x^i,\ v$, and that $X(L[u]) = L[W]$, use Eq. (230) and obtain

$$vL[W]) - WL^*[v] = D_i\mathsf{N}^i(vL[u]).$$

Letting here $W = u$ we arrive at the following generalization of Eq. (210)

$$vL[u] - uL^*[v] = D_i(\psi^i) \tag{232}$$

where ψ^i are defined as in (211) and (212)

$$\psi^i = \mathsf{N}^i(vL[u])\big|_{W=u} \equiv a^{ij}(x)[vu_i - uv_i] + [b^i(x) - D_i\big(a^{ij}(x)\big)]uv. \tag{233}$$

Equation (232) with ψ^i given by (233) is called *Lagrange's identity* (see [36], p. 80).

2.1.6 Application of the Operator Identity to Nonlinear Equations

Let us apply the constructions of Sect. 2.1.5 to nonlinear equation (6)

$$F_\alpha\big(x, u, u_{(1)}, \ldots, u_{(s)}\big) = 0, \qquad \alpha = 1, \ldots, m. \tag{234}$$

We write the operator (186) in the form

$$X = W^\alpha \frac{\partial}{\partial u^\alpha} + W_i^\alpha \frac{\partial}{\partial u_i^\alpha} + W_{ij}^\alpha \frac{\partial}{\partial u_{ij}^\alpha} + \cdots$$

where $W_i^\alpha = D_i(W^\alpha),\ W_{ij}^\alpha = D_i D_j(W^\alpha), \ldots$. Then the operator (183) is written

$$\mathsf{N}^i = W_j^\alpha \frac{\delta}{\delta u_i^\alpha} + W^\alpha \frac{\delta}{\delta u_{ij}^\alpha} + \cdots .$$

We act on $v^\beta F_\beta$ by both sides of the operator identity (179)

$$X = W^\alpha \frac{\delta}{\delta u^\alpha} + D_i \mathsf{N}^i$$

denote by $F_\alpha^*[v]$ the adjoint operator defined by Eq. (8) and obtain

$$v^\beta \hat{F}_\beta[W] - W^\alpha F_\alpha^*[v] = D_i(\Psi^i) \tag{235}$$

where

$$\Psi^i = \mathsf{N}^i(v^\beta F_\beta)$$

and $\hat{F}_\beta[W]$ is the *linear approximation* to F_β defined by (see also Sect. 1.1.3)

$$\hat{F}_\beta[W] = X(F_\beta) \equiv W^\alpha \frac{\partial F_\beta}{\partial u^\alpha} + W_i^\alpha \frac{\partial F_\beta}{\partial u_i^\alpha} + W_{ij}^\alpha \frac{\partial F_\beta}{\partial u_{ij}^\alpha} + \cdots .$$

Remark 14 Equation (235) shows that $F_\alpha^*[v] = \hat{F}_\beta^*[W]$, i.e., the adjoint operator F_α^* to nonlinear equation (234) is the usual adjoint operator $\hat{F}_\beta^*$ to the linear operator $\hat{F}_\beta[W]$ (see also [37]). But the linear self-adjointness of $\hat{F}_\beta[W]$ is not identical with the nonlinear self-adjointness of Eq. (234). For example, the KdV equation $F \equiv u_t - u_{xxx} - uu_x = 0$ is nonlinearly self-adjoint (see Example 2 in Sect. 1.1.6). But its linear approximation $\hat{F}[W] = W_t - W_{xxx} - uW_x - Wu_x$ is not a self-adjoint linear operator. Moreover, *all linear equations are nonlinearly self-adjoint.*

2.2 Conservation Laws: Generalities and Explicit Formula

2.2.1 Preliminaries

Let us consider a system of $\overline{m}$ differential equations

$$F_{\bar{\alpha}}\left(x, u, u_{(1)}, \ldots, u_{(s)}\right) = 0, \qquad \bar{\alpha} = 1, \ldots, \overline{m} \tag{236}$$

with m dependent variables $u^1, \ldots, u^m$ and n independent variables $x^1, \ldots, x^n$.

A conservation law for Eq. (236) is written

$$\left[D_i(C^i)\right]_{(236)} = 0. \tag{237}$$

The subscript $|_{(236)}$ means that the left-hand side of (237) is restricted on the solutions of Eq. (236). In practical calculations this restriction can be achieved by solving Eq. (236) with respect to certain derivatives of u and eliminating these derivatives from the left-hand side of (237). For example, if (236) is an evolution equation

$$u_t = \Phi(t, x, u, u_x, u_{xx})$$

the restriction $|_{(236)}$ can be understood as the elimination of u_t. The n-dimensional vector

$$C = (C^1, \ldots, C^n) \tag{238}$$

satisfying Eq. (237) is called a *conserved vector* for the system (236). If its components are functions $C^i = C^i(x, u, u_{(1)}, \ldots)$ of x, u and derivatives $u_{(1)}, \ldots$ of a finite order, the conserved vector (238) is called a *local conserved vector.*

Since the conservation equation (237) is linear with respect to C^i, any linear combination with constant coefficients of a finite number of conserved vectors is again a conserved vector. It is obvious that if the divergence of a vector (238) vanishes identically, it is a conserved vector for any system of differential equations. This is a *trivial* conserved vectors for all differential equations. Another type of *trivial conserved vectors* for Eq. (236) are provided by those vectors whose components C^i vanish on the solutions of the system (236). One ignores both types of trivial conserved vectors. In other words, conserved vectors (238) are simplified by considering them up to addition of these trivial conserved vectors.

The following less trivial operation with conserved vectors is particularly useful in practice. Let

$$C^1|_{(236)} = \widetilde{C}^1 + D_2(H^2) + \cdots + D_n(H^n) \tag{239}$$

the conserved vector (238) can be replaced with the equivalent conserved vector

$$\widetilde{C} = (\widetilde{C}^1, \widetilde{C}^2, \ldots, \widetilde{C}^n) = 0 \tag{240}$$

with the components

$$\widetilde{C}^1, \quad \widetilde{C}^2 = C^2 + D_1(H^2), \quad \ldots, \quad \widetilde{C}^n = C^n + D_1(H^n). \tag{241}$$

The passage from (238) to the vector (240) is based on the commutativity of the total differentiations. Namely, we have

$$D_1 D_2(H^2) = D_2 D_t(H^2), \qquad D_1 D_n(H^n) = D_n D_t(H^n)$$

and therefore the conservation Eq. (237) for the vector (238) is equivalent to the conservation equation

$$\left[D_i(\widetilde{C}^i)\right]_{(236)} = 0$$

for the vector (240). If $n \geq 3$, the simplification (241) of the conserved vector can be iterated: if $\widetilde{C}^2$ contains the terms

$$D_3(\widetilde{H}^3) + \cdots + D_n(\widetilde{H}^n)$$

one can subtract them from $\widetilde{C}^2$ and add to $\widetilde{C}^3, \ldots, \widetilde{C}^n$ the corresponding terms

$$D_2(\widetilde{H}^3), \ldots, D_2(\widetilde{H}^n).$$

Note that the conservation law (237) for equations (236) can be written in the form

$$D_i(C^i) = \mu^{\bar{\alpha}} F_{\bar{\alpha}}\big(x, u, u_{(1)}, \ldots, u_{(s)}\big) \tag{242}$$

with undetermined coefficients $\mu^{\bar{\alpha}} = \mu^{\bar{\alpha}}(x, u, u_{(1)}, \ldots)$ depending on a finite number of variables $x, u, u_{(1)}, \ldots$. If C^i depend on higher-order derivatives, Eq. (242) is replaced with

$$D_i(C^i) = \mu^{\bar{\alpha}} F_{\bar{\alpha}} + \mu^{i\bar{\alpha}} D_i\big(F_{\bar{\alpha}}\big) + \mu^{ij\bar{\alpha}} D_i D_j\big(F_{\bar{\alpha}}\big) + \cdots. \tag{243}$$

It is manifest from Eq. (242) or Eq. (243) that the total differentiations of a conserved vector (238) provide again conserved vectors. Therefore, e.g., the vector

$$D_1(C) = \big(D_1(C^1), \ldots, D_1(C^n)\big) \tag{244}$$

obtained from the known vector (238) is not considered as a new conserved vector.

If one of the independent variables is time, e.g., $x^1 = t$, then the conservation Eq. (237) is often written, using the divergence theorem, in the integral form

$$\frac{\mathrm{d}}{\mathrm{d}t}\int_{\mathbb{R}^{n-1}} C^1\, \mathrm{d}x^2 \cdots \mathrm{d}x^n = 0. \tag{245}$$

But the differential form (237) of conservation laws carries, in general, more information than the integral form (245). Using the integral form (245) one may even lose some nontrivial conservation laws. As an example, consider the two-dimensional Boussinesq equations

$$\begin{aligned} \Delta\psi_t - g\rho_x - f v_z &= \psi_x \Delta\psi_z - \psi_z \Delta\psi_x \\ v_t + f\psi_z &= \psi_x v_z - \psi_z v_x \\ \rho_t + \frac{N^2}{g}\psi_x &= \psi_x \rho_z - \psi_z \rho_x \end{aligned} \tag{246}$$

used in geophysical fluid dynamics for investigating uniformly stratified incompressible fluid flows in the ocean. Here Δ is the two-dimensional Laplacian

$$\Delta = \frac{\partial^2}{\partial x^2} + \frac{\partial^2}{\partial z^2}$$

and ψ is the stream function so that the x, z- components u, w of the velocity (u, v, w) of the fluid are given by

$$u = \psi_z, \qquad w = -\psi_x. \tag{247}$$

Equation (246) involve the physical constants: g is the gravitational acceleration, f is the Coriolis parameter, and N is responsible for the density stratification of the fluid. Each equation of the system (246) has the conservation form (237), namely

$$\begin{aligned} D_t(\Delta\psi) + D_x(-g\rho + \psi_z\Delta\psi) + D_z(-f v - \psi_x \Delta\psi) &= 0 \\ D_t(v) + D_x(v\psi_z) + D_z(f\psi - v\psi_x) &= 0 \\ D_t(\rho) + D_x\left(\frac{N^2}{g}\psi + \rho\psi_z\right) + D_z(-\rho\psi_x) &= 0. \end{aligned} \tag{248}$$

In the integral form (245) these conservation laws are written

$$\frac{\mathrm{d}}{\mathrm{d}t}\iint \Delta\psi \mathrm{d}x dz = 0, \quad \frac{\mathrm{d}}{\mathrm{d}t}\iint v \mathrm{d}x dz = 0, \quad \frac{\mathrm{d}}{\mathrm{d}t}\iint \rho \mathrm{d}x dz = 0. \tag{249}$$

We can rewrite the differential conservation equations (248) in an equivalent form by using the operations (239)–(241) of the conserved vectors. Namely, let us apply these operations to the first Eq. (248), i.e., to the conserved vector

$$C^1 = \Delta\psi, \qquad C^2 = -g\rho + \psi_z\Delta\psi, \qquad C^3 = -f v - \psi_x\Delta\psi. \tag{250}$$

Noting that

$$C^1 = D_x(\psi_x) + D_z(\psi_z)$$

and using the operations (239)–(241) we transform the vector (250) to the form

$$\widetilde{C}^1 = 0, \quad \widetilde{C}^2 = -g\rho + \psi_{tx} + \psi_z \Delta\psi, \quad \widetilde{C}^3 = -f v + \psi_{tz} - \psi_x \Delta\psi. \tag{251}$$

The integral conservation equation (245) for the vector in (251) is trivial, $0 = 0$. Thus, after the transformation of the conserved vector (250) to the equivalent form (251) we have lost the first integral conservation law in (249). But it does not mean that the conserved vector (251) has no physical significance. Indeed, if we write the differential conservation equation with the vector (251), we again obtain the first equation of the system (246)

$$D_x(\widetilde{C}^2) + D_z(\widetilde{C}^3) = \Delta\psi_t - g\rho_x - f v_z - \psi_x \Delta\psi_z + \psi_z \Delta\psi_x\,.$$

Let us assume that Eq. (236) have a nontrivial local conserved vector satisfying Eq. (242). Then not all $\mu^{\bar{\beta}}$ vanish simultaneously due to non-triviality of the conserved vector. Furthermore, since $\mu^{\bar{\beta}} F_{\bar{\beta}}$ depends on x, u and a finite number of derivatives $u_{(1)}, u_{(2)}, \ldots$ (i.e., it is a *differential function*) and has a divergence form, the following equations hold (for a detailed discussion see [4], Sect. 8.4.1)

$$\frac{\delta}{\delta \alpha}\Big[\mu^{\bar{\beta}} F_{\bar{\beta}}\big(x, u, u_{(1)}, \ldots, u_{(s)}\big)\Big] = 0, \qquad \alpha = 1, \ldots, m. \tag{252}$$

Note that Eq. (252) are identical with Eq. (69) where the differential substitution (76) is made with $\varphi^{\bar{\alpha}} = \mu^{\bar{\alpha}}$. Hence, the system (236) is nonlinearly self-adjoint. I formulate this simple observation as a theorem since it is useful in applications (see Sect. 2.5).

Theorem 4 *Any system of differential equation* (236) *having a nontrivial local conserved vector satisfying Eq.* (242) *is nonlinearly self-adjoint.*

2.2.2 Explicit Formula for Conserved Vectors

Using Definition 4 of nonlinear self-adjointness and the theorem on conservation laws proved in [1] by using the operator identity (179), we obtain the explicit formula for constructing conservation laws associated with symmetries of any nonlinearly self-adjoint system of equations. The method is applicable independently on the number of equations in the system and the number of dependent variables. The result is as follows.

Theorem 5 *Let the system of differential equation* (236) *be nonlinearly self-adjoint. Specifically, let the adjoint system* (69)–(236) *be satisfied for all solutions of Eq.* (236) *upon a substitution* (70)

$$v^{\bar{\alpha}} = \varphi^{\bar{\alpha}}(x, u), \qquad \bar{\alpha} = 1, \ldots, \overline{m}. \tag{253}$$

Then any Lie point, contact or Lie-Bäcklund symmetry

$$X = \xi^i(x, u, u_{(1)}, \ldots)\frac{\partial}{\partial x^i} + \eta^\alpha(x, u, u_{(1)}, \ldots)\frac{\partial}{\partial u^\alpha} \tag{254}$$

as well as a nonlocal symmetry of Eq. (236) *leads to a conservation law* (237) *constructed by the following formula*

$$\begin{aligned} C^i &= \xi^i \mathcal{L} + W^\alpha \left[\frac{\partial \mathcal{L}}{\partial u_i^\alpha} - D_j\left(\frac{\partial \mathcal{L}}{\partial u_{ij}^\alpha}\right) + D_j D_k\left(\frac{\partial \mathcal{L}}{\partial u_{ijk}^\alpha}\right) - \cdots\right] \\ &+ D_j\left(W^\alpha\right)\left[\frac{\partial \mathcal{L}}{\partial u_{ij}^\alpha} - D_k\left(\frac{\partial \mathcal{L}}{\partial u_{ijk}^\alpha}\right) + \cdots\right] + D_j D_k\left(W^\alpha\right)\left[\frac{\partial \mathcal{L}}{\partial u_{ijk}^\alpha} - \cdots\right] \end{aligned} \tag{255}$$

where

$$W^\alpha = \eta^\alpha - \xi^j u_j^\alpha \tag{256}$$

and $\mathcal{L}$ is the formal Lagrangian for the system (236)

$$\mathcal{L} = v^{\bar{\beta}} F_{\bar{\beta}}. \tag{257}$$

In (255) *the formal Lagrangian $\mathcal{L}$ should be written in the symmetric form with respect to all mixed derivatives u_{ij}^α, u_{ijk}^α, … and the "non-physical variables" $v^{\bar{\alpha}}$ should be eliminated via Eq.* (253).

One can omit in (255) the term $\xi^i \mathcal{L}$ when it is convenient. This term provides a trivial conserved vector mentioned in Sect. 2.2.1 because $\mathcal{L}$ vanishes on the solutions of Eq. (236). Thus, the conserved vector (255) can be taken in the following form

$$\begin{aligned} C^i &= W^\alpha \left[\frac{\partial \mathcal{L}}{\partial u_i^\alpha} - D_j\left(\frac{\partial \mathcal{L}}{\partial u_{ij}^\alpha}\right) + D_j D_k\left(\frac{\partial \mathcal{L}}{\partial u_{ijk}^\alpha}\right) - \cdots\right] \\ &+ D_j\left(W^\alpha\right)\left[\frac{\partial \mathcal{L}}{\partial u_{ij}^\alpha} - D_k\left(\frac{\partial \mathcal{L}}{\partial u_{ijk}^\alpha}\right) + \cdots\right] + D_j D_k\left(W^\alpha\right)\left[\frac{\partial \mathcal{L}}{\partial u_{ijk}^\alpha} - \cdots\right]. \end{aligned} \tag{258}$$

Remark 15 One can use Eq. (258) for constructing conserved vectors even if the system (236) is not self-adjoint, in particular, if one cannot find explicit formulae (253) or (76) for point or differential substitutions, respectively. The resulting conserved vectors will be nonlocal in the sense that they involve the variables v connected with the physical variables u via differential equations, namely, adjoint equations to (236).

Remark 16 Theorem 5, unlike Nother's theorem 3, does not require additional restrictions such as the invariance condition (189) or the divergence condition mentioned in Remark 12.

2.3 A Nonlinearly Self-Adjoint Irrigation System

Let us apply Theorem 5 to Eq. (171) satisfying the condition (173)

$$C(\psi)\psi_t = [K(\psi)\psi_x]_x + \left[K(\psi)\,(\psi_z - 1)\right]_z - S(\psi) \tag{259}$$

$$S'(\psi) = aC(\psi), \qquad a = \text{const}. \tag{260}$$

The formal Lagrangian (257) for Eq. (259) has the form

$$\mathcal{L} = \Big[-C(\psi)\psi_t + K(\psi)(\psi_{xx} + \psi_{zz}) + K'(\psi)(\psi_x^2 + \psi_z^2 - \psi_z) - S(\psi)\Big]\, v. \tag{261}$$

We will use the substitution (175) of the particular form

$$v = \mathrm{e}^{at}. \tag{262}$$

Denoting $t = x^1, x = x^2, z = x^3$ we write the conservation equation (237) in the form

$$D_t(C^1) + D_x(C^2) + D_z(C^3) = 0. \tag{263}$$

This equation should be satisfied on the solutions of Eq. (259).

The formal Lagrangian (261) does not contain derivatives of order higher than two. Therefore in our case Eq. (258) take the simple form

$$C^i = W\left[\frac{\partial\mathcal{L}}{\partial\psi_i} - D_j\left(\frac{\partial\mathcal{L}}{\partial\psi_{ij}}\right)\right] + D_j(W)\frac{\partial\mathcal{L}}{\partial\psi_{ij}} \tag{264}$$

and yield

$$\begin{aligned}
C^1 &= W\frac{\partial\mathcal{L}}{\partial\psi_t}\\
C^2 &= W\left[\frac{\partial\mathcal{L}}{\partial\psi_x} - D_x\left(\frac{\partial\mathcal{L}}{\partial\psi_{xx}}\right)\right] + D_x(W)\frac{\partial\mathcal{L}}{\partial\psi_{xx}}\\
C^3 &= W\left[\frac{\partial\mathcal{L}}{\partial\psi_z} - D_z\left(\frac{\partial\mathcal{L}}{\partial\psi_{zz}}\right)\right] + D_z(W)\frac{\partial\mathcal{L}}{\partial\psi_{zz}}\cdot
\end{aligned}$$

Substituting here the expression (261) for $\mathcal{L}$ we obtain

$$C^1 = -WC(\psi)v$$
$$C^2 = W[2K'(\psi)v\psi_x - D_x(K(\psi)v)] + D_x(W)K(\psi)v$$
$$C^3 = W[K'(\psi)v(2\psi_z - 1) - D_z(K(\psi)v)] + D_z(W)K(\psi)v$$

where v should be eliminated by means of the substitution (262). So, we have

$$C^1 = -WC(\psi)\mathrm{e}^{at}$$
$$C^2 = [WK'(\psi)\psi_x + D_x(W)K(\psi)]\mathrm{e}^{at} \tag{265}$$
$$C^3 = [WK'(\psi)(\psi_z - 1) + D_z(W)K(\psi)]\mathrm{e}^{at}.$$

Since Eq. (259) does not explicitly involve the independent variables t, x, z, it is invariant under the translations of these variables. Let us construct the conserved vector (265) corresponding to the time translation group with the generator

$$X = \frac{\partial}{\partial t} \cdot \tag{266}$$

For this operator Eq. (256) yields

$$W = -\psi_t. \tag{267}$$

Substituting (267) in Eq. (265) we obtain

$$C^1 = C(\psi)\psi_t\mathrm{e}^{at}$$
$$C^2 = -[K'(\psi)\psi_t\psi_x + K(\psi)\psi_{tx}]\mathrm{e}^{at} \tag{268}$$
$$C^3 = -[K'(\psi)\psi_t(\psi_z - 1) + K(\psi)\psi_{tz}]\mathrm{e}^{at}.$$

Now we replace in C^1 the term $C(\psi)\psi_t$ by the right-hand side of Eq. (259) to obtain

$$C^1 = -S(\psi)\mathrm{e}^{at} + D_x\left(K(\psi)\psi_x\mathrm{e}^{at}\right) + D_z\left(K(\psi)(\psi_z - 1)\mathrm{e}^{at}\right).$$

When we substitute this expression in the conservation Eq. (263), we can write

$$D_t\left(D_x\left(K(\psi)\psi_x\mathrm{e}^{at}\right)\right) = D_x\left(D_t\left(K(\psi)\psi_x\mathrm{e}^{at}\right)\right).$$

Therefore we can transfer the terms $D_x(\ldots)$ and $D_z(\ldots)$ from C^1 to C^2 and C^3, respectively (see 241). Thus, we rewrite the vector (268), changing its sign, as follows

$$C^1 = S(\psi)\mathrm{e}^{at}$$
$$C^2 = [K'(\psi)\psi_t\psi_x + K(\psi)\psi_{tx}]\mathrm{e}^{at} - D_t\left(K(\psi)\psi_x\mathrm{e}^{at}\right)$$
$$C^3 = [K'(\psi)\psi_t(\psi_z - 1) + K(\psi)\psi_{tz}]\mathrm{e}^{at} - D_t\left(K(\psi)(\psi_z - 1)\mathrm{e}^{at}\right).$$

Working out the differentiation D_t in the last terms of C^2 and C^3 we finally arrive at the following vector

$$C^1 = S(\psi)e^{at}, \quad C^2 = aK(\psi)\psi_x e^{at}, \quad C^3 = aK(\psi)(\psi_z - 1)e^{at}. \tag{269}$$

The reckoning shows that the vector (269) satisfies the conservation equation (263) due to the condition (260). Note that C^1 is the *density* of the conserved vector (269).

The use of the general substitution (175) instead of its particular case (262) leads to the conserved vector with the density

$$C^1 = S(\psi)(bx + l)e^{at}.$$

This approach opens a new possibility to find a variety of conservation laws for the irrigation model (171) by considering other self-adjoint cases of the model and using the extensions of symmetry Lie algebras (see [15–17], vol. 2, Sect. 9.8).

2.4 Utilization of Differential Substitutions

2.4.1 Equation $u_{xy} = \sin u$

We return to Sect. 1.3.2 and calculate the conservation laws for Eq. (78)

$$u_{xy} = \sin u \tag{270}$$

using the differential substitution (79)

$$v = A_1[xu_x - yu_y] + A_2u_x + A_3u_y \tag{271}$$

and the admitted three-dimensional Lie algebra with the basis

$$X_1 = \frac{\partial}{\partial x}, \quad X_2 = \frac{\partial}{\partial y}, \quad X_3 = x\frac{\partial}{\partial x} - y\frac{\partial}{\partial y} \cdot \tag{272}$$

The conservation equation for Eq. (270) will be written in the form

$$D_x(C^1) + D_y(C^2) = 0.$$

We write the formal Lagrangian for Eq. (270) in the symmetric form

$$\mathcal{L} = \left(\frac{1}{2}u_{xy} + \frac{1}{2}u_{yx} - \sin u\right)v. \tag{273}$$

Equation (258) yield

$$C^1 = \frac{1}{2} D_y(W)v - \frac{1}{2} W v_y, \quad C^2 = \frac{1}{2} D_x(W)v - \frac{1}{2} W v_x \tag{274}$$

where we have to eliminate the variable v via the differential substitution (271).

Substituting in (274) $W = -u_x$ corresponding to the operator X_1 from (272), replacing v with (271) and u_{xy} with $\sin u$, then transferring the terms of the form $D_y(\ldots)$ from C^1 to C^2 (see the simplification (241)) we obtain

$$C^1 = A_1 \cos u, \quad C^2 = \frac{1}{2} A_1 u_x^2.$$

We let $A_1 = 1$ and conclude that the application of Theorem 5 to the symmetry X_1 yields the conserved vector

$$C^1 = \cos u, \quad C^2 = \frac{1}{2} u_x^2 \cdot \tag{275}$$

The similar calculations with the operator X_2 from (272) lead to the conserved vector

$$C^1 = \frac{1}{2} u_y^2, \quad C^2 = \cos u. \tag{276}$$

The third symmetry, X_3 from (272), does not lead to a new conserved vector. Indeed, in this case $W = yu_y - xu_x$. Substituting it in the first formula (274) we obtain after simple calculations

$$C^1 = \frac{1}{2} A_3 u_y^2 - A_2 \cos u + D_y \left[(A_2 y + A_3 x) \left(\frac{1}{2} u_x u_y + \cos u \right) \right].$$

Hence, upon transferring the term $D_y(\ldots)$ from C^1 to C^2 the resulting C^1 will be a linear combination with constant coefficients of the components C^1 of the conserved vectors (275) and (276). The same will be true for C^2. Therefore the conserved vector provided by the symmetry X_3 will be a linear combination with constant coefficients of the conserved vectors (275) and (276).

One can also use the Noether theorem because Eq. (270) has the classical Lagrangian, namely

$$L = -\frac{1}{2} u_x u_y + \cos u. \tag{277}$$

Then the symmetries X_1 and X_2 provide again the conserved vectors (275) and (276), respectively. But now we obtain one more conserved vector using X_3, namely

$$C^1 = x \cos u - \frac{y}{2} u_y^2, \quad C^2 = \frac{x}{2} u_x^2 - y \cos u. \tag{278}$$

2.4.2 Short Pulse Equation

The differential equation (up to notation and appropriate scaling the physical variables)

$$D_t D_x(u) = u + \frac{1}{6} D_x^2(u^3) \tag{279}$$

was suggested in [38] (see there Eq. (11), also [39]) as a mathematical model for the propagation of ultra-short light pulses in media with nonlinearities, e.g., in silica fibers. The mathematical model is derived in [38] by considering the propagation of linearly polarized light in a one-dimensional medium and assuming that the light propagates in the infrared range. The final step in construction of the model is based on the method of multiple scales.

Equation (279) is connected with Eq. (270) by a non-point transformation which is constructed in [40] as a chain of differential substitutions (given also in [39] by Eq. (2)). Using this connection, an exact solitary wave solution (a *pulse solution*) to Eq. (279) is constructed in [39]. One can also find in [40] a Lax pair and a recursion operator for Eq. (279).

Note that Eq. (279) does not have a conservation form. I will find a conservation law of equation (279) thus showing that it can be rewritten in a conservation form. A significance of this possibility is commonly known and is not discussed here.

We write the *short pulse Eq.* (279) in the expanded form

$$u_{xt} = u + \frac{1}{2} u^2 u_{xx} + u u_x^2 \tag{280}$$

so that the formal Lagrangian is written

$$\mathcal{L} = v \left[u_{xt} - u - \frac{1}{2} u^2 u_{xx} - u u_x^2 \right]. \tag{281}$$

Substituting (281) in (69) we obtain the following *adjoint equation* to equation (280)

$$v_{xt} = v + \frac{1}{2} u^2 v_{xx}. \tag{282}$$

We first demonstrate the following statement.

Proposition 12 *Equation* (279) *is not nonlinearly self-adjoint with a substitution*

$$v = \varphi(t, x, u) \tag{283}$$

but it is nonlinearly self-adjoint with the differential substitution

$$v = u_t - \frac{1}{2} u^2 u_x. \tag{284}$$

Proof We write the nonlinear self-adjointness condition (72)

$$\left[v_{xt} - v - \frac{1}{2}u^2 v_{xx}\right]_{(283)} = \lambda[u_{xt} - u - \frac{1}{2}u^2 u_{xx} - uu_x^2]$$

substitute here the expression (283) for v and its derivatives

$$\begin{aligned} v_{xx} &= \varphi_u u_{xx} + \varphi_{uu} u_x^2 + 2\varphi_{xu} u_x + \varphi_{xx} \\ v_{xt} &= \varphi_u u_{xt} + \varphi_{uu} u_x u_t + \varphi_{xu} u_t + \varphi_{tu} u_x + \varphi_{xt} \end{aligned} \tag{285}$$

and first obtain $\lambda = \varphi_u$ by comparing the terms with the second-order derivatives of u. This reduces the nonlinear self-adjointness condition to the following equation

$$\begin{aligned} &\varphi_{uu} u_x u_t + \varphi_{xu} u_t + \varphi_{tu} u_x + \varphi_{xt} - \varphi + \varphi_{xx}) \\ &\qquad - \frac{1}{2}u^2(\varphi_{uu} u_x^2 + 2\varphi_{xu} u_x = -\varphi_u[u + uu_x^2]. \end{aligned} \tag{286}$$

The terms with u_t in Eq. (286) yield $\varphi_{uu} = \varphi_{xu} = 0$. Then we take the term with u_x^2 and obtain $\varphi_u = 0$. Hence

$$\varphi = a(t, x).$$

Now Eq. (286) gives $a_{xx} = 0,\ a_{xt} - a = 0$, whence $a = 0$. Thus

$$\varphi = 0$$

i.e., the substitution (283) is trivial. This proves the first part of Proposition 12. Its second part is proved by similar calculations with the substitution

$$v = \varphi(t, x, u, u_x, u_t).$$ □

I will not reproduce these rather lengthy calculations, but instead we will verify that the substitution (284) maps any solution of Eq. (270) into a solution of the adjoint Eq. (282). First we calculate

$$v_x = u_{xt} - \frac{1}{2}u^2 u_{xx} - uu_x^2$$

and see that on the solutions of Eq. (270) we have $v_x = u$. Now we calculate other derivatives and verify that on the solutions of Eq. (270) the following equations hold

$$v_x = u, \quad v_t = u_{tt} - \frac{1}{2}u^2 u_{xt} - uu_x u_t, \quad v_{xt} = u_t, \quad v_{xx} = u_x. \tag{287}$$

It is easily seen that Eq. (282) is satisfied. Namely, using (284) and (287) we have

$$v_{xt} - v - \frac{1}{2}u^2 v_{xx} = u_t - \left(u_t - \frac{1}{2}u^2 u_x\right) - \frac{1}{2}u^2 u_x = 0.$$

The maximal Lie algebra of point symmetries of Eq. (279) is the three-dimensional algebra spanned by the operators

$$X_1 = \frac{\partial}{\partial t}, \quad X_2 = \frac{\partial}{\partial x}, \quad X_3 = u\frac{\partial}{\partial u} + x\frac{\partial}{\partial x} - t\frac{\partial}{\partial t}. \tag{288}$$

Let us construct the conservation laws

$$D_t(C^1) + D_x(C^2) = 0 \tag{289}$$

for the basis operators (288).

Since the formal Lagrangian (281) does not contain derivatives of order higher than two, Eq. (258) are written (see 264)

$$C^i = W\left[\frac{\partial \mathcal{L}}{\partial u_i} - D_j\left(\frac{\partial \mathcal{L}}{\partial u_{ij}}\right)\right] + D_j(W)\frac{\partial \mathcal{L}}{\partial u_{ij}}.$$

In our case we have

$$\begin{aligned} C^1 &= - WD_x\left(\frac{\partial \mathcal{L}}{\partial u_{tx}}\right) + D_x(W)\frac{\partial \mathcal{L}}{\partial u_{tx}} \\ C^2 &= W\left[\frac{\partial \mathcal{L}}{\partial u_x} - D_t\left(\frac{\partial \mathcal{L}}{\partial u_{xt}}\right) - D_x\left(\frac{\partial \mathcal{L}}{\partial u_{xx}}\right)\right] \\ &\quad + D_t(W)\frac{\partial \mathcal{L}}{\partial u_{xt}} + D_x(W)\frac{\partial \mathcal{L}}{\partial u_{xx}}. \end{aligned} \tag{290}$$

Substituting in (290) the expression (281) for $\mathcal{L}$ written in the symmetric form

$$\mathcal{L} = v\left[\frac{1}{2}u_{tx} + \frac{1}{2}u_{xt} - u - \frac{1}{2}u^2 u_{xx} - uu_x^2\right] \tag{291}$$

we obtain

$$\begin{aligned} C^1 &= -\frac{1}{2}Wv_x + \frac{1}{2}vD_x(W) \\ C^2 &= -W\left[uvu_x + \frac{1}{2}v_t - \frac{1}{2}u^2 v_x\right] + \frac{1}{2}vD_t(W) - \frac{1}{2}u^2 vD_x(W). \end{aligned} \tag{292}$$

Since v should be eliminated via the differential substitution (284), we further simplify this vector by replacing v_x with u according to the first Eq. (287) and obtain

$$
\begin{aligned}
C^1 &= -\frac{1}{2}Wu + \frac{1}{2}\,vD_x(W)\\
C^2 &= -W\Big[uvu_x + \frac{1}{2}\,v_t - \frac{1}{2}\,u^3\Big] + \frac{1}{2}\,vD_t(W) - \frac{1}{2}\,u^2vD_x(W)
\end{aligned}
\tag{293}
$$

where v and v_t should be replaced with their values given in Eqs. (284) and (287).

Let us construct the conserved vectors using the symmetries (288). Their commutators are

$$
[X_1, X_3] = -X_1, \qquad [X_2, X_3] = X_2.
$$

Hence, according to [10], Sect. 22.4, the operator X_3 plays a distinguished role. Namely, the conserved vectors associated with X_1 and X_2 can be obtained from the conserved vector provided by X_3 using the adjoint actions of the operators X_1 and X_2, respectively. Therefore we start with X_3. Substituting in (293) the expression

$$
W = u + tu_t - xu_x
$$

corresponding to the symmetry X_3, eliminating the terms of the form $D_x(A)$ from C^1 and adding them to C^2 in the form $D_t(A)$ according to the simplification (241), we obtain after routine calculations the following conserved vector

$$
C^1 = u^2, \qquad C^2 = u^2u_xu_t - u_t^2 - \frac{1}{4}u^4 - \frac{1}{4}u^4u_x^2. \tag{294}
$$

The conservation Eq. (289) for the vector (294) holds in the form

$$
D_t(C^1) + D_x(C^2) = 2\Big(u_t - \frac{1}{2}\,u^2u_x\Big)\Big(u + \frac{1}{2}\,u^2u_{xx} + uu_x^2 - u_{xt}\Big). \tag{295}
$$

Let us turn now to the operators X_1 and X_2 from (288). To simplify the calculations it is useful to modify Eq. (293) as follows. Noting that

$$
vD_x(W) = D_x(vW) - Wv_x
$$

we rewrite the vector (292) in the form

$$
C^1 = -Wv_x, \qquad C^2 = -W\Big[uvu_x - \frac{1}{2}\,u^2v_x\Big] + vD_t(W) - \frac{1}{2}\,u^2vD_x(W).
$$

Then (293) is replaced with

$$
\begin{aligned}
C^1 &= -uW\\
C^2 &= -W\Big[uvu_x - \frac{1}{2}\,u^3\Big] + vD_t(W) - \frac{1}{2}\,u^2vD_x(W).
\end{aligned}
\tag{296}
$$

Substituting in the first formula (296) to expression $W = -u_t$ corresponding the operator X_1 we obtain $C^1 = uu_t$. This is the time derivative of C^1 from (294). Hence the symmetry X_1 leads to a trivial conserved vector obtained from the vector (294) by the differentiation D_t, in accordance with [10]. Likewise, it is manifest from (296) that the operator X_2 leads to a trivial conserved vector obtained from the conserved vector (294) by the differentiation D_x. Thus we have demonstrated the following statement.

Proposition 13 *The Lie point symmetries* (288) *of Eq.* (280) *yield one non-trivial conserved vector* (294). *Accordingly, the short pulse Eq.* (280) *can be written in the following conservation form*

$$D_t\left(u^2\right) + D_x\left(u^2 u_x u_t - u_t^2 - \frac{1}{4}u^4 - \frac{1}{4}u^4 u_x^2\right) = 0. \tag{297}$$

2.5 Gas Dynamics

2.5.1 Classical Symmetries and Conservation Laws

Let us consider the polytropic gasdynamic equations

$$\begin{aligned} &\boldsymbol{v}_t + (\boldsymbol{v}\cdot\nabla)\boldsymbol{v} + \frac{1}{\rho}\nabla p = 0 \\ &\rho_t + \boldsymbol{v}\cdot\nabla\rho + \rho\nabla\cdot\boldsymbol{v} = 0 \\ &p_t + \boldsymbol{v}\cdot\nabla p + \gamma p\nabla\cdot\boldsymbol{v} = 0 \end{aligned} \tag{298}$$

where γ is a constant known as the polytropic (or adiabatic) exponent. The independent variables are the time and the space coordinates

$$t, \qquad \boldsymbol{x} = (x^1, \ldots, x^n), \qquad n \le 3. \tag{299}$$

The dependent variables are the velocity, the density and the pressure

$$\boldsymbol{v} = (v^1, \ldots, v^n), \qquad \rho, \qquad p. \tag{300}$$

Equation (298) with arbitrary γ have the Lie algebra of point symmetries spanned by

$$\begin{aligned} &X_0 = \frac{\partial}{\partial t}, \quad X_i = \frac{\partial}{\partial x^i}, \quad Y_0 = t\frac{\partial}{\partial t} + x^i\frac{\partial}{\partial x^i}, \quad Y_i = t\frac{\partial}{\partial x^i} + \frac{\partial}{\partial v^i} \\ &X_{ij} = x^j\frac{\partial}{\partial x^i} - x^i\frac{\partial}{\partial x^j} + v^j\frac{\partial}{\partial v^i} - v^i\frac{\partial}{\partial v^j}, \qquad i < j \\ &Z_0 = \rho\frac{\partial}{\partial \rho} + p\frac{\partial}{\partial p}, \quad Z_1 = t\frac{\partial}{\partial t} - v^i\frac{\partial}{\partial v^i} + 2\rho\frac{\partial}{\partial \rho}, \qquad i, j = 1, \ldots, n \end{aligned} \tag{301}$$

and the following classical conservation laws

$$\frac{\mathrm{d}}{\mathrm{d}t}\int\limits_{\Omega(t)} \rho \mathrm{d}\omega = 0 \qquad \text{– Conservation of mass}$$

$$\frac{\mathrm{d}}{\mathrm{d}t}\int\limits_{\Omega(t)} \left(\frac{1}{2}\rho|\boldsymbol{v}|^2 + \frac{p}{\gamma-1}\right)\mathrm{d}\omega = -\int\limits_{S(t)} p\,\boldsymbol{v}\cdot\boldsymbol{\nu} dS \qquad \text{– Energy}$$

$$\frac{\mathrm{d}}{\mathrm{d}t}\int\limits_{\Omega(t)} \rho\boldsymbol{v}\mathrm{d}\omega = -\int\limits_{S(t)} p\,\boldsymbol{\nu}\mathrm{d}S \qquad \text{– Momentum}$$

$$\frac{\mathrm{d}}{\mathrm{d}t}\int\limits_{\Omega(t)} \rho(\boldsymbol{x}\times\boldsymbol{v})\mathrm{d}\omega = -\int\limits_{S(t)} p(\boldsymbol{x}\times\boldsymbol{\nu})\mathrm{d}S \qquad \text{– Angular momentum}$$

$$\frac{\mathrm{d}}{\mathrm{d}t}\int\limits_{\Omega(t)} \rho(t\boldsymbol{v}-\boldsymbol{x})\mathrm{d}\omega = -\int\limits_{S(t)} tp\,\boldsymbol{\nu}\mathrm{d}S \qquad \text{– Center-of-mass.}$$

The conservation laws are written in the integral form by using the standard symbols

$\Omega(t)$ – arbitrary n-dimensional volume, moving with fluid
$S(t)$ – boundary of the volume $\Omega(t)$
$\boldsymbol{\nu}$ – unit (outer) normal vector to the surface $S(t)$.

If we write the above conservation laws in the general form

$$\frac{\mathrm{d}}{\mathrm{d}t}\int\limits_{\Omega(t)} T\mathrm{d}\omega = -\int\limits_{S(t)} (\chi\cdot\boldsymbol{\nu})\mathrm{d}S \tag{302}$$

then the differential form of these conservation laws will be

$$D_t(T) + \nabla\cdot(\chi + T\boldsymbol{v}) = 0. \tag{303}$$

2.5.2 Adjoint Equations and Self-Adjointness When $n = 1$

Theorem 4 from Sect. 2.2.1 shows that the system of gasdynamic Eq. (298) is non-linearly self-adjoint. Let us illustrate this statement in the one-dimensional case

$$v_t + vv_x + \frac{1}{\rho}p_x = 0, \qquad \rho_t + v\rho_x + \rho v_x = 0, \qquad p_t + vp_x + \gamma p v_x = 0. \tag{304}$$

We write the formal Lagrangian in the form

$$\mathcal{L} = U\Big(v_t + vv_x + \frac{1}{\rho}p_x\Big) + R(\rho_t + v\rho_x + \rho v_x) + P(p_t + vp_x + \gamma p v_x) \tag{305}$$

and obtain the following adjoint system for the new dependent variables U, R, P

$$
\begin{aligned}
\frac{\delta \mathcal{L}}{\delta v} &\equiv -U_t - vU_x - \rho R_x + (1-\gamma)P p_x - \gamma p P_x = 0\\
\frac{\delta \mathcal{L}}{\delta \rho} &\equiv -R_t - vR_x - \frac{1}{\rho^2}\, U p_x = 0\\
\frac{\delta \mathcal{L}}{\delta p} &\equiv -P_t - \frac{1}{\rho}\, U_x + \frac{1}{\rho^2}\, U \rho_x + (\gamma-1)P v_x - vP_x = 0.
\end{aligned}
\tag{306}
$$

Let us take, e.g., the conservation of energy from Sect. 2.2.1. Then we have

$$
T = \frac{1}{2}\,\rho v^2 + \frac{p}{\gamma-1}\,, \qquad \chi = pv
$$

and using the differential form (303) of the energy conservation we obtain following the Eq. (242)

$$
\begin{aligned}
&D_t\left(\frac{1}{2}\,\rho v^2 + \frac{p}{\gamma-1}\right) + D_x\left(\frac{1}{2}\,\rho v^3 + \frac{\gamma}{\gamma-1}\,pv\right)\\
&= \rho v\Big(v_t + vv_x + \frac{1}{\rho}\,p_x\Big) + \frac{v^2}{2}(\rho_t + v\rho_x + \rho v_x) + \frac{1}{\gamma-1}(p_t + vp_x + \gamma p v_x).
\end{aligned}
\tag{307}
$$

Hence, the adjoint Eq. (306) are satisfied for all solutions of the gasdynamic Eq. (298) upon the substitution

$$
U = \rho v, \qquad R = \frac{v^2}{2}\,, \qquad P = \frac{1}{\gamma-1}\,\cdot
\tag{308}
$$

This conclusion can be easily verified by the direct substitution of (308) in the adjoint system (306). Namely, we have

$$
\begin{aligned}
\frac{\delta \mathcal{L}}{\delta v}\bigg|_{(308)} &= -\rho\Big(v_t + vv_x + \frac{1}{\rho}\,p_x\Big) - v(\rho_t + v\rho_x + \rho v_x)\\
\frac{\delta \mathcal{L}}{\delta \rho}\bigg|_{(308)} &= -v\Big(v_t + vv_x + \frac{1}{\rho}\,p_x\Big)\\
\frac{\delta \mathcal{L}}{\delta p}\bigg|_{(308} &= 0.
\end{aligned}
\tag{309}
$$

2.5.3 Adjoint System to Equations (298) with $n \geq 2$

For gasdynamic Eq. (298) with two and three space variables x^i the formal Lagrangian (305) is replaced by

$$\mathcal{L} = \boldsymbol{U} \cdot \left(\boldsymbol{v}_t + (\boldsymbol{v} \cdot \nabla)\boldsymbol{v} + \frac{1}{\rho} \nabla p\right) + R(\rho_t + \boldsymbol{v} \cdot \nabla \rho + \rho \nabla \cdot \boldsymbol{v})$$
$$+ P(p_t + \boldsymbol{v} \cdot \nabla p + \gamma p \nabla \cdot \boldsymbol{v}) \tag{310}$$

where the vector $\boldsymbol{U} = (U^1, \ldots, U^n)$ and the scalars R, P are new dependent variables. Using this formal Lagrangian, we obtain the following adjoint system instead of (306)

$$\begin{aligned}
\frac{\delta \mathcal{L}}{\delta \boldsymbol{v}} &\equiv - \boldsymbol{U}_t - (\boldsymbol{v} \cdot \nabla)\boldsymbol{U} + (\boldsymbol{U} \cdot \nabla)\boldsymbol{v} - (\nabla \cdot \boldsymbol{v})\boldsymbol{U} \\
&\quad - \rho \nabla R + (1 - \gamma) P \nabla p - \gamma p \nabla P = 0 \\
\frac{\delta \mathcal{L}}{\delta \rho} &\equiv - R_t - \boldsymbol{v} \cdot \nabla R - \frac{1}{\rho^2} \boldsymbol{U} \cdot \nabla p = 0 \\
\frac{\delta \mathcal{L}}{\delta p} &\equiv - P_t - \frac{1}{\rho} (\nabla \cdot \boldsymbol{U}) + \frac{1}{\rho^2} \boldsymbol{U} \cdot \nabla \rho + (\gamma - 1) P (\nabla \cdot \boldsymbol{v}) - \boldsymbol{v} \cdot \nabla P = 0.
\end{aligned} \tag{311}$$

The nonlinear self-adjointness of the system (298) can be demonstrated as in the one-dimensional case discussed in Sect. 2.5.2.

2.5.4 Application to Nonlocal Symmetries of the Chaplygin Gas

The Chaplygin gas is described by the one-dimensional gasdynamic Eq. (304) with $\gamma = -1$, i.e.,

$$v_t + v v_x + \frac{1}{\rho} p_x = 0, \quad \rho_t + v \rho_x + \rho v_x = 0, \quad p_t + v p_x - p v_x = 0. \tag{312}$$

Equation (312) have the same maximal Lie algebra of Lie point symmetries as Eq. (304) with arbitrary γ. This algebra is spanned by the symmetries (301) in the one-dimensional case, namely

$$\begin{aligned}
&X_1 = \frac{\partial}{\partial t}, \quad X_2 = \frac{\partial}{\partial x}, \quad X_3 = t \frac{\partial}{\partial x} + \frac{\partial}{\partial v}, \quad X_4 = t \frac{\partial}{\partial t} + x \frac{\partial}{\partial x} \\
&X_5 = \rho \frac{\partial}{\partial \rho} + p \frac{\partial}{\partial p}, \qquad X_6 = t \frac{\partial}{\partial t} - v \frac{\partial}{\partial v} + 2\rho \frac{\partial}{\partial \rho} \cdot
\end{aligned} \tag{313}$$

But the Chaplygin gas has more symmetries than an arbitrary one-dimensional polytropic gas upon rewriting it in Lagrange's variables obtained by replacing x and ρ with τ and q, respectively, obtained by the following *nonlocal transformation*

$$\tau = \int \rho \mathrm{d}x, \qquad q = \frac{1}{\rho}\cdot \tag{314}$$

Then the system (312) becomes

$$q_t - v_\tau = 0, \qquad v_t + p_\tau = 0, \qquad p_t - \frac{p}{q}\, v_\tau = 0 \tag{315}$$

and admits the 8-dimensional Lie algebra with the basis

$$\begin{aligned}
&Y_1 = \frac{\partial}{\partial t}, \qquad Y_2 = \frac{\partial}{\partial \tau}, \qquad Y_3 = \frac{\partial}{\partial v}, \qquad Y_4 = t\frac{\partial}{\partial t} + \tau\frac{\partial}{\partial \tau}\\
&Y_5 = \tau\frac{\partial}{\partial \tau} + p\frac{\partial}{\partial p} - q\frac{\partial}{\partial q}, \qquad Y_6 = v\frac{\partial}{\partial v} + p\frac{\partial}{\partial p} + q\frac{\partial}{\partial q}\\
&Y_7 = \frac{\partial}{\partial p} + \frac{q}{p}\frac{\partial}{\partial q}, \qquad Y_8 = t\frac{\partial}{\partial v} - y\frac{\partial}{\partial p} - \frac{yq}{p}\frac{\partial}{\partial q}\cdot
\end{aligned} \tag{316}$$

It is shown in [24] that the operators Y_7, Y_8 from (316) lead to the following *nonlocal symmetries* for Eq. (312)

$$\begin{aligned}
X_7 &= \sigma\frac{\partial}{\partial x} - \frac{\partial}{\partial p} + \frac{\rho}{p}\frac{\partial}{\partial \rho}\\
X_8 &= \left(\frac{t^2}{2} + s\right)\frac{\partial}{\partial x} + t\frac{\partial}{\partial v} - \tau\frac{\partial}{\partial p} + \frac{\rho\tau}{p}\frac{\partial}{\partial \rho}
\end{aligned} \tag{317}$$

where τ, s, σ are the following *nonlocal variables*

$$\tau = \int \rho \mathrm{d}x, \qquad s = -\int \frac{\tau}{p}\mathrm{d}x, \qquad \sigma = -\int \frac{\mathrm{d}x}{p}\cdot \tag{318}$$

They can be equivalently defined by the compatible over-determined systems

$$\begin{aligned}
\tau_x &= \rho, & \tau_t + v\tau_x &= 0\\
s_x &= -\frac{\tau}{p}, & s_t + vs_x &= 0\\
\sigma_x &= -\frac{1}{p}, & \sigma_t + v\sigma_x &= 0
\end{aligned} \tag{319}$$

or

$$\begin{aligned}
\tau_x &= \rho, & \tau_t &= -v\rho \\
s_x &= -\frac{\tau}{p}, & s_t &= \frac{v\tau}{p} \\
\sigma_x &= -\frac{1}{p}, & \sigma_t &= \frac{v}{p}\,.
\end{aligned} \tag{320}$$

Let us verify that the operator X_7 is admitted by Eq. (312). Its first prolongation is obtained by applying the usual prolongation procedure and eliminating the partial derivatives σ_x and σ_t via Eq. (320). It has the form

$$\begin{aligned}
X_7 = \sigma\frac{\partial}{\partial x} - \frac{\partial}{\partial p} + \frac{\rho}{p}\frac{\partial}{\partial \rho} - \frac{v v_x}{p}\frac{\partial}{\partial v_t} + \frac{v_x}{p}\frac{\partial}{\partial v_x} - \frac{v p_x}{p}\frac{\partial}{\partial p_t} + \frac{p_x}{p}\frac{\partial}{\partial p_x} \\
+ \left(\frac{\rho_t}{p} - \frac{\rho p_t}{p^2} - \frac{v\rho_x}{p}\right)\frac{\partial}{\partial \rho_t} + \left(2\frac{\rho_x}{p} - \frac{\rho p_x}{p^2}\right)\frac{\partial}{\partial \rho_x}\,.
\end{aligned} \tag{321}$$

The calculation shows that the invariance condition is satisfied in the following form:

$$\begin{aligned}
&X_7\left(v_t + v v_x + \frac{1}{\rho}\,p_x\right) = 0 \\
&X_7(\rho_t + v\rho_x + \rho v_x) = \frac{1}{p}\,(\rho_t + v\rho_x + \rho v_x) - \frac{\rho}{p^2}\,(p_t + v p_x - p v_x) \\
&X_7(p_t + v p_x - p v_x) = 0.
\end{aligned}$$

One can verify likewise that the invariance test for the operator X_8 is satisfied in the following form

$$\begin{aligned}
&X_8(v_t + v v_x + \frac{1}{\rho}\,p_x) = 0 \\
&X_8(\rho_t + v\rho_x + \rho v_x) = \frac{\tau}{p}\,(\rho_t + v\rho_x + \rho v_x) - \frac{\rho\tau}{p^2}\,(p_t + v p_x - p v_x) \\
&X_8(p_t + v p_x - p v_x) = 0.
\end{aligned}$$

The operators $Y_1, \ldots, Y_6$ from (316) do not add to the operators (313) new symmetries of the system (312).

Thus, the Chaplygin gas described by Eq. (312) admits the eight-dimensional vector space spanned by the operators (313) and (317). However this vector space is not a Lie algebra. Namely, the commutators of the dilation generators X_4, X_5, X_6 from (313) with the operators (317) are not linear combinations of the operators (313) and (317) with constants coefficients. The reason is that the operators X_4, X_5, X_6 are not admitted by the differential equation (319) for the nonlocal variables τ, s, σ. Therefore I will extend the action of the dilation generators to τ, s, σ so that the extended operators will be admitted by Eq. (319).

Let us take the operator X_4. We write it in the extended form

$$X_4' = t\frac{\partial}{\partial t} + x\frac{\partial}{\partial x} + \alpha\frac{\partial}{\partial \tau} + \beta\frac{\partial}{\partial s} + \mu\frac{\partial}{\partial \sigma}$$

where α, β, μ are unknown functions of $t, x, v, \rho, p, \tau, s, \sigma$. Then we make the prolongation of X_4' to the first-order partial derivatives of the nonlocal variables with respect to t and x by treating τ, s, σ as new dependent variables and obtain

$$\begin{aligned} X_4' = {} & t\frac{\partial}{\partial t} + x\frac{\partial}{\partial x} + \alpha\frac{\partial}{\partial \tau} + \beta\frac{\partial}{\partial s} + \mu\frac{\partial}{\partial \sigma} \\ & + [D_t(\alpha) - \tau_t]\frac{\partial}{\partial \tau_t} + [D_x(\alpha) - \tau_x]\frac{\partial}{\partial \tau_x} \\ & + [D_t(\beta) - s_t]\frac{\partial}{\partial s_t} + [D_x(\beta) - s_x]\frac{\partial}{\partial s_x} \\ & + [D_t(\mu) - \sigma_t]\frac{\partial}{\partial \sigma_t} + [D_x(\mu) - \sigma_x]\frac{\partial}{\partial \sigma_x}\cdot \end{aligned}$$

Now we require the invariance of the Eq. (319)

$$\begin{aligned} X_4'(\tau_x - \rho) &= 0, \qquad & X_4'(\tau_t + v\tau_x) &= 0 \\ X_4'\left(s_x + \frac{\tau}{p}\right) &= 0, \qquad & X_4'(s_t + vs_x) &= 0 \\ X_4'\left(\sigma_x + \frac{1}{p}\right) &= 0, \qquad & X_4'(\sigma_t + v\sigma_x) &= 0. \end{aligned} \tag{322}$$

As usual, Eq. (322) should be satisfied on the solutions of Eq. (319). Let us solve the equations $X_4'(\tau_x - \rho) = 0, \quad X_4'(\tau_t + v\tau_x) = 0$. They are written

$$[D_x(\alpha) - \tau_x]_{(319)} = 0, \quad [D_t(\alpha) - \tau_t + v\,(D_x(\alpha) - \tau_x)]_{(319)} = 0. \tag{323}$$

Since $\tau_x = D_x(\alpha)$, the first equation in (323) is satisfied if we take

$$\alpha = \tau.$$

With this α the second equation in (323) is also satisfied because $\tau_t + v\tau_x = 0$. Now the first equation in the second line of Eq. (322) becomes

$$\left[D_x(\beta) - s_x + \frac{\tau}{p}\right]_{(319)} = D_x(\beta) - 2s_x = 0$$

and yields

$$\beta = 2s.$$

Table 1 The structure of the Lie algebra L_8

	X_1	X_2	X_3	X_4'	X_5'	X_6'	X_7	X_8
X_1	0	0	X_2	X_1	0	X_1	0	X_3
X_2	0	0	0	X_2	0	0	0	0
X_3	$-X_2$	0	0	0	0	$-X_3$	0	0
X_4'	$-X_1$	$-X_2$	0	0	0	0	0	X_8
X_5'	0	0	0	0	0	0	$-X_7$	0
X_6'	$-X_1$	0	X_3	0	0	0	0	$2X_8$
X_7	0	0	0	0	X_7	0	0	0
X_8	$-X_3$	0	0	$-X_8$	0	$-2X_8$	0	0

The second equation in the second line of Eq. (322) is also satisfied with this β. Applying the same approach to the third line of Eq. (322) we obtain

$$\mu = \sigma.$$

After similar calculations with X_5 and X_6 we obtain the following extensions of the dilation generators

$$\begin{aligned} X_4' &= t\frac{\partial}{\partial t} + x\frac{\partial}{\partial x} + \tau\frac{\partial}{\partial \tau} + 2s\frac{\partial}{\partial s} + \sigma\frac{\partial}{\partial \sigma} \\ X_5' &= \rho\frac{\partial}{\partial \rho} + p\frac{\partial}{\partial p} + \tau\frac{\partial}{\partial \tau} - \sigma\frac{\partial}{\partial \sigma} \\ X_6' &= t\frac{\partial}{\partial t} - v\frac{\partial}{\partial v} + 2\rho\frac{\partial}{\partial \rho} + 2\tau\frac{\partial}{\partial \tau} + 2s\frac{\partial}{\partial s}\,. \end{aligned} \tag{324}$$

The operators (317) and (324) together with the operators X_1, X_2, X_3 from (313) span the eight-dimensional Lie algebra L_8 admitted by Eqs. (312) and (319). The algebra L_8 has the following commutator Table 1.

Let us apply Theorem 5 to the nonlocal symmetries (317) of the Chaplygin gas. The formal Lagrangian (305) for Eq. (312) has the form

$$\mathcal{L} = U\left(v_t + vv_x + \frac{1}{\rho}p_x\right) + R(\rho_t + v\rho_x + \rho v_x) + P(p_t + vp_x - pv_x). \tag{325}$$

Accordingly, the adjoint system (306) for the Chaplygin gas is written

$$\begin{aligned} \frac{\delta\mathcal{L}}{\delta v} &\equiv -U_t - vU_x - \rho R_x + 2Pp_x + pP_x = 0 \\ \frac{\delta\mathcal{L}}{\delta \rho} &\equiv -R_t - vR_x - \frac{1}{\rho^2}Up_x = 0 \\ \frac{\delta\mathcal{L}}{\delta p} &\equiv -P_t - \frac{1}{\rho}U_x + \frac{1}{\rho^2}U\rho_x - 2Pv_x - vP_x = 0. \end{aligned} \tag{326}$$

Let us proceed as in Sect. 2.5.2. Namely, let us first construct solutions to the adjoint system (326) by using the known conservation laws given in Sect. 2.2.1. Since the one-dimensional equation does not have the conservation of angular momentum, we use the conservation of mass, energy, momentum and center-of-mass and obtain the respective differential conservation equations (see the derivation of Eq. (307))

$$D_t(\rho) + D_x(\rho v) = \rho_t + v\rho_x + \rho v_x \tag{327}$$

$$\begin{aligned} D_t(\rho v^2 - p) + D_x(pv + \rho v^3) &= 2\rho v\Big(v_t + vv_x + \frac{1}{\rho}\,p_x\Big) \\ &+ v^2(\rho_t + v\rho_x + \rho v_x) - (p_t + vp_x - pv_x) \end{aligned} \tag{328}$$

$$\begin{aligned} D_t(\rho v) + D_x(p + \rho v^2) &= \rho\Big(v_t + vv_x + \frac{1}{\rho}\,p_x\Big) \\ &+ v(\rho_t + v\rho_x + \rho v_x) \end{aligned} \tag{329}$$

$$\begin{aligned} &D_t(t\rho v - x\rho) + D_x(tp + t\rho v^2 - x\rho v) \\ &= t\rho\Big(v_t + vv_x + \frac{1}{\rho}\,p_x\Big) + (tv - x)(\rho_t + v\rho_x + \rho v_x). \end{aligned} \tag{330}$$

Equations (327)–(330) give the following solutions to the system of the adjoint equation (326)

$$U = 0, \quad R = 1, \quad P = 0 \tag{331}$$

$$U = 2\rho v, \quad R = v^2, \quad P = -1 \tag{332}$$

$$U = \rho, \quad R = v, \quad P = 0 \tag{333}$$

$$U = t\rho, \quad R = tv - x, \quad P = 0. \tag{334}$$

The formal Lagrangian (325) contains the derivatives only of the first order. Therefore Eq. (258) for calculating the conserved vectors take the simple form

$$C^i = W^\alpha\,\frac{\partial \mathcal{L}}{\partial u_i^\alpha}, \qquad i = 1, 2. \tag{335}$$

We denote

$$t = x^1, \qquad x = x^2, \qquad v = u^1, \qquad \rho = u^2, \qquad p = u^3.$$

In this notation conservation Eq. (237) will be written in the form

$$\Big[D_t(C^1) + D_x(C^2)\Big]_{(312)} = 0. \tag{336}$$

Writing (335) in the form

$$C^1 = W^1 \frac{\partial \mathcal{L}}{\partial v_t} + W^2 \frac{\partial \mathcal{L}}{\partial \rho_t} + W^3 \frac{\partial \mathcal{L}}{\partial p_t}$$
$$C^2 = W^1 \frac{\partial \mathcal{L}}{\partial v_x} + W^2 \frac{\partial \mathcal{L}}{\partial \rho_x} + W^3 \frac{\partial \mathcal{L}}{\partial p_x}$$

and substituting the expression (325) for $\mathcal{L}$ we obtain the following final expressions for computing the components of conserved vectors

$$C^1 = UW^1 + RW^2 + PW^3 \tag{337}$$

$$C^2 = (vU + \rho R - pP)W^1 + vRW^2 + \left(\frac{1}{\rho}U + vP\right)W^3 \tag{338}$$

where

$$W^\alpha = \eta^\alpha - \xi^i u_i^\alpha, \qquad \alpha = 1, 2, 3. \tag{339}$$

We will apply Eqs. (337) and (338) to the nonlocal symmetries (317). First we write the expressions (339) for the operator X_7 from (317)

$$W^1 = -\sigma v_x, \qquad W^2 = \frac{\rho}{p} - \sigma \rho_x, \qquad W^3 = -(1 + \sigma p_x). \tag{340}$$

Then we substitute (340) in (337) and (338) and obtain four conserved vectors by replacing U, R, P with each of four different solutions (331)–(334) of the adjoint system (326). Some of these conserved vectors may be trivial. We select only the nontrivial ones.

Let us calculate the conserved vector obtained by eliminating U, R, P by using the solution (331), $U = 0,\ R = 1,\ P = 0$. In this case (337) and (338) and (340) yield

$$C^1 = W^2 = \frac{\rho}{p} - \sigma \rho_x \tag{341}$$
$$C^2 = \rho W^1 + vW^2 = -\sigma \rho v_x + \frac{\rho}{p} v - \sigma v \rho_x.$$

We write

$$\sigma \rho_x = D_x(\sigma \rho) + \rho \sigma_x$$

replace σ_x with $-1/p$ according to Eq. (319) and obtain

$$C^1 = -D_x(\sigma \rho).$$

Therefore application of the operations (239)–(241) yields $\widetilde{C}^1 = 0$ and

$$\begin{aligned}\widetilde{C}^2 &= -\sigma\rho v_x + \frac{\rho}{p}\,v - \sigma v\rho_x - D_t(\sigma\rho)\\ &= -\sigma\rho v_x + \frac{\rho}{p}\,v - \sigma v\rho_x - \sigma\rho_t - \sigma_t\rho\\ &= -\sigma(\rho_t + v\rho_x + \rho v_x).\end{aligned}$$

We have replaced σ_t with v/p according to Eq. (320). The above expression for $\widetilde{C}^2$ vanishes on Eq. (312). Hence, the conserved vector (341) is trivial.

Utilization of the solutions (332) and (333) also leads to trivial conserved vectors only. Finally, using the solution (334)

$$U = t\rho, \qquad R = tv - x, \qquad P = 0$$

we obtain, upon simplifying by using the operations (239)–(241), the following nontrivial conserved vector

$$C^1 = \sigma\rho, \qquad C^2 = \sigma\rho v + t. \tag{342}$$

The conservation Eq. (336) is satisfied in the following form

$$D_t(C^1) + D_x(C^2) = \sigma(\rho_t + v\rho_x + \rho v_x). \tag{343}$$

Note that we can write C^2 in (342) without t since it adds only the trivial conserved vector with the components $C^1 = 0,\ C^2 = t$. Thus, removing t in (342) and using the definition of σ given in (318) we formulate the result.

Proposition 14 *The nonlocal symmetry X_7 of the Chaplygin gas gives the following nonlocal conserved vector*

$$C^1 = -\rho\int\frac{\mathrm{d}x}{p}, \qquad C^2 = -\rho v\int\frac{\mathrm{d}x}{p}. \tag{344}$$

Mow we use the operator X_8 from (317). In this case

$$\begin{aligned}W^1 &= t - \left(\frac{t^2}{2} + s\right)v_x\\ W^2 &= \frac{\rho\tau}{p} - \left(\frac{t^2}{2} + s\right)\rho_x\\ W^3 &= -\tau - \left(\frac{t^2}{2} + s\right)p_x.\end{aligned} \tag{345}$$

Substituting in (337) and (338) the expressions (345) and the solution (331) of the adjoint system, i.e., letting $U = 0,\ R = 1,\ P = 0$, we obtain

$$C^1 = W^2 = \frac{\rho\tau}{p} - \left(\frac{t^2}{2} + s\right)\rho_x$$

$$C^2 = \rho W^1 + vW^2 = t\rho + \frac{\rho v\tau}{p} - \left(\frac{t^2}{2} + s\right)(\rho v_x + v\rho_x).$$

Noting that

$$-\left(\frac{t^2}{2} + s\right)\rho_x = -\frac{\rho\tau}{p} - D_x\left(\frac{t^2}{2}\rho + \rho s\right)$$

we reduce the above vector to the trivial conserved vector $\widetilde{C}^1 = 0,\ \widetilde{C}^2 = 0$.

Taking the solution (332) of the adjoint system, i.e., letting

$$U = 2\rho v, \qquad R = v^2, \qquad P = -1$$

we obtain

$$\begin{aligned} C^1 &= 2\rho v W^1 + v^2 W^2 - W^3 \\ &= 2t\rho v + \frac{\rho\tau v^2}{p} + \tau - \left(\frac{t^2}{2} + s\right) D_x\left(\rho v^2 - p\right) \\ C^2 &= (3\rho v^2 + p)W^1 + v^3 W^2 + vW^3 \\ &= t(3\rho v^2 + p) + \frac{\rho\tau v^3}{p} - v\tau \\ &\quad - \left(\frac{t^2}{2} + s\right)(3\rho v^2 v_x + v^3\rho_x + p v_x + v p_x). \end{aligned}$$

Then, upon rewriting C^1 in the form

$$C^1 = 2t\rho v + 2\tau - D_x\left[\left(\frac{t^2}{2} + s\right)(\rho v^2 - p)\right]$$

and applying the operations (239)–(241) we arrive at the following conserved vector

$$C^1 = t\rho v + \tau, \qquad C^2 = t(\rho v^2 + p). \tag{346}$$

The conservation Eq. (336) is satisfied for (346) in the following form

$$D_t(C^1) + D_x(C^2) = t\rho\left(v_t + vv_x + \frac{1}{\rho}p_x\right) + tv(\rho_t + v\rho_x + \rho v_x). \tag{347}$$

Taking the solution (333) of the adjoint system, i.e., letting

$$U = \rho, \qquad R = v, \qquad P = 0$$

we obtain

$$C^1 = \rho W^1 + v W^2, \qquad C^2 = 2\rho v W^1 + v^2 W^2 + W^3.$$

Substituting the expressions (345) for W^1, W^2, W^3 and simplifying as in the previous case we obtain the conserved vector

$$C^1 = t\rho, \qquad C^2 = t\rho v - \tau. \tag{348}$$

The conservation Eq. (336) is satisfied for (346) in the following form

$$D_t(C^1) + D_x(C^2) = t(\rho_t + v\rho_x + \rho v_x). \tag{349}$$

Finally, we take the solution (334), $U = t\rho,\ R = tv - x,\ P = 0$, and obtain

$$C^1 = t\rho W^1 + (tv - x)W^2, \qquad C^2(2t\rho v - x\rho)W^1 + (tv^2 - xv)W^2 + tW^3.$$

Simplifying as above, we arrive at the conserved vector

$$C^1 = \left(\frac{t^2}{2} - s\right)\rho, \qquad C^2 = \left(\frac{t^2}{2} - s\right)\rho v - t\tau. \tag{350}$$

The conservation Eq. (336) is satisfied for (346) in the following form

$$D_t(C^1) + D_x(C^2) = \left(\frac{t^2}{2} - s\right)(\rho_t + v\rho_x + \rho v_x). \tag{351}$$

Substituting in the conserved vectors (346), (348) and (350) the definition (318) of the nonlocal variables we formulate the result.

Proposition 15 *The nonlocal symmetry X_8 of the Chaplygin gas gives the following nonlocal conserved vectors*

$$C^1 = t\rho v + \int \rho \mathrm{d}x, \qquad C^2 = t(\rho v^2 + p) \tag{352}$$

$$C^1 = t\rho, \qquad C^2 = t\rho v - \int \rho \mathrm{d}x \tag{353}$$

$$C^1 = \left[\frac{t^2}{2} + \int \frac{1}{p}\left(\int \rho \mathrm{d}x\right)\mathrm{d}x\right]\rho$$
$$C^2 = \left[\frac{t^2}{2} + \int \frac{1}{p}\left(\int \rho \mathrm{d}x\right)\mathrm{d}x\right]\rho v - t\int \rho \mathrm{d}x. \tag{354}$$

Theorem 6 *Application of* Theorem 5 *to* two *nonlocal symmetries* (317) *gives* four *nonlocal conservation laws* (344) and (352)–(354) *for the Chaplygin gas* (312).

2.5.5 Steady Two-Dimensional Gas Flow

Consider the steady two-dimensional equations of gasdynamics

$$(\rho u)_x + (\rho v)_y = 0, \qquad u s_x + v s_y = 0$$
$$p_x + \rho(u u_x + v u_y) = 0, \qquad p_y + \rho(u v_x + v v_y) = 0 \tag{355}$$

where s denotes the specific entropy. It is shown in [41] that the Bateman-type reciprocal transformations provide the following nonlocal symmetry of equations (355)

$$X = \alpha\frac{\partial}{\partial x} + \sigma\frac{\partial}{\partial y} - p^2\frac{\partial}{\partial p} - pu\frac{\partial}{\partial u} - pv\frac{\partial}{\partial v} - (u^2+v^2)\rho^2\frac{\partial}{\partial \rho}$$

where α and σ are nonlocal variables determined by the equations

$$\alpha_x = p + \rho v^2, \qquad \alpha_y = -\rho u v$$

and

$$\sigma_x = -\rho u v, \qquad \sigma_y = p + \rho u^2$$

respectively. Using this nonlocal symmetry and the nonlinear self-adjointness of the system (355), the following nonlocal conserved vector is constructed in [41]

$$C^1 = \rho p u + u(u^2+v^2)\rho^2 + \alpha(\rho u)_x + \sigma(\rho u)_y$$
$$C^2 = \rho p v + v(u^2+v^2)\rho^2 + \alpha(\rho v)_x + \sigma(\rho v)_y. \tag{356}$$

The vector (356) satisfies the conservation law in the following form

$$\begin{aligned} D_x(C^1) + D_y(C^2) = {} & 2[p + \rho(u^2+v^2)][(\rho u)_x + (\rho v)_y] \\ & + \rho u[p_x + \rho(u u_x + v u_y)] + \rho v[p_y + \rho(u v_x + v v_y)] \\ & + \alpha[(\rho u)_x + (\rho v)_y]_x + \sigma[(\rho u)_x + (\rho v)_y]_y \end{aligned}$$

and the nonlocal conservation law that this symmetry

2.5.6 The Operator Identity for Nonlocal Symmetries

Example 18 Let us verify that the operator identity (179) is satisfied for the nonlocal symmetry X_7 of the Chaplygin gas. To this end, we have to verify that the coefficients of

$$\frac{\partial}{\partial v}, \quad \frac{\partial}{\partial \rho}, \quad \frac{\partial}{\partial p}, \quad \frac{\partial}{\partial v_t}, \quad \frac{\partial}{\partial v_x}, \quad \frac{\partial}{\partial \rho_t}, \quad \frac{\partial}{\partial \rho_x}, \quad \frac{\partial}{\partial p_t}, \quad \frac{\partial}{\partial p_x} \tag{357}$$

in both sides of (179) are equal. Using the first prolongation (321) of X_7 and the definition of the nonlocal variable σ given in Eq. (320) we see that the left-hand side of the identity (179) is written

$$\begin{aligned} X_7 + D_i(\xi^i) = {} & \sigma\frac{\partial}{\partial x} - \frac{\partial}{\partial p} + \frac{\rho}{p}\frac{\partial}{\partial \rho} - \frac{v v_x}{p}\frac{\partial}{\partial v_t} + \frac{v_x}{p}\frac{\partial}{\partial v_x} - \frac{v p_x}{p}\frac{\partial}{\partial p_t} \\ & + \frac{p_x}{p}\frac{\partial}{\partial p_x} + \left(\frac{\rho_t}{p} - \frac{\rho p_t}{p^2} - \frac{v\rho_x}{p}\right)\frac{\partial}{\partial \rho_t} + \left(2\frac{\rho_x}{p} - \frac{\rho p_x}{p^2}\right)\frac{\partial}{\partial \rho_x} - \frac{1}{p}\cdot \end{aligned} \tag{358}$$

Then we use the expressions (340) of W^α for the operator X_7, substitute them in the definition (183) of N^i and obtain in our approximation

$$\begin{aligned} \mathsf{N}^1 &= -\sigma v_x \frac{\partial}{\partial v_t} + \left(\frac{\rho}{p} - \sigma\rho_x\right)\frac{\partial}{\partial \rho_t} - (1 + \sigma p_x)\frac{\partial}{\partial p_t} \\ \mathsf{N}^2 &= \sigma - \sigma v_x \frac{\partial}{\partial v_x} + \left(\frac{\rho}{p} - \sigma\rho_x\right)\frac{\partial}{\partial \rho_x} - (1 + \sigma p_x)\frac{\partial}{\partial p_t}\cdot \end{aligned}$$

Now the right-hand side of (179) is written

$$\begin{aligned} & W^1\frac{\delta}{\delta v} + W^2\frac{\delta}{\delta \rho} + W^3\frac{\delta}{\delta p} + D_t\mathsf{N}^1 + D_x\mathsf{N}^2 \\ & = -\sigma v_x\left[\frac{\partial}{\partial v} - D_t\frac{\partial}{\partial v_t} - D_t\frac{\partial}{\partial v_x}\right] \\ & + \left(\frac{\rho}{p} - \sigma\rho_x\right)\left[\frac{\partial}{\partial \rho} - D_t\frac{\partial}{\partial \rho_t} - D_x\frac{\partial}{\partial \rho_x}\right] \\ & - (1 + \sigma p_x)\left[\frac{\partial}{\partial p} - D_t\frac{\partial}{\partial p_t} - D_x\frac{\partial}{\partial p_x}\right] \\ & + D_t\left[-\sigma v_x\frac{\partial}{\partial v_t} + \left(\frac{\rho}{p} - \sigma\rho_x\right)\frac{\partial}{\partial \rho_t} - (1 + \sigma p_x)\frac{\partial}{\partial p_t}\right] \\ & + D_x\left[\sigma - \sigma v_x\frac{\partial}{\partial v_x} + \left(\frac{\rho}{p} - \sigma\rho_x\right)\frac{\partial}{\partial \rho_x} - (1 + \sigma p_x)\frac{\partial}{\partial p_t}\right]\cdot \end{aligned} \tag{359}$$

Making the changes in two last lines of Eq. (359) such as

$$
\begin{aligned}
D_t\left[-\sigma v_x \frac{\partial}{\partial v_t}\right] &= -\sigma v_x D_t \frac{\partial}{\partial v_t} - D_t(\sigma v_x)\frac{\partial}{\partial v_t} \\
&= -\sigma v_x D_t \frac{\partial}{\partial v_t} - \left(\frac{v}{p}\, v_x + \sigma v_{tx}\right)\frac{\partial}{\partial v_t}
\end{aligned}
$$

one can see that the coefficients of the differentiations (357) in (358) and (359) coincide. Inspection of the coefficients of the differentiations in higher derivatives $v_{tt}, v_{tx}, v_{xx}, \ldots$ requires the higher-order prolongations of the operator X_7.

Exercise 7 Verify that the operator identity (179) is satisfied in the same approximation as in Example 18 for the nonlocal symmetry operator X_8 from (317).

2.6 Comparison With the "Direct Method"

2.6.1 General Discussion

Theorem 5 allows to construct conservation laws for equations with known symmetries simply by substituting in Eq. (258) the expressions W^α and $\mathcal{L}$ given by Eqs. (256) and (256), respectively.

The "direct method" means the determination of the conserved vectors (238) by solving Eq. (237) for C^i. Upon restricting the highest order of derivatives of u involved in C^i, Eq. (237) splits into several equations. If one can solve the resulting system, one obtains the desired conserved vectors. Existence of symmetries is not required.

To the best of my knowledge, the direct method was used for the first time in 1798 by Laplace [42]. He applied the method to Kepler's problem in celestial mechanics and found a new vector-valued conserved quantity (see [42], Book II, Chap. III, equations (P)) known as Laplace's vector.

The application of the direct method to the gasdynamic Eq. (298) allowed to demonstrate in [43] that all conservation laws involving only the independent and dependent variables (299) and (300) were provided by the classical conservation laws (mass, energy, momentum, angular momentum and center-of-mass) given in Sect. 2.5.1 and the following two special conservation laws

$$
\frac{\mathrm{d}}{\mathrm{d}t}\int\limits_{\Omega(t)} \left\{t(\rho|\boldsymbol{v}|^2 + np) - \rho\boldsymbol{x}\cdot\boldsymbol{v}\right\}\mathrm{d}\omega = -\int\limits_{S(t)} p\,(2t\boldsymbol{v} - \boldsymbol{x})\cdot\boldsymbol{\nu}\mathrm{d}S
$$

$$
\frac{\mathrm{d}}{\mathrm{d}t}\int\limits_{\Omega(t)} \left\{t^2(\rho|\boldsymbol{v}|^2 + np) - \rho\boldsymbol{x}\cdot(2t\boldsymbol{v} - \boldsymbol{x}\right\}\mathrm{d}\omega = -\int\limits_{S(t)} 2tp\,(t\boldsymbol{v} - \boldsymbol{x})\cdot\boldsymbol{\nu}\mathrm{d}S
$$

that were found in [44] in the case $\gamma = (n+2)/n$ by using the symmetry ideas.

All local conservation laws for the heat equation $u_t - u_{xx} = 0$ have been found by the direct method in [45] (see [15–17, vol. 1, Sect. 10.1] and [46]). Namely it has been shown by considering the conservation equations of the form

$$D_t[\tau(t,x,u,u_x,u_{xx},\ldots)] + D_x[\psi(t,x,u,u_x,u_{xx},\ldots)] = 0$$

that all such conservation laws are given by

$$D_t[\varphi(t,x)u] + D_x[u\varphi_x(t,x) - \varphi(t,x)u_x] = 0$$

where $v = \varphi(t,x)$ is an arbitrary solution of the adjoint equation $v_t + v_{xx} = 0$ to the heat equation. Similar result can be obtained by applying Theorem 5 for any linear equation, e.g., for the heat equation $u_t - \Delta u = 0$ with any number of spatial variables $x = (x^1, \ldots, x^n)$. Namely, applying formula (258) to the scaling symmetry $X = u\partial/\partial u$ we obtain the conservation law

$$D_t[\varphi(t,x)u] + \nabla \cdot [u\nabla\varphi(t,x) - \varphi(t,x)\nabla u] = 0$$

where $v = \varphi(t,x)$ is an arbitrary solution of the adjoint equation $v_t + \Delta v = 0$ to the heat equation. This conservation law embraces the conservation laws associated with all other symmetries of the heat equation.

Various mathematical models for describing the geological process of segregation and migration of large volumes of molten rock were proposed in the geophysical literature (see the papers [47–51] and the references therein). One of them is known as the *generalized magma equation* and has the form

$$u_t + D_z\left[u^n - u^n D_z\left(u^{-m}u_t\right)\right] = 0, \qquad n, m = \text{const.} \tag{360}$$

It is accepted as a reasonable mathematical model for describing melt migration through the Earth's mantle. Several conservation laws for this model have been calculated by the direct method in [47, 48] and interpreted from symmetry point of view in [49]. It is shown in [52] that Eq. (360) is quasi self-adjoint with the substitution (34) given by $v = u^{1-n-m}$ if $m + n \neq 1$ and $v = \ln|u|$ if $m + n = 1$. These substitutions show that Eq. (360) is strictly self-adjoint (Definition 2) if $m + n = 0$. Using the quasi self-adjointness, the conservation laws are easily computed in [52].

Some simplification of the direct method was suggested in [37]. Namely, one writes the conservation equation in the form (242)

$$D_i(C^i) = \mu^{\bar{\alpha}} F_{\bar{\alpha}}\big(x, u, u_{(1)}, \ldots, u_{(s)}\big) \tag{361}$$

and first finds the undetermined coefficients $\mu^{\bar{\alpha}}$ by satisfying the integrability condition of Eq. (361), i.e., by solving the equations (see Proposition 8 in Sect. 2.1.2)

$$\frac{\delta}{\delta u^\alpha}\left[\mu^{\bar{\beta}}(x,u,u_{(1)},\ldots)\, F_{\bar{\beta}}\big(x,u,u_{(1)},\ldots,u_{(s)}\big)\right] = 0, \quad \alpha = 1,\ldots,m. \tag{362}$$

Then, for each solution $\mu^{\bar{\alpha}}$ of Eq. (362), the components C^i of the corresponding conserved vector are computed from Eq. (361). In simple situations C^i can be detected merely by looking at the right-had side of Eq. (361), see further Example 19.

Remark 17 Note that Eq. (362) should be satisfied on the solutions of Eq. (236). Then the left-hand side of (362) can be written as

$$F_\alpha^*\big(x, u, v, \ldots, u_{(s)}, v_{(s)}\big)\Big|_{v=\mu(x,u,u_{(1)},\ldots)}$$

with F_α^* defined by Eq. (69).

The reader can find a detailed discussion of the direct method in the recent book [14]. I will compare two methods by considering few examples and exercises.

2.6.2 Examples and Exercises

Example 19 (See [14, Sect. 1.3]). Let us consider the KdV Eq. (73)

$$u_t = u_{xxx} + uu_x \tag{73}$$

and write the condition (362) for $\mu = \mu(t, x, u)$. We have

$$\begin{aligned}
&\frac{\delta}{\delta u}\,[\mu(t,x,u)(u_t - u_{xxx} - uu_x)]\\
&\quad = -D_t(\mu) + D_x^3(\mu) + D_x(u\mu) - \mu u_x + (u_t - u_{xxx} - uu_x)\frac{\partial \mu}{\partial u}\\
&\quad = -D_t(\mu) + D_x^3(\mu) + uD_x(\mu) + (u_t - u_{xxx} - uu_x)\frac{\partial \mu}{\partial u}\,.
\end{aligned}$$

In accordance with Remark 17, we consider this expression on the solutions of the KdV equation and see that Eq. (362) coincides with the adjoint Eqs. (73) and (74)

$$D_t(\mu) = D_x^3(\mu) + uD_x(\mu). \tag{363}$$

Its solution is given in Example 11 and has the form (75)

$$\mu = A_1 + A_2u + A_3(x + tu), \qquad A_1, A_2, A_3 = \text{const}.$$

Thus, we have the following three linearly independent solutions of Eq. (363)

$$\mu_1 = 1, \qquad \mu_2 = u, \qquad \mu_3 = (x + tu).$$

and the corresponding three Eq. (361)

$$D_t(C^1) + D_x(C^2) = u_t - u_{xxx} - uu_x \tag{364}$$
$$D_t(C^1) + D_x(C^2) = u(u_t - u_{xxx} - uu_x) \tag{365}$$
$$D_t(C^1) + D_x(C^2) = (x + tu)(u_t - u_{xxx} - uu_x). \tag{366}$$

In this simple example the components C^1, C^2 of the conserved vector can be easily seen from the right-hand sides of Eqs. (364)–(366). In the case of (364) and (365) it is obvious. Therefore let us consider the right-hand side of Eq. (366). We see that

$$\begin{aligned}
(x + tu)u_t &= D_t\left(xu + \frac{1}{2}tu^2\right) - \frac{1}{2}u^2 \\
-(x + tu)uu_x &= -D_x\left(\frac{1}{2}xu^2 + \frac{1}{3}tu^3\right) + \frac{1}{2}u^2 \\
-(x + tu)u_{xxx} &= -D_x\left(xu_{xx} + tuu_{xx}\right) + u_{xx} + tu_x u_{xx} \\
&= D_x\left(u_x + \frac{1}{2}tu_x^2 - xu_{xx} - tuu_{xx}\right).
\end{aligned}$$

Hence, the right-hand side of Eq. (366) can be written in the divergence form

$$\begin{aligned}
&(x + tu)(u_t - u_{xxx} - uu_x) \\
&= D_t\left(t\frac{u^2}{2} + xu\right) + D_x\left[u_x + t\left(\frac{u_x^2}{2} - uu_{xx} - \frac{u^3}{3}\right) - x\left(\frac{u^2}{2} + u_{xx}\right)\right].
\end{aligned}$$

The expressions under $D_t(\cdots)$ and $D_x(\cdots)$ give C^1 and C^2, respectively, in (366). Note that the corresponding conservation law

$$D_t\left(t\frac{u^2}{2} + xu\right) + D_x\left[u_x + t\left(\frac{u_x^2}{2} - uu_{xx} - \frac{u^3}{3}\right) - x\left(\frac{u^2}{2} + u_{xx}\right)\right] = 0 \tag{367}$$

was derived from the Galilean invariance of the KdV equation (see [10, Sect. 22.5]) and by the direct method (see [14, Sect. 1.3.5]).

The similar treatment of the right-hand sides of the Eqs. (364) and (365) leads to Eq. (73) and to the conservation law

$$D_t(u^2) + D_x\left(u_x^2 - 2uu_{xx} - \frac{2}{3}u^3\right) = 0 \tag{368}$$

respectively. Theorem 5 associates the conservation law (368) with the scaling symmetry of the KdV equation.

Exercise 8 Apply the direct method to the short pulse Eq. (280) using the differential substitution (284). In this case Eq. (361) is written

$$D_t(C^1) + D_x(C^2) = u_t u_{xt} - \frac{1}{2} u^2 u_x u_{xt}$$
$$- \left(u + \frac{1}{2} u^2 u_{xx} + u u_x^2 \right) u_t + \frac{1}{2} u^3 u_x + \frac{1}{4} u^4 u_x u_{xx} + \frac{1}{2} u^3 u_x^3. \tag{369}$$

Exercise 9 Consider the Boussinesq Eq. (246). Taking its formal Lagrangian

$$\mathcal{L} = \omega \left[\Delta\psi_t - g\rho_x - f v_z - \psi_x \Delta\psi_z + \psi_z \Delta\psi_x \right]$$
$$+ \mu \left[v_t + f\psi_z - \psi_x v_z + \psi_z v_x \right] + r \left[\rho_t + (N^2/g)\, \psi_x - \psi_x \rho_z + \psi_z \rho_x \right]$$

where ω, μ, r are new dependent variables, we obtain the adjoint system to Eq. (246)

$$\frac{\delta \mathcal{L}}{\delta \psi} = 0, \qquad \frac{\delta \mathcal{L}}{\delta v} = 0, \qquad \frac{\delta \mathcal{L}}{\delta \rho} = 0. \tag{370}$$

It is shown in [53] that the system (246) is self-adjoint. Namely, the substitution

$$\omega = \psi, \qquad \mu = -v, \qquad r = -(g^2/N^2)\, \rho \tag{371}$$

maps the adjoint system (370) into the system (246). Using the so eatablished self-adjointness, nontrivial conservation laws were constructed via Theorem 5. Apply the direct method to the system (246). Note that knowledge of the substitution (371) gives the following Eq. (361)

$$D_t(C^1) + D_x(C^2) + D_z(C^3) = \psi\big[\psi_{txx} + \psi_{tzz} - g\rho_x - f v_z$$
$$- \psi_x \left(\psi_{zxx} + \psi_{zzz}\right) + \psi_z \left(\psi_{xxx} + \psi_{xzz}\right)\big] - v \left[v_t + f\psi_z - \psi_x v_z + \psi_z v_x \right] \tag{372}$$
$$- \frac{g^2}{N^2}\, \rho \Big[\rho_t + \frac{N^2}{g}\, \psi_x - \psi_x \rho_z + \psi_z \rho_x \Big].$$

Example 20 Let us consider the conservation Eq. (343)

$$D_t(C^1) + D_x(C^2) = \sigma(\rho_t + v\rho_x + \rho v_x)$$

where σ is connected with the velocity v and the pressure p of the Chaplygin gas by Eq. (319)

$$\sigma_x = -\frac{1}{p}, \qquad \sigma_t + v\sigma_x = 0.$$

In this example Eq. (362) are not satisfied. Indeed, we have

$$\frac{\delta}{\delta v}\left[\sigma(\rho_t + v\rho_x + \rho v_x)\right] = \sigma\rho_x - D_x(\sigma\rho) = -\rho\sigma = \rho\int\frac{\mathrm{d}x}{p} \neq 0$$
$$\frac{\delta}{\delta \rho}\left[\sigma(\rho_t + v\rho_x + \rho v_x)\right] = \sigma_t - D_x(\sigma v) + \sigma v_x = -(\sigma_t + v\sigma_x) = 0$$
$$\frac{\delta}{\delta p}\left[\sigma(\rho_t + v\rho_x + \rho v_x)\right] = 0.$$

Example 21 Let us consider the conservation Eq. (347)

$$D_t(C^1) + D_x(C^2) = t\rho\left(v_t + vv_x + \frac{1}{\rho}\,p_x\right) + tv(\rho_t + v\rho_x + \rho v_x).$$

Here Eq. (362) are not satisfied. Namely, writing

$$t\rho\left(v_t + vv_x + \frac{1}{\rho}\,p_x\right) + tv(\rho_t + v\rho_x + \rho v_x)$$
$$= t\rho v_t + 2t\rho v v_x + tp_x + tv\rho_t + tv^2\rho_x$$

we obtain

$$\frac{\delta}{\delta v}\left[t\rho v_t + 2t\rho v v_x + tp_x + tv\rho_t + tv^2\rho_x\right] = -\rho$$
$$\frac{\delta}{\delta \rho}\left[t\rho v_t + 2t\rho v v_x + tp_x + tv\rho_t + tv^2\rho_x\right] = -v$$
$$\frac{\delta}{\delta p}\left[t\rho v_t + 2t\rho v v_x + tp_x + tv\rho_t + tv^2\rho_x\right] = 0.$$

Exercise 10 Check if Eq. (362) are satisfied for the conservation Eqs. (349) and (351).

3 Utilization of Conservation Laws for Constructing Solutions of PDEs

In this part we will discuss a method for constructing exact solutions for systems of nonlinear partial differential equations. The method is based on knowledge of conservation laws of the equations under consideration and therefore it is called the *method of conservation laws.*

Application of the method to the Chaplygin gas allowed to construct new solutions containing several arbitrary parameters. One can verify that these solutions cannot be obtained as group invariant solutions.

3.1 General Discussion of the Method

As mentioned in Sect. 2.1.4, one can integrate or reduce the order of linear ordinary differential equations by rewriting them in a conservation form (217). Likewise one can integrate or reduce the order of a nonlinear ordinary differential equation as well as a system of ordinary differential equations using their conservation laws. Namely, a conservation law

$$D_x\left(\psi(x, y, y', \ldots, y^{(s-1)})\right) = 0 \tag{373}$$

for a nonlinear ordinary differential equation

$$F(x, y, y', \ldots, y^{(s)}) = 0 \tag{374}$$

yields the first integral

$$\psi(x, y, y', \ldots, y^{(s-1)}) = C_1. \tag{375}$$

We will discuss now an extension of this idea to partial differential equations. Namely, we will apply conservation laws for constructing particular exact solutions of systems of partial differential equations. Detailed calculations are given in examples considered in the next sections.

Let us assume that the system (236)

$$F_{\bar{\alpha}}\left(x, u, u_{(1)}, \ldots, u_{(s)}\right) = 0, \qquad \bar{\alpha} = 1, \ldots, \overline{m} \tag{376}$$

has a conservation law (237)

$$\left[D_i(C^i)\right]_{(376)} = 0 \tag{377}$$

with a known conserved vector

$$C = \left(C^1, \ \ldots, \ C^n\right) \tag{378}$$

where

$$C^i = C^i\left(x, u, u_{(1)}, \ldots\right), \qquad i = 1, \ldots, n.$$

We write the conservation Eq. (377) in the form (361)

$$D_i(C^i) = \mu^{\bar{\alpha}} F_{\bar{\alpha}}\big(x, u, u_{(1)}, \ldots, u_{(s)}\big). \tag{379}$$

For a given conserved vector (378) the coefficients $\mu^{\bar{\alpha}}$ in Eq. (379) are known functions $\mu^{\bar{\alpha}} = \mu^{\bar{\alpha}}(x, u, u_{(1)}, \ldots)$.

We will construct particular solutions of the system (376) by requiring that *on these solutions the vector* (378) *reduces to the following trivial conserved vector*

$$C = \left(C^1(x^2, \ldots, x^n), \quad \ldots, \quad C^n(x^1, \ldots, x^{n-1})\right). \tag{380}$$

In other words, we look for particular solutions of the system (376) by adding to Eq. (376) the *differential constraints*

$$\begin{aligned} C^1\left(x, u, u_{(1)}, \ldots\right) &= h^1(x^2, x^3, \ldots, x^n) \\ C^2\left(x, u, u_{(1)}, \ldots\right) &= h^2(x^1, x^3, \ldots, x^n) \\ \ldots\ldots\ldots\ldots\ldots\ldots\ldots\ldots\ldots\ldots\ldots \\ C^n\left(x, u, u_{(1)}, \ldots\right) &= h^n(x^1, \ldots, x^{n-1}) \end{aligned} \tag{381}$$

where $C^i\left(x, u, u_{(1)}, \ldots\right)$ are the components of the known conserved vector (378). Due to the constraints (381), the left-hand side of Eq. (379) vanishes identically. Hence the number of equations in the system (376) will be reduced by one.

The differential constraints (381) can be equivalently written as follows

$$\begin{aligned} D_1\left[C^1\left(x, u, u_{(1)}, \ldots\right)\right] &= 0 \\ D_2\left[C^2\left(x, u, u_{(1)}, \ldots\right)\right] &= 0 \\ \ldots\ldots\ldots\ldots\ldots\ldots\ldots\ldots \\ D_n\left[C^n\left(x, u, u_{(1)}, \ldots\right)\right] &= 0. \end{aligned} \tag{382}$$

Remark 18 The overdetermined system of $\overline{m} + n$ Eqs. (376) and (382) reduces to $\overline{m} + n - 1$ equations due to the conservation law (377).

3.2 Application to the Chaplygin Gas

3.2.1 Detailed Discussion of One Case

Let us apply the method to the Chaplygin gas equations (312)

$$\begin{aligned} v_t + v v_x + \frac{1}{\rho} p_x &= 0 \\ \rho_t + v \rho_x + \rho v_x &= 0 \\ p_t + v p_x - p v_x &= 0. \end{aligned} \tag{383}$$

We will construct a particular solution of the system (383) using the simplest conservation law (327)

$$D_t(\rho) + D_x(\rho v) = \rho_t + v\rho_x + \rho v_x. \tag{384}$$

The conservation Eq. (384) is written in the form (379) with the following conserved vector (378)

$$C^1 = \rho, \qquad C^2 = \rho v. \tag{385}$$

The differential constraints (381) are written as follows

$$\rho = g(x), \qquad \rho v = h(t). \tag{386}$$

Thus we look for solutions of the form

$$\rho = g(x), \qquad v = \frac{h(t)}{g(x)} \cdot \tag{387}$$

The functions (387) solve the second equation in (383) because the conservation law (384) coincides with the second Eq. (383) (see Remark 18). Therefore it remains to substitute (387) in the first and third equations of the system (383). The result of this substitution can be solved for the derivatives of p

$$p_x = -h' + \frac{h^2 g'}{g^2}, \qquad p_t = -\frac{hg'}{g^2}\,p + \frac{hh'}{g} - \frac{h^3 g'}{g^3} \cdot \tag{388}$$

The compatibility condition $p_{xt} = p_{tx}$ of the system (388) gives the equation

$$\left(g'' - 2\frac{g'^2}{g}\right) p = g^2\frac{h''}{h} - 2g'h' - h^2\frac{g''}{g} + 2h^2\frac{g'^2}{g^2} \cdot \tag{389}$$

For illustration purposes I will simplify further calculations by considering the particular case when the coefficient in front of p in Eq. (389) vanishes

$$g'' - 2\frac{g'^2}{g} = 0. \tag{390}$$

Thc solution of Eq. (390) is

$$g(x) = \frac{1}{ax + b}, \qquad a, b = \text{const}. \tag{391}$$

Substituting (391) in Eq. (389) we obtain

$$h'' + 2ahh' = 0 \tag{392}$$

whence

$$h(t) = k\tan(c - akt) \tag{393}$$

if $a \neq 0$, and

$$h(t) = At + B \tag{394}$$

if $a = 0$.

If the constant a in (391) does not vanish, we substitute (391) and (393) in Eq. (388), integrate them and obtain

$$p = k^2(ax + b) + Q\cos(c - akt), \qquad Q = \text{const.} \tag{395}$$

In the case $a = 0$ the similar calculations yield

$$p = -Ax + \frac{b}{2}A^2t^2 + ABbt + Q, \qquad Q = \text{const.} \tag{396}$$

Thus, using the conservation law (384) we have arrived at the solutions

$$\begin{aligned} \rho &= \frac{1}{ax + b} \\ v &= k(ax + b)\tan(c - akt) \\ p &= k^2(ax + b) + Q\cos(c - akt) \end{aligned} \tag{397}$$

and

$$\begin{aligned} \rho &= \frac{1}{b} \\ v &= b(At + B) \\ p &= -Ax + \frac{b}{2}A^2t^2 + ABbt + Q. \end{aligned} \tag{398}$$

3.2.2 Differential Constraints Provided by Other Conserved Vectors

The conservation laws (328) and (330) give the following differential constraints (381)

$$\rho v^2 - p = g(x), \qquad pv + \rho v^3 = h(t) \tag{399}$$

$$\rho v = g(x), \qquad p + \rho v^2 = h(t) \tag{400}$$

$$t\rho v - x\rho = g(x), \qquad tp + t\rho v^2 - x\rho v = h(t). \tag{401}$$

The nonlocal conserved vectors (346), (348) and (350) lead to the following differential constraints (381)

$$t\rho v + \tau = g(x), \qquad p + \rho v^2 = h(t) \tag{402}$$

$$t\rho = g(x), \qquad t\rho v - \tau = h(t) \tag{403}$$

$$\left(\frac{t^2}{2} - s\right)\rho = g(x), \quad \left(\frac{t^2}{2} - s\right)\rho v - t\tau = h(t). \tag{404}$$

The constraints (402) are not essentially different from the constraints (400). It is manifest if we write them in the form (382).

3.3 *Application to Nonlinear Equation Describing an Irrigation System*

The method of Sect. 3.1 can be used for constructing particular solutions not only of a system, but of a single partial differential equations as well.

Let us consider the nonlinear Eq. (171)

$$C(\psi)\psi_t = [K(\psi)\psi_x]_x + [K(\psi)(\psi_z - 1)]_z - S(\psi) \tag{405}$$

satisfying the nonlinear self-adjointness condition (173)

$$S'(\psi) = aC(\psi), \qquad a = \text{const.} \tag{406}$$

and apply the method of Sect. 3.1 to the conserved vector (269)

$$C^1 = S(\psi)\mathrm{e}^{at}, \quad C^2 = aK(\psi)\psi_x\mathrm{e}^{at}, \quad C^3 = aK(\psi)(\psi_z - 1)\mathrm{e}^{at}. \tag{407}$$

The conditions (381) are written

$$S(\psi)\mathrm{e}^{at} = f(x,z), \quad aK(\psi)\psi_x\mathrm{e}^{at} = g(t,z), \quad aK(\psi)(\psi_z - 1)\mathrm{e}^{at} = h(t,x).$$

These conditions mean that the left-hand sides of the first, second and third equation do not depend on t, x and z, respectively. Therefore they can be equivalently written as the following differential constraints (see Eq. 382)

$$aS(\psi) + S'(\psi)\psi_t = 0, \quad [K(\psi)\psi_x]_x = 0, \quad [K(\psi)(\psi_z - 1)]_z = 0. \tag{408}$$

The constraints (408) reduce Eqs. (405) and (406). Hence, the particular solutions of Eq. (405) provided by the conserved vector (407) are described by the system

$$
\begin{aligned}
aC(\psi) - S'(\psi) &= 0, \qquad aS(\psi) + S'(\psi)\psi_t = 0 \\
[K(\psi)\psi_x]_x &= 0, \qquad \left[K(\psi)\,(\psi_z - 1)\right]_z = 0.
\end{aligned} \tag{409}
$$

4 Approximate Self-Adjointness and Approximate Conservation Laws

The methods developed in this paper can be extended to differential equations with a small parameter in order to construct approximate conservation laws using approximate symmetries. I will illustrate this possibility by examples. The reader interested in approximate symmetries can find enough material in [15–17], vol. 3, Chaps. 2 and 9.

4.1 The Van Der Pol Equation

The van der Pol equation has the form

$$
F \equiv y'' + y + \varepsilon(y'^3 - y') = 0, \qquad \varepsilon = \text{const.} \neq 0. \tag{410}
$$

4.1.1 Approximately Adjoint Equation

We have

$$
\frac{\delta}{\delta y}\left\{z\left[y'' + y + \varepsilon\left(y'^3 - y'\right)\right]\right\} = z'' + z + \varepsilon D_x\left(z - 3zy'^2\right).
$$

Thus, the adjoint equation to the van der Pol equation is

$$
F^* \equiv z'' + z + \varepsilon\left(z' - 3z'y'^2 - 6zy'y''\right) = 0.
$$

We eliminate here y'' by using Eq. (410), consider ε as a small parameter and write F^* in the first order of precision with respect to ε. In other words, we write

$$
y'' \approx -y. \tag{411}
$$

Then we obtain the following *approximately adjoint equation* to equation (410)

$$
F^* \equiv z'' + z + \varepsilon\left(z' - 3z'y'^2 + 6zyy'\right) = 0. \tag{412}
$$

4.1.2 Approximate Self-Adjointness

Let us investigate Eq. (410) for approximate self-adjointness. Specifically, I will call Eq. (410) *approximately self-adjoint* if there exists a non-trivial (not vanishing identically) approximate substitution

$$z \approx f(x, y, y') + \varepsilon g(x, y, y') \tag{413}$$

such that F given by Eq. (410) and F^* defined by Eq. (412) approximately satisfy the condition (72) of nonlinear self-adjointness. In other words, the following equation is satisfied in the first-order of precision in ε

$$F^*\big|_{z=f+\varepsilon g} = \lambda F. \tag{414}$$

Note, that the unperturbed equation $y''+y = 0$ is nonlinearly self-adjoint. Namely it coincides with the adjoint equation $z'' + z = 0$ upon the substitution

$$z = \alpha y + \beta \cos x + \gamma \sin x, \qquad \alpha, \beta, \gamma = \text{const.} \tag{415}$$

Therefore we will consider the substitution (413) of the following restricted form

$$z \approx f(x, y) + \varepsilon g(x, y, y'). \tag{416}$$

In differentiating $g(x, y, y')$ we will use Eq. (411) because we make out calculations in the first order of precision with respect to ε. Then we obtain

$$\begin{aligned} z' &= D_x(f) + \varepsilon D_x(g)\big|_{y''=-y} \equiv f_x + y' f_y + \varepsilon(g_x + y' g_y - y g_{y'}) \\ z'' &= D_x^2(f) + \varepsilon D_x^2(g)\big|_{y''=-y} \equiv f_{xx} + 2y' f_{xy} + y'^2 f_{yy} + y'' f_y \\ &\quad + \varepsilon(g_{xx} + 2y' g_{xy} - 2y g_{xy'} + y'^2 g_{yy} \\ &\quad - 2yy' g_{yy'} + y^2 g_{y'y'} - y g_y - y' g_{y'}). \end{aligned} \tag{417}$$

Substituting (417) in (412) and solving Eq. (414) with $\varepsilon = 0$ we see that f is given by Eq. (415). Then $\lambda = C$ and the terms with ε in Eq. (414) give the following second-order linear partial differential equation for $g(x, y, y')$

$$\begin{aligned} g + D_x^2(g)\big|_{y''=-y} &= \alpha \left(4y'^3 - 6y^2 y' - 2y'\right) \\ &\quad + \beta \left(\sin x - 3y'^2 \sin x - 6yy' \cos x\right) \\ &\quad + \gamma \left(3y'^2 \cos x - \cos x - 6yy' \sin x\right). \end{aligned} \tag{418}$$

The standard existence theorem guarantees that Eq. (418) has a solution. It is manifest that the solution does not vanish because $g = 0$ does not satisfy Eq. (418). We conclude that the van der Pol equation (410) with a small parameter ε is approximately self-adjoint. The substitution (416) satisfying the approximate self-adjointness condition (414) has the form

$$z \approx \alpha y + \beta \cos x + \gamma \sin x + \varepsilon g(x, y, y') \tag{419}$$

where $\alpha,\ \beta,\ \gamma$ are arbitrary constants and $g(x, y, y')$ solves Eq. (418).

4.1.3 Exact and Approximate Symmetries

If ε is treated as an arbitrary constant, Eq. (410) has only one point symmetry, namely the one-parameter group of translations of the independent variable x. We will write the generator $X_1 = \partial/\partial x$ of this group in the form (186)

$$X_1 = y' \frac{\partial}{\partial y} \cdot \tag{420}$$

If ε is a small parameter, then Eq. (410) has, along with the exact symmetry (420), the following 7 approximate symmetries ([15–17], vol. 3, Sect. 9.1.3.3)

$$\begin{aligned}
X_2 &= \left\{4y - \varepsilon\left[y^2 y' + 3xy\left(y^2 + y'^2\right)\right]\right\}\frac{\partial}{\partial y}\\
X_3 &= \left\{8\cos x + \varepsilon\left[\left(4 - 3y'^2 - 9y^2\right)x\cos x + 3(xy^2)'\sin x\right]\right\}\frac{\partial}{\partial y}\\
X_4 &= \left\{8\sin x + \varepsilon\left[\left(4 - 3y'^2 - 9y^2\right)x\sin x - 3(xy^2)'\cos x\right]\right\}\frac{\partial}{\partial y}\\
X_5 &= \{24y^2\cos x - 24yy'\sin x + \varepsilon\big[(12yy' + 9yy'^3 + 9y^3y')x\sin x\\
&\quad + \left(12y^2 - 9y^2y'^2 - 6y^4\right)\sin x - \left(12y^2 - 9y^2y'^2 - 9y^4\right)x\cos x\\
&\quad - 3y^3y'\cos x\big]\}\frac{\partial}{\partial y} \qquad (421)\\
X_6 &= \{24y^2\sin x + 24yy'\cos x - \varepsilon\big[(12yy' + 9yy'^3 + 9y^3y')x\cos x\\
&\quad + \left(12y^2 - 9y^2y'^2 - 6y^4\right)\cos x + \left(12y^2 + 9y^2y'^2 + 9y^4\right)x\sin x\\
&\quad + 3y^3y'\sin x\big]\}\frac{\partial}{\partial y}\\
X_7 &= \{4y\cos 2x - 4y'\sin 2x + \varepsilon\big[3(yy'^2 - y^3)x\cos 2x\\
&\quad - 3y^2y'\cos 2x + 6y^2y'x\sin 2x + 2(y - y^3)\sin 2x\big]\}\frac{\partial}{\partial y}\\
X_8 &= \{4y\sin 2x + 4y'\cos 2x - \varepsilon\big[3(y^3 - yy'^2)x\sin 2x
\end{aligned}$$

$$+ 3y^2y' \sin 2x + 6y^2y'x \cos 2x + 2(y - y^3) \cos 2x]\}\frac{\partial}{\partial y}\cdot$$

4.1.4 Approximate Conservation Laws

We can construct now approximate conserved quantities for the van der Pol equation using the formula (258) and the approximate substitution (419). Inserting in (258) the formal Lagrangian

$$\mathcal{L} = z\left[y'' + y + \varepsilon\left(y'^3 - y'\right)\right]$$

we obtain

$$C = W\left[-z' + \varepsilon\left(3y'^2 z - z\right)\right] + W'z. \tag{422}$$

Let us calculate the conserved quantity (422) for the operator X_1 given by Eq. (420). In this case $W = y',\ W' = y''$, and therefore (422) has the form

$$C = -y'z' + \varepsilon\left(3y'^3 - y'\right)z + y''z.$$

We eliminate here y'' via Eq. (410), use the approximate substitution (419) and obtain (in the first order of precision with respect to ε) the following approximate conserved quantity

$$\begin{aligned} C = &- \alpha\left(y^2 + y'^2\right) + \beta\left(y' \sin x - y \cos x\right) - \gamma\left(y' \cos x + y \sin x\right) \\ &+ \varepsilon\left(2\alpha y y'^3 + 2\beta y'^3 \cos x + 2\gamma y'^3 \sin x - yg - y' D_x(g)\big|_{y''=-y}\right). \end{aligned} \tag{423}$$

Differentiating it and using the Eqs. (410) and (411) we obtain

$$\begin{aligned} D_x(C) =& \varepsilon y'\Big[\alpha\left(4y'^3 - 6y^2y' - 2y'\right) + \beta\left(\sin x - 3y'^2 \sin x - 6yy' \cos x\right) \\ &+ \gamma\left(3y'^2 \cos x - \cos x - 6yy' \sin x\right) - g - D_x^2(g)\big|_{y''=-y}\Big] + o(\varepsilon) \end{aligned} \tag{424}$$

where $o(\varepsilon)$ denotes the higher-order terms in ε. The Eqs. (418) and (424) show that the quantity (423) satisfies the approximate conservation law

$$D_x(C)\big|_{(410)} \approx 0. \tag{425}$$

Let us consider the operator X_2 from (421). In this case we have

$$W = 4y - \varepsilon\Big[y^2 y' + 3xy\left(y^2 + y'^2\right)\Big]$$
$$W' \approx 4y' - \varepsilon\Big[2y^3 + 5yy'^2 + 3x\left(y^2 y' + y'^3\right)\Big]. \tag{426}$$

Proceeding as above we obtain the following approximate conserved quantity

$$\begin{aligned} C = {} & 4y'(\beta\cos x + \gamma\sin x) - 4y(\gamma\cos x - \beta\sin x) \\ & + \varepsilon\Big\{2\alpha y^2\left(4y'^2 - y^2 - 2\right) + 4y'g - 4yD_x(g)\big|_{y''=-y} \\ & + \Big[7yy'^2 - 3xy'(y^2 + y'^2) - 2y^3 - 4y\Big](\beta\cos x + \gamma\sin x) \\ & + \Big[y^2 y' + 3xy(y^2 + y'^2)\Big](\gamma\cos x - \beta\sin x)\Big\}. \end{aligned} \tag{427}$$

The calculation shows that the quantity (427) satisfies the approximate conservation law (425) in the following form

$$\begin{aligned} D_x(C) = {} & 4(\beta\cos x + \gamma\sin x)\Big[y'' + y + \varepsilon\left(y'^3 - y'\right)\Big] \\ & + 4\varepsilon y\Big[\alpha\left(4y'^3 - 6y^2 y' - 2y'\right) + \beta\left(\sin x - 3y'^2\sin x - 6yy'\cos x\right) \\ & + \gamma\left(3y'^2\cos x - \cos x - 6yy'\sin x\right) - g - D_x^2(g)\big|_{y''=-y}\Big] + o(\varepsilon). \end{aligned} \tag{428}$$

Continuing this procedure, one can construct approximate conservation laws for the remaining approximate symmetries (421).

4.2 *Perturbed* KdV *Equation*

Let us consider again the KdV equation (73)

$$u_t = u_{xxx} + uu_x$$

and the following perturbed equation

$$F \equiv u_t - u_{xxx} - uu_x - \varepsilon u = 0. \tag{429}$$

We will follow the procedure described in Sect. 4.1.

4.2.1 Approximately Adjoint Equation

Let us write the formal Lagrangian for Eq. (429) in the form

$$\mathcal{L} = v\,[-u_t + u_{xxx} + uu_x + \varepsilon u]\,. \tag{430}$$

Then

$$\frac{\delta\mathcal{L}}{\delta u} = v_t - v_{xxx} - D_x(uv) + vu_x + \varepsilon v = v_t - v_{xxx} - uv_x + \varepsilon v.$$

Hence, the approximately adjoint equation to equation (429) has the form

$$F^* \equiv v_t - v_{xxx} - uv_x + \varepsilon v = 0. \tag{431}$$

4.2.2 Approximate Self-Adjointness

As mentioned in Sect. 1.3.1, Example 11, the KdV Eq. (73) is nonlinearly self-adjoint with the substitution (75)

$$v = A_1 + A_2 u + A_3(x + tu). \tag{75}$$

Therefore in the case of the perturbed Eq. (429) we look for the substitution

$$v = \phi(t, x, u) + \varepsilon\psi(t, x, u)$$

satisfying the nonlinear self-adjointness condition

$$F^*\big|_{v=\phi+\varepsilon\psi} = \lambda F \tag{432}$$

in the first-order of precision in ε, in the following form

$$v = A_1 + A_2 u + A_3(x + tu) + \varepsilon\psi(t, x, u). \tag{433}$$

When we substitute the expression (433) in the definition (431) of F^*, the terms without ε in Eq. (432) disappear by construction of the substitution (75) and give $\lambda = A_2 + A_3 t$. Then we write Eq. (432), rearranging the terms, in the form

$$\begin{aligned}&\varepsilon\psi_u[u_t - u_{xxx} - uu_x] - 3\varepsilon u_{xx}[u_x\psi_{uu} + \psi_{xu}]\\ &\quad - \varepsilon u_x[u_x^2\psi_{uuu} + 3u_x\psi_{xuu} + 3\psi_{xxu}] + \varepsilon[\psi_t - \psi_{xxx} - u\psi_x\\ &\quad + A_1 + A_2 u + A_3(x + tu)] = -\varepsilon(A_2 + A_3 t)u.\end{aligned} \tag{434}$$

In view Eq. (429), the first term in the first line of Eq. (434) is written $\varepsilon^2 u\psi_u$. Hence, this term vanishes in our approximation. The terms with u_{xx} in the first line of Eq. (434) yield

$$\psi_{uu} = 0, \qquad \psi_{xu} = 0$$

whence

$$\psi = f(t)u + g(t, x).$$

The third bracket in the first line of Eq. (434) vanishes, and Eq. (434) becomes

$$[f'(t) - g_x(t, x)]u + g_t(t, x) - g_{xxx}(t, x) + 2[A_2 + A_3 t]u + A_1 + A_3 x = 0.$$

After rather simple calculations we solve this equation and obtain

$$g(t, x) = A_4 - A_1 t + (A_5 + 2A_2 - A_3 t)x, \qquad f(t) = A_6 + A_5 t - \frac{3}{2} A_3 t^2.$$

We conclude that the perturbed KdV equation (429) is approximately self-adjoint. The approximate substitution (433) has the following form

$$v \approx A_1 + A_2 u + A_3(x + tu) \tag{435}$$
$$+ \varepsilon\left[\left(A_6 + A_5 t - \frac{3}{2} A_3 t^2\right) u + A_4 - A_1 t + (A_5 + 2A_2 - A_3 t)x\right].$$

4.2.3 Approximate Symmetries

Recall that the Lie algebra of point symmetries of the KdV equation (73) is spanned by the following operators

$$X_1 = \frac{\partial}{\partial t}, \qquad X_2 = \frac{\partial}{\partial x}, \qquad X_3 = t\frac{\partial}{\partial x} - \frac{\partial}{\partial u}$$
$$X_4 = 3t\frac{\partial}{\partial t} + x\frac{\partial}{\partial x} - 2u\frac{\partial}{\partial u} \cdot \tag{436}$$

Following the method for calculating approximate symmetries and using the terminology presented in [15–17], vol. 3, Chap. 2, we can prove that all symmetries (436) are stable. Namely the perturbed equation (429) inherits the symmetries (436) of the KdV equation in the form of the following approximate symmetries

$$X_1 = \frac{\partial}{\partial t}, \quad X_2 = \frac{\partial}{\partial x}, \quad X_3 = t\frac{\partial}{\partial x} - \frac{\partial}{\partial u} + \varepsilon\left(\frac{1}{2}t^2\frac{\partial}{\partial x} - t\frac{\partial}{\partial u}\right)$$
$$X_4 = 3t\frac{\partial}{\partial t} + x\frac{\partial}{\partial x} - 2u\frac{\partial}{\partial u} - \varepsilon\left[\frac{9}{2}t^2\frac{\partial}{\partial t} + 3tx\frac{\partial}{\partial x} - (6tu + 3x)\frac{\partial}{\partial u}\right]. \quad (437)$$

4.2.4 Approximate Conservation Laws

We can construct now the approximate conservation laws

$$\left[D_t(C^1) + D_x(C^2)\right]_{(429)} \approx 0 \quad (438)$$

for the perturbed KdV equation (429) using its approximate symmetries (437), the general formula (258) and the approximate substitution (435). Inserting in (258) the formal Lagrangian (430) we obtain

$$C^1 = -Wv, \qquad C^2 = W\,[uv + v_{xx}] - v_x D_x(W) + vD_x^2(W). \quad (439)$$

I will calculate here the conserved vector (439) for the operator X_4 from (437). In this case we have

$$W = -2u - 3tu_t - xu_x + \varepsilon\left(6tu + 3x + \frac{9}{2}t^2u_t + 3txu_x\right). \quad (440)$$

We further simplify the calculations by taking the particular substitution (435) with $A_2 = 1,\ A_1 = A_3 = \cdots = A_6 = 0$. Then

$$v = u + 2\varepsilon x. \quad (441)$$

Substituting (440) and (441) in the first component of the vector (439) and then eliminating u_t via Eq. (429) we obtain

$$C^1 \approx (2u + 3tu_t + xu_x)(u + 2\varepsilon x) - \varepsilon\left(6tu + 3x + \frac{9}{2}t^2u_t + 3txu_x\right)u$$
$$= 2u^2 + 3tuu_{xxx} + 3tu^2u_x + xuu_x + \varepsilon\left(xu + 6txu_{xxx} + 3txuu_x\right.$$
$$\left. + 2x^2u_x - 3tu^2 - \frac{9}{2}t^2uu_{xxx} - \frac{9}{2}t^2u^2u_x\right).$$

Upon singling out the total derivatives in x, it is written

$$C^1 \approx \frac{3}{2}u^2 - 3\varepsilon\left(xu + \frac{3}{2}tu^2\right) + D_x\left[\frac{1}{2}xu^2 + tu^3 - \frac{3}{2}tu_x^2 + 3tuu_{xx}\right. \quad (442)$$
$$\left. + \varepsilon\left(2x^2u + \frac{3}{2}txu^2 - \frac{3}{2}t^2u^3 - 6tu_x + 6txu_{xx} + \frac{9}{4}t^2u_x^2 - \frac{9}{2}t^2uu_{xx}\right)\right].$$

Then we substitute (440) and (441) in the second component of the vector (423), transfer the term $D_x(\ldots)$ from C^1 to C^2, multiply the resulting vector (C^1, C^2) by 2/3 and arrive at the following vector

$$C^1 = u^2 - 2\varepsilon\Big[xu + \frac{3}{2}tu^2\Big]$$
$$C^2 = u_x^2 - \frac{2}{3}u^3 - 2uu_{xx} + \varepsilon\Big[xu^2 - 2u_x + 2xu_{xx} + 2tu^3 - 3tu_x^2 + 6tuu_{xx}\Big]. \tag{443}$$

The approximate conservation law (438) for the vector (443) is satisfied in the following form

$$D_t(C^1) + D_x(C^2) = 2u(u_t - u_{xxx} - uu_x - \varepsilon u)$$
$$- 2\varepsilon(x + 3tu)(u_t - u_{xxx} - uu_x) + o(\varepsilon).$$

Acknowledgments I am cordially grateful to my wife Raisa for reading the manuscript carefully. Without her help and questions the work would be less palatable and contain essentially more misprints. My thanks are due to Sergey Svirshchevskii for a discussion of the material presented in Sect. 2.1. I acknowledge a financial support of the Government of Russian Federation through Resolution # 220, Agreement # 11.G34.31.0042.

References

1. Ibragimov, N.: A new conservation theorem. J. Math. Anal. Appl. **333**, 311–328 (2007). doi:10.1016/j.jmaa.2006.10.078
2. Ibragimov, N.: Integrating factors, adjoint equations and lagrangians. J. Math. Anal. Appl. **318**, 742–757 (2006). doi:10.1016/j.jmaa.2005.11.012
3. Ibragimov, N.: Quasi-self-adjoint differential equations. Arch. ALGA **4**, 55–60 (2007)
4. Ibragimov, N.: Elementary lie Group Analysis and Ordinary Differential Equations. Wiley, New York (1999)
5. Atherton, R., Homsy, G.: On the existence and formulation of variational principles for nonlinear differential equations. Stuides Appl. Math. **54**, 31–60 (1975)
6. Ibragimov N., Torrisi M., Tracina R.: Quasi self-adjoint nonlinear wave equations. J. Phys. A: Math. Theor. **43** (2010). doi:10.1088/1751-8113/43/44/442001
7. Fornberg, B., Whitham, G.: A numerical and theoretical study of certain nonlinear wave phenomena. Philos. Trans. R. Soc. **289**, 373–404 (1978)
8. Rosen, G.: Nonlinear heat conduction in solid H_2. Phys. Rev. **19**, 2398–2399 (1979)
9. Bluman, G., Kumei, S.: On the remarkable nonlinear diffusion equation $\frac{\partial}{\partial x}\left[a(u+b)^{-2}\frac{\partial u}{\partial x}\right] - \frac{\partial u}{\partial t} = 0$. J. Math. Phys. **21**, 1019–1023 (1980)
10. Ibragimov, N.: Transformation Groups in Mathematical Physics. Nauka, Moscow (1983) (English trans.: Transformation Groups Applied to Mathematical Physics. Reidel, Dordrecht (1985))
11. Rogers, C.: Application of a reciprocal transformation to a two phase stefan problem. J. Phys. A: Math. Gen. **18**, L105–L109 (1985)
12. Rogers, C.: On a class of moving boundary value problems in nonlinear heat conduction: application of a Bäcklund transformation. Int. J. Nonlinear Mech. **21**, 249–256 (1986)
13. Ibragimov, N.: Nonlinear self-adjointness in constructing conservation laws, Arch. ALGA **7**(8), 1–99 (2010–2011)

14. Bluman, G., Cheviakov, A., Anco, S.: Applications of Symmetry Methods to Partial Differential Equations. Springer, New York (2010)
15. Ibragimov N. (ed.): CRC handbook of lie group analysis of differential equations. vol. 1 1994, vol. 2 1995, vol. 3 1996, CRC Press, Boca Raton
16. Ibragimov, N. (ed.): CRC Handbook of Lie Group Analysis of Differential Equations, vol. 2. CRC Press, Boca Raton (1995)
17. Ibragimov, N. (ed.): CRC Handbook of Lie Group Analysis of Differential Equations, vol. 3. CRC Press, Boca Raton (1996)
18. Kompaneets A.: The establishment of thermal equilibrium between quanta and electrons, Zh. Eksp. Teor. Fiz **31**, 876–885 (1956). English translations. Soviet Phys. JETP **4**, 730–885 (1957)
19. Weymann, R.: Diffusion approximation for a photon gas interacting with a plasma via the compton effect. Phys. Fluids **8**, 2112–2114 (1965)
20. Dreicer, H.: Kinetic theory of an electron-photon gas. Phys. Fluids **7**, 735–753 (1964)
21. Syunyaev, R.: Induced compton scattering by thermal electrons and low-frequency spectrum of radio sourses. Astronomicheski Zhurnal **48** 244–252 (1971) (English transl., Soviet Astronomy - AJ **15** 190–196 (1971))
22. Weymann, R.: The energy spectrum of radiation in the expanding universe. Astrophys. J. **145**, 560–571 (1966)
23. Chluba, J., Sunyaev, R.: Evolution of low-frequency features in the CMB spectrum due to stimulated compton scattering and doppler broadening. Astron. Astrophys. **488**, 861–865 (2008)
24. Akhatov, I., Gazizov, R., Ibragimov, N.: Nonlocal symmetries: heuristic approach, Itogi Nauki i Tekhniki. Sovremennie Probl. Matematiki **34** (1989) 3–84, English translations. J. Soviet Math. **55**, 1401–1450 (1991)
25. Ibragimov, N., Torrisi, M., Valenti, A.: Preliminary group classification of equations $v_{tt} = f(x, v_x)v_{xx} + g(x, v_x)$. J. Math. Phys. **32**, 2988–2995 (1991). Reprinted in: N.H. Ibragimov, Selected Work, vol. II, ALGA Publications, Karlskrona 2006, Paper 7
26. Tsyganov, M., Biktashev, V., Brindley, J., Holden, A., Ivanitsky G.: Waves in systems with cross-diffusion as a new class of nonlinear waves. Uspekhi Fizicheskikh Nauk **177** 275–300 (2007) (English transl., Physics - Uspekhi **50** 263–286 (2007))
27. Krichever, I., Novikov, S.: Curves, holomorphic bundles over algebraic, equations, nonlinear, Nauk, Uspekhi Matematicheskikh **35**, 47–68. English translations. Russian Math. Surveys **35**, 53–80 (1980)
28. Galiakberova, L., Ibragimov, N.: Nonlinear self-adjointness of the Krichever-Novikov equation. Commun. Nonlinear Sci. Numer. Simul. **19**, 361–363 (2013)
29. Noether, E.: Invariante Variationsprobleme, Königliche Gesellschaft der Wissenschaften zu Göttingen, Nachrichten. Mathematisch-Physikalische Klasse, Heft 2: (1918) 235–257. English transl. Transp. Theory Stat. Phy. **1**, 186–207 (1971)
30. Ibragimov, N.: Invariant variational problems and conservation laws. Teoreticheskaya i Matematicheskaya Fizika **1**, 350–359 (1969) (English trans.: Theor. Math. Phys. **1** 267–276 (1969). Reprinted in: N.H. Ibragimov, Selected Works, vol. I, ALGA Publications, Karlskrona 2006, Paper 8)
31. Ibragimov, N.: The Noether Identity (In Russian), Continuum Dynamics, vol. 38, pp. 26–32. Institute of Hydrodynamics, USSR Academy Science, Siberian Branch, Novosibirsk (1979)
32. Rosen, J.: Some Properties of the Euler-Lagrange Operators, Preprint TAUP-269-72. Tel-Aviv University, Tel-Aviv (1972)
33. Courant, R., Hilbert, D.: Methods of Mathematical Physics, vol. I. Wiley, New York (1989)
34. Stepanov, V.: A Course of Differential Equations (In Russian), 7th Edn. State Publisher of Phys.- Math. Literature, Moscow (1958)
35. Ibragimov, N.: Classical and new results on integrating factors. Arch. ALGA **5**, 63–120 (2008)
36. Dennemeyer, R.: Introduction to Partial Differential Equations and Boundary Value Problems. McGraw-Hill Book Company, New York (1986)
37. Anco, S., Bluman, G.: Direct construction of conservation laws from field equations. Phys. Rev. Lett. **78**, 2869–2873 (1997)

38. Schäfer, T., Wayne, C.: Propagation of ultra-short optical pulses in cubic nonlinear media. Phys. D **196**, 90–105 (2004)
39. Sakovich, A., Sakovich, S.: Solitary wave solutions of the short pulse equation. J. Phys A: Math. Gen. **39**, L361–367 (2006)
40. Sakovich, A., Sakovich, S.: The short pulse equation is integrable. J. Phys. Soc. Jpn. **74**, 239–241 (2005)
41. Ibragimov, N., Rogers, C.: On infinitesimal reciprocal-type transformations in gasdynamics. Lie group connections and nonlinear self-adjointness. Ufa Math. J. **4**, 191–202 (2012)
42. Laplace, P.: Traité de Méchanique Céleste. Paris: vol. 1, 1798. Reprinted in P. S. Laplace, Ouvres complétes, vol. I, Gauthier-Villars, Paris 1878; English trans.: New York 1966
43. Terentyev, E., Shmyglevskii, J.: A complete system of equations in divergence form for the dynamics of an ideal gas. Zh. Vychisl. Mat. i Mat. Fiz. **15** 1535–1544 (1975) (English transl., USSR Comput. Math. and Math. Phys. **15** (1975) 167–176)
44. Ibragimov, N.: Conservation Laws in Hydrodynamics, Dokl. Akad. Nauk SSSR **210**, 1307–1309 (1973). English trans.: Soviet Physics Dokl. 18 1973/74. Original unabridged version see in: N.H. Ibragimov, Selected Works, vol. I, ALGA Publications, Karlskrona 2006, Paper 12
45. Dorodnitsyn V. and Svirshchevskii S., On Lie-Bäcklund Groups Admitted by the Heat Equation with a Source, Preprint 101, Inst. Applied Math., USSR Acad. Sci., Moscow 1983.
46. Steinberg, S., Wolf, K.: Symmetry, conserved quantities and moments in diffusive equations. J. Math. Anal. Appl. **80**, 36–45 (1981)
47. Barcilon, V., Richter, F.: Nonlinear waves in compacting media. J. Fluid Mech. **164**, 429–448 (1986)
48. Harris, S.: Conservation laws for a nonlinear wave equation. Nonlinearity **9**, 187–208 (1996)
49. Maluleke, G., Mason, D.: Derivation of conservation laws for a nonlinear wave equation modelling melt migration using lie point symmetry generators. Commun. Nonlinear Sci. Numer. Simul. **12**, 423–433 (2007)
50. Scott, D., Stevenson, D.: Magma solitons. Geophys. Res. Lett. **11**, 1161–1164 (1984)
51. Scott, D., Stevenson, D.: Mechanics of fluid-rock systems. Ann. Rev. Fluid Mech. **23**, 305–339 (1991)
52. Khamitova, R.: Symmetries and nonlocal conservation laws of the general magma equation. Commun. Nonlinear Sci. Numer. Simul. **14**, 3754–3769 (2009). doi:10.1016/j.cnsns.2008.08.009
53. Ibragimov, N., Ibragimov, R.N.: Group Analysis of Nonlinear Internal Waves in Oceans I: Lagrangian, Conservation laws, Invariant Solutions. Arch. ALGA **6**, 19–44 (2009). arXiv: math-ph 1108–1877v1, 28 pp. 2011

Symbolic Computation of Nonlocal Symmetries and Nonlocal Conservation Laws of Partial Differential Equations Using the GeM Package for Maple

Alexei F. Cheviakov

Abstract The use of the symbolic software package `GeM` for `Maple` is illustrated with examples of computation of nonlocal symmetries and nonlocal conservation laws of nonlinear partial differential equations. In the considered examples, the nonlocal symmetries and conservation laws arise as local symmetries and conservation laws of potential systems. Full `Maple` code with detailed comments is presented. Examples of automated symmetry and conservation law classification are included.

1 Introduction

The majority of contemporary mathematical models involving partial and ordinary differential equations (PDE, ODE) are essentially nonlinear. The analysis of such models often proceeds using approximate, numerical, and/or problem-specific methods. In particular, the efficiency and precision of numerical solutions is commonly restricted by nonlinear effects, which limit mesh sizes and boost computation times, as well as by extra large data structures arising in discretizations of multi-dimensional problems.

Methods based on the framework of symmetry and conservation law analysis can be systematically applied to wide classes of PDE and ODE models. This research area, pioneered by Sophus Lie and Emmy Noether, has been recently developed in various directions, having become a set of interrelated methods that can provide essential analytical information about the underlying equations. For further details, an interested reader is referred to [6, 9, 10, 15, 26, 33].

For ODEs, seeking conservation laws is equivalent to seeking integrating factors; conserved quantities (first integrals) lead to the reduction of order. Conser-

A. F. Cheviakov (✉)
Department of Mathematics and Statistics, University of Saskatchewan,
Saskatoon, SK S7N 5E6, Canada
e-mail: cheviakov@math.usask.ca

J.-F. Ganghoffer and I. Mladenov (eds.), *Similarity and Symmetry Methods*, Lecture Notes in Applied and Computational Mechanics 73, DOI: 10.1007/978-3-319-08296-7_3,

vation laws (divergence forms) of governing PDEs yield local densities conserved by the process, as well as global conserved quantities under appropriate boundary conditions. Local conservation laws of PDEs are also used in existence, uniqueness and stability analysis (e.g., [4, 27, 31]). An important application area of local conservation laws of PDE systems is numerical simulation. Many modern numerical methods, such as finite volume, finite element, discontinuous Galerkin methods, etc., (see, e.g., [25, 28, 29]) rely on the divergence forms of the given equations.

Local symmetries of ODEs lead to the reduction of order, and can be used for the construction of particular symmetry-invariant solutions (see, e.g., [7, 12]). Depending on the structure of the symmetry Lie algebra, the knowledge of an r-parameter Lie group of point symmetries of an ODE can lead to the reduction of order by up to r.

One of the most important applications of local symmetries to nonlinear PDEs is the construction of exact solutions. This includes obtaining new solutions from known ones through the symmetry mapping, and the construction of symmetry-invariant solutions, in particular, physically important traveling wave and self-similar solutions. Additional exact solutions can be obtained using nonlocal symmetries, when they are known. Multiple examples can be found in [10] and references therein.

If a PDE system has an infinite set of local symmetries and/or local conservation laws involving arbitrary functions, it can sometimes be mapped into a linear PDE system by an invertible transformation [3, 13]. Similarly, infinite families of nonlocal symmetries and/or nonlocal conservation laws admitted by a PDE system may be used to construct respective non-invertible mappings [10, 14]. An infinite countable set of local conservation laws may be associated with integrability.

An important application of local conservation laws is the construction of potential systems, nonlocally related to a given one, through the introduction of nonlocal potential variables. Other types of nonlocally related systems, in particular, nonlocally related subsystems, can also arise. The resulting framework of nonlocally related PDE systems [8, 10, 11] has been successfully used in multiple applications, yielding nonlocal symmetries and conservation laws, nonlocal linearizations, and new classes of exact solutions of various PDE systems (see, e.g., [10] and references therein).

The systematic computation of symmetries and conservation laws of PDE systems, especially symmetry and conservation law *classifications* and case splitting for systems involving arbitrary functions or constant parameters, may present a significant computational challenge. Indeed, systems of symmetry and conservation law determining equations can involve thousands of linear PDEs. Symbolic computation software is routinely used to carry out such computations. A number of symbolic software packages have been written for local symmetry and conservation law computations in various computer algebra systems. In the current paper, the use of `GeM` package for `Maple`, developed by the author, is discussed ([22–24]). The current version 32.02 of the `GeM` package has been tested to work with `Maple` versions 14–18.

The present contribution is devoted to practical aspects of computation of nonlocal symmetries and nonlocal conservation laws of nonlinear PDEs. After the general introduction and definitions of Sect. 2, in Sect. 3, we present basic detailed examples of the use of GeM package to compute nonlocal symmetries and nonlocal conservation laws of nonlinear PDEs through the local symmetry and conservation law computations applied to potential systems. In particular, a nonlocal symmetry for a specific nonlinear wave equation is derived; nonlocal symmetries of a class of nonlinear telegraph equations are classified; nonlocal conservation laws are sought for a class of diffusion-convection equations.

The paper is concluded with Sect. 4 containing a discussion and further remarks.

2 Nonlocal Symmetries and Nonlocal Conservation Laws

Consider a system $\mathbf{R}\{x;\, u\}$ of N differential equations of order k, with n independent variables $x = (x^1, \ldots, x^n)$ and m dependent variables $u(x) = (u^1(x), \ldots, u^m(x))$, given by

$$R^\sigma[u] \equiv R^\sigma(x, u, \partial u, \ldots, \partial^k u) = 0, \qquad \sigma = 1, \ldots, N. \tag{1}$$

Here and below, the notation $f[u]$ denotes a differential function depending on x, u and the derivatives of u up to some finite order,

$$\partial u \equiv \partial^1 u = \left(u^1_1(x), \ldots, u^1_n(x), \ldots, u^m_1(x), \ldots, u^m_n(x)\right)$$

denotes the set of all first-order partial derivatives, and

$$\begin{aligned}\partial^p u &= \left\{u^\mu_{i_1\ldots i_p}\,;\;\; \mu = 1, \ldots, m;\;\; i_1, \ldots, i_p = 1, \ldots, n\right\}\\ &= \left\{\frac{\partial^p u^\mu(x)}{\partial x^{i_1}\ldots \partial x^{i_p}}\,;\; \mu = 1, \ldots, m; i_1, \ldots, i_p = 1, \ldots, n\right\}\end{aligned}$$

denote higher-order derivatives. Summation in any pair of repeated indices is assumed below. Subscripts are used to denote partial derivatives: $u_x \equiv \partial u/\partial x$, etc.

2.1 Lie Point Symmetries

Consider a one-parameter Lie group of point transformations

$$\begin{aligned}(x^*)^i &= f^i(x, u; \varepsilon) = x^i + \varepsilon\xi^i(x, u) + O(\varepsilon^2), \qquad i = 1, \ldots, n\\ (u^*)^\mu &= g^\mu(x, u; \varepsilon) = u^\mu + \varepsilon\eta^\mu(x, u) + O(\varepsilon^2), \qquad \mu = 1, \ldots, m\end{aligned} \tag{2}$$

with the corresponding infinitesimal generator

$$\mathrm{X} = \xi^i(x,u)\frac{\partial}{\partial x^i} + \eta^\mu(x,u)\frac{\partial}{\partial u^\mu}. \tag{3}$$

Definition 1 The one-parameter Lie group of point transformations (2) *leaves the* DE *system* (1) *invariant* if it maps any family of solution surfaces $u = u(x)$ of the DE system (1) into another family of solution surfaces $u^* = u^*(x^*)$ of DE system (1). In this case, the transformation (2) are referred to as a *point symmetry* of the DE system (1).

The Lie's algorithm for finding the point symmetries of a DE system (1) written in a solved form in terms of a set of leading derivatives is based on the following theorem (for details, see, e.g., [10, 15, 33]).

Theorem 1 *Let* (3) *be the infinitesimal generator of a one-parameter Lie group of point transformations* (2)*, and* $\mathrm{X}^{(k)}$ *its* kth *extension. Then the transformation* (2) *is a point symmetry of the* DE *system* (1) *if and only if for each* $\alpha = 1, \ldots, N$

$$\mathrm{X}^{(k)} R^\alpha(x, u, \partial u, \ldots, \partial^k u) = 0 \tag{4}$$

when

$$R^\sigma(x, u, \partial u, \ldots, \partial^k u) = 0, \qquad \sigma = 1, \ldots, N. \tag{5}$$

In (4), it is assumed that (5) and the differential consequences of (5) are taken into account.

Other types of local symmetries, including contact and higher-order symmetries, can be computed in a similar manner, when they exist. For such extensions, the symmetry components may depend on derivatives (e.g., [10, 15, 33]).

Remark 1 It is important to mention that some PDE systems have an infinite number of local symmetries, with symmetry components involving arbitrary functions of one or more variables. In particular, linear PDEs always admit "trivial" symmetries

$$\mathrm{X}_\infty = \sigma^\mu \frac{\partial}{\partial u^\mu}$$

where a set of functions $\sigma(x) = (\sigma^1(x), \ldots, \sigma^m(x))$ is an arbitrary solution of the homogeneous version of the given linear equations.

Conversely, if a given PDE system has a sufficiently large infinite set of local symmetries, it can be mapped into a linear system with a point transformation. For details on necessary and sufficient conditions for the existence of such mappings, see [5, 10, 13].

In practical symmetry computations for linear PDEs, the presence of the "trivial" infinite-dimensional symmetry groups poses certain difficulties; for details and techniques of such computations, see [24].

2.2 Local Conservation Laws

Definition 2 *A local divergence-type conservation law* of a PDE system (1) is a divergence expression of the form

$$\mathrm{D}_i\Phi^i(x, u, \partial u, \ldots, \partial^r u) = 0 \tag{6}$$

in terms of total derivative operators

$$\mathrm{D}_i = \frac{\partial}{\partial x^i} + u_i^{\mu}\frac{\partial}{\partial u^{\mu}} + u_{ii_1}^{\mu}\frac{\partial}{\partial u_{i_1}^{\mu}} + u_{ii_1i_2}^{\mu}\frac{\partial}{\partial u_{i_1i_2}^{\mu}} + \cdots \tag{7}$$

holding on solutions of (1).

In the 1+1-dimensional situation, with $x = (x, t)$, the conservation law (6) has the form

$$\mathrm{D}_t\Theta + \mathrm{D}_x\Psi = 0 \tag{8}$$

where the density Θ and the spatial flux Ψ can depend on independent and dependent variables of the given equations, as well as their derivatives.

Remark 2 In practice, one is interested in finding sets of non-trivial, non-equivalent, linearly independent conservation laws. A trivial conservation law of a normal PDE system is a divergence expression that vanishes identically, or if its density and fluxes vanish on solutions of the given PDE system. For further details, see, e.g., [10, 33].

Local conservation laws (6), (8) are systematically sought by applying the *direct conservation law construction method* [2]. The method consists in finding sets of multipliers $\{\Lambda_\sigma[U]\}_{\sigma=1}^N = \{\Lambda_\sigma(x, U, \partial U, \ldots, \partial^\ell U)\}_{\sigma=1}^N$, depending on some prescribed independent and dependent variables and possibly their derivatives to some finite order ℓ, which, taken in linear combinations with the given PDEs, yield a divergence expression

$$\Lambda_\sigma[U]R^\sigma[U] \equiv \mathrm{D}_i\Phi^i[U] \tag{9}$$

holding for arbitrary functions U. Then on solutions $U = u(x)$ of the PDE system (1), one has a local conservation law

$$\Lambda_\sigma[u]R^\sigma[u] = \mathrm{D}_i\Phi^i[u] = 0. \tag{10}$$

Determining equations for the multipliers are obtained from the fact that an expression $F(U)$ is annihilated by Euler operators

$$\begin{aligned} \mathrm{E}_{U^j} &= \frac{\partial}{\partial U^j} - \mathrm{D}_i\frac{\partial}{\partial U_i^j} + \cdots + (-1)^s\mathrm{D}_{i_1}\ldots\mathrm{D}_{i_s}\frac{\partial}{\partial U_{i_1\ldots i_s}^j} + \cdots \\ i, i_q &= 1, \ldots, n, \qquad j = 1, \ldots, m \end{aligned} \tag{11}$$

if and only if $F(U)$ is a divergence expression (e.g., [10, 33]). Hence the local conservation law multiplier determining equations are given by

$$\mathrm{E}_{U^j}\big(\Lambda_\sigma[U]R^\sigma[U]\big) = 0, \qquad j = 1, \ldots, m. \tag{12}$$

After the linear equations (12) are solved for the multipliers $\{\Lambda_\sigma[U]\}_{\sigma=1}^{N}$, the conservation law fluxes and/or density is calculated using (9) (see, e.g., [23]).

Remark 3 Some PDE systems admit an infinite number of independent local conservation laws. In such cases, multipliers may involve arbitrary functions of one or several variables. This happens for both nonlinear and linear PDEs. In particular, linear PDEs always admit an infinite number of conservation laws; the corresponding conservation law multipliers are solutions of a linear system of PDEs adjoint to the given linear system (see, e.g., [5, 10]).

When a given PDE system admits a sufficiently large infinite set of local conservation laws, it can be mapped into a linear system with a point transformation, see [3, 10].

Remark 4 Local variational symmetries and local conservation laws of self-adjoint (variational) PDEs are related through the Noether's theorem. For non-variational PDE systems, this relation generally does not hold. The direct conservation law construction method described above is applicable to both variational and non-variational PDE systems [9, 10, 17, 33].

2.3 Nonlocally Related PDE Systems

Consider a PDE system $\mathbf{R}\{x, t; u\}$ with two independent variables (x, t) and m dependent variables $u = (u^1, \ldots, u^m)$ given by

$$R^\sigma[u] = R^\sigma(x, t, u, \partial u, \partial^2 u, \ldots, \partial^l u) = 0, \quad \sigma = 1, \ldots, s. \tag{13}$$

Suppose that the PDE system (13) has one or more nontrivial conservation laws (8). For each such conservation law, one can introduce a potential variable v satisfying

$$v_x = \Theta[u], \qquad v_t = -\Psi[u]. \tag{14}$$

The potential variable v is a *nonlocal variable* of the PDE system (13), i.e., v cannot be expressed as a local function of the variables in the PDE system (13) and their derivatives [10].

A *potential system* is obtained by appending one or more sets of potential equations (14) to the given PDE system (13). We denote a potential system involving q potential variables by $\mathbf{S}\{x, t; u, v\}$, $v = (v^1, \ldots, v^q)$.

Remark 5 In the case of PDE systems involving $n \geq 3$ independent variables, the application of divergence-type local conservation laws to the construction of potential systems is less straightforward. In particular, the corresponding potential system is underdetermined. Overdetermined potential systems cannot yield nonlocal symmetries [1], but can yield nonlocal conservation laws [1, 21]. The gauge freedom may be eliminated using a gauge constraint, however, finding an"optimal" gauge for a specific PDE system—conservation law pair remains an open problem.

2.4 Nonlocal Symmetries and Nonlocal Conservation Laws

Consider a given PDE system $\mathbf{R}\{x, t; u\}$ (13) and its potential system $\mathbf{S}\{x, t; u, v\}$ involving a single potential variable: (13), (14). Point symmetries of the potential system $\mathbf{S}\{x, t; u, v\}$ are given by infinitesimal generators

$$\mathrm{X} = \xi^x(x, t, u, v)\frac{\partial}{\partial x} + \xi^t(x, t, u, v)\frac{\partial}{\partial t} + \sum_{i=1}^{m} \eta^{u^i}(x, t, u, v)\frac{\partial}{\partial u^i} + \eta^v(x, t, u, v)\frac{\partial}{\partial v}. \tag{15}$$

Definition 3 A generator (15) corresponds to a *nonlocal symmetry* of the given PDE system $\mathbf{R}\{x, t; u\}$ (13) if it does not yield a local symmetry of (13) when projected on the space of its variables.

The criterion for the symmetry (15) to be a nonlocal symmetry of the system $\mathbf{R}\{x, t; u\}$ (13) is provided by the following theorem (e.g, [10, 15, 16]).

Theorem 2 *The point symmetry* (15) *of the potential system* $\mathbf{S}\{x, t; u, v\}$ *yields a nonlocal symmetry* (potential symmetry) *of the given PDE system* (13) *if and only if one or more of the infinitesimals* ($\xi^x(x, t, u, v)$, $\xi^t(x, t, u, v)$, $\eta^{u^1}(x, t, u, v)$, $\ldots$, $\eta^{u^m}(x, t, u, v)$) *depend explicitly on the potential variable* v, *i.e.,*

$$\left(\frac{\partial \xi^x}{\partial v}\right)^2 + \left(\frac{\partial \xi^t}{\partial v}\right)^2 + \sum_{i=1}^{m}\left(\frac{\partial \eta^{u^i}}{\partial v}\right)^2 > 0.$$

Remark 6 Nonlocal symmetries can also arise from nonlocally related subsystems obtained by differential exclusions of dependent variables, and from other PDE systems in the trees of nonlocally related PDE systems. For details, examples, and applications, see [10] and references therein.

Now consider a local conservation law

$$\mathrm{D}_t \Theta[u, v] + \mathrm{D}_x \Psi[u, v] = 0 \tag{16}$$

of the potential system $\mathbf{S}\{x, t; u, v\}$.

Definition 4 A nontrivial local conservation law (16) of the potential system $\mathbf{S}\{x, t; u, v\}$ is called a *nonlocal conservation law of the given PDE system* $\mathbf{R}\{x, t; u\}$ if it is not equivalent to any linear combination of local conservation laws of $\mathbf{R}\{x, t; u\}$ and trivial conservation laws, i.e., the flux and/or density in (16) have an essential dependence on the components of the potential variable v.

The following fundamental theorem [11, 32] holds.

Theorem 3 *Each conservation law of any potential system* $\mathbf{S}\{x, t; u, v\}$*, arising from multipliers that do not essentially depend on the potential variable* v*, is equivalent to a local conservation law of the given system* $\mathbf{R}\{x, t; u\}$ (13).

It follows that in order to construct nonlocal conservation laws of the original system using the direct method, one must consider multipliers that essentially involve potential variable(s). A similar theorem holds for equations with three or more independent variables [10].

The procedure of construction of *an extended tree of nonlocally related* PDE *systems*, starting from a given PDE system (13), is presented in [8, 10, 11]. It is based on the systematic construction, or a given system (13) on local conservation laws, potential systems, further local and nonlocal conservation laws, further potential systems, subsystems, and so on. Similar constructs in multi-dimensions are discussed in [10, 20, 21].

Remark 7 It is important to note that in practice, nonlocal symmetries and nonlocal conservation laws usually arise in classifications when given systems involve arbitrary (constitutive) functions or constant parameters, for special cases of those constitutive functions/parameters. Many examples of such classifications can be found in [10].

Remark 8 Similarly to local symmetries and conservation laws, infinite sets of nonlocal symmetries and conservation laws can lead to a linearization by a nonlocal transformation (e.g., [10, 14]). For example, this is the case for all 1+1-dimensional nonlinear wave equations $u_{tt} = (c^2(u)u_x)_x$, whose basic potential system is linearizable by a hodograph transformation, and for the specific instances of the nonlinear telegraph equation considered in Sect. 3.2 below.

3 Symbolic Computations of Nonlocal Symmetries and Nonlocal Conservation Laws

3.1 Example 1: Local and Nonlocal Symmetry Analysis of a Nonlinear Wave Equation

Consider a nonlinear wave equation on $u = u(x, t)$, denoted by $\mathbf{R}\{x, t; u\}$

$$u_{tt} = (c^2(u)u_x)_x. \tag{17}$$

For simplicity, we restrict to a specific case

$$c^2(u) = \frac{1}{u^2+1}. \tag{18}$$

for which nonlocal symmetries are known to arise (see, e.g., [10], Sec. 4.2.2).
(A) Point symmetries. We start from point symmetry analysis of the wave equation (17) with (18). The command sequence for the GeM package, version 32.02, and the output proceeds as follows.

All variables are cleared GeM package is initialized with the command

```
restart; read("d:/gem32_02.mpl"):
```

Independent and dependent variables are put together, for convenience, by commands

```
ind:=x,t; all_dep:=U(ind);
```

In the absence of arbitrary constants and/or functions in the given equation, the variables are declared as follows:

```
gem_decl_vars(indeps=[ind], deps=[all_dep]);
```

The wave speed is defined by

```
c(U(ind)):=1/(U(ind)^2+1);
```

and the given PDE is further defined, in the solved form, as follows:

```
gem_decl_eqs([diff(U(ind),t,t)=diff(c(U(ind))^2*diff(U(ind),x),x)],
          solve_for=[diff(U(ind),t,t)]);
```

The split system of linear symmetry determining equations, where the symmetry components depend on all independent and dependent variables, is generated by the function

```
det_eqs:=gem_symm_det_eqs([ind, all_dep]):
```

yielding 16 determining equations. A variable containing the unknown symmetry components will be needed for the further computations. It is initialized using the function

```
sym_components:=gem_symm_components();
```

The output value is

sym_components := [xi_t(x, t, U), xi_x(x, t, U), eta_U(x, t, U)]

where the three quantities correspond to the symmetry components for t, x, u respectively.

The simplification and differential elimination of redundant determining equations is obtained by calling the Maple rifsimp routine, as follows

```
symm_det_eqs:=DEtools[rifsimp](det_eqs, sym_components, mindim=1);
```

The mindim=1 option will force rifsimp to output the dimension of the solution space, i.e., the number of independent point symmetries of the PDE (17). In this case, the returned dimension is three. The system of 16 determining equations is reduced to seven equations and is stored in symm_det_eqs[Solved]. The final solution is performed using the standard Maple pdsolve routine,

```
symm_sol:=pdsolve(symm_det_eqs[Solved],sym_components);
```

returning

*symm_sol:=eta_U(x, t, U) = 0, xi_t(x, t, U) = _C1*t+_C3, xi_x(x, t, U)=_C1*x+_C2*

This final solution involves three arbitrary constants *_C1, _C2, _C3*, which agrees with the dimension of the solution space returned by rifsimp.

Finally, the three independent symmetry generators are output using the command

```
gem_output_symm(symm_sol);
```

which yields the canonical forms of the point symmetries of the PDE (17) with the wave speed (18)

$$\mathrm{X}_1 = \frac{\partial}{\partial t}, \qquad \mathrm{X}_2 = \frac{\partial}{\partial x}, \qquad \mathrm{X}_3 = t\frac{\partial}{\partial t} + x\frac{\partial}{\partial x}.$$

The equation is thus invariant under $t-$ and $x-$ translations and a scaling.

(B) A nonlocal symmetry computation. For an arbitrary $c(u)$, the PDE (17) has four zeroth-order conservation laws with multipliers $\Lambda = 1, t, xt, x$. In this example, we use the second one. The conservation law is given by

$$\mathrm{D}_t(tu_t - u) - \mathrm{D}_x(tc^2(u)u_x) = 0$$

and the resulting potential system $\mathbf{S}\{x, t; u, w\}$ is

$$w_x = tu_t - u, \quad w_t = tc^2(u)u_x. \tag{19}$$

We study local symmetries of (19) to seek nonlocal symmetries of the PDE (17) with the wave speed (18). The program proceeds in a fashion similar to the above example.

```
restart; read("d:/gem32_02.mpl"):
ind:=x,t; all_dep:=U(ind),W(ind);
gem_decl_vars(indeps=[ind], deps=[all_dep]);
c(U(ind)):=1/(U(ind)^2+1);
```

```
gem_decl_eqs([diff(W(ind),x)=t*diff(U(ind),t)-U(ind),
             diff(W(ind),t)=t*c(U(ind))^2*diff(U(ind),x)],
             solve_for=[diff(W(ind),x),diff(W(ind),t)]);

det_eqs:=gem_symm_det_eqs([ind, all_dep]):
sym_components:=gem_symm_components();
symm_det_eqs:=DEtools[rifsimp](det_eqs, sym_components, mindim=1);
symm_sol:=pdsolve(symm_det_eqs[Solved],sym_components);
gem_output_symm(symm_sol);
```

Here the dimension of the solution space is four and the output contains four symmetry generators

$$Y_1 = \frac{\partial}{\partial w}, \qquad Y_2 = \frac{\partial}{\partial x}, \qquad Y_3 = t\frac{\partial}{\partial t} + x\frac{\partial}{\partial x} + w\frac{\partial}{\partial w}$$

$$Y_4 = tu\frac{\partial}{\partial t} + w\frac{\partial}{\partial x} + (u^2+1)\frac{\partial}{\partial w} - x\frac{\partial}{\partial w}.$$

The symmetry Y_4 yields a nonlocal symmetry of the given nonlinear wave equation (17), since the component $\xi^x = w$ involves the potential variable (cf. Definition 3).

3.2 Example 2: A Potential Symmetry Classification for the Nonlinear Telegraph Equation

Let $\mathbf{R}\{x, t; u\}$ denote the nonlinear telegraph equation with the unknown function $u = u(x, t)$, given by

$$u_{tt} = (F(u)u_x)_x + (G(u))_x. \tag{20}$$

The complete point symmetry classification of the PDE (20) with respect to the constitutive functions $F(u)$ and $G(u)$ can be found in Ref. [30].

The PDE (20) is a conservation law as it stands, hence one can introduce a potential $v(x, t)$ to obtain a potential system

$$u_t = v_x, \qquad v_t = F(u)u_x + G(u). \tag{21}$$

The point symmetry classification of the PDE system (21) has been performed in [18]. In particular, it has been shown that in the cases $F(u) = u^{-2}$, $G(u) = u^{-1}$, and $F(u)$ arbitrary, $G(u) = \text{const}$, the potential system (21) has an infinite number of point symmetries (nonlocal symmetries of the PDE (20)), and moreover, is linearizable by a point transformation. (Thus the corresponding NLT equations (20) are linearizable by a nonlocal transformation).

The first equation of the system (21) is also a conservation law. Introducing a second potential w accordingly, one has a potential system $\mathbf{S}\{x, t; u, v, w\}$ for three dependent variables $u(x, t)$, $v(x, t)$, $w(x, t)$, given by three PDEs

$$w_t = v, \qquad w_x = u, \qquad v_t = F(u)u_x + G(u). \tag{22}$$

In our computations, for the simplicity of presentation, we will be avoiding detailed calculations for the linearization cases, since they yield infinite sets of symmetries.

For the current example, we are interested in finding point symmetries of the potential system $\mathbf{S}\{x, t; u, v, w\}$ (22) that correspond to nonlocal symmetries of the original PDE (20). For brevity, we will restrict ourselves to the case of power nonlinearities,

$$F(u) = u^{\alpha}, \quad G(u) = u^{\beta}. \tag{23}$$

The classification will thus be performed with respect to two constitutive parameters $\alpha \neq 0$, $\beta \neq 0$. (For the complete classification, see [10], Sect. 4.2.)

Point symmetry generators of the potential system $\mathbf{S}\{x, t; u, v, w\}$ (22) are of the form

$$\begin{aligned} \mathrm{Z} = {} & \xi(x, t, u, v, w)\frac{\partial}{\partial x} + \tau(x, t, u, v, w)\frac{\partial}{\partial t} + \eta^u(x, t, u, v, w)\frac{\partial}{\partial u} \\ & + \eta^v(x, t, u, v, w)\frac{\partial}{\partial v} + \eta^w(x, t, u, v, w)\frac{\partial}{\partial w}. \end{aligned} \tag{24}$$

In order to find symmetries (24) that correspond to nonlocal symmetries of the given PDE (20), one requires that at least one of the following six conditions is satisfied

$$\begin{aligned} &\frac{\partial \xi^t}{\partial v} \neq 0, \quad \frac{\partial \xi^t}{\partial w} \neq 0, \quad \frac{\partial \xi^x}{\partial v} \neq 0 \\ &\frac{\partial \xi^x}{\partial w} \neq 0, \quad \frac{\partial \eta^u}{\partial v} \neq 0, \quad \frac{\partial \eta^u}{\partial w} \neq 0. \end{aligned} \tag{25}$$

The `Maple` code for the symmetry classification proceeds as follows.

```
restart; read("d:/gem32_02.mpl"):

ind:=x,t; all_dep:=U(ind),V(ind),W(ind);

gem_decl_vars(indeps=[ind], deps=[all_dep],
        freeconst=[],freefunc=[F(U(ind)),G(U(ind))]);
```

Here $F(u)$ and $G(u)$ are defined as arbitrary functions. It is important to do so even though we are going to consider only power nonlinearities. The reason is that in order to do the case splitting and simplification of the determining equations, the `Maple` function `rifsimp` is used; the latter can only handle polynomial nonlinearities. Hence the code will proceed as follows

- Generate symmetry determining equations treating the nonlinear functions as arbitrary (free) at the initial stage.
- In order to use the desired form of the "arbitrary functions", employ the `Maple` `dpolyform` function. This function converts a specified condition into the differential polynomial form. An analogous operation can be performed by hand. E.g., if we need to have $H(u) = Ae^{ku}$, $A, k = \text{const}$, the linear ODE and conditions on $H(u)$ can be

$$\frac{dH(u)}{du} = kH(u), \qquad k \neq 0, \qquad H(u) \neq 0.$$

- When case splitting with `rifsimp` is performed, the system of determining equations should be appended with the conditions defining the arbitrary functions.

The same approach should be used for any nonlinearities, including logarithms, exponents, trigonometric functions, etc.

In our case, the conditions of $F(u)$ and $G(u)$ being power nonlinearities (23) can be generated as follows. (We note that in the determining equations in `GeM`, dependent variables of the given equations are treated as simple variables, not functions; hence in the determining equations, `F(U)` not `F(U(ind))` should be used.)

```
cond_FG_powers:={F(U)=U^n, G(U)=U^m};

cond_F_G:=PDEtools[dpolyform](cond_FG_powers,no_Fn);

cond_F_G_full:=convert(cond_F_G,list)[1][], F(U)<>0, G(U)<>0,
                                              m<>0, n<>0;
```

The resulting set of conditions is given by a `Maple` set-type variable

*cond_F_G_full := {diff(F(U), U) = F(U)*n/U, diff(G(U), U) = G(U)*m/U,*

m<>0, n<>0, F(U) <> 0, G(U) <> 0 }

The equations are declared as follows.

```
gem_decl_eqs([diff(W(x,t),t)=V(x,t),
             diff(W(x,t),x)=U(x,t),
             diff(V(x,t),t)=F(U(x,t))*diff(U(x,t),x) +G(U(x,t))],
             solve_for=[diff(W(x,t),t), diff(W(x,t),x),
             diff(V(x,t),t)]);
```

Then the symmetry determining equations are generated and a variable of symmetry components is initialized

```
det_eqs:=gem_symm_det_eqs([ind, all_dep]):
```

```
sym_components:=gem_symm_components();
```

The next step is to unite the determining equations and the conditions `cond_F_G_full` on the functions $F(u)$ and $G(u)$ to be power nonlinearities. A set union is used:

```
det_eqs:=det_eqs union cond_F_G_full:
```

We now perform six rounds the symmetry classification and case splitting, using, one by one, the conditions (25) for the symmetry to be essentially nonlocal. For the classification, it is essential to use the `casesplit` option in the call to `rifsimp`.

Round 1: $\partial\xi^t/\partial v \neq 0$.

```
symm_det_eqs:=DEtools[rifsimp](det_eqs
                union diff{xi_t(x, t, U, V, W),V)<>0},
                sym_components, mindim=1, casesplit);
```

The result contains only one case, with $m = -1, n = -2$, with the solution space of $dimension = \infty$. As remarked above, we will not go into detailed computations for this linearization case.

Round 2: $\partial\xi^t/\partial w \neq 0$.

```
symm_det_eqs:=DEtools[rifsimp](det_eqs
                union {diff(xi_t(x, t, U, V, W),W)<>0},
                sym_components, mindim=1, casesplit);
```

This computation yields two cases: the linearization case $m = -1, n = -2$ and another case $m = 3, n = 2$. (A case tree with pivots may be plotted using the command `DEtools[caseplot](symm_det_eqs,pivots);`).

Let us compute all symmetries for the case $m = 3, n = 2$. The solution space dimension (number of linearly independent symmetries) is equal to six. To obtain a general symmetry generator, one uses the following commands.

```
symm_sol:=pdsolve(
        subs({m=3,n=2,F(U)=A*U^2,G(U)=B*U^3}
        symm_det_eqs[1][Solved]), sym_components);
```

The six symmetries can be output separately using

```
gem_output_symm(symm_sol);
```

The resulting set of symmetries is given by

$$Z_1 = \frac{\partial}{\partial x}, \qquad Z_2 = \frac{\partial}{\partial t}, \qquad Z_3 = \frac{\partial}{\partial w}, \qquad Z_4 = t\frac{\partial}{\partial w} + \frac{\partial}{\partial v} + w\frac{\partial}{\partial w}$$

$$Z_5 = t\frac{\partial}{\partial t} - u\frac{\partial}{\partial u} - 2v\frac{\partial}{\partial v} - w\frac{\partial}{\partial w}$$

$$Z_6 = (Au + Bw)\frac{\partial}{\partial t} + Av\frac{\partial}{\partial x} - Buv\frac{\partial}{\partial u} - Bv^2\frac{\partial}{\partial v} + Auv\frac{\partial}{\partial w}.$$

The symmetry Z_6 yields the nonlocal symmetry of the original NLT equation (20).
Round 3: $\partial\xi^x/\partial v \neq 0$.

```
symm_det_eqs:=DEtools[rifsimp](det_eqs
              union {diff(xi_x(x, t, U, V, W),V)<>0},
              sym_components, mindim=1, casesplit);
```

This computation results in the same two cases: $m = -1$, $n = -2$ and $m = 3$, $n = 2$, no new symmetries arise.
Round 4: $\partial\xi^x/\partial w \neq 0$.

```
symm_det_eqs:=DEtools[rifsimp](det_eqs
              union {diff(xi_x(x, t, U,V,W),W)<>0},
              sym_components, mindim=1, casesplit);
```

The output of the above command is

symm_det_eqs := *table*([*status* = "system is inconsistent"])

which means that there are no point symmetries (24) of the potential system $\mathbf{S}\{x, t; u, v, w\}$ (22) that satisfy $\partial\xi^x/\partial w \neq 0$.
Round 5: $\partial\eta^u/\partial v \neq 0$.

```
symm_det_eqs:=DEtools[rifsimp](det_eqs
              union {diff(eta_U(x, t, U, V, W),V)<>0},
              sym_components, mindim=1, casesplit);
```

The computation again yields the same two cases: $m = -1$, $n = -2$ and $m = 3$, $n = 2$.
Round 6: $\partial\eta^u/\partial w \neq 0$.

```
symm_det_eqs:=DEtools[rifsimp](det_eqs
              union {diff(eta_U(x, t, U, V, W),W)<>0},
              sym_components, mindim=1, casesplit);
```

The result of this computations is the same as for Round 4 above: no point symmetries of the potential system $\mathbf{S}\{x, t; u, v, w\}$ (22) satisfying $\partial\eta^u/\partial w \neq 0$ exist.

3.3 Example 3: A Nonlocal Conservation Law Classification for the Nonlinear Telegraph Equation

Consider a class of diffusion-convection equations $\mathbf{R}\{x, t; u\}$ of the form

$$u_t = (A(u)u_x)_x + (B(u))_x \tag{26}$$

where $A(u)$ and $B(u)$ are arbitrary constitutive functions, and $A(u) \neq 0$. The break linear case $A = 1$, $B =$ const is excluded. The complete classification of linearly independent local conservation laws for (26) yields the following results [34].

1. For arbitrary $A(u)$, $B(u)$, the only local conservation law of (26) is given by

$$\mathrm{D}_t(u) - \mathrm{D}_x\Big(A(u)u_x + B(u)\Big) = 0. \tag{27}$$

2. For arbitrary $A(u)$, and $B(u) = 0$, there are two local conservation laws of (26).
3. For arbitrary $A(u)$, and $B(u) = A(u)$, the PDE (26) has four local conservation laws.

We employ the conservation law (27) to construct the potential system $\mathbf{S}\{x, t; u, v\}$

$$v_x = u, \qquad v_t = A(u)u_x + B(u). \tag{28}$$

We wish to perform the local conservation law classification of the potential system (28) and find conservation laws that yield nonlocal conservation laws of the given PDE (26).

Here we restrict to zeroth-order multipliers $\Lambda_1(x, t, U, V)$, $\Lambda_2(x, t, U, V)$. Moreover, for simplicity of computation, we specify

$$A(u) := u^4$$

and perform the nonlocal conservation law classification with respect to the remaining arbitrary function $B(u)$. The full `Maple` program for the computation is given below.

First, the package is initialized, and variables and the free function $B(u)$ are declared.

```
restart; read("d:/gem32_02.mpl"):
ind:=x,t; all_dep:=U(ind),V(ind);
gem_decl_vars(indeps=[ind], deps=[all_dep],freefunc=[B(U(ind))]);
```

Second, the function $A(u)$ is specialized, and the PDEs (28) are declared.

```
A(U(ind)):= (U(ind)^4);
```

```
gem_decl_eqs([diff(V(ind),x)=U(ind),
             diff(V(ind),t)=A(U(ind))*diff(U(ind),x) +B(U(ind))],
             solve_for=[diff(V(ind),x),diff(V(ind),t)]);
```

The conservation law determining equations are obtained, in the split form, by calling the functon

```
det_eqs:=gem_conslaw_det_eqs([ind, all_dep]):
```

where the list in `[...]` determines the dependence of the multipliers. This yields 6 determining equations.

The multiplier variables are accessed by calling

```
CL_multipliers:=gem_conslaw_multipliers();
```

Further, we perform the case splitting, assuming that $B(u) \neq 0$, and seeking conservation laws where at least one multiplier essentially involves the potential v (cf. Theorem 3).

Round 1: $\partial\Lambda_1/\partial V \neq 0$. Here one has

```
simplified_eqs:=DEtools[rifsimp]( {det_eqs[]} union {B(U)<>0}
               union {diff(Lambda1(x, t, U, V),V)<>0},
               CL_multipliers, casesplit, mindim=1);
```

The only nontrivial case returned has dimension one, i.e., it yields a single conservation law of the required type. The condition on the function $B(u)$ within `simplified_eqs[Solved]` is

diff(B(U), U, U, U, U) = (6(diff(B(U), U, U, U))*U−12*(diff(B(U), U, U)))/U^2*

Using `dsolve`, one readily finds that $B(U)$ must have the form

$$B(U) = M_1U^6 + M_2U^5 + M_3U + M_4, \qquad M_1, \ldots, M_4 = \text{const}.$$

For the subsequent computations, it is more straightforward to initialize the function $B(U)$ to the above expression, and then again perform the simplification of determining equations

```
B(U):=M1*U^6+M2*U^5+M3*U+M4;
```

```
simplified_eqs:=DEtools[rifsimp](det_eqs, CL_multipliers, mindim=1);
```

One then solves for the multipliers

```
multipliers_sol:=pdsolve(simplified_eqs[Solved], CL_multipliers);
```

to obtain

multipliers_sol :=
*{Lambda1(x, t, U, V) = -_C1*exp(−(5*(M1*M4-M2*M3))*t)*exp(5*M1*V)*
**exp(5*M2*x)*(M1*U^5+M3)*
*Lambda2(x, t, U, V) =_C1*exp(-(5*(M1*M4-M2*M3))*t)*
**exp(5*M1*V)*exp(5*M2*x)}*

Here $\_C1$ is the only arbitrary constant, so indeed, one conservation law is obtained. In the regular form

$$\begin{aligned} \Lambda_1(x,t,U,V) &= -C_1(M_1U^5+M_3)\mathrm{e}^{5(M_2x+M_1V+(M_2M_3-M_1M_4)t)} \\ \Lambda_2(x,t,U,V) &= C_1\mathrm{e}^{5(M_2x+M_1V+(M_2M_3-M_1M_4)t)}. \end{aligned} \tag{29}$$

Finally, the conservation law density and flux are computed using the function

```
gem_get_CL_fluxes(multipliers_sol);
```

(Other flux computation methods are available; see [10, 23].)

The newly computed conservation law is given by

$$\begin{aligned} &D_t\mathrm{e}^{5(M_2x+M_1v+(M_2M_3-M_1M_4)t)} \\ &\quad -D_x\left((M_1u^5+M_3)\mathrm{e}^{5(M_2x+M_1v+(M_2M_3-M_1M_4)t)}\right)=0. \end{aligned} \tag{30}$$

This is a nonlocal conservation law of the diffusion-convection PDE (26) since it is not equivalent to any of its local conservation laws [10, 34].

Round 2: $\partial\Lambda_2/\partial V \neq 0$. For this case, one obtains exactly the same result, i.e., a single pair of potential-dependent multipliers (29). Indeed, both multipliers there have a similar exponential dependence on v.

4 Discussion

The symbolic software package GeM for Maple, in conjunction with standard Maple routines like rifsimp and dsolve/pdsolve, offers convenient ways to compute symmetries and conservation laws of systems of differential equations, and importantly, perform symmetry and conservation law classifications with respect to arbitrary functions and parameters.

Such computations can be applied to potential systems, as shown in Sect. 3.1 where local symmetry analysis of a potential system for a nonlinear wave equation was used to compute a nonlocal symmetry of that equation.

It is rather straightforward, by appending extra conditions, to restrict the computations to seek specifically nonlocal symmetries or new (local or nonlocal) conservation laws arising as local ones for the potential system. In Sect. 3.2, the nonlinear telegraph equation (20) and its potential system (22) involving two potential variables were considered. By using the conditions (25) in symbolic symmetry computations for the potential system, the two cases of power nonlinearities $m = -1, n = -2$ and $m = 3, n = 2$ were isolated, for which nonlocal symmetries of the original equation arise.

In symmetry and conservation law computations in `Maple` that use `rifsimp`, nonlinearities have to be restricted to differential polynomial ones. Hence if a given DE system contains nonlinearities of other types, one has to recast the nonlinear functions into "arbitrary" functions that satisfy additional linear or differential polynomial equations. An example for power nonlinearities was presented in Sect. 3.2.

In cases when a PDE system under consideration has an infinite set of symmetries and/or conservation laws, the `rifsimp` routine returns *dimension* $= \infty$ for each such case. It is important to exercise care in the analysis of determining equations in the `rifsimp` output in such situations. In particular, `Maple` `pdsolve` may return incomplete results.

For PDE systems involving $n \geq 3$ independent variables, computations generally proceed the same way. Many specific aspects of construction of nonlocally related systems and nonlocal symmetry and conservation law computations were discussed in [10, 20, 21].

In terms of the further development of symbolic software for nonlocal and local symmetry/conservation law analysis and classification, the future research directions will naturally include the detailed development of more examples, and the automation of case-by-case consideration in the classifications.

Acknowledgments The author is thankful to NSERC for the research support.

References

1. Anco, S.C., Bluman, G.W.: Nonlocal symmetries and nonlocal conservation laws of Maxwell's equations. J. Math. Phys. **38**, 3508–3532 (1997)
2. Anco, S.C., Bluman, G.W.: Direct construction of conservation laws from field equations. Phys. Rev. Lett. **78**, 2869–2873 (1997)
3. Anco, S.C., Bluman, G.W., Wolf, T.: Invertible mappings of nonlinear PDEs to linear PDEs through admitted conservation laws. Acta Appl. Math. **101**, 21–38 (2008)
4. Benjamin, T.B.: The stability of solitary waves. Proc. Roy. Soc. Lond. A **328**, 153–183 (1972)
5. Bluman, G.W.: Use of group methods for relating linear and nonlinear partial differential equations. In: Proceedings of Symposium on Symmetry, Similarity and Group Theoretic Methods in Mechanics, pp. 203–218, Calgary AB (1974)
6. Bluman, G.: Connections Between Symmetries and Conservation Laws, Symm. Integr. & Geom.: Meth. Appl. (SIGMA) 1 (2005) paper 011, 16 pages
7. Bluman, G.W., Anco, S.C.: Symmetry and Integration Methods for Differential Equations for Differential Equations, Applied Mathematical Sciences. Springer, New York (2005)
8. Bluman, G., Cheviakov, A.F.: Framework for potential systems and nonlocal symmetries: algorithmic approach. J. Math. Phys. **46**, 123506 (2005)
9. Bluman, G., Cheviakov, A.F., Anco, S.: Construction of conservation laws: how the direct method generalizes Noether's theorem. In: Proceedings of 4th Workshop "Group Analysis of Differential Equations & Integrability, vol. 1, pp. 1–23 (2009)

10. Bluman, G.W., Cheviakov, A.F., Anco, S.C.: Applications of Symmetry Methods to Partial Differential Equations. Applied Mathematical Sciences. Springer, New York (2010)
11. Bluman, G., Cheviakov, A.F., Ivanova, N.M.: Framework for nonlocally related PDE systems and nonlocal symmetries: extension, simplification, and examples. J. Math. Phys. **47**, 113505 (2006)
12. Bluman, G.W., Cheviakov, A.F., Senthilvelan, M.: Solution and asymptotic/blow-up behaviour of a class of nonlinear dissipative systems. J. Math. Anal. Appl. **339**, 1199–1209 (2008)
13. Bluman, G.W., Kumei, S.: Symmetry-based algorithms to relate partial differential equations: I. local symmetries. Eur. J. Appl. Math. **1**, 189–216 (1990)
14. Bluman, G.W., Kumei, S.: Symmetry-based algorithms to relate partial differential equations: II. linearization by nonlocal symmetries. Eur. J. Appl. Math. **1**, 217–223 (1990)
15. Bluman, G.W., Kumei, S.: Symmetries and Differential Equations, Applied Mathematical Sciences. Springer, New York (1989)
16. Bluman, G.W., Kumei, S., Reid, G.J.: Classes, new, of symmetries for partial differential equations. J. Math. Phys. **29**, 806–811 (Erratum. J. Math. Phys. **29**(1988), 2320 (1988))
17. Bluman, G.W., Temuerchaolu.: Comparing symmetries and conservation laws of nonlinear telegraph equations. J. Math. Phys. **46**, 073513 (2005)
18. Bluman, G.W., Temuerchaolu, Sahadevan, R.: Local and nonlocal symmetries for nonlinear telegraph equations. J. Math. Phys. **46**, 023505 (2005)
19. Cheviakov, A.F.: Comp. Phys. Comm. **176**(1), 48–61 (2007). (In the current paper, we used a new version of `"GeM"` software, which is scheduled for public release in 2009. See http://math.usask.ca/cheviakov/gem/)
20. Cheviakov, A.F., Bluman, G.W.: Multidimensional partial differential equation systems: generating new systems via conservation laws, potentials, gauges, subsystems. J. Math. Phys. **51**, 103521 (2010)
21. Cheviakov, A.F., Bluman, G.W.: Multidimensional partial differential equation systems: nonlocal symmetries, nonlocal conservation laws, exact solutions. J. Math. Phys. **51**, 103522 (2010)
22. Cheviakov, A.F.: GeM software package for computation of symmetries and conservation laws of differential equations. Comput. Phys. Commun. **176**, 48–61 (2007)
23. Cheviakov, A.F.: Computation of fluxes of conservation laws. J. Eng. Math. **66**, 153–173 (2010)
24. Cheviakov, A.F.: Symbolic computation of local symmetries of nonlinear and linear partial and ordinary differential equations. Math. Comp. Sci. **4**, 203–222 (2010)
25. Godlewski, E., Raviart, P.-A.: Numerical Approximation of Hyperbolic Systems of Conservation Laws. Springer, Berlin (1996)
26. Hydon, P.E.: SSymmetry Methods for Differential Equations: A Beginner's Guide. Cambridge University Press, Cambridge (2000)
27. Lax, P.D.: Integrals of nonlinear equations of evolution and solitary waves. Comm. Pure Appl. Math. **21**, 467–490 (1968)
28. LeVeque, R.J.: Numerical Methods for Conservation Laws. Birkhäuser, Basel (1992)
29. Kallendorf, C., Cheviakov, A.F., Oberlack, M., Wang, Y.: Conservation laws of surfactant transport equations. Phys. Fluids **24**(10), 102105 (2012)
30. Kingston, G., Sophocleous, C.: Symmetries and form-preserving transformations of one-dimensional wave equations with dissipation. J. Non-Linear Mech. **36**, 987–997 (2001)
31. Knops, R.J., Stuart, C.A.: Quasiconvexity and uniqueness of equilibrium solutions in nonlinear elasticity. Arch. Rat. Mech. Anal. **86**, 234–249 (1984)
32. Kunzinger, M., Popovych, R.O.: Potential conservation laws. J. Math. Phys. **49**, 103506 (2008)
33. Olver, P.J.: Applications of Lie Groups to Differential Equations. Graduate Texts in Mathematics. Springer, New York (1986)
34. Popovych, R.O., Ivanova, N.M.: Hierarchy of conservation laws of diffusion-convection equations. J. Math. Phys. **46**, 043502 (2005)
35. Wolf, T.: A comparison of four approaches to the calculation of conservation laws. Europ. J. Appl. Math. **13**(2), 129–152 (2002)
36. Wolf, T.: Crack, LiePDE, ApplySym and ConLaw, section 4.3.5 and computer program on CD-ROM. In: Grabmeier J., Kaltofen E., Weispfenning, V. (eds.) Computer Algebra Handbook. pp. 465–468, Springer (2002)

Symmetries of Hamiltonian Systems on Symplectic and Poisson Manifolds

Charles-Michel Marle

Abstract This text presents some basic notions in symplectic geometry, Poisson geometry, Hamiltonian systems, Lie algebras and Lie groups actions on symplectic or Poisson manifolds, momentum maps and their use for the reduction of Hamiltonian systems. It should be accessible to readers with a general knowledge of basic notions in differential geometry. Full proofs of many results are provided.

1 Introduction

This text presents some basic notions in symplectic geometry, Poisson geometry, Hamiltonian systems, Lie algebras and Lie groups actions on symplectic or Poisson manifolds, momentum maps and their use for the reduction of Hamiltonian systems. It should be accessible to readers with a general knowledge of basic notions in differential geometry. Full proofs of many results are provided.

1.1 Contents of the Paper

Symplectic and Poisson manifolds are defined in Sects. 2 and 3, where their basic properties are given, often with detailed proofs. Darboux theorem and the main results about the local structure of Poisson manifolds, however, are given without proof. Actions of a Lie group or of a Lie algebra on a smooth manifold and, when this manifold is endowed with a symplectic or a Poisson structure, symplectic, Poisson and Hamiltonian actions are introduced in Sect. 4. For Hamiltonian actions of a Lie group on a connected symplectic manifold, the equivariance of the momentum map with respect to an affine action of the group on the dual of its Lie algebra is

C.-M. Marle (✉)
Institut de Mathématiques de Jussieu, Université Pierre et Marie Curie, Paris, France
e-mail: cmm1934@orange.fr

J.-F. Ganghoffer and I. Mladenov (eds.), *Similarity and Symmetry Methods*, Lecture Notes in Applied and Computational Mechanics 73, DOI: 10.1007/978-3-319-08296-7_4,

proven, and the notion of symplectic cocycle is introduced. We prove (4.2.13) that given a Lie algebra symplectic cocycle, there exists on the associated connected and simply connected Lie group a unique corresponding Lie group symplectic cocycle. The Hamiltonian actions of a Lie group on its cotangent bundle obtained by lifting the actions of the group on itself by translations on the left and on the right are fully discussed in Sect. 4.5. We prove that there exists a two-parameter family of deformations of these actions into a pair of mutually symplectically orthogonal Hamiltonian actions whose momentum maps are equivariant with respect to an affine action involving any given Lie group symplectic cocycle (4.5.4). The use of first integrals and, more generally, of momentum maps for the resolution of Hamiltonian dynamical systems, is discussed in Sect. 5. For a system whose Hamiltonian is invariant under a Hamiltonian Lie algebra action, the Marsden–Weinstein reduction procedure can be used: through the use of Noether's theorem, this procedure leads to a reduced symplectic manifold on which a reduced Hamiltonian system can be solved in a first step. Another way of using the symmetries of the system rests on the use of the Euler–Poincaré equation. This equation can be written for classical Lagrangian mechanical systems when there exists a locally transitive Lie algebra action on their configuration space, or for the corresponding Hamiltonian systems when the Lagrangian is hyper-regular. However, the Euler–Poincaré equation does not always lead to a reduction of the system: such a reduction occurs mainly when the Hamiltonian can be expressed as the momentum map composed with a smooth function defined on the dual of the Lie algebra; the Euler–Poincaré equation is then equivalent to the Hamilton equation written on the dual of the Lie algebra. Finally in Sect. 6 three classical examples are considered: the spherical pendulum, the motion of a rigid body around a fixed point and the Kepler problem. For each example the Euler–Poincaré equation is derived (for the Kepler problem a transitive Lie algebra action is obtained by adding the Lie algebra of the group of positive homotheties to the Lie algebra of the group of rotations around the attractive centre), the first integrals linked to symmetries are given. In this Section, the classical concepts of vector calculus on an Euclidean three-dimensional vector space (scalar, vector and mixed products) are used and their interpretation in terms of concepts such as the adjoint or coadjoint action of the group of rotations are explained.

1.2 Further Reading

Of course this text is just an introduction. Several important parts of the theory of Hamiltonian systems are not discussed here, for example completely integrable systems (although the three examples presented belong to that class of systems), action-angle coordinates, monodromy, singular reduction, the Kolmogorov-Arnold-Moser theorem, symplectic methods in Hydrodynamics, ... To extend his knowledge of the subject, the reader can consult the books by Abraham and Marsden [1], Arnold [2],

Arnold and Khesin [3], Iglesias-Zemmour [4], Camille Laurent-Gengoux, Pichereau and Vanhaecke [5], Yvette Kosmann-Schwarzbach (editor) [6] on both the scientific and historical aspects of the development of modern Poisson geometry, Vaisman [7].

1.3 Notations

Our notations are those which today are generally used in differential geometry. For example, the tangent and the cotangent bundles to a smooth n-dimensional manifold M are denoted, respectively, by TM and by T^*M, and their canonical projections on M by $\tau_M : TM \to M$ and by $\pi_M : T^*M \to M$. The space of differential forms of degree p, i.e. the space of smooth sections of $\bigwedge^p(T^*M)$, the p-th exterior power of the cotangent bundle, is denoted by $\Omega^p(M)$. Similarly the space of multivectors of degree p, i.e. the space of smooth sections of $\bigwedge^p(TM)$, the p-th exterior power of the tangent bundle, is denoted by $A^p(M)$. By convention $\Omega^p(M) = A^p(M) = 0$ for $p < 0$ or $p > n$, and $\Omega^0(M) = A^0(M) = C^\infty(M, \mathbb{R})$. The exterior algebras of differential forms and of multivectors are, respectively, $\Omega(M) = \oplus_{p=0}^n \Omega^p(M)$ and $A(M) = \oplus_{p=0}^n A^p(M)$. Their main properties are briefly recalled in Sect. 3.2.

When $f : M \to N$ is a smooth map between two smooth manifolds M and N, the natural lift of f to the tangent bundles is denoted by $Tf : TM \to TN$. The same notation $Tf : \bigwedge^p TM \to \bigwedge^p TN$ is used to denote its natural prolongation to the p-th exterior power of TM. The pull-back by f of a smooth differential form $\alpha \in \Omega(N)$ is denoted by $f^*\alpha$.

When $f : M \to N$ is a smooth diffeomorphism, the push-forward f_*X of a smooth vector field $X \in A^1(M)$ is the vector field $f_*X \in A^1(N)$ defined by

$$f_*X(y) = Tf\Bigl(X\bigl(f^{-1}(y)\bigr)\Bigr), \quad y \in N.$$

Similarly, the pull-back of a smooth vector field $Y \in A^1(N)$ is the vector field $f^*Y \in A^1(M)$ defined by

$$f^*Y(x) = Tf^{-1}\Bigl(Y\bigl(f(x)\bigr)\Bigr), \quad x \in M.$$

The same notation is used for the push-forward of any smooth tensor field on M and the pull-back of any smooth tensor field on N.

2 Symplectic Manifolds

2.1 Definition and Elementary Properties

Definition 2.1.1 A *symplectic form* on a smooth manifold M is a bilinear skew-symmetric differential form ω on that manifold which satisfies the following two properties:

- the form ω is closed; it means that its exterior differential $\mathrm{d}\omega$ vanishes: $\mathrm{d}\omega = 0$;
- the rank of ω is everywhere equal to the dimension of M; it means that for each point $x \in M$ and each vector $v \in T_xM, v \neq 0$, there exists another vector $w \in T_xM$ such that $\omega(x)(v, w) \neq 0$.

Equipped with the symplectic form ω, the manifold M is called a *symplectic manifold* and denoted (M, ω). One says also that ω determines a *symplectic structure* on the manifold M.

2.1.1 Elementary Properties of Symplectic Manifolds

Let (M, ω) be a symplectic manifold.
1. For each $x \in M$ and each $v \in T_xM$ we denote by $\mathrm{i}(v)\omega : T_xM \to \mathbb{R}$ the map $w \mapsto \omega(x)(v, w)$; it is a linear form on the vector space T_xM, in other words an element of the cotangent space T^*_xM. Saying that the rank of ω is everywhere equal to the dimension of M amounts to say that the map $v \mapsto \mathrm{i}(v)\omega$ is an isomorphism of the tangent bundle TM onto the cotangent bundle T^*M.
2. Let V be a finite-dimensional vector space, and $\eta : V \times V \to \mathbb{R}$ be a skew-symmetric bilinear form. As above, $v \mapsto \mathrm{i}(v)\eta$ is a linear map defined on V, with values in its dual space V^*. The *rank* of η is the dimension of the image of that map. An easy result in linear algebra is that the rank of a skew-symmetric bilinear form is always an even integer. When (M, ω) is a symplectic manifold, for each $x \in M$ that result can be applied to the bilinear form $\omega(x) : T_xM \times T_xM \to \mathbb{R}$, and we see that the dimension of M must be an even integer $2n$.
3. The *Darboux theorem*, due to the French mathematician Gaston Darboux (1842–1917), states that in a $2n$-dimensional symplectic manifold (M, ω) any point has a neighbourhood on which there exists local coordinates $(x^1, \dots, x^{2n})$ in which the $(2n) \times (2n)$-matrix (ω_{ij}) $(1 \leq i, j \leq 2n)$ of components of ω is a constant, skew-symmetric invertible matrix. We recall that

$$\omega_{ij} = \omega\left(\frac{\partial}{\partial x^i}, \frac{\partial}{\partial x^j}\right).$$

These local coordinates can even be chosen in such a way that

$$\omega_{ij} = \begin{cases} 1 & \text{if } i - j = n, \\ -1 & \text{if } i - j = -n, \\ 0 & \text{if } |i - j| \neq n, \end{cases} \qquad 1 \leq i, j \leq 2n.$$

Local coordinates which satisfy this property are called *Darboux local coordinates*.
4. On the $2n$-dimensional symplectic manifold (M, ω), the $2n$-form ω^n (the n-th exterior power of ω) is a *volume form* (it means that it is everywhere $\neq 0$). Therefore a symplectic manifold always is *orientable*.

2.2 Examples of Symplectic Manifolds

2.2.1 Surfaces

A smooth orientable surface embedded in an Euclidean 3-dimensional affine space, endowed with the area form determined by the Euclidean metric, is a symplectic manifold.

More generally, any 2-dimensional orientable manifold, equipped with a nowhere vanishing area form, is a symplectic manifold.

2.2.2 Symplectic Vector Spaces

A *symplectic vector space* is a finite-dimensional real vector space E equipped with a skew-symmetric bilinear form $\omega : E \times E \to \mathbb{R}$ of rank equal to the dimension of E; therefore $\dim E$ is an even integer $2n$. Considered as a constant differential two-form on E, η is symplectic, which allows us to consider (E, η) as a symplectic manifold.

The canonical example of a symplectic vector space is the following. Let V be a real n-dimensional vector space and let V^* be its dual space. There exists on the direct sum $V \oplus V^*$ a natural skew-symmetric bilinear form

$$\eta\bigl((x_1, \zeta_1), (x_2, \zeta_2)\bigr) = \langle \zeta_1, x_2 \rangle - \langle \zeta_2, x_1 \rangle .$$

The rank of η being $2n$, $(V \oplus V^*, \eta)$ is a symplectic vector space.

Conversely, any $2n$-dimensional symplectic vector space (E, ω) can be identified with the direct sum of any of its n-dimensional vector subspaces V such that the symplectic form ω vanishes identically on $V \times V$, with its dual space V^*. In this identification, the symplectic form ω on E becomes identified with the above-defined symplectic form η on $V \oplus V^*$.

2.2.3 Cotangent Bundles

Let N be a smooth n-dimensional manifold. With the notations of Sect. 1.3 for the canonical projections of tangent or cotangent bundles onto their base manifold and for prolongation to vectors of a smooth map, we recall that the diagram

$$\begin{array}{ccc} T(T^*N) & \xrightarrow{T\pi_N} & TN \\ \Big\downarrow{\scriptstyle \tau_{T^*N}} & & \Big\downarrow{\scriptstyle \tau_N} \\ T^*N & \xrightarrow{\pi_N} & N \end{array}$$

is commutative. For each $w \in T(T^*N)$, we can therefore write

$$\eta_N(w) = \bigl\langle \tau_{T^*N}(w), T\pi_N(w) \bigr\rangle .$$

This formula defines a differential 1-form η_N on the manifold T^*N, called the *Liouville 1-form*. Its exterior differential $d\eta_N$ is a symplectic form, called the *canonical symplectic form* on the cotangent bundle T^*N.

Let $(x^1, \dots, x^n)$ be a system of local coordinates on N, $(x^1, \dots, x^n, p_1, \dots, p_n)$ be the corresponding system of local coordinates on T^*N. The local expressions of the Liouville form η_N and of its exterior differential $\mathrm{d}\eta_N$ are

$$\eta_N = \sum_{i=1}^{n} p_i \mathrm{d}x^i \,, \quad \mathrm{d}\eta_N = \sum_{i=1}^{n} \mathrm{d}p_i \wedge \mathrm{d}x^i \,.$$

We see that $(x^1, \dots, x^n, p_1, \dots, p_n)$ is a system of Darboux local coordinates. Therefore any symplectic manifold is locally isomorphic to a cotangent bundle.

2.2.4 The Complex Plane

The complex plane $\mathbb{C}$ is naturally endowed with a Hermitian form

$$\eta(z_1, z_2) = z_1 \overline{z_2} \,, \quad z_1 \text{and}\, z_2 \in \mathbb{C} \,,$$

where $\overline{z_2}$ is the conjugate of the complex number z_2. Let us write $z_1 = x_1 + iy_1$, $z_2 = x_2 + iy_2$, where $x_1,\ y_1,\ x_2,\ y_2$ are real, and separate the real and imaginary parts of $\eta(z_1, z_2)$. We get

$$\eta(z_1, z_2) = (x_1 x_2 + y_1 y_2) + i(y_1 x_2 - y_2 x_1).$$

The complex plane $\mathbb{C}$ has an underlying structure of real, 2-dimensional vector space, which can be identified with $\mathbb{R}^2$, each complex number $z = x + iy \in \mathbb{C}$ being identified with $(x, y) \in \mathbb{R}^2$. The real and imaginary parts of the Hermitian form η on $\mathbb{C}$ are, respectively, the Euclidean scalar product g and the symplectic form ω on $\mathbb{R}^2$ such that

$$\begin{aligned}\eta(z_1, z_2) &= (x_1 x_2 + y_1 y_2) + i(y_1 x_2 - y_2 x_1) \\ &= g\big((x_1, y_1), (x_2, y_2)\big) + i\omega\big((x_1, y_1), (x_2, y_2)\big) \,.\end{aligned}$$

2.2.5 Kähler Manifolds

More generally, a n-dimensional *Kähler manifold* (i.e. a complex manifold of complex dimension n endowed with a Hermitian form whose imaginary part is a closed two-form), when considered as a real $2n$-dimensional manifold, is automatically endowed with a Riemannian metric and a symplectic form given, respectively, by the real and the imaginary parts of the Hermitian form.

Conversely, it is not always possible to endow a symplectic manifold with a complex structure and a Hermitian form of which the given symplectic form is the imaginary part. However, it is always possible to define, on a symplectic manifold, an *almost complex structure* and an *almost complex 2-form* with which the properties of the symplectic manifold become similar to those of a Kähler manifold (but with change of chart functions which are not holomorphic functions). This possibility was used by Gromov [8] in his theory of *pseudo-holomorphic curves*.

2.3 Remarkable Submanifolds of a Symplectic Manifold

Definitions 2.3.1 Let (V, ω) be a symplectic vector space, and W be a vector subspace of V. The *symplectic orthogonal* of W is the vector subspace

$$\text{orth}\, W = \{\, v \in V\, ;\ \omega(v, w) = 0 \text{ for all } w \in W\,\}\,.$$

The vector subspace W is said to be

- *isotropic* if $W \subset \text{orth}\, W$,
- *coisotropic* if $W \supset \text{orth}\, w$,
- *Lagrangian* if $W = \text{orth}\, W$,
- *symplectic* if $W \oplus \text{orth}\, W = V$.

2.3.1 Properties of Symplectic Orthogonality

The properties stated below are easily consequences of the above definitions
1. For any vector subspace W of the symplectic vector space (W, ω), we have $\text{orth}(\text{orth}\, W) = W$.
2. Let $\dim V = 2n$. For any vector subspace W of V, we have $\dim(\text{orth}\, W) = \dim V - \dim W = 2n - \dim W$. Therefore, if W is isotropic, $\dim W \leq n$; if W is coisotropic, $\dim W \geq n$; and if W is Lagrangian, $\dim W = n$.
3. Let W be an isotropic vector subspace of V. The restriction to $W \times W$ of the symplectic form ω vanishes identically. Conversely, if W is a vector subspace such that the restriction of ω to $W \times W$ vanishes identically, W is isotropic.
4. A Lagrangian vector subspace of V is an isotropic subspace whose dimension is the highest possible, equal to half the dimension of V.
5. Let W be a symplectic vector subspace of V. Since $W \cap \text{orth}\, W = \{0\}$, the rank of the restriction to $W \times W$ of the form ω is equal to $\dim W$; therefore $\dim W$ is even and, equipped with the restriction of ω, W is a symplectic vector space. Conversely if, when equipped with the restriction of ω, a vector subspace W of V is a symplectic vector space, we have $W \oplus \text{orth}\, W = V$, and W is a symplectic vector subspace of V in the sense of the above definition.
6. A vector subspace W of V is symplectic if and only if $\text{orth}\, W$ is symplectic.

Definitions 2.3.2 Let (M, ω) be a symplectic manifold. For each $x \in M$, $((T_x M, \omega(x))$ is a symplectic vector space. A submanifold N of M is said to be

- *isotropic* if for each $x \in N$, $T_x N$ is an isotropic vector subspace of the symplectic vector space $\big(T_x M, \omega(x)\big)$,
- *coisotropic* if for each $x \in N$, $T_x N$ is a coisotropic vector subspace of $\big(T_x M, \omega(x)\big)$,
- *Lagrangian* if for each $x \in N$, $T_x N$ is a Lagrangian vector subspace of $\big(T_x M, \omega(x)\big)$,
- *symplectic* if for each $x \in N$, $T_x N$ is a symplectic vector subspace of $\big(T_x M, \omega(x)\big)$.

2.4 Hamiltonian Vector Fields on a Symplectic Manifold

Let (M, ω) be a symplectic manifold. We have seen that the map which associates to each vector $v \in TM$ the covector $\mathrm{i}(v)\omega$ is an isomorphism from TM onto T^*M. So, for any given differential one-form α, there exists a unique vector field X such that $\mathrm{i}(X)\omega = \alpha$. We are therefore allowed to state the following definitions.

Definitions 2.4.1 Let (M, ω) be a symplectic manifold and $f : M \to \mathbb{R}$ be a smooth function. The vector field X_f which satisfies

$$\mathrm{i}(X_f)\omega = -\mathrm{d}f$$

is called the *Hamiltonian vector field* associated to f. The function f is called a *Hamiltonian* for the Hamiltonian vector field X_f.

A vector field X on M such that the one-form $\mathrm{i}(X)\omega$ is closed,

$$\mathrm{d}\,\mathrm{i}(X)\omega = 0\,,$$

is said to be *locally Hamiltonian.*

Remarks 2.4.2 The function f is not the unique Hamiltonian of the Hamiltonian vector field X_f: any function g such that $\mathrm{i}(X_f)\omega = -\mathrm{d}g$ is another Hamiltonian for X_f. Given a Hamiltonian f of X_f, a function g is another Hamiltonian for X_f if and only if $\mathrm{d}(f - g) = 0$, or in other words if and only if $f - g$ keeps a constant value on each connected component of M

Of course, a Hamiltonian vector field is locally Hamiltonian. The converse is not true when the cohomology space $H^1(M, \mathbb{R})$ is not trivial.

Proposition 2.4.3 *On a symplectic manifold (M, ω), a vector field X is locally Hamiltonian if and only if the Lie derivative $\mathcal{L}(X)\omega$ of the symplectic form ω with respect to X vanishes:*

$$\mathcal{L}(X)\omega = 0\,.$$

The bracket $[X, Y]$ of two locally Hamiltonian vector fields X and Y is Hamiltonian, and has as a Hamiltonian the function $\omega(X, Y)$.

Proof The well known formula which relates the exterior differential d, the interior product $\mathrm{i}(X)$ and the Lie derivative $\mathcal{L}(X)$ with respect to the vector field X

$$\mathcal{L}(X) = \mathrm{i}(X)\mathrm{d} + \mathrm{d}\mathrm{i}(X)$$

proves that when X is a vector field on a symplectic manifold (M, ω)

$$\mathcal{L}(X)\omega = \mathrm{d}\mathrm{i}(X)\omega\,,$$

since $\mathrm{d}\,\omega = 0$. Therefore $\mathrm{i}(X)\omega$ is closed if and only if $\mathcal{L}(X)\omega = 0$.

Let X and Y be two locally Hamiltonian vector fields. We have

$$\begin{aligned}\mathrm{i}\big([X,Y]\big)\omega &= \mathcal{L}(X)\mathrm{i}(Y)\omega - \mathrm{i}(Y)\mathcal{L}(X)\omega\\ &= \mathcal{L}(X)\mathrm{i}(Y)\omega\\ &= \big(\mathrm{i}(X)\mathrm{d} + \mathrm{d}\,\mathrm{i}(X)\big)\mathrm{i}(Y)\omega\\ &= \mathrm{d}\,\mathrm{i}(X)\mathrm{i}(Y)\omega\\ &= -d\big(\omega(X,Y)\big)\,,\end{aligned}$$

which proves that $\omega(X, Y)$ is a Hamiltonian for $[X, Y]$. □

2.4.1 Expression in a System of Darboux Local Coordinates

Let $(x^1, \ldots, x^{2n})$ be a system of Darboux local coordinates. The symplectic form ω can be locally writen as

$$\omega = \sum_{i=1}^{n} \mathrm{d}\,x^{n+i} \wedge \mathrm{d}\,x^i\,,$$

so we see that the Hamiltonian vector field X_f associated to a smooth function f can be locally written as

$$X_f = \sum_{i=1}^{n} \frac{\partial f}{\partial x^{n+i}}\frac{\partial}{\partial x^i} - \frac{\partial f}{\partial x^i}\frac{\partial}{\partial x^{n+i}}\,.$$

A smooth curve φ drawn in M parametrized by the real variable t is said to be a *solution* of the differential equation determined by X_f, or an *integral curve* of X_f, if it satisfies the equation, called the *Hamilton equation* for the Hamiltonian f,

$$\frac{\mathrm{d}\varphi(t)}{\mathrm{d}t} = X_f\big(\varphi(t)\big)\,.$$

Its local expression in the considered system of Darboux local coordinates is

$$\begin{cases} \dfrac{\mathrm{d}x^i}{\mathrm{d}t} = \dfrac{\partial f}{\partial x^{n+i}}\,, \\ \dfrac{\mathrm{d}x^{n+i}}{\mathrm{d}t} = -\dfrac{\partial f}{\partial x^i}\,, \end{cases} \qquad (1 \le i \le n).$$

Definition 2.4.4 Let $\Phi{:}N \to N$ be a diffeomorphism of a smooth manifold N onto itself. The *canonical lift* of Φ to the cotangent bundle is the transpose of the vector bundles isomorphism $T(\Phi^{-1}) = (T\Phi)^{-1}{:}TN \to TN$. In other words, denoting by $\widehat{\Phi}$ the canonical lift of Φ to the cotangent bundle, we have for all $x \in N$, $\xi \in T^*_xN$ and $v \in T_{\Phi(x)}N$,

$$\bigl\langle \widehat{\Phi}(\xi), v \bigr\rangle = \bigl\langle \xi, (T\Phi)^{-1}(v) \bigr\rangle .$$

Remark 2.4.5 With the notations of Definition 2.4.4, we have $\pi_N \circ \widehat{\Phi} = \Phi \circ \pi_N$.

2.4.2 The Flow of a Vector Field

Let X be a smooth vector field on a smooth manifold M. We recall that the *reduced flow* of X is the map Φ, defined on an open subset Ω of $\mathbb{R} \times M$ and taking its values in M, such that for each $x \in M$ the parametrized curve $t \mapsto \varphi(t) = \Phi(t, x)$ is the maximal integral curve of the differential equation

$$\frac{\mathrm{d}\varphi(t)}{\mathrm{d}t} = X\bigl(\varphi(t)\bigr)$$

which satisfies $\varphi(0) = x$. For each $t \in \mathbb{R}$, the set $D_t = \{x \in M; (t, x) \in \Omega\}$ is an open subset of M and when D_t is not empty the map $x \mapsto \Phi_t(x) = \Phi(t, x)$ is a diffeomorphism of D_t onto D_{-t}.

Definitions 2.4.6 Let N be a smooth manifold, TN and T^*N be its tangent and cotangent bundles, $\tau_N : TN \to N$ and $\pi_N : T^*N \to N$ be their canonical projections. Let X be a smooth vector field on N and $\{\Phi^X_t\,; t \in \mathbb{R}\}$ be its reduced flow.

1. The *canonical lift* of X to the tangent bundle TN is the unique vector field $\overline{X}$ on TM whose reduced flow $\{\Phi^{\overline{X}}_t\,; t \in \mathbb{R}\}$ is the prolongation to vectors of the reduced flow of X. In other words, for each $t \in \mathbb{R}$,

$$\Phi^{\overline{X}}_t = T\Phi^X_t\,,$$

therefore, for each $v \in TN$,

$$\overline{X}(v) = \frac{\mathrm{d}}{\mathrm{d}t}\bigl(T\Phi^X_t(v)\bigr)\bigr|_{t=0}\,.$$

2. The *canonical lift* of X to the cotangent bundle T^*N is the unique vector field $\widehat{X}$ on T^*M whose reduced flow $\{\Phi^{\widehat{X}}_t\,; t \in \mathbb{R}\}$ is the lift to the cotangent bundle of the reduced flow $\{\Phi^X_t\,; t \in \mathbb{R}\}$ of X. In other words, for each $t \in \mathbb{R}$,

$$\Phi_t^{\widehat{X}} = \widehat{\Phi_t^X}\,,$$

therefore, for each $\xi \in T^*N$,

$$\widehat{X}(\xi) = \frac{\mathrm{d}}{\mathrm{d}t}\big(\widehat{\Phi_t^X}(\xi)\big)\,\big|_{t=0}\,.$$

Remark 2.4.7 Let X be a smooth vector field defined on a smooth manifold N. Its canonical lift $\overline{X}$ to the tangent bundle TN (2.4.6) is related to the prolongation to vectors $TX : TN \to T(TN)$ by the formula

$$\overline{X} = \kappa_N \circ TX\,,$$

where $\kappa_N : T(TN) \to T(TN)$ is the canonical involution of the tangent bundle to TN (see [9]).

Proposition 2.4.8 *Let $\Phi : N \to N$ be a diffeomorphism of a smooth manifold N onto itself and $\widehat{\Phi} : T^*N \to T^*N$ the canonical lift of Φ to the cotangent bundle. Let η_N be the Liouville form on T^*N. We have*

$$\widehat{\Phi}^*\eta_N = \eta_N\,.$$

Let X be a smooth vector field on N, and $\widehat{X}$ be the canonical lift of X to the cotangent bundle. We have

$$\mathcal{L}(\widehat{X})(\eta_N) = 0\,.$$

Proof Let $\xi \in T^*N$ and $v \in T_\xi(T^*N)$. We have

$$\widehat{\Phi}^*\eta_N(v) = \eta_N\big(T\widehat{\Phi}(v)\big) = \big\langle \tau_{T^*N} \circ T\widehat{\Phi}(v), T\pi_N \circ T\widehat{\Phi}(v)\big\rangle\,.$$

But $\tau_{T^*N} \circ T\widehat{\Phi} = \widehat{\Phi} \circ \tau_{T^*N}$ and $T\pi_N \circ T\widehat{\Phi} = T(\pi_N \circ \widehat{\Phi}) = T(\Phi \circ \pi_N)$. Therefore

$$\widehat{\Phi}^*\eta_N(v) = \big\langle \widehat{\Phi} \circ \tau_{T^*N}(v), T(\Phi \circ \pi_N)(v)\big\rangle = \big\langle \tau_{T^*N}(v), T\pi_N(v)\big\rangle = \eta_N(v)$$

since $\widehat{\Phi} = (T\Phi^{-1})^T$. Now let X be a smooth vector field on N, $\{\Phi_t^X\,;\, t \in \mathbb{R}\}$ be its reduced flow, and $\widehat{X}$ be the canonical lift of X to the cotangent bundle. We know that the reduced flow of $\widehat{X}$ is $\{\widehat{\Phi_t^X}\,;\, t \in \mathbb{R}\}$, so we can write

$$\mathcal{L}(\widehat{X})\eta_N = \frac{\mathrm{d}}{\mathrm{d}t}\big(\widehat{\Phi_t^X}^*\eta_N\big)\,\big|_{t=0}\,.$$

Since $\widehat{\Phi_t^X}^*\eta_N = \eta_N$ does not depend on t, $\mathcal{L}(\widehat{X})\eta_N = 0$. □

The following Proposition, which presents an important example of Hamiltonian vector field on a cotangent bundle, will be used when we will consider Hamiltonian actions of a Lie group on its cotangent bundle.

Proposition 2.4.9 *Let N be a smooth manifold, T^*N be its cotangent bundle, η_N be the Liouville form and $\mathrm{d}\eta_N$ be the canonical symplectic form on T^*N. Let X be a smooth vector field on N and $f_X : T^*N \to \mathbb{R}$ be the smooth function defined by*

$$f_X(\xi) = \Big\langle \xi, X\big(\pi_N(\xi)\big)\Big\rangle , \quad \xi \in T^*N .$$

On the symplectic manifold $(T^*N, \mathrm{d}\eta_N)$, the vector field $\widehat{X}$, canonical lift to T^*N of the vector field X on N in the sense defined above (2.4.6), is a Hamiltonian field which has the function f_X as a Hamiltonian. In other words

$$\mathrm{i}(\widehat{X})\mathrm{d}\eta_N = -\mathrm{d}f_X .$$

Moreover,

$$f_X = \mathrm{i}(\widehat{X})\eta_N .$$

Proof We have seen (Proposition 2.4.8) that $\mathcal{L}(\widehat{X})\eta_N = 0$. Therefore

$$\mathrm{i}(\widehat{X})\mathrm{d}\eta_N = \mathcal{L}(\widehat{X})\eta_N - \mathrm{d}\mathrm{i}(\widehat{X})\eta_N = -\mathrm{d}\mathrm{i}(\widehat{X})\eta_N ,$$

which proves that $\widehat{X}$ is Hamiltonian and admits $\mathrm{i}(\widehat{X})\eta_N$ as Hamiltonian. For each $\xi \in T^*N$

$$\mathrm{i}(\widehat{X})\eta_N(\xi) = \eta_N(\widehat{X})(\xi) = \Big\langle \xi, T\pi_N\big(\widehat{X}(\xi)\big)\Big\rangle = \Big\langle \xi, X\big(\pi_N(\xi)\big)\Big\rangle = f_X(\xi) .$$ □

2.5 The Poisson Bracket

Definition 2.5.1 The *Poisson bracket* of an ordered pair (f, g) of smooth functions defined on the symplectic manifold (M, ω) is the smooth function $\{f, g\}$ defined by the equivalent formulae

$$\{f, g\} = \mathrm{i}(X_f)\,\mathrm{d}g = -\mathrm{i}(X_g)\,\mathrm{d}f = \omega(X_f, X_g) ,$$

where X_f and X_g are the Hamiltonian vector fields on M with, respectively, the functions f and g as Hamiltonian.

Lemma 2.5.2 *Let (M, ω) be a symplectic manifold, let f and g be two smooth functions on M and let X_f and X_g be the associated Hamiltonian vector fields. The bracket $[X_f, X_g]$ is a Hamiltonian vector field which admits $\{f, g\}$ as Hamiltonian.*

Proof This result is an immediate consequence of Proposition 2.4.3. □

Proposition 2.5.3 *Let (M, ω) be a symplectic manifold. The Poisson bracket is a bilinear composition law on the space $C^\infty(M, \mathbb{R})$ of smooth functions on M, which satisfies the following properties.*

1. *It is skew-symmetric:* $\{g,f\} = -\{f,g\}$.
2. *It satisfies the Leibniz identity with respect to the ordinary product of functions:*

$$\{f, gh\} = \{f, g\}h + g\{f, h\}\,.$$

3. *It satisfies the Jacobi identity, which is a kind of Leibniz identity with respect to the Poisson bracket itself:*

$$\bigl\{f, \{g, h\}\bigr\} = \bigl\{\{f, g\}, h\bigr\} + \bigl\{g, \{f, h\}\bigr\}\,,$$

which can also be written, when the skew-symmetry of the Poisson bracket is taken into account,

$$\bigl\{\{f, g\}, h\bigr\} + \bigl\{\{g, h\}, f\bigr\} + \bigl\{\{h, f\}, g\bigr\} = 0\,.$$

Proof The proofs of Properties (i) and (ii) are very easy and left to the reader. Let us proove Property (iii).

We have

$$\bigl\{\{f, g\}, h\bigr\} = \omega(X_{\{f,g\}}, X_h) = -\mathrm{i}(X_{\{f,g\}})\mathrm{i}(X_h)\omega = \mathrm{i}(X_{\{f,g\}})\mathrm{d}h\,.$$

By Lemma 2.5.2, $X_{\{f,g\}} = [X_f, X_g]$ so we have

$$\bigl\{\{f, g\}, h\bigr\} = \mathrm{i}\bigl([X_f, X_g]\bigr)\mathrm{d}h = \mathcal{L}\bigl([X_f, X_g]\bigr)h\,.$$

We also have

$$\bigl\{\{g, h\}, f\bigr\} = -\mathcal{L}(X_f) \circ \mathcal{L}(X_g)h\,, \quad \bigl\{\{h, f\}, g\bigr\} = \mathcal{L}(X_g) \circ \mathcal{L}(X_f)h\,.$$

Taking the sum of these three terms, and taking into account the identity

$$\mathcal{L}\bigl([X_f, X_g]\bigr) = \mathcal{L}(X_f) \circ \mathcal{L}(X_g) - \mathcal{L}(X_g) \circ \mathcal{L}(X_f)\,,$$

we see that the Jacobi identity is satisfied. □

Remarks 2.5.4 **1.** In a system of Darboux local coordinates $(x^1, \ldots, x^{2n})$, the Poisson bracket can be written

$$\{f, g\} = \sum_{i=1}^{n} \left(\frac{\partial f}{\partial x^{n+i}} \frac{\partial g}{\partial x^i} - \frac{\partial f}{\partial x^i} \frac{\partial g}{\partial x^{n+i}} \right)\,.$$

2. Let H be a smooth function on the symplectic manifold (M, ω), and X_H be the associaled Hamiltonian vector field. By using the Poisson bracket, one can write in a very concise way the Hamilton equation for X_H. Let $t \mapsto \varphi(t)$ be any integral curve of X_H. Then for any smooth function $f : M \to \mathbb{R}$

$$\frac{\mathrm{d}f\bigl(\varphi(t)\bigr)}{\mathrm{d}t} = \{H,f\}\bigl(\varphi(t)\bigr).$$

By successively taking for f the coordinate functions $x^1, \ldots, x^{2n}$ of a system of Darboux local coordinates, we recover the equations

$$\begin{cases} \dfrac{\mathrm{d}x^i}{\mathrm{d}t} = \dfrac{\partial H}{\partial x^{n+i}}, \\ \dfrac{\mathrm{d}x^{n+i}}{\mathrm{d}t} = -\dfrac{\partial H}{\partial x^i}, \end{cases} \qquad (1 \leq i \leq n).$$

3 Poisson Manifolds

3.1 The Inception of Poisson Manifolds

Around the middle of the XX-th century, several scientists felt the need of a frame in which Hamiltonian differential equations could be considered, more general than that of symplectic manifolds. Paul Dirac for example proposed such a frame in his famous 1950 paper *Generalized Hamiltonian dynamics* [10, 11].

In many applications in which, starting from a symplectic manifold, another manifold is built by a combination of processes (products, quotients, restriction to a submanifold, ...), there exists on that manifold a structure, more general than a symplectic structure, with which a vector field can be associated to each smooth function, and the bracket of two smooth functions can be defined. It was also known that on a (odd-dimensional) contact manifold one can define the bracket of two smooth functions.

Several generalizations of symplectic manifolds were defined and investigated by André Lichnerowicz during the years 1975–1980. He gave several names to these generalizations: *canonical, Poisson, Jacobi* and *locally conformally symplectic* manifolds [12, 13].

In 1976 Alexander Kirillov published a paper entitled *Local Lie algebras* [14] in which he determined all the possible structures on a manifold allowing the definition of a bracket with which the space of smooth functions becomes a local Lie algebra. *Local* means that the value taken by the bracket of two smooth functions at each point only depends of the values taken by these functions on an arbitrarily small neighbourhood of that point. The only such structures are those called by Lichnerowicz *Poisson structures, Jacobi structures* and *locally conformally symplectic structures.*

In what follows we will mainly consider Poisson manifolds.

3.2 Definition and Structure of Poisson Manifolds

Definition 3.2.1 A *Poisson structure* on a smooth manifold M is the structure determined by a bilinear, skew-symmetric composition law on the space of smooth functions, called the *Poisson bracket* and denoted by $(f, g) \mapsto \{f, g\}$, satisfying the Leibniz identity

$$\{f, gh\} = \{f, g\}h + g\{f, h\}$$

and the Jacobi identity

$$\bigl\{\{f, g\}, h\bigr\} + \bigl\{\{g, h\}, f\bigr\} + \bigl\{\{h, f\}, g\bigr\} = 0\,.$$

A manifold endowed with a Poisson structure is called a *Poisson manifold.*

Proposition 3.2.2 *On a Poisson manifold M, there exists a unique smooth bivector field Λ, called the* Poisson bivector field *of M, such that for any pair (f, g) of smooth functions defined on M, the Poisson bracket $\{f, g\}$ is given by the formula*

$$\{f, g\} = \Lambda(\mathrm{d}f, \mathrm{d}g)\,.$$

Proof The existence, uniqueness and skew-symmetry of Λ are easy consequences of the Leibniz identity and of the skew-symmetry of the Poisson bracket. It does not depend on the Jacobi identity. □

Remark 3.2.3 The Poisson bivector field Λ determines the Poisson structure of M, since it allows the calculation of the Poisson bracket of any pair of smooth functions. For this reason a Poisson manifold M is often denoted by (M, Λ).

Definition 3.2.4 Let (M, Λ) be a Poisson manifold. We denote by $\Lambda^\sharp : T^*M \to TM$ the vector bundle homomorphism such that, for each $x \in M$ and each $\alpha \in T^*_xM$, $\Lambda^\sharp(\alpha)$ is the unique element in T_xM such that, for any $\beta \in T^*_xM$,

$$\bigl\langle\beta, \Lambda^\sharp(\alpha)\bigr\rangle = \Lambda(\alpha, \beta)\,.$$

The subset $C = \Lambda^\sharp(T^*M)$ of the tangent bundle TM is called the *characteristic field* of the Poisson manifold (M, Λ).

The following theorem, due to Alan Weinstein [15, 16], proves that, loosely speaking, a Poisson manifold is the disjoint union of symplectic manifolds, arranged in such a way that the union is endowed with a differentiable structure.

Theorem 3.2.5 *Let (M, Λ) be a Poisson manifold. Its characteristic field C is a completely integrable generalized distribution on M. It means that M is the disjoint union of immersed connected submanifolds, called the* symplectic leaves *of (M, Λ), with the following properties: a leaf S is such that, for each $x \in S$, $T_xS = T_xM \cap C$; moreover, S is maximal in the sense that any immersed connected submanifold S' containing S and such that for each $x \in S'$, $T_xS' = T_xM \cap C$, is equal to S.*

Moreover, the Poisson structure of M determines, on each leaf S, a symplectic form ω_S, such that the restriction to S of the Poisson bracket of two smooth functions defined on M only depends on the restrictions of these functions to S, and can be calculated as the Poisson bracket of these restrictions, using the symplectic form ω_S.

The reader may look at [15, 16] or at [17] for a proof of this theorem.

3.2.1 The Schouten–Nijenhuis Bracket

Let M be a smooth n-dimensional manifold. We recall that the exterior algebra $\Omega(M)$ of differential forms on M is endowed with an associative composition law, the *exterior product*, which associates to a pair (η, ζ), with $\eta \in \Omega^p(M)$ and $\zeta \in \Omega^q(M)$ the form $\eta \wedge \zeta \in \Omega^{p+q}(M)$, with the following properties.
1. When $p = 0$, $\eta \in \Omega^0(M) \equiv C^\infty(M, \mathbb{R})$; the exterior product $\eta \wedge \zeta$ is the usual product $\eta\zeta$ of the differential form ζ of degree q by the function η.
2. The exterior product satisfies

$$\zeta \wedge \eta = (-1)^{pq} \eta \wedge \zeta \,.$$

3. When $p \geq 1$ and $q \geq 1$, $\eta \wedge \zeta$ evaluated on the $p+q$ vector fields $v_i \in A^1(M)$ $(1 \leq i \leq p+q)$ is expressed as

$$\eta \wedge \zeta(v_1, \ldots, v_{p+q}) = \sum_{\sigma \in \mathcal{S}_{(p,q)}} \varepsilon(\sigma) \eta(v_{\sigma(1)}, \ldots, v_{\sigma(p)}) \zeta(v_{\sigma(p+1)}, \ldots, v_{\sigma(p+q)}) \,.$$

We have denoted by $\mathcal{S}_{(p,q)}$ the set of permutations σ of $\{1, \ldots, p+q\}$ which satisfy

$$\sigma(1) < \sigma(2) < \cdots < \sigma(p) \quad \text{and} \quad \sigma(p+1) < \sigma(p+2) < \cdots < \sigma(p+q) \,,$$

and set

$$\epsilon(\sigma) = \begin{cases} 1 & \text{if } \sigma \text{ is even,} \\ -1 & \text{if } \sigma \text{ is odd.} \end{cases}$$

The exterior algebra $\Omega(M)$ is endowed with a linear map $\mathrm{d} : \Omega(M) \to \Omega(M)$ called the *exterior differential*, with the following properties.
1. The exterior differential d is a graded map of degree 1, which means that $\mathrm{d}\big(\Omega^p(M)\big) \subset \Omega^{p+1}(M)$.
2. It is a *derivation* of the exterior algebra $\Omega(M)$, which means that when $\eta \in \Omega^p(M)$ and $\zeta \in \Omega^q(M)$,

$$\mathrm{d}(\eta \wedge \zeta) = (\mathrm{d}\eta) \wedge \zeta + (-1)^p \eta \wedge \mathrm{d}\zeta \,.$$

3. It satisfies

$$\mathrm{d} \circ \mathrm{d} = 0 \,.$$

Similarly, the exterior algebra $A(M)$ of smooth multivector fields on M is endowed with an associative composition law, the *exterior product*, which associates to a pair (P, Q), with $P \in A^p(M)$ and $Q \in A^q(M)$, the multivector field $P \wedge Q \in A^{p+q}(M)$. It is defined by the formulae given above for the exterior product of differential forms, the only change being the exchange of the roles of $\Omega^p(M)$ and $A^p(M)$. Its properties are essentially the same as those of the exterior product of differential forms.

There is a natural pairing of elements of same degree in $A(M)$ and in $\Omega(M)$. It is first defined for decomposable elements: let $\eta = \eta_1 \wedge \cdots \wedge \eta_p \in \Omega^p(M)$ and $P = X_1 \wedge \cdots \wedge X_p \in A^p(M)$. We set

$$\langle \eta, P \rangle = \det\bigl(\langle \eta_i, X_j \rangle\bigr) .$$

Then this pairing can be uniquely extended to $\Omega^p(M) \times A^p(M)$ by bilinearity.

With any $P \in A^p(M)$ we can associate a graded endomorphism $\mathrm{i}(P)$ of the exterior algebra of differential forms $\Omega(M)$, of degree $-p$, which means that when $\eta \in \Omega^q(M)$, $\mathrm{i}(P)\eta \in \Omega^{q-p}(M)$. This endomorphism, which extends to multivector fields the interior product of forms with a vector field, is determined by the formula, in which $P \in A^p(M)$, $\eta \in \Omega^q(M)$ and $R \in A^{q-p}(M)$,

$$\bigl\langle \mathrm{i}(P)\eta, R \bigr\rangle = (-1)^{(p-1)p/2} \langle \eta, P \wedge Q \rangle .$$

Besides the exterior product, there exists on the graded vector space $A(M)$ of multivector fields another bilinear composition law, which naturally extends to multivector fields the Lie bracket of vector fields. It associates to $P \in A^p(M)$ and $Q \in A^q(M)$ an element denoted $[P, Q] \in A^{p+q-1}(M)$, called the *Schouten-Nijenhuis bracket* of P and Q. The Schouten-Nijenhuis bracket $[P, Q]$ is defined by the following formula, which gives the expression of the corresponding graded endomorphism of $\Omega(M)$,

$$\mathrm{i}\bigl([P, Q]\bigr) = \Bigl[\bigl[\mathrm{i}(P), \mathrm{d}\bigr], \mathrm{i}(Q)\Bigr] .$$

The brackets in the right hand side of this formula are the *graded commutators* of graded endomorphisms of $\Omega(M)$. Let us recall that if E_1 and E_2 are graded endomorphisms of $\Omega(M)$ of degrees e_1 and e_2 respectively, their graded commutator is

$$[E_1, E_2] = E_1 \circ E_2 - (-1)^{e_1 e_2} E_2 \circ E_1 .$$

The following properties of the Schouten-Nijenhuis bracket can be deduced from the above formulae.

1. For f and $g \in A^0(M) = C^\infty(M, \mathbb{R})$, $[f, g] = 0$.
2. For a vector field $V \in A^1(M)$, $q \in \mathbb{Z}$ and $Q \in A^q(M)$, the Schouten-Nijenhuis bracket $[V, Q]$ is the Lie derivative $\mathcal{L}(V)(Q)$.
3. For two vector fields V and $W \in A^1(M, E)$, the Schouten-Nijenhuis bracket $[V, W]$ is the usual Lie bracket of these vector fields.
4. For all p and $q \in \mathbb{Z}$, $P \in A^p(M)$, $Q \in A^q(M)$,

$$[P,Q] = -(-1)^{(p-1)(q-1)}[Q,P]\,.$$

5. Let $p \in \mathbb{Z}$, $P \in A^p(M)$. The map $Q \mapsto [P,Q]$ is a derivation of degree $p-1$ of the graded exterior algebra $A(M)$. In other words, for q_1 and $q_2 \in \mathbb{Z}$, $Q_1 \in A^{q_1}(M)$ and $Q_2 \in A^{q_2}(M)$,

$$[P, Q_1 \wedge Q_2] = [P,Q_1] \wedge Q_2 + (-1)^{(p-1)q_1} Q_1 \wedge [P,Q_2]\,.$$

6. Let p, q and $r \in \mathbb{Z}$, $P \in A^p(M)$, $Q \in A^q(M)$ and $R \in A^r(M)$. The Schouten-Nijenhuis bracket satisfies the *graded Jacobi identity*

$$\begin{aligned}(-1)^{(p-1)(r-1)}\big[[P,Q],R\big] &+ (-1)^{(q-1)(p-1)}\big[[Q,R],P\big]\\ &+ (-1)^{(r-1)(q-1)}\big[[R,P],Q\big]\\ &= 0\,.\end{aligned}$$

For more information about the Schouten-Nijenhuis bracket, the reader may look at [18] or [19].

Proposition 3.2.6 *Let Λ be a smooth bivector field on a smooth manifold M. Then Λ is a Poisson bivector field (and (M,Λ) is a Poisson manifold) if and only if $[\Lambda,\Lambda] = 0$.*

Proof We define the vector bundle homomorphism $\Lambda^\sharp : T^*M \to TM$ by setting, for all $x \in M$, α and $\beta \in T^*_xM$,

$$\big\langle \beta, \Lambda^\sharp(\alpha)\big\rangle = \Lambda(\alpha,\beta)\,.$$

For any pair (f,g) of smooth functions we set

$$X_f = \Lambda^\sharp(\mathrm{d}f)\,,\quad \{f,g\} = \mathrm{i}(X_f)(\mathrm{d}g) = \Lambda(\mathrm{d}f,\mathrm{d}g)\,.$$

This bracket is a bilinear skew-symmetric composition law on $C^\infty(M,\mathbb{R})$ which satisfies the Leibniz identity. Therefore Λ is a Poisson bivector field if and only if the above defined bracket of functions satisfies the Jacobi identity.

Let f, g and h be three smooth functions on M. We easily see that X_f and $\{f,g\}$ can be expressed in terms of the Schouten-Nijenhuis bracket. Indeed we have

$$X_f = -[\Lambda,f] = -[f,\Lambda]\,,\quad \{f,g\} = \big[[\Lambda,f],g\big]\,.$$

Therefore

$$\big\{\{f,g\},h\big\} = \Big[\Big[\Lambda,\big[[\Lambda,f],g\big]\Big],h\Big]\,.$$

By using the graded Jacobi identity satisfied by Schouten-Nijenhuis bracket, we see that

$$\Big[\Lambda, \big[[\Lambda, f], g\big]\Big] = -\big[[g, \Lambda], [f, \Lambda]\big] + 2\Big[\big[[\Lambda, \Lambda], f\big], g\Big].$$

Using the equalities $X_f = -[\Lambda, f] = -[f, \Lambda]$ and $X_g = -[\Lambda, g] = -[g, \Lambda]$ we obtain

$$\begin{aligned}\big\{\{f, g\}, h\big\} &= \big[[X_f, X_g], h\big] + 2\bigg[\Big[\big[[\Lambda, \Lambda], f\big], g\Big], h\bigg]\\ &= \mathcal{L}\big([X_f, X_g]\big)h + 2\bigg[\Big[\big[[\Lambda, \Lambda], f\big], g\Big], h\bigg].\end{aligned}$$

On the other hand, we have

$$\big\{\{g, h\}, f\big\} = -\mathcal{L}(X_f) \circ \mathcal{L}(X_g)h\,, \quad \big\{\{h, f\}, g\big\} = \mathcal{L}(X_g) \circ \mathcal{L}(X_f)h\,.$$

Taking into account the equality

$$\mathcal{L}\big([X_f, X_g]\big) = \mathcal{L}(X_f) \circ \mathcal{L}(X_g) - \mathcal{L}(X_g) \circ \mathcal{L}(X_f)$$

we obtain

$$\big\{\{f, g\}, h\big\} + \big\{\{g, h\}, f\big\} + \big\{\{h, f\}, g\big\} = 2\bigg[\Big[\big[[\Lambda, \Lambda], f\big], g\Big], h\bigg].$$

By using the formula which defines the Schouten-Nijenhuis bracket, we check that for any $P \in A^3(M)$

$$\Big[\big[[P, f], g\big], h\Big] = P(\mathrm{d}f, \mathrm{d}g, \mathrm{d}h)\,.$$

Therefore

$$\big\{\{f, g\}, h\big\} + \big\{\{g, h\}, f\big\} + \big\{\{h, f\}, g\big\} = 2[\Lambda, \Lambda](\mathrm{d}f, \mathrm{d}g, \mathrm{d}h)\,,$$

so Λ is a Poisson bivector field if and only if $[\Lambda, \Lambda] = 0$. □

3.3 Some Properties of Poisson Manifolds

Definitions 3.3.1 Let (M, Λ) be a Poisson manifold.

1. The *Hamiltonian vector field* associated to a smooth function $f \in C^\infty(M, \mathbb{R})$ is the vector field X_f on M defined by

$$X_f = \Lambda^\sharp(\mathrm{d}f)\,.$$

The function f is called a *Hamiltonian* for the Hamiltonian vector field X_f.
2. A *Poisson vector field* is a vector field X which satisfies

$$\mathcal{L}(X)\Lambda = 0\,.$$

Example 3.3.2 On a symplectic manifold (M,ω) we have defined the Poisson bracket of smooth functions. That bracket endows M with a Poisson structure, said to be *associated* to its symplectic structure. The Poisson bivector field Λ is related to the symplectic form ω by

$$\Lambda(\mathrm{d}f,\mathrm{d}g) = \omega(X_f,X_g)\,,\quad f \text{ and } g \in C^\infty(M,\mathbb{R})\,.$$

The map $\Lambda^\sharp : T^*M \to TM$ such that, for any $x \in M$, α and $\beta \in T^*_xM$,

$$\langle \beta, \Lambda^\sharp(\alpha) = \Lambda(\alpha,\beta)$$

is therefore the inverse of the map $\omega^\flat : TM \to T^*M$ such that, for any $x \in M$, v and $w \in T_xM$,

$$\langle \omega^\flat(v), w\rangle = -\langle i(v)\omega, w\rangle = \omega(w,v)\,.$$

Hamiltonian vector fields for the symplectic structure of M coincide with Hamiltonian vector fields for its Poisson structure. The Poisson vector fields on the symplectic manifold (M,ω) are the locally Hamiltonian vector fields. However, on a general Poisson manifold, Poisson vector fields are more general than locally Hamiltonian vector fields: even restricted to an arbitrary small neighbourhood of a point, a Poisson vector field may not be Hamiltonian.

Remarks 3.3.3 **1.** Another way in which the Hamiltonian vector field X_f associated to a smooth function f can be defined is by saying that, for any other smooth function g on the Poisson manifold (M,Λ),

$$\mathrm{i}(X_f)(\mathrm{d}g) = \{f,g\}\,.$$

2. A smooth function g defined on the Poisson manifold (M,Λ) is said to be a *Casimir* if for any other smooth function h, we have $\{g,h\} = 0$. In other words, a Casimir is a smooth function g whose associated Hamiltonian vector field is $X_g = 0$. On a general Poisson manifold, there may exist Casimirs other than the locally constant functions.
3. A smooth vector field X on the Poisson manifold (M,Λ) is a Poisson vector field if and only if, for any pair (f,g) of smooth functions,

$$\mathcal{L}(X)\bigl(\{f,g\}\bigr) = \bigl\{\mathcal{L}(X)f,g\bigr\} + \bigl\{f,\mathcal{L}(X)g\bigr\}\,.$$

Indeed we have

$$\begin{aligned}\mathcal{L}(X)\bigl(\{f,g\}\bigr) &= \mathcal{L}(X)\bigl(\Lambda(\mathrm{d}f,\mathrm{d}g)\bigr)\\ &= \bigl(\mathcal{L}(X(\Lambda)\bigr)(\mathrm{d}f,\mathrm{d}g) + \Lambda\bigl(\mathcal{L}(X)(\mathrm{d}f),\mathrm{d}g\bigr) + \Lambda\bigl(\mathrm{d}f,\mathcal{L}(X)(\mathrm{d}g)\bigr)\\ &= \bigl(\mathcal{L}(X)(\Lambda)\bigr)(\mathrm{d}f,\mathrm{d}g) + \bigl\{\mathcal{L}(X)f,g\bigr\} + \bigl\{f,\mathcal{L}(X)g\bigr\}\,.\end{aligned}$$

3. Any Hamiltonian vector field X_f is a Poisson vector field. Indeed, if f is a Hamiltonian for X_f, g and h two other smooth functions,we have according to the Jacobi identity

$$\begin{aligned}\mathcal{L}(X_f)\bigl(\{g,h\}\bigr) &= \bigl\{f,\{g,h\}\bigr\} = \bigl\{\{f,g\},h\bigr\} + \bigl\{g,\{f,h\}\bigr\}\\ &= \bigl\{\mathcal{L}(X_f)g,h\bigr\} + \bigl\{g,\mathcal{L}(X_f)h\bigr\}\,.\end{aligned}$$

4. Since the characteristic field of the Poisson manifold (M,Λ) is generated by the Hamiltonian vector fields, any Hamiltonian vector field is everywhere tangent to the symplectic foliation. A Poisson vector field may not be tangent to that foliation.

Proposition 3.3.4 *Let (M,Λ) be a Poisson manifold, $H\in C^\infty(M,\mathbb{R})$ be a smooth function and $X_H=\Lambda^\sharp(\mathrm{d}H)\in A^1(M)$ be the associated Hamiltonian vector field. A smooth function $g\in C^\infty(M,\mathbb{R})$ keeps a constant value on each integral curve of X_H if and only if $\{H,g\}=0$ identically. Such a function g is said to be a* first integral *of X_H.*

A specially important first integral of X_H, called the energy first integral, *is the Hamiltonian H.*

Proof Let $\varphi:I\to M$ be an integral curve of X_H, defined on an open interval I of $\mathbb{R}$. For each $t\in I$

$$\frac{\mathrm{d}\varphi(t)}{\mathrm{d}t} = X_H\bigl(\varphi(t)\bigr)\,.$$

The function $g\circ\varphi$ is smooth and satisfies

$$\frac{\mathrm{d}(g\circ\varphi)(t)}{\mathrm{d}t} = \mathrm{i}(X_H)(\mathrm{d}g)\bigl(\varphi(t)\bigr) = \{H,g\}\bigl(\varphi(t)\bigr)\,.$$

Since I is connected, $g\circ\varphi$ keeps a constant value if and only if, for each $t\in I$, $\dfrac{\mathrm{d}(g\circ\varphi)(t)}{\mathrm{d}t}=0$, and the above equality proves that such is the case if and only if $\{H,g\}\bigl(\varphi(t)\bigr)=0$. The indicated result follows from the fact that for any point $x\in M$, there exists an integral curve $\varphi:I\to M$ of X_H and an element t in I such that $\varphi(t)=x$.

The skew-symmetry of the Poisson bracket implies $\{H,H\}=0$, therefore the Hamiltonian H is a first integral of X_H. □

Remark 3.3.5 Some Hamiltonian mechanical systems encountered in Mechanics, defined on a Poisson manifold (M,Λ), have as Hamiltonian a smooth function H

defined on $\mathbb{R} \times M$ rather than on the manifold M. Such a function H is said to be a *time-dependent Hamiltonian*. The associated Hamiltonian vector field X_H is no more an ordinary vector field on M, i.e. a smooth map $M \to TM$ wich associates to each $x \in M$ an element in T_xM, but rather a *time-dependent vector field*, i.e. a smooth map $X_H : \mathbb{R} \times M \to TM$ such that, for each $t \in \mathbb{R}$ and each $x \in M$ $X_H(t, x) \in T_xM$. For each fixed value of $t \in \mathbb{R}$, the map $x \mapsto X_H(t, x)$ is the Hamiltonian vector field on M whose Hamiltonian is the function $H_t : M \to \mathbb{R}$, defined by

$$H_t(x) = H(t, x) \,, \quad x \in M \,.$$

Therefore

$$X_H(t, x) = \Lambda^\sharp\bigl(\mathrm{d}H_t\bigr)(x) \,, \quad x \in M \,,\ t \in \mathbb{R} \,.$$

A smooth parametrized curve $\varphi : I \to M$, defined on an open interval I of $\mathbb{R}$, is an integral curve of the time-dependent vector field X_H if for each $t \in I$ it satisfies the non-autonomous differential equation

$$\frac{\mathrm{d}\varphi(t)}{\mathrm{d}t} = X_H\bigl(t, \varphi(t)\bigr) \,.$$

The time-dependent Hamiltonian $H : \mathbb{R} \times M \to \mathbb{R}$ is no more a first integral of X_H since, for each integral curve $\varphi : I \to M$ of X_H and each $t \in I$,

$$\frac{\mathrm{d}(H \circ \varphi)(t)}{\mathrm{d}t} = \frac{\partial H\bigl(t, \varphi(t)\bigr)}{\partial t} \,.$$

Proposition 3.3.6 *Let (M_1, Λ_1) and (M_2, Λ_2) be two Poisson manifolds and let $\varphi : M_1 \to M_2$ be a smooth map. The following properties are equivalent.*

1. *For any pair (f, g) of smooth functions defined on M_2*

$$\{\varphi^* f, \varphi^* g\}_{M_1} = \varphi^* \{f, g\}_{M_2} \,.$$

2. *For any smooth function $f \in C^\infty(M_2, \mathbb{R})$ the Hamiltonian vector fields $\Lambda_2^\sharp(\mathrm{d}f)$ on M_2 and $\Lambda_1^\sharp\bigl(\mathrm{d}(f \circ \varphi)\bigr)$ on M_1 are φ-compatible, which means that for each $x \in M_1$*

$$T_x\varphi\Bigl(\Lambda_1^\sharp\bigl(\mathrm{d}(f \circ \varphi)(x)\bigr)\Bigr) = \Lambda_2^\sharp\Bigl(\mathrm{d}f\bigl(\varphi(x)\bigr)\Bigr) \,.$$

3. *The bivector fields Λ_1 on M_1 and Λ_2 on M_2 are φ-compatible, which means that for each $x \in M_1$*

$$T_x\varphi\bigl(\Lambda_1(x)\bigr) = \Lambda_2\bigl(\varphi(x)\bigr) \,.$$

A map $\varphi : M_1 \to M_2$ which satisfies these equivalent properties is called a Poisson map.

Proof Let f and g be two smooth functions defined on M_2. For each $x \in M_1$, we have

$$\{\varphi^*f, \varphi^*g\}_{M_1}(x) = \{f \circ \varphi, g \circ \varphi\}(x) = \Lambda_1(x)\big(\mathrm{d}(f \circ \varphi)(x), \mathrm{d}(g \circ \varphi)(x)\big)$$
$$= \Big\langle \mathrm{d}(g \circ \varphi)(x), \Lambda_1^\sharp\big(\mathrm{d}(f \circ \varphi(x))\big)\Big\rangle .$$

We have also

$$\varphi^*\{f, g\}_{M_2}(x) = \{f, g\}_{M_2}\big(\varphi(x)\big)$$
$$= \Big\langle \mathrm{d}g\big(\varphi(x)\big), \Lambda_2^*\big(\mathrm{d}f\big(\varphi(x)\big)\big)\Big\rangle .$$

These formulae show that Properties 1 and 2 are equivalent.

We recall that $T_x\varphi\big(\Lambda_1(x)\big)$ is, by its very definition, the bivector at $\varphi(x) \in M_2$ such that, for any pair (f, g) of smooth functions on M_2

$$T_x\varphi\big(\Lambda_1(x)\big)\Big(\mathrm{d}f\big(\varphi(x)\big), \mathrm{d}g\big(\varphi(x)\big)\Big) = \Lambda_1\big(\mathrm{d}(f \circ \varphi)(x), \mathrm{d}(g \circ \varphi)(x)\big) .$$

The above equalities therefore prove that Properties 2 and 3 are equivalent. □

Poisson manifolds often appear as quotients of symplectic manifolds, as indicated by the following Proposition, due to Paulette Libermann [20].

Proposition 3.3.7 *Let (M, ω) be a symplectic manifold and let $\varphi : M \to P$ be a surjective submersion of M onto a smooth manifold P whose fibres are connected (it means that for each $y \in P$, $\varphi^{-1}(y)$ is connected). The following properties are equivalent.*

1. *On the manifold M, the distribution* $\mathrm{orth}(\ker T\varphi)$ *is integrable.*
2. *For any pair (f, g) of smooth functions defined on P, the Poisson bracket $\{f \circ \varphi, g \circ \varphi\}$ is constant on each fibre $\varphi^{-1}(y)$ of the submersion φ (with $y \in P$).*

When these two equivalent properties are satisfied, there exists on P a unique Poisson structure for which $\varphi : M \to P$ is a Poisson map (the manifold M being endowed with the Poisson structure associated to its symplectic structure).

Proof On the manifold M, $\ker T\varphi$ is a an integrable distribution of rank $\dim M - \dim P$ whose integral submanifolds are the fibres of the submersion φ. Its symplectic orthog(onal $\mathrm{orth}(\ker T\varphi)$ is therefore a distribution of rank $\dim P$. Let f and g be two smooth functions defined on M_2. On M_1, the Hamiltonian vector fields $X_{f\circ\varphi}$ and $X_{g\circ\varphi}$ take their values in $\mathrm{orth}(\ker T\varphi)$. We have

$$[X_{f\circ\varphi}, X_{g\circ\varphi}] = X_{\{f\circ\varphi, g\circ\varphi\}} .$$

Therefore $[X_{f\circ\varphi}, X_{g\circ\varphi}]$ takes its values in $\mathrm{orth}(\ker T\varphi)$ if and only if the function $\{f \circ \varphi, g \circ \varphi\}$ is constant on each fibre $\varphi^{-1}(y)$ of the submersion φ. The equivalence of Properties 1 and 2 easily follows.

Let us now assume that the equivalent properties 1 and 2 are satisfied. Since the map $\varphi : M \to P$ is a submersion with connected fibres, the map which associates to each function $f \in C^\infty(M_2, \mathbb{R})$ the function $f \circ \varphi$ is an isomorphism of $C^\infty(M_2, \mathbb{R})$ onto the subspace of $C^\infty(M_1, \mathbb{R})$ made by smooth functions which are constant on each fibre of φ. The existence and unicity of a Poisson structure on M_2 for which φ is a Poisson map follows. □

Remark 3.3.8 Poisson manifolds obtained as quotients of symplectic manifolds often come by pairs. Let us assume indeed that (M, ω) is a symplectic manifold and that the above Proposition can be applied to a smooth surjective submersion with connected fibres $\varphi : M \to P$, and defines a Poisson structure on P for which φ is a Poisson map. Since $\operatorname{orth}(\ker T\varphi)$ is integrable, it defines a foliation of M, which is said to be *simple* when the set of leaves Q of that foliation has a smooth manifold structure such that the map $\psi : M \to Q$, which associates to each point in M the leaf through this point, is a submersion. Then the maps $\varphi : M \to P$ and $\psi : M \to Q$ play similar parts, so there exists on Q a unique Poisson structure for which ψ is a Poisson map. Alan Weinstein [15, 16] has determined the links which exist between the local structures of the two Poisson manifolds P and Q at corresponding points (that means, at points which are the images of the same point in M by the maps φ and ψ).

Several kinds of remarkable submanifolds of a Poisson manifold can be defined [15, 16]. The most important are the *coisotropic* submanifolds, defined below.

Definition 3.3.9 A submanifold N of a Poisson manifold (M, Λ) is said to be *coisotropic* if for any point $x \in N$ and any pair (f, g) of smooth functions defined on a neighbourhood U of x in M whose restrictions to $U \cap N$ are constants, the Poisson bracket $\{f, g\}$ vanishes on $U \cap N$.

3.4 Examples of Poisson Manifolds

3.4.1 Symplectic Manifolds

We have seen above that any symplectic manifold is a Poisson manifold.

3.4.2 Dual Spaces of Finite-Dimensional Lie algebras

Let $\mathcal{G}$ be a finite-dimensional Lie algebra, and $\mathcal{G}^*$ its dual space. The Lie algebra $\mathcal{G}$ can be considered as the dual of $\mathcal{G}^*$, that means as the space of linear functions on $\mathcal{G}^*$, and the bracket of the Lie algebra $\mathcal{G}$ is a composition law on this space of linear functions. This composition law can be extended to the space $C^\infty(\mathcal{G}^*, \mathbb{R})$ by setting

$$\{f, g\}(x) = \Big\langle x, \big[\mathrm{d}f(x), \mathrm{d}g(x)\big]\Big\rangle, \quad f \text{ and } g \in C^\infty(\mathcal{G}^*, \mathbb{R}), \quad x \in \mathcal{G}^*.$$

This bracket on $C^\infty(\mathcal{G}^*, \mathbb{R})$ defines a Poisson structure on $\mathcal{G}^*$, called its *canonical Poisson structure*. It implicitly appears in the works of Sophus Lie, and was rediscovered by Kirillov [21], Kostant [22] and Jean-Marie Souriau [23]. Its existence can be seen as an application of Proposition 3.3.7. Let indeed G be the connected and simply connected Lie group whose Lie algebra is $\mathcal{G}$. We know that the cotangent bundle T^*G has a canonical symplectic structure. One can check easily that for this symplectic structure, the Poisson bracket of two smooth functions defined on T^*G and invariant with respect to the lift to T^*G of the action of G on itself by left translations, is too invariant with respect to that action. Application of Proposition 3.3.7, the submersion $\varphi : T^*G \to \mathcal{G}^*$ being the left translation which, for each $g \in G$, maps T^*_gG onto $T^*_eG \equiv \mathcal{G}^*$, yields the above defined Poisson structure on $\mathcal{G}^*$. If instead of translations on the left, we use translation on the right, we obtain on $\mathcal{G}^*$ the opposite Poisson structure. This illustrates Remark 3.3.8, since, as we will see later, each one of the tangent spaces at a point $\xi \in T^*G$ to the orbits of that point by the lifts to T^*G of the actions of G on itself by translations on the left and on the right, is the symplectic orthogonal of the other.

The symplectic leaves of $\mathcal{G}^*$ equipped with the above defined Poisson structure are the coadjoint orbits.

3.4.3 Symplectic Cocycles

A *symplectic cocycle* of the Lie algebra $\mathcal{G}$ is a skew-symmetric bilinear map $\widetilde{\Theta} : \mathcal{G} \times \mathcal{G} \to \mathbb{R}$ which satisfies

$$\widetilde{\Theta}\big([X,Y],Z\big) + \widetilde{\Theta}\big([Y,Z],X\big) + \widetilde{\Theta}\big([Z,X],Y\big) = 0\,.$$

The above defined canonical Poisson structure on $\mathcal{G}^*$ can be modified by means of a symplectic cocycle $\widetilde{\Theta}$ by defining the new bracket (see for example [17])

$$\{f,g\}_{\widetilde{\Theta}}(x) = \Big\langle x, \big[\mathrm{d}f(x), \mathrm{d}g(x)\big]\Big\rangle - \widetilde{\Theta}\big(\mathrm{d}f(x), \mathrm{d}g(x)\big)\,,$$

where f and $g \in C^\infty(\mathcal{G}^*, \mathbb{R})$, $x \in \mathcal{G}^*$. This Poisson structure is called the *modified canonical Poisson structure* by means of the symplectic cocycle $\widetilde{\Theta}$. We will see in Sect. 4.5 that the symplectic leaves of $\mathcal{G}^*$ equipped with this Poisson structure are the orbits of an affine action whose linear part is the coadjoint action, with an additional term determined by $\widetilde{\Theta}$.

4 Symplectic, Poisson and Hamiltonian Actions

4.1 Actions on a Smooth Manifold

Let us first recall some definitions and facts about actions of a Lie algebra or of a Lie group on a smooth manifold.

Definition 4.1.1 An *action on the left* (resp. an *action on the right*) of a Lie group G on a smooth manifold M is a smooth map $\Phi : G \times M \to M$ (respectively, $\Psi : M \times G \to M$) such that, for any $x \in M$, g_1 and $g_2 \in G$, $e \in G$ being the neutral element,

- for an action on the left

$$\Phi\bigl(g_1, \Phi(g_2, x)\bigr) = \Phi(g_1 g_2, x)\,, \quad \Phi(e, x) = x\,,$$

- for an action on the right

$$\Psi\bigl(\Psi(x, g_1), g_2\bigr) = \Psi(x, g_1 g_2)\,, \quad \Psi(x, e) = x\,.$$

4.1.1 Consequences

Let $\Phi : G \times M \to M$ be an action on the left of the Lie group G on the smooth manifold M. For each $g \in G$, we denote by $\Phi_g : M \to M$ the map

$$\Phi_g(x) = \Phi(g, x)\,.$$

The map $g \mapsto \Phi_g$ is a groups homomorphism of G into the group of smooth diffeomorphisms of M. In other words, for each $g \in G$, Φ_g is a diffeomorphism of M, and we have

$$\Phi_g \circ \Phi_h = \Phi_{gh}\,, \quad (\Phi_g)^{-1} = \Phi_{g^{-1}}\,, \quad g \text{ and } h \in G\,.$$

Similarly, let $\Psi : M \times G \to M$ be an action on the right of the Lie group G on the smooth manifold M. For each $g \in G$, we denote by $\Psi_g : M \to M$ the map

$$\Psi_g(x) = \Psi(x, g)\,.$$

The map $g \mapsto \Psi_g$ is a groups anti-homomorphism of G into the group of smooth diffeomorphisms of M. In other words, for each $g \in G$, Ψ_g is a diffeomorphism of M, and we have

$$\Psi_g \circ \Psi_h = \Psi_{hg}\,, \quad (\Psi_g)^{-1} = \Psi_{g^{-1}}\,, \quad g \text{ and } h \in G\,.$$

Definition 4.1.2 Let $\Phi : G \times M \to M$ be an action on the left (resp. let $\Psi : M \times G \to M$ be an action of the right) of the Lie group G on the smooth manifold M. With each element $X \in \mathcal{G} \equiv T_e G$ (the tangent space to the Lie group G at the neutral element) we associate the vector field X_M on M defined by

$$X_M(x) = \begin{cases} \left.\dfrac{\mathrm{d}\Phi\bigl(\exp(sX), x\bigr)}{\mathrm{d}s}\right|_{s=0} & \text{if } \Phi \text{ is an action on the left,} \\ \left.\dfrac{\mathrm{d}\Psi\bigl(x, \exp(sX)\bigr)}{\mathrm{d}s}\right|_{s=0} & \text{if } \Psi \text{ is an action on the right.} \end{cases}$$

The vector field X_M is called the *fundamental vector field* on M associated to X.

Definition 4.1.3 An *action* of a Lie algebra $\mathcal{G}$ on a smooth manifold M is a Lie algebras homomorphism φ of $\mathcal{G}$ into the Lie algebra $A^1(M)$ of smooth vector fields on M (with the Lie bracket of vector fields as composition law). In other words, it is a linear map $\varphi : \mathcal{G} \to A^1(M)$ such that for each pair $(X, Y) \in \mathcal{G} \times \mathcal{G}$,

$$\varphi\bigl([X, Y]\bigr) = \bigl[\varphi(X), \varphi(Y)\bigr].$$

Remark 4.1.4 Let G be a Lie group. There are two natural ways in which the tangent space $T_eG \equiv \mathcal{G}$ to the Lie group G at the neutral element e can be endowed with a Lie algebra structure.

In the first way, we associate with each element $X \in T_eG$ the *left invariant* vector field X^L on G such that $X^L(e) = X$; its value at a point $g \in G$ is $X^L(g) = TL_g(X)$, where $L_g : G \to G$ is the map $h \mapsto L_g(h) = gh$. We observe that for any pair (X, Y) of elements in $\mathcal{G}$ the Lie bracket $[X^L, Y^L]$ of the vector fields X^L and Y^L on G is left invariant, and we define the bracket $[X, Y]$ by setting $[X, Y] = [X^L, Y^L](e)$. This Lie algebra structure on $\mathcal{G} \equiv T_eG$ will be called the Lie algebra structure of *left invariant vector fields* on G.

In the second way, we choose the *right invariant* vector fields on G X^R and Y^R, instead of the left invariant vector fields X^L and Y^L. Since $[X^R, Y^R](e) = -[X^L, Y^L](e)$, the Lie algebra structure on $\mathcal{G} \equiv T_eG$ obtained in this way, called the Lie algebra structure of *right invariant vector fields*, is the *opposite* of that of left invariant vector fields. We have therefore on T_eG two opposite Lie algebras structures, both equally natural. Fortunately, the choice of one rather than the other as the Lie algebra $\mathcal{G}$ of G does not matter because the map $X \mapsto -X$ is a Lie algebras isomorphism between these two structures.

Proposition 4.1.5 *Let $\Phi : G \times M \to M$ be an action on the left (resp. let $\Psi : M \times G \to M$ be an action on the right) of a Lie group G on a smooth manifold M. We endow $\mathcal{G} \equiv T_eG$ with the Lie algebra structure of right invariant vector fields on G (resp, with the Lie algebra structure of left invariant vector fields on G). The map $\varphi : \mathcal{G} \to A^1(M)$ (resp. $\psi : \mathcal{G} \to A^1(M)$) which associates to each element X of the Lie algebra $\mathcal{G}$ of G the corresponding fundamental vector field X_M, is an action of the Lie algebra $\mathcal{G}$ on the manifold M. This Lie algebra action is said to be* associated *to the Lie group action Φ (resp. Ψ).*

Proof Let us look at an action on the left Φ. Let $x \in M$, and let $\Phi^x : G \to M$ be the map $g \mapsto \Phi^x(g) = \Phi(g, x)$.For any $X \in T_eG$ and $g \in G$, we have

$$\begin{aligned} X_M\big(\Phi(g,x)\big) &= \frac{\mathrm{d}}{\mathrm{d}s}\,\Phi\big(\exp(sX),\Phi(g,x)\big)\Big|_{s=0} = \frac{\mathrm{d}}{\mathrm{d}s}\,\Phi\big(\exp(sX)g,x\big)\Big|_{s=0}\\ &= \frac{\mathrm{d}}{\mathrm{d}s}\,\Phi\Big(R_g\big(\exp(sX)\big),x\Big)\Big|_{s=0} = T\Phi^x\circ TR_g(X)\,. \end{aligned}$$

We see that for each $X \in T_xG$, the right invariant vector field X^R on G and the fundamental vector field X_M on M are compatible with respect to the map $\Phi^x : G \to M$. Therefore for any pair (X, Y) of elements in T_eG, we have $[X, Y]_M = [X_M, Y_M]$. In other words the map $X \mapsto X_M$ is an action of the Lie algebra $\mathcal{G} = T_eG$ (equipped with the Lie algebra structure of right invariant vector fields on G) on the manifold M.

For an action on the right Ψ, the proof is similar, $\mathcal{G} = T_eG$ being this time endowed with the Lie algebra structure of left invariant vector fields on G. □

Proposition 4.1.6 *Let $\Phi : G \times M \to M$ be an action on the left (resp. let $\Psi : M \times G \to M$ be an action on the right) of a Lie group G on a smooth manifold M. Let X_M be the fundamental vector field associated to an element $X \in \mathcal{G}$. For any $g \in G$, the direct image $(\Phi_g)_*(X_M)$ (resp. $(\Psi_g)_*(X_M)$) of the vector field X_M by the diffeomorphism $\Phi_g : M \to M$ (resp. $\Psi_g : M \to M$) is the fundamnetal vector field $(\mathrm{Ad}_g X)_M$ associated to $\mathrm{Ad}_g X$ (resp. the fundamental vector field $(\mathrm{Ad}_{g^{-1}} X)_M$ associated to $\mathrm{Ad}_{g^{-1}} X$).*

Proof For each $x \in M$

$$\begin{aligned} (\Phi_g)_*(X_M)(x) &= T\Phi_g\Big(X_M\big(\Phi(g^{-1},x)\big)\Big)\\ &= T\Phi_g\left(\frac{\mathrm{d}}{\mathrm{d}s}\Phi\big(\exp(sX)g^{-1},x\big)\Big|_{s=0}\right)\\ &= \frac{\mathrm{d}}{\mathrm{d}s}\Big(\Phi\big(g\exp(sX)g^{-1},x\big)\Big)\Big|_{s=0} = (\mathrm{Ad}_g X)_M(x)\,, \end{aligned}$$

since $g\exp(sX)g^{-1} = \exp(\mathrm{Ad}_g X)$. The proof for the action on the right Ψ is similar. □

4.2 Linear and Affine Representations

In this section, after recalling some results about linear and affine transformation groups, we discuss linear and affine representations of a Lie group or of a Lie algebra in a finite-dimensional vector space, which can be seen as special examples of actions.

4.2.1 Linear and Affine Transformation Groups and Their Lie Algebras

Let E be a finite-dimensional vector space. The set of linear isomorphisms $l : E \to E$ will be denoted by $\mathrm{GL}(E)$. We recall that equipped with the composition of maps

$$(l_1, l_2) \mapsto l_1 \circ l_2$$

as a composition law, $\mathrm{GL}(E)$ is a Lie group whose dimension is $(\dim E)^2$. Its Lie algebra, which will be denoted by $\mathfrak{gl}(E)$, is the set $\mathcal{L}(E, E)$ of linear maps $f : E \to E$, with the commutator

$$(f_1, f_2) \mapsto [f_1, f_2] = f_1 \circ f_2 - f_2 \circ f_1$$

as a composition law.

A map $a : E \to E$ is called an *affine map* if it can be written as

$$a(x) = l(x) + c\,, \quad x \in E\,,$$

the map $l : E \to E$ being linear, and $c \in E$ being a constant. The affine map a is invertible if and only if its linear part l is invertible, in other words if and only if $l \in \mathrm{GL}(E)$; when this condition is satisfied, its inverse is

$$a^{-1}(y) = l^{-1}(y - c)\,, \quad y \in E\,.$$

By identifying the invertible affine map a with the pair (l, c), with $l \in \mathrm{GL}(E)$ and $c \in E$, the set of invertible affine maps of E onto itself becomes identified with $\mathrm{GL}(E) \times E$. The composition law and the inverse map on this product (which is called the *semi-direct product of* $\mathrm{GL}(E)$ *with* E) are

$$(l_1, c_1), (l_2, c_2) \mapsto \bigl((l_1 \circ l_2, l_1(c_2) + c_1)\bigr)\,, \quad (l, c)^{-1} = \bigl(l^{-1}, -l^{-1}(c)\bigr)\,.$$

The semi-direct product $\mathrm{Aff}(E) = \mathrm{GL}(E) \times E$ is a Lie group whose dimension is $(\dim E)^2 + \dim E$; its Lie algebra is the product $\mathfrak{aff}(E) = \mathcal{L}(E, E) \times E$, with the composition law

$$\bigl((f_1, d_1), (f_2, d_2)\bigr) \mapsto \bigl[(f_1, d_1), (f_2, d_2)\bigr] = \bigl(f_1 \circ f_2 - f_2 \circ f_1, f_1(d_2) - f_2(d_1)\bigr)\,.$$

The adjoint representation is given by the formula

$$\mathrm{Ad}_{(l,c)}\bigl((f, d)\bigr) = \bigl(l \circ f \circ l^{-1}, l(d) - l \circ f \circ l^{-1}(d)\bigr)\,.$$

Remark 4.2.1 The finite-dimensional vector space E can be considered as a smooth manifold on which E itself transitively acts by translations. That action determines a natural trivialization of the tangent bundle TE, the tangent space T_xE at each point $x \in E$ being identified with E. An element $a \in \mathfrak{aff}(E)$, in other words an affine map $a : E \to E$, can therefore be considered as the vector field on E whose value, at each $x \in E$, is $a(x) \in T_xE \equiv E$. A question naturally arises: how the bracket of two elements a_1 and $a_2 \in \mathfrak{aff}(E)$, for the Lie algebra structure of $\mathfrak{aff}(E)$ defined in Sect. 4.2, compares with the bracket of these two elements when considered as vector fields on E? An easy calculation in local coordinates shows that the bracket

$[a_1, a_2]$ defined in Sect. 4.2 is the *opposite* of the bracket of these two elements when considered as vector fields on E. Remark 4.2.7 below will explain the reason of that change of sign.

Definitions 4.2.2 Let G be a Lie group, $\mathcal{G}$ a Lie algebra and E a finite-dimensional vector space.

1. A *linear representation* (respectively, an *affine representation*) of the Lie group G in the vector space E is a Lie groups homomorphism $R : G \to \mathrm{GL}(E)$ of G in the Lie group $\mathrm{GL}(E)$ of linear transformations of E (respectively, a Lie groups homomorphism $A : G \to \mathrm{Aff}(E)$ of G in the Lie group $\mathrm{Aff}(E)$ of affine transformations of E).

2. A *linear representation* (respectively, an *affine representation*) of the Lie algebra $\mathcal{G}$ in the vector space E is a Lie algebras homomorphism $r : \mathcal{G} \to \mathfrak{gl}(E)$ of the Lie algebra $\mathcal{G}$ in the Lie algebra $\mathfrak{gl}(E)$ of the group of linear transformations of E (resp, a Lie algebras homomorphism $a : \mathcal{G} \to \mathfrak{aff}(E)$ of the Lie algebra $\mathcal{G}$ in the Lie algebra $\mathfrak{aff}(E)$ of the group of affine transformations of E).

Examples 4.2.3 Let G be a Lie group. The *adjoint representation* of G is the linear representation of G in its Lie algebra $\mathcal{G}$ which associates, to each $g \in G$, the linear isomorphism $\mathrm{Ad}_g \in \mathrm{GL}(\mathcal{G})$

$$\mathrm{Ad}_g(X) = TL_g \circ TR_{g^{-1}}(X) , \quad (X \in \mathcal{G}) .$$

The *coadjoint representation* of G is the contragredient of the adjoint representation. It associates to each $g \in G$ the linear isomorphism $\mathrm{Ad}^*_{g^{-1}} \in \mathrm{GL}(\mathcal{G}^*)$, which satisfies, for each $\zeta \in \mathcal{G}^*$ and $X \in \mathcal{G}$,

$$\left\langle \mathrm{Ad}^*_{g^{-1}}(\zeta), X \right\rangle = \left\langle \zeta, \mathrm{Ad}_{g^{-1}}(X) \right\rangle .$$

The *adjoint representation* of the Lie algebra $\mathcal{G}$ is the linear representation of $\mathcal{G}$ into itself which associates, to each $X \in \mathcal{G}$, the linear map $\mathrm{ad}_X \in \mathfrak{gl}(\mathcal{G})$

$$\mathrm{ad}_X(Y) = [X, Y] , \quad (Y \in \mathcal{G}) .$$

The *coadjoint representation* of the Lie algebra $\mathcal{G}$ is the contragredient of the adjoint representation. It associates, to each $X \in \mathcal{G}$, the linear map $\mathrm{ad}^*_{-X} \in \mathfrak{gl}(\mathcal{G}^*)$ which satisfies, for each $\zeta \in \mathcal{G}^*$ and $X \in \mathcal{G}$,

$$\left\langle \mathrm{ad}^*_{-X}\, \zeta, Y \right\rangle = \left\langle \zeta, [-X, Y] \right\rangle .$$

The adjoint representation (respectively, the coadjoint representation) of $\mathcal{G}$ is the Lie algebra representation associated to the adjoint representation (respectively, the coadjoint representation) of the Lie group G, in the sense recalled below in the proof of Proposition 4.2.6.

Proposition 4.2.4 *Let G be a Lie group and E a finite-dimensional vector space. A map $A : G \to \mathrm{Aff}(E)$ always can be written as*

$$A(g)(x) = R(g)(x) + \theta(g)\,, \quad \text{with } g \in G\,,\ x \in E\,,$$

where the maps $R : G \to \mathrm{GL}(E)$ and $\theta : G \to E$ are determined by A. The map A is an affine representation of G in E if and only if the following two properties are satisfied:

- *$R : G \to \mathrm{GL}(E)$ is a linear representation of G in the vector space E,*
- *the map $\theta : G \to E$ is a one-cocycle of G with values in E, for the linear representation R; it means that θ is a smooth map which satisfies, for all g and $h \in G$,*
$$\theta(gh) = R(g)\bigl(\theta(h)\bigr) + \theta(g)\,.$$

When these two properties are satisfied, the linear representation R is called the linear part *of the affine representation A, and θ is called the* one-cocycle *of G associated to the affine representation A.*

Proof Since $\mathrm{Aff}(E) = \mathrm{GL}(E) \times E$, for each $g \in G$ and $x \in E$, we have

$$A(g)(x) = R(g)(x) + \theta(g)\,,$$

where the maps $R : G \to \mathrm{GL}(E)$ and $\theta : G \to E$ are determined by A. By comparing $A(gh)$ and $A(g) \circ A(h)$, for g and $h \in G$, using the composition law of $\mathrm{Aff}(E)$ recalled in Sect. 4.2, we easily check that A is an affine representation, which means that it is smooth and satisfies, for all g and $h \in G$, $A(gh) = A(g) \circ A(h)$, and $A(e) = \mathrm{id}_E$, if and only if the two above stated properties are satisfied. □

For linear and affine representations of a Lie algebra, we have the following infinitesimal analogue of Proposition 4.2.4.

Proposition 4.2.5 *Let $\mathcal{G}$ be a Lie algebra and E a finite-dimensional vector space. A linear map $a : \mathcal{G} \to \mathfrak{aff}(E)$ always can be written as*

$$a(X)(x) = r(X)(x) + \Theta(X)\,, \quad \text{with } X \in \mathcal{G}\,,\ x \in E\,,$$

where the linear maps $r : \mathcal{G} \to \mathfrak{gl}(E)$ and $\Theta : \mathcal{G} \to E$ are determined by a. The map a is an affine representation of G in E if and only if the following two properties are satisfied:

- *$r : \mathcal{G} \to \mathfrak{gl}(E)$ is a linear representation of the Lie algebra $\mathcal{G}$ in the vector space E,*
- *the linear map $\Theta : \mathcal{G} \to E$ is a one-cocycle of $\mathcal{G}$ with values in E, for the linear representation r; it means that Θ satisfies, for all X and $Y \in \mathcal{G}$,*

$$\Theta\bigl([X, Y]\bigr) = r(X)\bigl(\Theta(Y)\bigr) - r(Y)\bigl(\Theta(X)\bigr)\,.$$

When these two properties are satisfied, the linear representation r is called the linear part *of the affine representation a, and Θ is called the* one-cocycle *of $\mathcal{G}$ associated to the affine representation a.*

Proof Since $\mathfrak{aff}(E) = \mathfrak{gl}(E) \times E = \mathcal{L}(E,E) \times E$, for each $X \in \mathcal{G}$ and $x \in E$, we have

$$a(X)(x) = r(X)(x) + \Theta(X)\,,$$

where the linear maps $r : \mathcal{G} \to \mathfrak{gl}(E) = \mathcal{L}(E,E)$ and $\Theta : \mathcal{G} \to E$ are determined by a. By comparing $a\bigl([X,Y]\bigr)$ and $\bigl[a(X),a(Y)\bigr]$, for X and $Y \in \mathcal{G}$, using the expression of the bracket of $\mathfrak{aff}(E)$ recalled in Sect. 4.2, we easily check that A is an affine representation, which means that it is smooth and satisfies, for all X and $Y \in \mathcal{G}$, $a\bigl([X,Y]\bigr) = \bigl[a(X),a(Y)\bigr]$ if and only if the two above stated properties are satisfied. □

Proposition 4.2.6 *Let $A : G \to \mathrm{Aff}(E)$ be an affine representation of a Lie group G in a finite-dimensional vector space E, and $\mathcal{G}$ be the Lie algebra of G. Let $R : G \to \mathrm{GL}(E)$ and $\theta : G \to E$ be, respectively, the linear part and the associated cocycle of the affine representation A. Let $a : \mathcal{G} \to \mathfrak{aff}(E)$ be the affine representation of the Lie algebra $\mathcal{G}$ associated (in the sense recalled below in the proof) to the affine representation $A : G \to \mathrm{Aff}(E)$ of the Lie group G. The linear part of a is the linear representation $r : \mathcal{G} \to \mathfrak{gl}(E)$ associated to the linear representation $R : G \to \mathrm{GL}(E)$, and the associated cocycle $\Theta : \mathcal{G} \to E$ is related to the one-cocycle $\theta : G \to E$ by*

$$\Theta(X) = T_e\theta\bigl(X(e)\bigr)\,, \quad (X \in \mathcal{G})\,.$$

Proof We recall that when we have a Lie groups homomorphism $A : G \to H$ of a Lie group G into another Lie group H, the associated Lie algebras homomorphism $a : \mathcal{G} \to \mathcal{H}$ of Lie algebras associates, to each $X \in \mathcal{G}$ (seen as the space of left-invariant vector fields on G) the left-invariant vector field $a(X)$ on H whose value at the neutral element is $T_eA\bigl(X(e)\bigr)$. Let $X \in \mathcal{G}$. For each $t \in \mathbb{R}$ and $x \in E$, we have

$$A\bigl(\exp(tX)\bigr)(x) = R\bigl(\exp(tX)\bigr)(x) + \theta\bigl(\exp(tX)\bigr)\,.$$

By taking the derivative of both sides of this equality with respect to t, then setting $t = 0$, we get

$$a(X)(x) = r(X)(x) + T_e\theta(X)\,.$$

Therefore the affine representation a has r as linear part and $\Theta = T_e\theta$ as associated one-cocycle. □

Remark 4.2.7 Let $A : G \to \mathrm{Aff}(E)$ be an affine representation of a Lie group G in a finite-dimensional vector space E. The map $\widetilde{A} : G \times E \to E$,

$$\widetilde{A}(g,x) = A(g)(x)\,, \quad g \in G\,,\ x \in E\,,$$

is an action on the left of G on E. Proposition 4.2.6 shows that $a : \mathcal{G} \to \mathfrak{aff}(E)$ is a Lie algebras homomorphism, the Lie algebra structure of $\mathfrak{aff}(E)$ being the structure defined in Sect. 4.2. For each $X \in \mathcal{G}$, the element $a(X) \in \mathfrak{aff}(E)$, when considered as an affine vector field on E, is the fundamental vector field associated to X, for the action on the left $\widetilde{A}$ of G on E. We have seen (4.1.5) that for an action on the left of G on E, the map which associates to each $X \in \mathcal{G}$ the corresponding fundamental vector field on E is a Lie algebras homomorphism of the Lie algebra of *right invariant* vector fields on G into the Lie algebra of smooth vector fields on E. This explains why, as was observed in 4.2.1, the Lie algebra structure of $\mathfrak{aff}(E)$ defined in Sect. 4.2 is the opposite of the Lie algebra structure which exists on the space of affine vector fields on E. Of course, this remark is also valid for a linear representation $R : G \to \mathrm{GL}(E)$, since $\mathrm{GL}(E)$ is a Lie subgroup of $\mathrm{Aff}(E)$.

Definitions 4.2.8

1. Let $R : G \to \mathrm{GL}(E)$ be a linear representation of a Lie group G in a finite-dimensional vector space E. A *one-coboundary* of G with values in E, for the linear representation R, is a map $\theta : G \to E$ which can be expressed as

$$\theta(g) = R(g)(c) - c\,, \quad (g \in G)\,,$$

where c is a fixed element in E.

2. Ler $r : \mathcal{G} \to \mathfrak{gl}(E)$ be a linear representation of a Lie algebra $\mathcal{G}$ in a finite-dimensional vector space E. A *one-coboundary* of $\mathcal{G}$ with values in E, for the linear representation r, is a linear map $\Theta : \mathcal{G} \to E$ which can be expressed as

$$\Theta(X) = r(X)(c)\,, \quad (X \in \mathcal{G})\,,$$

where c is a fixed element in E.

Remark 4.2.9 The reader will easily check the following properties. A one-coboundary of a Lie group G with values in a finite-dimensional vector space E for a linear representation $R : G \to \mathrm{GL}(E)$, automatically is a one-cocycle in the sense of 4.2.4. Similarly, a one-coboundary of a Lie algebra $\mathcal{G}$ with values in E for the linear representation $r : \mathcal{G} \to \mathfrak{gl}(E)$, automatically is a one-cocycle of $\mathcal{G}$ in the sense of 4.2.5. When a Lie group one-cocycle $\theta : G \to E$ is in fact a one-coboundary, the associated Lie algebra one-cocycle $\Theta = T_e\theta$ is a Lie algebra one-coboundary.

Proposition 4.2.10 *Let $A : G \to \mathrm{Aff}(E)$ be an affine representation of a Lie group G in a finite-dimensional vector space E, $R : G \to \mathrm{GL}(E)$ be its linear part and $\theta : G \to E$ be the associated Lie group one-cocycle. The following properties are equivalent.*

1. *There exists an element $c \in E$ such that, for all $g \in G$ and $x \in E$,*

$$A(g)(x) = R(g)(x + c) - c\,.$$

2. *The one-cocycle $\theta : G \to E$ is in fact a* 1*-coboudary, whose expression is*

$$\theta(g) = R(g)(c) - c\,.$$

Proof Since for each $g \in G$ $R(g)$ is linear, Property 1 can be written

$$A(g)(x) = R(g)(x) + \bigl(R(g)(c) - c\bigr)\,.$$

Therefore Property 1 is true if and only if $\theta(g) = R(g)(c) - c$, in other words if and only if Property 2 is true. □

The following Proposition is the infinitesimal analogue, for affine representations of a Lie algebra, of Proposition 4.2.10.

Proposition 4.2.11 *Let $a : \mathcal{G} \to \mathfrak{aff}(E)$ be an affine representation of a Lie algebra $\mathcal{G}$ in a finite-dimensional vector space E, $r : \mathcal{G} \to \mathfrak{gl}(E)$ be its linear part and $\Theta : \mathcal{G} \to E$ be the associated Lie algebra one-cocycle. The following properties are equivalent.*

1. *There exists an element $c \in E$ such that, for all $X \in \mathcal{G}$ and $x \in E$,*

$$a(X)(x) = r(X)(x + c)\,.$$

2. *The one-cocycle $\Theta : \mathcal{G} \to E$ is in fact a* 1*-coboudary, whose expression is*

$$\Theta(X) = r(X)(c)\,.$$

Proof Since for each $X \in \mathcal{G}$ $r(X)$ is linear, Property (i) can be written

$$a(X)(x) = r(X)(x) + r(X)(c)\,.$$

Therefore Property 1 is true if and only if $\Theta(X) = r(X)(c)$, in other words if and only if Property 2 is true. □

Remark 4.2.12 Let us say that an affine representation $A : G \to \mathrm{Aff}(E)$ of a Lie group G in a finite-dimensional vector space E is equivalent to its linear part $R : G \to \mathrm{GL}(E)$ if there exists a translation $T : E \to E$ such that, for all $g \in G$ and $x \in E$,

$$A(g)(x) = T^{-1} \circ R(g) \circ T(x)\,.$$

Proposition 4.2.10 expresses the fact that the affine representation A is equivalent to its linear part R if and only if its associated Lie group cocycle θ is a one-coboundary. The reader will easily formulate a similar interpretation of Proposition 4.2.11.

Proposition 4.2.13 *Let G be a connected and simply connected Lie group, $R : G \to \mathrm{GL}(E)$ be a linear representation of G in a finite-dimensional vector space E, and*

$r : \mathcal{G} \to \mathfrak{gl}(E)$ be the associated linear representation of its Lie algebra $\mathcal{G}$. For any one-cocycle $\Theta : \mathcal{G} \to E$ of the Lie algebra $\mathcal{G}$ for the linear representation r, there exists a unique one-cocycle $\theta : G \to E$ of the Lie group G for the linear representation R such that $\Theta = T_e\theta$, in other words, which has Θ as associated Lie algebra one-cocycle. The Lie group one-cocycle θ is a Lie group one-coboundary if and only if the Lie algrebra one-cocycle Θ is a Lie algebra one-coboundary.

Proof If $\theta : G \to E$ is a Lie group one-cocycle such that $T_e\theta = \Theta$ we have, for any $g \in G$ and $X \in \mathcal{G}$,

$$\theta\bigl(g\exp(tX)\bigr) = \theta(g) + R(g)\Bigl(\theta\bigl(\exp(tX)\bigr)\Bigr).$$

By taking the derivative of both sides of this equality with respect to t, then setting $t = 0$, we see that

$$T_g\theta\bigl(TL_g(X)\bigr) = R(g)\bigl(\Theta(x)\bigr),$$

which proves that if it exists, the Lie group one-cocycle θ such that $T_e\theta = \Theta$ is unique.

For each $g \in G$ let $\eta(g) : T_gG \to E$ be the map

$$\eta(g)(X) = R(g) \circ \Theta \circ TL_{g^{-1}}(X), \quad X \in T_gG.$$

The map η is an E-valued differential one-form on G. Let us calculate its exterior differential $\mathrm{d}\eta$, which is an E-valued differential two-form on G (if the reader does not feel at ease with E-valued differential forms on G, he can consider separately the components of η in a basis of E, which are ordinary real-valued one-forms). Let X and Y be two left-invariant vector fields on G. We have, for each $g \in G$,

$$\mathrm{d}\eta(g)\bigl(X(g), Y(g)\bigr) = \mathcal{L}(X)\bigl(\langle\eta, Y\rangle(g)\bigr) - \mathcal{L}(Y)\bigl(\langle\eta, X\rangle(g)\bigr) - \bigl\langle\eta, [X, Y]\bigr\rangle(g).$$

But

$$\langle\eta, Y\rangle(g) = R(g) \circ \Theta(Y), \quad \langle\eta, X\rangle(g) = R(g) \circ \Theta(X),$$

therefore

$$\mathcal{L}(X)\bigl(\langle\eta, Y\rangle(g)\bigr) = \frac{\mathrm{d}}{\mathrm{d}t}\Bigl(R\bigl(g\exp(tX)\bigr) \circ \Theta(Y)\Bigr)\Big|_{t=0} = R(g) \circ r(X) \circ \Theta(Y).$$

Similarly

$$\mathcal{L}(Y)\bigl(\langle\eta, X\rangle(g)\bigr) = R(g) \circ r(Y) \circ \Theta(X),$$

and

$$\bigl\langle\eta, [X, Y]\bigr\rangle(g) = R(g) \circ \Theta\bigl([X, Y]\bigr),$$

Since the condition which expresses that Θ is a Lie algebra one-cocycle for the linear representation r asserts that

$$r(X)\circ\Theta(Y)-r(Y)\circ\Theta(X)-\Theta\bigl([X,Y]\bigr)=0\,,$$

we conclude that the one-form η is closed, i.e. satisfies $d\eta=0$. Since G is assumed to be simply connected, the one-form η is exact, and since G is assumed to be connected, for any g in G, there exists a smooth parametrized curve $\gamma:[0,T]\to G$ such that $\gamma(0)=e$ and $\gamma(T)=g$. Let us set

$$\theta(g)=\int_0^T\eta\left(\frac{\mathrm{d}\gamma(t)}{\mathrm{d}t}\right)dt\,.$$

Since η is exact, the right hand side of the above equality only depends on the end points $\gamma(0)=e$ and $\gamma(T)=g$ of the parametrized curve γ, which allows us to define $\theta(g)$ by that equality. So defined, $\gamma:G\to E$ is a smooth map. Its very definition shows that $T_e\theta=\Theta$. If g and h are two elements in G, let $\gamma:[0,T_2]\to G$ be a smooth parametrized curve such that $0<T_1<T_2$, $\gamma(0)=e$, $\gamma(T_1)=g$ and $\gamma(t_2)=gh$. We have

$$\theta(gh)=\int_0^{T_2}\eta\left(\frac{\mathrm{d}\gamma(t)}{\mathrm{d}t}\right)dt=\int_0^{T_1}\eta\left(\frac{\mathrm{d}\gamma(t)}{\mathrm{d}t}\right)dt+\int_{T_1}^{T_2}\eta\left(\frac{\mathrm{d}\gamma(t)}{\mathrm{d}t}\right)dt\,.$$

Observe that

$$\int_0^{T_2}\eta\left(\frac{\mathrm{d}\gamma(t)}{\mathrm{d}t}\right)dt=\theta(g)$$

and that

$$\int_{T_1}^{T_2}\eta\left(\frac{\mathrm{d}\gamma(t)}{\mathrm{d}t}\right)dt=R(g)\circ\int_{T_1}^{T_2}\eta\left(\frac{\mathrm{d}\bigl(L_{g^{-1}}\circ\gamma(t)\bigr)}{dt}\right)dt=R(g)\bigl(\theta(h)\bigr)\,,$$

which proves that θ is a Lie group one-cocycle.

We already know that if θ is a Lie group one-coboundary, $\Theta=T_e\theta$ is a Lie algebra coboundary. Conversely let us assume that Θ is a Lie algebra one-coboundary. We have, for each $X\in\mathcal{G}$,

$$\Theta(X)=r(X)(c)\,,$$

where c is a fixed element in E. Let $g\in G$ and let $\gamma:[0,T]\to G$ be a smooth parametrized curve in G such that $\gamma(0)=e$ and $\gamma(T)=g$. We have

$$\theta(g) = \int_0^T \eta\left(\frac{\mathrm{d}\gamma(t)}{\mathrm{d}t}\right) dt = \int_0^t R\bigl(\gamma(t)\bigr) \circ r\left(TL_{\left(\gamma(t)\right)^{-1}} \frac{\mathrm{d}\gamma(t)}{\mathrm{d}t}\right)(c)\, dt\,.$$

But by taking the derivative with respect to t of the two sides of the equality

$$R\bigl(g\exp(tX)\bigr) = R(g) \circ R\bigl(\exp(tX)\bigr)$$

and then setting $t = 0$, we see that

$$\frac{\mathrm{d}}{\mathrm{d}t} R\bigl(\gamma(t)\bigr) = R\bigl(\gamma(t)\bigr) \circ r\left(TL_{\left(\gamma(t)\right)^{-1}} \frac{\mathrm{d}\gamma(t)}{\mathrm{d}t}\right).$$

Therefore θ is a Lie group one-coboundary since we have

$$\theta(g) = \int_0^T \frac{\mathrm{d}R\bigl(\gamma(t)\bigr)}{\mathrm{d}t}(c)\, dt = R(g)(c) - R(e)(c) = R(g)(c) - c\,. \qquad \square$$

4.3 Poisson, Symplectic and Hamiltonian Actions

Definitions 4.3.1
1. An action φ of a Lie algebra $\mathcal{G}$ on a Poisson manifold (M, Λ) is called a *Poisson action* if for any $X \in \mathcal{G}$ the corresponding vector field $\varphi(X)$ is a Poisson vector field. When the Poisson manifold is in fact a symplectic manifold (M, ω), Poisson vector fields on M are locally Hamiltonian vector fields and a Poisson action is called a *symplectic action.*
2. An action Φ (either on the left or on the right) of a Lie group G on a Poisson manifold (M, Λ) is called a *Poisson action* when for each $g \in G$,

$$(\Phi_g)_*\Lambda = \Lambda\,.$$

When the Poisson manifold (M, Λ) is in fact a symplectic manifold (M, ω), a Poisson action is called a *symplectic action*; the fibre bundles isomorphism $\Lambda^\sharp : T^*M \to TM$ being the inverse of $\omega^\flat : TM \to T^*M$, we also can say that an action Φ of a Lie group G on a symplectic manifold (M, ω) is called a *symplectic action* when for each $g \in G$,

$$(\Phi_g)^*\omega = \omega\,.$$

Proposition 4.3.2 *We assume that G is a connected Lie group which acts by an action Φ, either on the left or on the right, on a Poisson manifold (M, Λ), in such a way that the corresponding action of its Lie algebra $\mathcal{G}$ is a Poisson action. Then the action Φ itself is a Poisson action.*

Proof Let $X \in \mathcal{G}$. For each $x \in M$, the parametrized curve $s \mapsto \Phi_{\exp(sX)}(x)$ is the integral curve of the fundamental vector field X_M which takes the value x for $s = 0$. In other words, the reduced flow of the vector field X_M is the map, defined on $\mathbb{R} \times M$ and taking its values in M,

$$(s, x) \mapsto \Phi_{\exp(sX)}(x)\,.$$

According to a formula which relates inverse images of multivectors or differential forms with respect to the flow of a vector field, with their Lie derivatives with respect to that vector field (see for example [17], Appendix 1, Sect. 3.4, page 351), for any $s_0 \in \mathbb{R}$

$$\frac{\mathrm{d}}{\mathrm{d}s}\Bigl(\bigl((\Phi_{\exp(sX)})^*(\Lambda)\bigr)(x)\Bigr)\Big|_{s=s_0} = \Bigl((\Phi_{\exp(s_0X)})^*\bigl(\mathcal{L}(X_M)\Lambda\bigr)\Bigr)(x) = 0\,,$$

since $\mathcal{L}(X_M)\Lambda = 0$. Therefore for any $s \in \mathbb{R}$,

$$(\Phi_{\exp(sX)})^*\Lambda = (\Phi_{\exp(-sX)})_*\Lambda = \Lambda\,.$$

The Lie group G being connected, any $g \in G$ is the product of a finite number of exponentials, so $(\Phi_g)_*\Lambda = \Lambda$. □

4.3.1 Other Characterizations of Poisson Actions

Let Φ be an action, either on the left or on the right, of a Lie group G on a Poisson manifold (M, Λ). The reader will easily prove that the following properties are equivalent. Therefore any of these properties can be used as the definition of a Poisson action.

1. For each $g \in G$,
$$(\Phi_g)_*\Lambda = \Lambda\,.$$
2. For each $g \in G$ and $f \in C^\infty(M, \mathbb{R})$,
$$(\Phi_g)_*(X_f) = X_{(\Phi_g)_*(f)}\,.$$
3. For each $g \in G$, $\Phi_g : M \to M$ is a *Poisson map*, which means that for each pair (f_1, f_2) of smooth functions on M,
$$\bigl\{(\Phi_g)^*f_1, (\Phi_g)^*f_2\bigr\} = (\Phi_g)^*\bigl(\{f_1, f_2\}\bigr)\,;$$
4. In the special case when the Poisson manifold (M, Λ) is in fact a symplectic manifold (M, ω), for each $g \in G$,
$$(\Phi_g)^*\omega = \omega\,.$$

The reader will easily prove that when these equivalent properties are satisfied, the action of the Lie algebra $\mathcal{G}$ of G which associates, to each $X \in \mathcal{G}$, the fundamental vector field X_M on M, is a Poisson action.

Definitions 4.3.3

1. An action φ of a Lie algebra $\mathcal{G}$ on a Poisson manifold (M, Λ) is called a *Hamiltonian action* if for every $X \in \mathcal{G}$ the corresponding vector field $\varphi(X)$ is a Hamiltonian vector field on M.

2. An action Φ (either on the left or on the right) of a Lie group G on a Poisson manifold (M, Λ) is called a *Hamiltonian action* if it is a Poisson action (or a symplectic action when the Poisson manifold (M, Λ) is in fact a symplectic manifold(M, ω)) and if, in addition, the associated action φ of its Lie algebra is a Hamiltonian action.

Remarks 4.3.4

1. A Hamiltonian action of a Lie algebra on a Poisson manifold is automatically a Poisson action.

2. An action Φ of a connected Lie group G on a Poisson manifold such that the corresponding action of its Lie algebra is Hamiltonian, automatially is a Hamiltonian action.

3. Very often, Hamiltonian actions of a Lie algebra (or of a Lie group) on the cotangent bundle T^*N to a smooth manifold N encountered in applications come from an action of this Lie algebra (or of this Lie group) on the manifold N itself. Proposition 4.3.5 explains how an action on N can be lifted to T^*N into a Hamiltonian action.

Proposition 4.3.5 *Let $\varphi : \mathcal{G} \to A^1(N)$ be an action of a finite-dimensional Lie algebra $\mathcal{G}$ on a smooth manifold N. Let $\widehat{\varphi} : \mathcal{G} \to A^1(T^*N)$ be the map wich associates to each $X \in \mathcal{G}$ the canonical lift to T^*N of the vector field $\varphi(X)$ on N (2.4.6). The map $\widehat{\varphi}$ is a Hamiltonian action of $\mathcal{G}$ on $(T^*N, \mathrm{d}\eta_N)$ (where η_N is the Liouville form and $\mathrm{d}\eta_N$ the canonical symplectic form on T^*N). For each $X \in \mathcal{G}$, the smooth function $f_X : T^*N \to \mathbb{R}$*

$$f_X(\xi) = \Big\langle \xi, \varphi(X)\big(\pi_N(\xi)\big)\Big\rangle = \mathrm{i}\big(\widehat{\varphi}(X)\big)\eta_N(\xi)\,, \quad \xi \in T^*N\,,$$

is a Hamiltonian for the vector field $\widehat{\varphi}(X)$. Moreover, for each pair (X, Y) of elements in $\mathcal{G}$,

$$\{f_X, f_Y\} = f_{[X,Y]}\,.$$

Proof Proposition 2.4.9 proves that for each $X \in \mathcal{G}$ the vector field $\widehat{\varphi}(X)$ is Hamiltonian and admits the function f_X as Hamiltonian. This Proposition also shows that f_X is given by the two equivalent expressions

$$f_X(\xi) = \Big\langle \xi, \varphi(X)\big(\pi_N(\xi)\big)\Big\rangle = \mathrm{i}\big(\widehat{\varphi}(X)\big)\eta_N(\xi)\,, \quad \xi \in T^*N\,.$$

Let (X, Y) be a pair of elements in $\mathcal{G}$. Since the vector fields $\widehat{\varphi}(X)$ and $\widehat{\varphi}(Y)$ admit f_X and f_Y as Hamiltonians, Lemma 2.5.2 shows that $\big[\widehat{\varphi}(X), \widehat{\varphi}(Y)\big]$ admits $\{f_X, f_Y\}$ as

Hamiltonian. We have

$$\{f_X, f_Y\} = \mathcal{L}\big(\widehat{\varphi}(X)\big)f_Y = \mathcal{L}\big(\widehat{\varphi}(X)\big) \circ \mathrm{i}\big(\widehat{\varphi}(Y)\big)\eta_N = \mathrm{i}\Big[\widehat{\varphi}(X), \widehat{\varphi}(Y)\Big]\eta_N$$

since, using 2.4.8, we see that $\mathcal{L}\big(\widehat{\varphi}(X)\big)\eta_N = 0$. Therefore, for each $\xi \in T^*N$,

$$\{f_X, f_Y\}(\xi) = \Big\langle \xi, T\pi_N\big([\widehat{\varphi}(X), \widehat{\varphi}(Y)](\xi)\big)\Big\rangle = \big\langle \xi, [X, Y] \circ \pi_N(\xi)\big\rangle = f_{[X,Y]}(\xi)$$

since $T\pi_N\big([\widehat{\varphi}(X), \widehat{\varphi}(Y)](\xi)\big) = [X, Y] \circ \pi_N(\xi)$. Since $\{f_X, f_Y\} = f_{[X,Y]}$, the corresponding Hamiltonian vector fields $\big[\widehat{\varphi}(X), \widehat{\varphi}(Y)\big]$ and $\widehat{\varphi}\big([X, Y]\big)$ are equal. In other words, $\widehat{\varphi}$ is a Lie algebra action of $\mathcal{G}$ on $(T^*N, \mathrm{d}\eta_N)$. □

Proposition 4.3.6 *Let φ be a Hamiltonian action of a Lie algebra $\mathcal{G}$ on a Poisson manifold (M, Λ). Let $\mathcal{G}^*$ be the dual space of $\mathcal{G}$. There exists a smooth map $J : M \to \mathcal{G}^*$ such that for each $X \in \mathcal{G}$ the corresponding Hamiltonian vector field X_M has the function $J_X : M \to \mathbb{R}$, defined by*

$$J_X(x) = \big\langle J(x), X\big\rangle, \quad \textit{with } x \in M,$$

as Hamiltonian.

Such a map $J : M \to \mathcal{G}^*$ is called a momentum map *for the Hamiltonian Lie algebra action φ. When φ is the Lie algebra action associated to a Hamiltonian action Φ of a Lie group G on the Poisson manifold (M, Λ), J is called a* momentum map *for the Hamiltonian Lie group action Φ.*

Proof Let $(e_1, \ldots, e_p)$ be a basis of the Lie algebra $\mathcal{G}$ and $(\varepsilon^1, \ldots, \varepsilon^p)$ be the dual basis of $\mathcal{G}^*$. Since φ is Hamiltonian, for each i $(1 \leq i \leq p)$ there exists a Hamiltonian $J_{e_i} : M \to \mathbb{R}$ for the Hamiltonian vector field $\varphi(e_i)$. The map $J : M \to \mathcal{G}$ defined by

$$J(x) = \sum_{i=1}^{p} J_{e_i}\varepsilon^i, \quad x \in M,$$

is a momentum map for φ.

The momentum map was introduced by Souriau [23] and, in the Lagrangian formalism, by Smale [24, 25]. □

4.4 Some Properties of Momentum Maps

Proposition 4.4.1 *Let φ be a Hamiltonian action of a Lie algebra $\mathcal{G}$ on a Poisson manifold (M, Λ), and $J : M \to \mathcal{G}^*$ be a momentum map for that action. For any pair $(X, Y) \in \mathcal{G} \times \mathcal{G}$, the smooth function $\widetilde{\Theta}(X, Y) : M \to \mathbb{R}$ defined by*

$$\widetilde{\Theta}(X, Y) = \{J_X, J_Y\} - J_{[X,Y]}$$

is a Casimir of the Poisson algebra $C^\infty(M,\mathbb{R})$*, which satisfies, for all* X*,* Y *and* $Z\in\mathcal{G}$*,*

$$\widetilde{\Theta}\big([X,Y],Z\big)+\widetilde{\Theta}\big([Y,Z],X\big)+\widetilde{\Theta}\big([Z,X],Y\big)=0\,. \tag{1}$$

When the Poisson manifold (M,Λ) *is in fact a connected symplectic manifold* (M,ω)*, for any pair* $(X,Y)\in\mathcal{G}\times\mathcal{G}$ *the function* $\widetilde{\Theta}(X,Y)$ *is constant on* M*, and the map* $\widetilde{\Theta}:\mathcal{G}\times\mathcal{G}\to\mathbb{R}$ *is a skew-symmetric bilinear form, which satisfies the above identity* (1).

Proof Since J_X and J_Y are Hamiltonians for the Hamiltonian vector fields $\varphi(X)$ and $\varphi(Y)$, the Poisson bracket $\{J_X,J_Y\}$ is a Hamiltonian for $\big[\varphi(X),\varphi(Y)\big]$. Since $\varphi:\mathcal{G}\to A^1(M)$ is a Lie algebras homomorphism, $\big[\varphi(X),\varphi(Y)\big]=\varphi\big([X,Y]\big)$, and $J_{[X,Y]}$ is a Hamiltonian for this vector field. We have two different Hamiltonians for the same Hamiltonian vector field. Their difference $\widetilde{\Theta}(X,Y)$ is therefore a Casimir of the Poisson algebra $C^\infty(M,\mathbb{R})$.

Let X, Y and Z be three elements in $\mathcal{G}$. We have

$$\begin{aligned}\widetilde{\Theta}\big([X,Y],Z\big)&=\{J_{[X,Y]},J_Z\}-J_{\big[[X,Y],Z\big]}\\&=\big\{\{J_X,J_Y\}-\widetilde{\Theta}(X,Y),J_Z\big\}-J_{\big[[X,Y],Z\big]}\\&=\big\{\{J_X,J_Y\},J_Z\big\}-J_{\big[[X,Y],Z\big]}\end{aligned}$$

since $\widetilde{\Theta}(X,Y)$ is a Casimir of the Poisson algebra $C^\infty(M,\mathbb{R})$. Similarly

$$\begin{aligned}\widetilde{\Theta}\big([Y,Z],X\big)&=\big\{\{J_Y,J_Z\},J_X\big\}-J_{\big[[Y,Z],X\big]}\,,\\\widetilde{\Theta}\big([Z,X],Y\big)&=\big\{\{J_Z,J_X\},J_Y\big\}-J_{\big[[Z,X],Y\big]}\,.\end{aligned}$$

Adding these three terms and using the fact that the Poisson bracket of functions and the bracket in the Lie algebra $\mathcal{G}$ both satisfy the Jacobi identity, we see that $\widetilde{\Theta}$ satisfies (1).

When (M,Λ) is in fact a connected symplectic manifold (M,ω), the only Casimirs of the Poisson algebra $C^\infty(M,\mathbb{R})$ are the constants, and $\widetilde{\Theta}$ becomes a bilinear skew-symmetric form on $\mathcal{G}$. □

Definition 4.4.2 Under the assumptions of Proposition 4.4.1, the skew-symmetric bilinear map $\widetilde{\Theta}$, defined on $\mathcal{G}\times\mathcal{G}$ and taking its values in the space of Casimirs of the Poisson algebra $C^\infty(M,\mathbb{R})$ (real-valued when the Poisson manifold (M,Λ) is in fact a connected symplectic manifold (M,ω)), is called the *symplectic cocycle of the Lie algebra* $\mathcal{G}$ associated to the momentum map J.

Remark 4.4.3 Under the assumptions of Proposition 4.4.1, let us assume in addition that the Poisson manifold (M,Λ) is in fact a connected symplectic manifold (M,ω). The symplectic cocycle $\widetilde{\Theta}$ is then a real-valued skew-symmetric bilinear form on $\mathcal{G}$. Therefore it is a symplectic cocycle in the sense of Sect. 3.4. Two different interpretations of this cocycle can be given.

1. Let $\Theta : \mathcal{G} \to \mathcal{G}^*$ be the map such that, for all X and $Y \in \mathcal{G}$

$$\langle \Theta(X), Y \rangle = \widetilde{\Theta}(X, Y) .$$

   Written for Θ, Eq. (1) of 4.4.1 becomes

$$\Theta\bigl([X, Y]\bigr) = \mathrm{ad}^*_{-X}\bigl(\Theta(Y)\bigr) - \mathrm{ad}^*_{-Y}\bigl(\Theta(X)\bigr) , \quad X \text{ and } Y \in \mathcal{G} .$$

   The map Θ is therefore the one-cocycle of the Lie algebra $\mathcal{G}$ with values in $\mathcal{G}^*$, for the coadjoint representation (4.2.3) $X \mapsto \mathrm{ad}^*_{-X}$ of $\mathcal{G}$, associated to the affine action of $\mathcal{G}$ on its dual

$$a_\Theta(X)(\zeta) = \mathrm{ad}^*_{-X}(\zeta) + \Theta(X) , \quad X \in \mathcal{G} , \ \zeta \in \mathcal{G}^* ,$$

   in the sense of 4.2.5. The reader is referred to the book [26] for a more thorough discussion of the cohomology theories of Lie groups and Lie algebras.
2. Let G be a Lie group whose Lie algebra is $\mathcal{G}$. The skew-symmetric bilinear form $\widetilde{\Theta}$ on $\mathcal{G} = T_eG$ can be extended, either by left translations or by right translations, into a left invariant (or a right invariant) closed differential two-form on G, since the identity (1) of 4.4.1 means that its exterior differential $\mathrm{d}\widetilde{\Theta}$ vanishes. In other words, $\widetilde{\Theta}$ is a 2-cocycle for the restriction of the de Rham cohomology of G to left (or right) invariant differential forms.

Proposition 4.4.4 *Let $\varphi : \mathcal{G} \to A^1(N)$ be an action of a finite-dimensional Lie algebra $\mathcal{G}$ on a smooth manifold N, and let $\widehat{\varphi} : \mathcal{G} \to A^1(T^*N)$ be the Hamiltonian action of $\mathcal{G}$ on $(T^*N, \mathrm{d}\eta_N)$ introduced in Proposition 4.3.5. The map $J : T^*N \to \mathcal{G}^*$ defined by*

$$\langle J(\xi), X \rangle = \mathrm{i}\bigl(\widehat{\varphi}(X)\bigr)\eta_N(\xi) , \quad X \in \mathcal{G} , \ \xi \in T^*N ,$$

is a momentum map for the action $\widehat{\varphi}$ which satisfies, for all X and $Y \in \mathcal{G}$,

$$\{J_X, J_Y\} = J_{[X,Y]} .$$

In other words, the symplectic cocycle of $\mathcal{G}$ associated to J, in the sense of 4.4.2, identically vanishes.

Proof These properties immediately follow from 4.3.5. □

Theorem 4.4.5 (First Emmy Noether's theorem in Hamiltonian form) *Let φ be a Hamiltonian action of a Lie algebra $\mathcal{G}$ on a Poisson manifold (M, Λ), $J : M \to \mathcal{G}^*$ be a momentum map for φ and $H : M \to \mathbb{R}$ be a smooth Hamiltonian. If the action φ leaves H invariant, that means if*

$$\mathcal{L}\bigl(\varphi(X)\bigr)H = 0 \quad \textit{for any } X \in \mathcal{G} ,$$

the momentum map J is a $\mathcal{G}^*$-valued first integral (3.3.4) of the Hamiltonian vector field $\Lambda^\sharp(\mathrm{d}H)$, which means that it keeps a constant value along each integral curve of that vector field.

Proof For any $X \in \mathcal{G}$, let $J_X : M \to \mathbb{R}$ be the function $x \mapsto \langle J(x), X\rangle$. Let $t \mapsto \psi(t)$ be an integral curve of the Hamiltonian vector field $\Lambda^\sharp(\mathrm{d}H)$. We have

$$\begin{aligned}\frac{\mathrm{d}}{\mathrm{d}t}\Bigl(J_X\bigl(\psi(t)\bigr)\Bigr) &= \mathcal{L}\bigl(\Lambda^\sharp(\mathrm{d}H)\bigr)\bigl(J_X\bigr)\bigl(\psi(t)\bigr) = \Lambda\bigl(\mathrm{d}H, \mathrm{d}J_X\bigr)(\psi(t))\\ &= -\mathcal{L}\Bigl(\Lambda^\sharp\bigl(\mathrm{d}J_X\bigr)\Bigr)H = -\mathcal{L}\bigl(\varphi(X)\bigr)H = 0\,.\end{aligned}$$

Therefore, for any $X \in \mathcal{G}$, the derivative of $\langle J, X\rangle\bigl(\psi(t)\bigr)$ with respect to the parameter t of the parametrized curve $t \mapsto \psi(t)$ vanishes identically, which means that J keeps a constant value along that curve. □

The reader will find in the book by Yvette Kosmann-Schwarzbach [27] a very nice exposition of the history and scientific applications of the Noether's theorems.

Proposition 4.4.6 *Let φ be a Hamiltonian action of a Lie algebra $\mathcal{G}$ on a Poisson manifold (M, Λ) and $J : M \to \mathcal{G}^*$ be a momentum map for that action. Let S be a symplectic leaf of (M, Λ) and ω_S be its symplectic form.*

1. *For each $x \in S$, in the symplectic vector space $\bigl(T_xS, \omega_S(x)\bigr)$, each of the two vector subspaces $T_xS \cap ker(T_xJ)$ and $\{\,\varphi(X)(x)\,;\ X \in \mathcal{G}\,\}$ is the symplectic orthogonal of the other.*

2. *For each $x \in S$, $T_xJ(T_xS)$ is the annihilator of the isotropy subalgebra $\mathcal{G}_x = \bigl\{X \in \mathcal{G};\ \phi(X)(x) = 0\bigr\}$ of x.*

Proof Let $v \in T_xS$. For each $X \in \mathcal{G}$ we have

$$\omega_S\bigl(v, \varphi(X)(x)\bigr) = \bigl\langle\, d\langle J, X\rangle(x), v\bigr\rangle = \bigl\langle T_xJ(v), X\bigr\rangle\,.$$

Therefore a vector $v \in T_xS$ belongs to $\mathrm{orth}\bigl\{\,\varphi(X)(x)\,;\ X \in \mathcal{G}\,\bigr\}$ if and only if $T_xJ(v) = 0$. In other words, in the symplectic vector space $\bigl(T_xS, \omega_S(x)\bigr)$, $T_xS \cap \ker(T_xJ)$ is the symplectic orthogonal of $\bigl\{\,\varphi(X)(x)\,;\ X \in \mathcal{G}\,\bigr\}$. Of course, conversely $\bigl\{\,\varphi(X)(x)\,;\ X \in \mathcal{G}\,\bigr\}$ is the symplectic orthogonal of $T_xS \cap \ker(T_xJ)$.

The same formula shows that $\bigl\langle T_xJ(v), X\bigr\rangle = 0$ for all $v \in T_xS$ if and only if $X \in \mathcal{G}_x$. □

Remark 4.4.7 Under the assumptions of 4.4.6, when φ is the Lie algebra action associated to a Hamiltonian action Φ of a Lie group G, the vector space $\bigl\{\,\varphi(X)(x)\,;\ X \in \mathcal{G}\,\bigr\}$ is the space tangent at x to the G-orbit of this point.

Corollary 4.4.8 *Let φ be a Hamiltonian action of a Lie algebra $\mathcal{G}$ on a symplectic manifold (M, ω) and $J : M \to \mathcal{G}^*$ be a momentum map for that action.*

1. *For each $x \in M$, in the symplectic vector space $\bigl(T_xM, \omega(x)\bigr)$ each of the two vector subspaces* $\ker(T_xJ)$ *and $\bigl\{\,\varphi(X)(x)\,;\ X \in \mathcal{G}\,\bigr\}$ is the symplectic orthogonal of the other.*

2. *For each* $x \in M$, $T_xJ(T_xM)$ *is the annihilator of the isotropy subalgebra* $\mathcal{G}_x = \{X \in \mathcal{G};\, \varphi(X)(x) = 0\}$ *of* x.

Proof These assertions both follow immediately from 4.4.6 since the symplectic leaves of (M, ω) are its connected components. □

Proposition 4.4.9 *Let* Φ *be a Hamiltonian action of a Lie group* G *on a connected symplectic manifold* (M, ω) *and* $J : M \to \mathcal{G}^*$ *be a momentum map for that action. There exists a unique action* A *of the Lie group* G *on the dual* $\mathcal{G}^*$ *of its Lie algebra for which the momentum map* J *is equivariant, that means satisfies for each* $x \in M$ *and* $g \in G$

$$J\bigl(\Phi_g(x)\bigr) = A_g\bigl(J(x)\bigr)\,.$$

The action A *is an action on the left (respectively, on the right) if* Φ *is an action on the left (respectively, on the right), and its expression is*

$$\begin{cases} A(g,\xi) = \mathrm{Ad}^*_{g^{-1}}(\xi) + \theta(g) & \text{if } \Phi \text{ is an action on the left,} \\ A(\xi,g) = \mathrm{Ad}^*_{g}(\xi) - \theta(g^{-1}) & \text{if } \Phi \text{ is an action on the right,} \end{cases} \qquad g \in G\,,\ \xi \in \mathcal{G}^*\,.$$

The map $\theta : G \to \mathcal{G}^*$ *is called the* symplectic cocycle of the Lie group G *associated to the momentum map* J.

Proof Let us first assume that Φ is an action on the left. For each $X \in \mathcal{G}$ the associated fundamental vector field X_M is Hamiltonian and the function $J_X : M \to \mathbb{R}$ defined by

$$J_X(x) = \bigl\langle J(x), X\bigr\rangle\,, \quad x \in M\,,$$

is a Hamiltonian for X_M. We know by the characterizations Sect. 4.3 of Poisson actions that $(\Phi_{g^{-1}})_*(X_M)$, the direct image of X_M by the diffeomorphism $\Phi_{g^{-1}}$, is a Hamiltonian vector field for which the function $J_X \circ \Phi_g$ is a Hamiltonian. Proposition 4.1.6 shows that $(\Phi_{g^{-1}})_*(X_M)$ is the fundamental vector field associated to $\mathrm{Ad}_{g^{-1}}(X)$, therefore has the function

$$x \mapsto \bigl\langle J(x), \mathrm{Ad}_{g^{-1}}(X)\bigr\rangle = \bigl\langle \mathrm{Ad}^*_{g^{-1}} \circ J(x), X\bigr\rangle$$

as a Hamiltonian. The difference between these two Hamiltonians for the same Hamiltonian vector field is a constant since M is assumed to be connected. Therefore the expression

$$\bigl\langle J \circ \Phi_g(x) - \mathrm{Ad}^*_{g^{-1}} \circ J(x), X\bigr\rangle$$

does not depend on $x \in M$, and depends linearly on $X \in \mathcal{G}$ (and of course smoothly depends on $g \in G$). We can therefore define a smooth map $\theta : G \to \mathcal{G}^*$ by setting

$$\theta(g) = J \circ \Phi_g - \mathrm{Ad}^*_{g^{-1}} \circ J\,, \quad g \in G\,.$$

It follows that the map $a : G \times \mathcal{G}^* \to \mathcal{G}^*$,

$$a(g, \xi) = \mathrm{Ad}^*_{g^{-1}}(\xi) + \theta(g)$$

is an action on the left of the Lie group G on the dual $\mathcal{G}^*$ of its Lie algebra, which renders the momentum map J equivariant.

The case when Φ is an action on the right easily follows by observing that $(g, x) \mapsto \Phi(x, g^{-1})$ is a Hamiltonian action on the left whose momentum map is the opposite of that of Φ. □

Proposition 4.4.10 *Under the same assumptions as those of Proposition 4.4.9, the map $\theta : G \to \mathcal{G}^*$ satisfies, for all g and $h \in G$,*

$$\theta(gh) = \theta(g) + \mathrm{Ad}^*_{g^{-1}}\big(\theta(h)\big)\,.$$

Proof In Proposition 4.4.9, the cocycle θ introduced for an action on the right $\Psi : M \times G \to M$ was the cocycle of the corresponding action on the left $\Phi : G \times M \to M$ defined by $\Phi(g, x) = \Psi(x, g^{-1})$. We can therefore consider only the case when Φ is an action on the left.

Let g and $h \in G$. We have

$$\begin{aligned}
\theta(gh) &= J\big(\Phi(gh, x)\big) - \mathrm{Ad}^*_{(gh)^{-1}} J(x) \\
&= J\Big(\Phi\big(g, \Phi(h, x)\big)\Big) - \mathrm{Ad}^*_{g^{-1}} \circ \mathrm{Ad}^*_{h^{-1}} J(x) \\
&= \theta(g) + \mathrm{Ad}^*_{g^{-1}}\Big(J\big(\Phi(h, x)\big) - \mathrm{Ad}^*_{h^{-1}} J(x)\Big) \\
&= \theta(g) + \mathrm{Ad}^*_{g^{-1}} \theta(h)\,.
\end{aligned}$$

□

Proposition 4.4.11 *Let Φ be a Hamiltonian action of a Lie group G on a connected symplectic manifold (M, ω) and $J : M \to \mathcal{G}^*$ be a momentum map for that action. The symplectic cocycle $\theta : G \to \mathcal{G}^*$ of the Lie group G introduced in Proposition 4.4.9 and the symplectic cocycle $\Theta : \mathcal{G} \to \mathcal{G}^*$ of its Lie algebra $\mathcal{G}$ introduced in Definition 4.4.2 and Remark 4.4.3 are related by*

$$\Theta = T_e\theta\,,$$

where e is the neutral element of G, the Lie algebra $\mathcal{G}$ being identified with T_eG and the tangent space at $\mathcal{G}^*$ at its origin being identified with $\mathcal{G}^*$. Moreover J is a Poisson map when $\mathcal{G}^*$ is endowed with

- *its canonical Poisson structure modified by the symplectic cocycle Θ (defined in Sect. 3.4) if Φ is an action on the right,*
- *the opposite of this Poisson structure if Φ is an action on the left.*

Proof As in the proof of Proposition 4.4.10, we have only to consider the case when Φ is an action on the left. The map which associates to each $X \in \mathcal{G}$ the fundamental vector field X_M is a Lie algebras homomorphism when $\mathcal{G}$ is endowed with the Lie algebra structure of *right invariant vector fields* on the Lie group G. We will follow here the more common convention, in which $\mathcal{G}$ is endowed with the Lie algebra structure of *left invariant vector fields* on G. With this convention the map $X \mapsto X_M$ is a Lie algebras antihomomorphism and we must change a sign in the definition of $\widetilde{\Theta}$ given in Proposition 4.4.1 and take

$$\widetilde{\Theta}(X, Y) = \bigl\langle \Theta(X), Y \bigr\rangle = \{J_X, J_Y\} + J_{[X,Y]}, \quad X \text{ and } Y \in \mathcal{G}.$$

We have, for any $x \in M$,

$$\begin{aligned} \{J_X, J_Y\}(x) &= \omega(X_M, Y_M)(x) = i(X_M)d\bigl(\langle J, Y\rangle\bigr)(x) \\ &= \frac{\mathrm{d}}{\mathrm{d}t}\Bigl\langle J\bigl(\Phi(\exp(tX), x\bigr), Y\Bigr\rangle\Big|_{t=0} \\ &= \frac{\mathrm{d}}{\mathrm{d}t}\Bigl\langle \mathrm{Ad}^*_{\exp(-tX)} J(x) + \theta\bigl(\exp(tX)\bigr), Y\Bigr\rangle\Big|_{t=0} \\ &= \bigl\langle J(x), -[X, Y]\bigr\rangle + \bigl\langle T_e\theta(X), Y\bigr\rangle \\ &= -J_{[X,Y]}(x) + \bigl\langle T_e\theta(X), Y\bigr\rangle . \end{aligned}$$

We see that $\Theta = T_e\theta$. Moreover, the elements X and Y in $\mathcal{G}$ can be considered as linear functions on $\mathcal{G}^*$. Their Poisson bracket, when $\mathcal{G}^*$ is equipped with its canonical Poisson structure modified by $\widetilde{\Theta}$, is

$$\{X, Y\}_{\widetilde{\Theta}}(\xi) = \bigl\langle \xi, [X, Y]\bigr\rangle - \widetilde{\Theta}(X, Y) .$$

The formula $\{J_X, J_Y\}(x) = -J_{[X,Y]}(x) + \widetilde{\Theta}(X, Y)$ can be read as

$$\{X \circ J, Y \circ J\}(x) = -\{X, Y\}_{\widetilde{\Theta}} \circ J(x) .$$

Since the value taken at a point by the Poisson bracket of two functions only depends on the values of the differentials of these two functions at that point, this result proves that J is a Poisson map when $\mathcal{G}^*$ is equipped with the opposite of the Poisson bracket $\{\ ,\ \}_{\widetilde{\Theta}}$. □

Remarks 4.4.12 Let Φ be a Hamiltonian action on the left of a Lie group G on a connected symplectic manifold (M, ω), $J : M \to \mathcal{G}^*$ be a momentum map for that action and $\theta : G \to \mathcal{G}^*$ be the symplectic cocycle of the Lie group G introduced in Proposition 4.4.9.

1. The symplectic cocycle $\theta : G \to \mathcal{G}^*$ is the Lie group one-cocycle with values in $\mathcal{G}^*$, for the coadjoint representation, associated to the affine representation $A : G \to \mathrm{Aff}(\mathcal{G}^*)$,

$$A(g)(\zeta) = \mathrm{Ad}^*_{g^{-1}}(\zeta) + \theta(g)\,, \quad \zeta \in \mathcal{G}^*\,,$$

in the sense of 4.2.4.

2. If instead of J we take for momentum map

$$J'(x) = J(x) - c\,, \quad x \in M\,,$$

where $c \in \mathcal{G}^*$ is constant, the symplectic cocycle θ is replaced by

$$\theta'(g) = \theta(g) + \mathrm{Ad}^*_{g^{-1}}(c) - c\,.$$

The map $\theta' - \theta$ is a one-coboundary of G with values in $\mathcal{G}^*$ for the coadjoint representation (4.2.8). Therefore the cohomology class of the symplectic cocycle θ only depends on the Hamiltonian action Φ, not on the choice of its momentum map J. This property is used by Souriau ([23], chapter III, p. 153) to offer a very nice cohomological interpretation of the total mass of a classical (non-relativistic) isolated mechanical system. He proves that the space of all possible motions of the system is a symplectic manifold on which the Galilean group acts by a Hamiltonian action. The dimension of the symplectic cohomology space of the Galilean group (the quotient of the space of symplectic one-cocycles by the space of symplectic one-coboundaries) is equal to 1. The cohomology class of the symplectic cocycle associated to a momentum map of the action of the Galilean group on the space of motions of the system is interpreted as the *total mass* of the system.

4.4.1 Other Properties of the Momentum Map

The momentum map has several other very remarkable properties. Atiyah [28], Guillemin and Sternberg [29, 30] have shown that the image of the momentum map of a Hamiltonian action of a torus on a compact symplectic manifold is a convex polytope. Kirwan [31] adapted this result when the torus is replaced by any compact Lie group. Delzant [32] has shown that the convex polytope which is the image of a Hamiltonian action of a torus on a compact symplectic manifold determines this manifold.

4.5 Actions of a Lie Group on Its Cotangent Bundle

In this section G is a Lie group, $\mathcal{G}$ is its Lie algebra and $\mathcal{G}^*$ is the dual space of $\mathcal{G}$. The Liouville one-form on T^*G is denoted by η_G.

The group composition law $m : G \times G \to G$, $m(g,h) = gh$, can be seen as an action of G on itself either on the left, or on the right. For each $g \in G$ we will denote by $L_g : G \to G$ and $R_g : G \to G$ the diffeomorphisms

$$L_g(h) = gh\,, \quad R_g(h) = hg\,, \quad h \in G\,.$$

called, respectively, the *left translation* and the *right translation* of G by g.

Definitions 4.5.1 The *canonical lifts* to the tangent bundle TG of the actions of G on itself by left translations (respectively, by right translations) are, repectively, the maps $\overline{L} : G \times TG \to TG$ and $\overline{R} : TG \times G \to TG$

$$\overline{L}(g, v) = TL_g(v)\,, \quad \overline{R}(v, g) = TR_g(v)\,, \quad g \in G\,, \quad v \in TG\,.$$

The *canonical lifts* to the cotangent bundle T^*G of the actions of G on itself by left translations (respectively, by right translations) are, respectively, the maps $\widehat{L}$: $G \times T^*G \to T^*G$ and $\widehat{R} : T^*G \times G \to T^*G$

$$\widehat{L}(g, \xi) = \left(TL_{g^{-1}}\right)^T(\xi)\,, \quad \widehat{R}(\xi, g) = \left(TR_{g^{-1}}\right)^T(\xi)\,, \quad g \in G\,, \quad \xi \in T^*G\,.$$

We have denoted by $\left(TL_{g^{-1}}\right)^T$ and $\left(TR_{g^{-1}}\right)^T$ the transposes of the vector bundles morphisms $TL_{g^{-1}}$ and $TR_{g^{-1}}$, respectively.

Proposition 4.5.2 *The canonical lifts to the tangent bundle and to the cotangent bundle of the actions of the Lie group G on itself by left translations (respectively, by right translations) are actions on the left (respectively, on the right) of G on its tangent bundle and on its cotangent bundle, which project onto the actions of G on itself by left translations (respectively, by right translations). It means that for all $g \in G$ and $v \in TG$*

$$\tau_G\big(\overline{L}(g, v)\big) = L_g\big(\tau_G(v)\big)\,, \quad \tau_G\big(\overline{R}(v, g)\big) = R_g\big(\tau_G(v)\big)\,,$$

and that for all $g \in G$ and $\xi \in T^*G$

$$\pi_G\big(\widehat{L}(g, \xi)\big) = L_g\big(\pi_G(\xi)\big)\,, \quad \pi_G\big(\widehat{R}(\xi, g)\big) = R_g\big(\pi_G(\xi)\big)\,.$$

Proof It is an easy verification that the properties of actions are indeed satisfied by the maps $\overline{L}$, $\overline{R}$, $\widehat{L}$ and $\widehat{R}$, which is left to the reader. □

Theorem 4.5.3 *The canonical lifts to the cotangent bundle $\widehat{L}$ and $\widehat{R}$ of the actions of the Lie group G on itself by translations on the left and on the right are two Hamiltonian actions of G on the symplectic manifold $(T^*G, \mathrm{d}\eta_G)$. The maps J^L : $T^*G \to \mathcal{G}^*$ and $J^R : T^*G \to \mathcal{G}^*$ defined, for each $\xi \in T^*G$, by*

$$J^L(\xi) = \widehat{R}\big(\xi, \pi_G(\xi)^{-1}\big)\,, \quad J^R(\xi) = \widehat{L}\big(\pi_G(\xi)^{-1}, \xi\big)$$

are momentum maps for the actions $\widehat{L}$ and $\widehat{R}$, respectively.

Moreover, the map J^L is constant on each orbit of the action $\widehat{R}$, the map J^R is constant on each orbit of the action $\widehat{L}$ and for each $\xi \in T^*G$ each of the tangent spaces at ξ to the orbits $\widehat{L}(G, \xi)$ and $\widehat{R}(\xi, G)$ is the symplectic orthogonal of the other. The maps $J^L : T^*G \to \mathcal{G}^*$ and $J^R : T^*G \to \mathcal{G}^*$ are Poisson maps when

T^*G is equipped with the Poisson structure associated to its canonical symplectic structure and when $\mathcal{G}^*$ is equipped, respectively, with its canonical Poisson structure (Sect. 3.4) and with the opposite of its canonical Poisson structure.

Proof For each $X \in \mathcal{G}$, let X_G^L and X_G^R be the fundamental vector fields on G associated to X for the actions of G on itself, respectively by left and by right translations. Similarly, let $X_{T^*G}^L$ and $X_{T^*G}^R$ be the fundamental vector fields on T^*G associated to X for the actions $\widehat{L}$ and $\widehat{R}$ of G on T^*G defined in 4.5.1. The reduced flows of X^L and of X^R are the maps

$$\Phi^{X^L}(t, g) = \exp(tX)g\,, \quad \Phi^{X^R}(t, g) = g\exp(tX)\,, \quad t \in \mathbb{R}\,, \ g \in G\,.$$

Therefore

$$X^L(g) = TR_g(X)\,, \quad X^R(g) = TL_g(X)\,, \quad g \in G\,,$$

and we see that the fundamental vector fields $X_{T^*G}^L$ and $X_{T^*G}^R$ on T^*G are the canonical lifts to the cotangent bundle of the vector fields X_G^L and X_G^R on the Lie group G. Proposition 2.4.9 proves that $X_{T^*G}^L$ and $X_{T^*G}^R$ are Hamiltonian vector fields which admit as Hamiltonians, respectively, the maps

$$J_X^L(\xi) = \Bigl\langle \xi, X_G^L\bigl(\pi_G(\xi)\bigr)\Bigr\rangle\,, \quad J_X^R(\xi) = \Bigl\langle \xi, X_G^R\bigl(\pi_G(\xi)\bigr)\Bigr\rangle\,, \quad \xi \in T^*G\,.$$

Replacing X_G^L and X_G^R by their expressions given above and using the definitions of $\widehat{R}$ and $\widehat{L}$, we easily get the stated expressions for J^L and J^R. These expressions prove that J^L is constant on each orbit of the action $\widehat{R}$, and that J^R is constant on each orbit of the action $\widehat{L}$.

The actions $\widehat{L}$ and $\widehat{R}$ being free, each of their orbits is a smooth submanifold of T^*G of dimension dim G. The ranks of the maps J^L and J^R are everywhere equal to dim G since their restrictions to each fibre of T^*G is a diffeomorphism of that fibre onto $\mathcal{G}^*$. Therefore, for each $\xi \in T^*G$,

$$\ker T_\xi J^L = T_\xi\bigl(\widehat{R}(\xi, G)\bigr)\,, \quad \ker T_\xi J^R = T_\xi\bigl(\widehat{L}(\xi, G)\bigr)\,.$$

Corollary 4.4.8 proves that for each $\xi \in T^*G$ each of the two vector subspaces of $T_\xi(T^*G)$:

$$T_\xi\bigl(\widehat{L}(G, \xi)\bigr) \quad \text{and} \quad T_\xi\bigl(\widehat{R}(\xi, G)\bigr)$$

is the symplectic orthogonal of the other.

Finally, the fact that J^L and J^R are Poisson maps when $\mathcal{G}$ is equipped with its canonical Poisson structure or its opposite is an easy consequence of Proposition 3.3.7. □

In [17], Chapter IV, Sect. 4, we proposed a generalization of Proposition 4.5.3 taking into account a symplectic cocycle $\theta : G \to \mathcal{G}^*$ in which the action $\widehat{L} : G \times T^*G \to T^*G$ remained unchanged while the action $\widehat{R} : T^*G \times G \to T^*G$ was modified. Below we propose a more general and more symmetrical generalization. The symplectic form on T^*G will be the sum of its canonical symplectic form $d\eta_G$ and of the pull-back by the canonical projection $\pi_G : T^*G \to G$ of a suitable closed two-form on G, deduced from θ. The actions $\widehat{L} : G \times T^*G \to T^*G$ and $\widehat{R} : T^*G \times G \to T^*G$ will be modified in the following way: for each $g \in G$, the map $\widehat{L}_g : T^*G \to T^*G$ will be composed with a translation in the fibres of T^*G, determined by addition of a *right-invariant* one-form on G depending of the element $g \in G$, deduced from θ; similarly, the map $\widehat{R}_g : T^*G \to T^*G$ will be composed with a translation in the fibres of T^*G, determined by addition of a *left-invariant* one-form on G depending of the element $g \in G$, deduced from θ. As the reader will see, it is possible to modify the action $\widehat{L}$ and to keep $\widehat{R}$ unchanged, or to modify the action $\widehat{R}$ and to keep $\widehat{L}$ unchanged; in the first case, the momentum map $J^L : T^*G \to \mathcal{G}^*$ remains unchanged, while $J^R : T^*G \to \mathcal{G}^*$ must be modified; in the second case, it is $J^R : T^*G \to \mathcal{G}^*$ which remains unchanged while $J^L : T^*G \to \mathcal{G}^*$ must be modified. It is even possible to simultaneously modify both the actions $\widehat{L}$ and $\widehat{R}$; then we get a pair of actions of G on T^*G depending on two real parameters.

Theorem 4.5.4 *Let G be a Lie group, $\theta : G \to \mathcal{G}^*$ be a symplectic cocycle of G, $\Theta = T_e\theta : \mathcal{G} \to \mathcal{G}^*$ be the associated symplectic cocycle of its Lie algebra $\mathcal{G}$, and $\widetilde{\Theta} : \mathcal{G} \times \mathcal{G} \to \mathbb{R}$ be the skew-symmetric bilinear form $\widetilde{\Theta}(X, Y) = \langle \Theta(X), Y \rangle$. Let $\widetilde{\Theta}_L$ and $\widetilde{\Theta}_R$ be the differential two-forms on G, respectively left-invariant and right-invariant, whose value at the neutral element is $\widetilde{\Theta}$. The differential two-form on T^*G*

$$\omega_{T^*G} = \mathrm{d}\eta_G + \pi_G^*(\lambda_L\widetilde{\Theta}_L - \lambda_R\widetilde{\Theta}_R)\,,$$

where λ_L and λ_R are real constants and where η_G is the Liouville form on T^*G, is a symplectic form on T^*G. The formulae, in which $g \in G$, $\xi \in T^*G$,

$$\begin{aligned}\Phi^L(g, \xi) &= \widehat{L}_g(\xi) + \lambda_R\widehat{R}_{g\pi_G(\xi)}\big(\theta(g)\big)\,,\\ \Phi^R(\xi, g) &= \widehat{R}_g(\xi) + \lambda_L\widehat{L}_{\pi_G(\xi)g}\big(\theta(g^{-1})\big)\end{aligned}$$

define two Hamiltonian actions $\Phi^L : G \times T^*G \to T^*G$ and $\Phi^R : T^*G \times G \to T^*G$ of G on the symplectic manifold (T^*G, ω_{T^*G}), respectively on the left and on the right. The maps $J^{L,\lambda_L} : T^*G \to \mathcal{G}^*$ and $J^{R,\lambda_R} : T^*G \to \mathcal{G}^*$ defined, for each $\xi \in T^*G$, by

$$\begin{aligned}J^{L,\lambda_L}(\xi) &= \widehat{R}_{\left(\pi_G(\xi)\right)^{-1}}(\xi) + \lambda_L\theta\big(\pi_G(\xi)\big)\,,\\ J^{R,\lambda_R}(\xi) &= \widehat{L}_{\left(\pi_G(\xi)\right)^{-1}}(\xi) + \lambda_R\theta\Big(\big(\pi_G(\xi)\big)^{-1}\Big)\end{aligned}$$

are momentum maps for the actions Φ^L and Φ^R, respectively.

Moreover, the map J^{L,λ_L} is constant on each orbit of the action Φ^R, the map J^{R,λ_R} is constant on each orbit of the action Φ^L and for each $\xi \in T^*G$ each of the tangent spaces at ξ to the orbits $\Phi^L(G, \xi)$ and $\Phi^R(\xi, G)$ is the symplectic orthogonal *of the other (with respect to the symplectic form ω_{T^*G}). The maps $J^{L,\lambda_L} : T^*G \to \mathcal{G}^*$ and $J^{R,\lambda_R} : T^*G \to \mathcal{G}^*$ are Poisson maps when T^*G is equipped with the Poisson structure associated to the symplectic form ω_{T^*G} and when $\mathcal{G}^*$ is equipped, respectively, with its canonical Poisson structure modified by the cocycle $(\lambda_L+\lambda_R)\widetilde{\Theta}$ (Sect. 3.4)*

$$\{f,g\}_{(\lambda_L+\lambda_R)\widetilde{\Theta}}(\zeta) = \Big\langle \zeta, \big[\mathrm{d}f(\zeta), \mathrm{d}g(\zeta)\big]\Big\rangle - (\lambda_L+\lambda_R)\widetilde{\Theta}\big(\mathrm{d}f(\zeta), \mathrm{d}g(\zeta)\big)$$

and with the opposite of this Poisson structure.

Proof The sum of the canonical symplectic form on T^*G with the pull-back of any closed two-form on G always is nondegenerate, therefore symplectic. So ω_{T^*G} is symplectic. For g and $h \in G$, $\xi \in T^*G$, let us calculate

$$\Phi^L\big(g, \Phi^L(h,\xi)\big) - \Phi^L(gh,\xi) \quad \text{and} \quad \Phi^R\big(\Phi^R(\xi,g),h\big) - \Phi^R(\xi,gh)\,.$$

We get

$$\begin{aligned}\Phi^L\big(g, \Phi^L(h,\xi)\big) - \Phi^L(gh,\xi) &= \lambda_R \widehat{R}_{gh\pi_G(\xi)}\Big(\mathrm{Ad}^*_{g^{-1}}\big(\theta(h)\big) + \theta(g) - \theta(gh)\Big)\\ &= 0\end{aligned}$$

since θ is a one-cocycle. The map Φ^L is therefore an action on the left of G on T^*G. Similarly

$$\begin{aligned}\Phi^R\big(\Phi^R(\xi,g),h\big) - \Phi^R(\xi,gh) &= \lambda_L \widehat{L}_{\pi_G(\xi)gh}\Big(\mathrm{Ad}^*_h\,\theta(g^{-1}) + \theta(h^{-1}) - \theta(h^{-1}g^{-1})\Big)\\ &= 0\end{aligned}$$

for the same reason. The map Φ^R is therefore an action on the right of G on T^*G.

Let $X \in \mathcal{G}$ and $\xi = T^*G$. By calculating the derivative with respect to t of $\Phi^L\big(\exp(tX),\xi\big)$ and of $\Phi^R\big(\xi,\exp(tX)\big)$, then setting $t = 0$, we get the following expressions for the fundamental vector fields on T^*G associated to the actions Φ^L and Φ^R:

$$\begin{aligned}X^{L,\lambda_R}_{T^*G}(\xi) &= X^L_{T^*G}(\xi) + \lambda_R T\widehat{R}_{\pi_G(\xi)}\Theta(X)\,,\\ X^{R,\lambda_L}_{T^*G}(\zeta) &= X^R_{T^*G}(\zeta) - \lambda_L T\widehat{L}_{\pi_G(\xi)}\Theta(X)\,,\end{aligned}$$

the vector fields $X^L_{T^*G}$ and $X^R_{T^*G}$ being, as in the proof of Theorem 4.5.3, the canonical lifts to T^*G of the fundamental vector fields X^L and X^R on G, for the actions of G on itself by translations on the left and on the right, respectively. Using these expressions, we easily check that

$$\mathrm{i}(X_{T^*G}^{L,\lambda_R})\omega_{T^*G} = -\mathrm{d}J_X^{L,\lambda_L}\,, \quad \mathrm{i}(X_{T^*G}^{R,\lambda_L})\omega_{T^*G} = -\mathrm{d}J_X^{R,\lambda_R}\,,$$

which means that the actions Φ^L and Φ^R are Hamiltonian and have, respectively, J^{L,λ_L} and J^{R,λ_R} as momentum maps.

The facts that J^{R,λ_R} is constant on each orbit of Φ^L and that J^{L,λ_L} is constant on each orbit of Φ^R directly follow from the expressions of Φ^L, Φ^R, J^{L,λ_L} and J^{R,λ_R}.

Finally, let X and $Y \in \mathcal{G}$. When considered as linear functions on $\mathcal{G}^*$, their Poisson bracket for the Poisson structure on $\mathcal{G}^*$ for which J^{L,λ_L} is a Poisson map is easily determined by calculating the Poisson bracket $\{J^{L,\lambda_L} \circ X, J^{L,\lambda_L} \circ Y\} = \{J_X^{L,\lambda_L}, J_Y^{L,\lambda_L}\}$, for the Poisson structure on T^*G associated to the symplectic form ω_{T^*G}. This calculation fully determines the Poisson structure on $\mathcal{G}^*$ for which J^{L,λ_L} is a Poisson map, and proves that it is indeed the canonical Poisson structure on T^*G modified by the symplectic cocycle $(\lambda_L + \lambda_R)\widetilde{\Theta}$, in the sense of Sect. 3.4. A similar calculation shows that J^{R,λ_R} is a Poisson map when $\mathcal{G}^*$ is equipped with the opposite Poisson structure. □

Proposition 4.5.5 *Under the assumptions and with the notations of 4.5.4, the momentum map* $J^{L,\lambda_L} : T^*G \to \mathcal{G}^*$ *is equivariant when G acts on the left on* T^*G *by the action* Φ^L *and on* $\mathcal{G}^*$ *by the action*

$$(g,\zeta) \mapsto \mathrm{Ad}^*_{g^{-1}}(\zeta) + \theta(g)\,, \quad (g,\zeta) \in G \times \mathcal{G}^*\,.$$

Similarly, the momentum map $J^{R,\lambda_R} : T^*G \to \mathcal{G}^*$ *is equivariant when G acts on the right on* T^*G *by the action* Φ^R *and on* $\mathcal{G}^*$ *by the action*

$$(\zeta, g) \mapsto \mathrm{Ad}^*_{g}(\zeta) + \theta(g^{-1})\,, \quad (\zeta, g) \in \mathcal{G}^* \times G\,.$$

Proof Let $g \in G$ and $\xi \in T^*G$. Using the expressions of J^{L,λ_L} and of Φ^L, we obtain

$$J^{L,\lambda_L}\big(\Phi^L(g,\xi)\big) = \mathrm{Ad}^*_{g^{-1}}\big(J^{L,\lambda_L}(\xi)\big) + (\lambda_L + \lambda_R)\theta(g)\,,$$
$$J^{R,\lambda_R}\big(\Phi^R(\xi,g)\big) = \mathrm{Ad}^*_{g}\big(J^{R,\lambda_R}(\xi)\big) + (\lambda_L + \lambda_R)\theta(g^{-1})\,,$$

which proves that J^{L,λ_L} and J^{R,λ_R} are equivariant with respect to the indicated actions, respectively on the left and on the right, of G on T^*G and on $\mathcal{G}^*$. □

5 Reduction of Hamiltonian Systems with Symmetries

Very early, many scientists (Lagrange, Jacobi, Poincaré, ...) used first integrals to facilitate the determination of integral curves of Hamiltonian systems. It was observed that the knowledge of *one* real-valued first integral often allows the reduction by *two* units of the dimension of the phase space in which solutions are searched for.

Sniatycki and Tulczyjew [33] and, when first integrals come from the momentum map of a Lie group action, Meyer [34], Marsden and Weinstein [35], developed a geometric presentation of this reduction procedure, widely known now under the name "Marsden–Weinstein reduction".

Another way in which symmetries of a Hamiltonian system can be used to facilitate the determination of its integral curves was discovered around 1750 by Leonard Euler (1707–1783) when he derived the equations of motion of a rigid body around a fixed point. In a short Note published in 1901 [36], Henri Poincaré formalized and generalized this reduction procedure, often called today, rather improperly, "Lagrangian reduction" while the equations obtained by its application are called the "Euler–Poincaré equations" [37, 38].

We present in the following sections these two reduction procedures.

5.1 The Marsden–Weinstein Reduction Procedure

Theorem 5.1.1 *Let (M, ω) be a connected symplectic manifold on which a Lie group G acts by a Hamiltonian action Φ, with a momentum map $J : M \to \mathcal{G}^*$. Let $\xi \in J(M) \subset \mathcal{G}^*$ be a possible value of J. The subset G_ξ of elements $g \in G$ such that $\Phi_g\bigl(J^{-1}(\xi)\bigr) = J^{-1}(\xi)$ is a closed Lie subgroup of G.*

If in addition ξ is a weakly regular value of J in the sense of Bott [39], $J^{-1}(\xi)$ is a submanifold of M on which G_ξ acts, by the action Φ restricted to G_ξ and to $J^{-1}(\xi)$, in such a way that all orbits are of the same dimension. For each $x \in J^{-1}(\xi)$ the kernel of the two-form induced by ω on $J^{-1}(\xi)$ is the space tangent at this point to its G_ξ-orbit. Let $M_\xi = J^{-1}(\xi)/G_\xi$ be the set of all these orbits. When M_ξ has a smooth manifold structure for which the canonical projection $\pi_\xi : J^{-1}(\xi) \to M_\xi$ is a submersion, there exists on M_ξ a unique symplectic form ω_ξ such that $\pi_\xi^*\omega_\xi$ is the two-form induced on $J^{-1}(\xi)$ by ω. The symplectic manifold (M_ξ, ω_ξ) is called the reduced symplectic manifold *(in the sense of Marsden an Weinstein) for the value ξ of the momentum map.*

Proof Proposition 4.4.9 shows that there exists an affine action a of G on $\mathcal{G}^*$ for which the momentum map J is equivariant. The subset G_ξ of G is therefore the isotropy subgroup of ξ for the action a, which proves that it is indeed a closed subgroup of G. A well known theorem due to Élie Cartan allows us to state that G_ξ is a Lie subgroup of G.

When ξ is a weakly regular value of J, $J^{-1}(\xi)$ is a submanifold of M and, for each $x \in J^{-1}(\xi)$, the tangent space at x to this submanifold is $\ker T_xJ$ (it is the definition of a weakly regular value in the sense of Bott). Let $N = J^{-1}(\xi)$ and let $i_N : N \to M$ be the canonical injection. For all $x \in N$, the vector spaces $\ker T_xJ$ all are of the same dimension $\dim N$, and $\dim\bigl(T_xJ(T_xM)\bigr) = \dim M - \dim N$. Corollary 4.4.8 shows that $T_xJ\bigl(T_xM\bigr)$ is the annihilator of $\mathcal{G}_x$. Therefore for all $x \in N$ the isotropy subalgebras $\mathcal{G}_x$ are of the same dimension $\dim G - \dim M + \dim N$. The G_ξ-orbits of all points $x \in N$ are all of the same dimension $\dim G_\xi - \dim G_x$.

Corollary 4.4.8 also shows that $\mathrm{orth}(\ker T_xJ) = \mathrm{orth}(T_xN) = T_x\big(\Phi(G,x)\big)$. Therefore, for each $x \in N$,

$$\ker(i_N^*\omega)(x) = T_xN \cap \mathrm{orth}(T_xN) = T_xN \cap T_x\big(\Phi(G,x)\big) = T_x\big(\Phi(G_\xi,x)\big)\,.$$

It is indeed the space tangent at this point to its G_ξ-orbit. When $M_\xi = N/G_\xi$ has a smooth manifold structure such that the canonical projection $\pi_\xi : N \to M_\xi$ is a submersion, for each $x \in N$ the kernel of $T_x\pi_\xi$ is $\ker(i_N^*\omega)(x)$, and the existence on M_ξ of a symplectic form ω_ξ such that $\pi_\xi^*(\omega_\xi) = i_N^*\omega$ easily follows. □

Proposition 5.1.2 *The assumptions made here are the strongest of those made in Theorem 5.1.1: the set $J^{-1}(\xi)/G_\xi$ has a smooth manifold structure such that the canonical projection $\pi_\xi : J^{-1}(\xi)/G_\xi$ is a submersion. Let $H : M \to \mathbb{R}$ be a smooth Hamiltonian, invariant under the action Φ. There exists an unique smooth function $H_\xi : M_\xi \to \mathbb{R}$ such that $H_\xi \circ \pi_\xi$ is equal to the restriction of H to $J^{-1}(\xi)$. Each integral curve $t \mapsto \varphi(t)$ of the Hamiltonian vector field X_H which meets $J^{-1}(\xi)$ is entirely contained in $J^{-1}(\xi)$, and in the reduced symplectic manifold (M_ξ, ω_ξ) the parametrized curve $t \mapsto \pi_\xi \circ \varphi(t)$ is an integral curve of X_{H_ξ}.*

Proof As in the proof of Theorem 5.1.1, we set $N = J^{-1}(\xi)$ and denote by $i_N : N \to M$ the canonical injection. Let $\omega_N = i_N^*\omega$. Since H is invariant under the action Φ, it keeps a constant value on each orbit of G_ξ contained in N, so there exists on M_ξ an unique function H_ξ such that $H_\xi \circ \pi_\xi = H \circ i_N$. The projection π_ξ being a surjective submersion, H_ξ is smooth. Noether's theorem (4.4.5) proves that the momentum map J remains constant on each integral curve of the Hamiltonian vector field X_H. So if one of these integral curves meets N it is entirely contained in N, and we see that the Hamiltonian vector field X_H is tangent to N. We have, for each $x \in N$,

$$\begin{aligned}\pi_\xi^*\Big(\mathrm{i}\big(T_x\pi_\xi\big(X_H(x)\big)\big)\omega_\xi\big(\pi_\xi(x)\big)\Big) &= i\big(X_H(x)\big)\big(i_N^*\omega(x)\big) = -\mathrm{d}(i_N^*H)(x)\\ &= -\pi_\xi^*\big(\mathrm{d}H_\xi\big)(x) = \pi_\xi^*\big(\mathrm{i}(X_{H_\xi})\omega_\xi\big)(x)\,.\end{aligned}$$

Since π_ξ is a submersion and ω_ξ a non-degenerate two-form, this implies that for each $x \in N$, $T_x\pi_\xi\big(X_H(x)\big) = X_{H_\xi}\big(\pi_\xi(x)\big)$. The restriction of X_H to N and X_{H_ξ} are therefore two vector fields compatible with respect to the map $\pi_\xi : N \to M_\xi$, which implies the stated result. □

Remark 5.1.3 Theorem 5.1.1 and Proposition 5.1.2 still hold when instead of the Lie group action Φ we have an action φ of a finite-dimensional Lie algebra. The proof of the fact that the G_ξ-orbits in $J^{-1}(\xi)$ all are of the same dimension can easily be adapted to prove that for all $x \in J^{-1}(\xi)$, the vector spaces $\{\varphi(X)(x);\, X \in \mathcal{G}_\xi\}$ all are of the same dimension and determine a foliation of $J^{-1}(\xi)$. We have then only to replace the G_ξ-orbits by the leaves of this foliation.

5.1.1 Use of the Marsden–Weinstein Reduction Procedure

Theorem 5.1.1 and Proposition 5.1.2 are used to determine the integral curves of the Hamiltonian vector field X_H contained in $J^{-1}(\xi)$ in two steps:

- their projections on M_ξ are first determined: they are integral curves of the Hamiltonian vector field X_{H_ξ}; this step is often much easier than the full determination of the integral curves of X_H, since the dimension of the reduced symplectic manifold M_ξ is smaller than the dimension of M;
- then these curves themselves are determined; this second step, called *reconstruction*, involves the resolution of a differential equation on the Lie group G_ξ.

Many scientists (T. Ratiu, R. Cushman, J. Sniatycki, L. Bates, J.-P. Ortega, ...) generalized this reduction procedure in several ways: when M is a Poisson manifold instead of a symplectic manifold, when ξ is not a weakly regular value of J, ... The reader will find more results on the subject in the recent book *Momentum maps and Hamiltonian reduction* by Ortega and Ratiu [40].

Reduced symplectic manifolds occur in many applications other than the determination of integral curves of Hamiltonian systems. The reader will find such applications in the book *Symplectic techniques in Physics* by Guillemin and Sternberg [41] and in the papers on the phase space of a particle in a Yang-Mills field [42, 43].

5.2 The Euler–Poincaré Equation

In his Note [36], Henri Poincaré writes the equations of motion of a Lagrangian mechanical system when a finite-dimensional Lie algebra acts on its configuration space by a locally transitive action. Below we adapt his results to the Hamiltonian formalism.

Proposition 5.2.1 *Let $\mathcal{G}$ be a finite-dimensional Lie algebra which acts, by an action $\varphi : \mathcal{G} \to A^1(N)$, on a smooth manifold N. The action φ is assumed to be locally transitive, which means that for each $x \in N$, $\{\varphi(X)(x)\,;X \in \mathcal{G}\} = T_xN$. Let $\widehat{\varphi} : \mathcal{G} \to A^1(T^*N)$ be the Hamiltonian action of $\mathcal{G}$ on $(T^*N, \mathrm{d}\eta_N)$ which associates, to each $X \in \mathcal{G}$, the canonical lift to T^*N of the vector field $\varphi(X)$ on N (4.3.5), and let $J : T^*N \to \mathcal{G}^*$ be the momentum map of $\widehat{\varphi}$ given by the formula (4.4.4)*

$$\langle J(\xi), X\rangle = \mathrm{i}\big(\widehat{\varphi}(X)\big)\eta_N(\xi)\,, \quad X \in \mathcal{G}\,,\ \xi \in T^*N\,.$$

Let $H : T^*N \to \mathbb{R}$ be a smooth Hamiltonian, which comes from a hyper-regular Lagrangian $L : TN \to \mathbb{R}$ (hyper-regular means that the associated Legendre map $\mathcal{L} : TN \to T^*N$is a diffeomorphism). Let $\psi : I \to T^*N$ be an integral curve of the Hamiltonian vector field X_H defined on an open interval I and $V : I \to \mathcal{G}$ be a smooth parametrized curve in $\mathcal{G}$ which satisfies, for each $t \in I$,

$$\varphi\big(V(t)\big)\big(\pi_N\circ\psi(t)\big)=\frac{d\big(\pi_N\circ\psi(t)\big)}{\mathrm{d}t}\,. \tag{1}$$

The curve $J\circ\psi : I\to\mathcal{G}^*$, obtained by composition with J of the integral curve ψ of the Hamiltonian vector field X_H, satisfies the differential equation in $\mathcal{G}^*$

$$\left(\frac{\mathrm{d}}{\mathrm{d}t}-\mathrm{ad}^*_{V(t)}\right)\big(J\circ\psi(t)\big)=J\Big(d_1\overline{L}\big(\pi_N\circ\psi(t),V(t)\big)\Big)\,. \tag{2}$$

We have denoted by $\overline{L}: N\times\mathcal{G}\to\mathbb{R}$ the map

$$(x,X)\mapsto\overline{L}(x,X)=L\big(\varphi(X)(x)\big)\,,\quad x\in N\,,\ X\in\mathcal{G}\,,$$

and by $d_1\overline{L}: N\times\mathcal{G}\to T^*N$ the partial differential of $\overline{L}$ with respect to its first variable.

Equation (2) *is called the* Euler–Poincaré equation, *while Eq.* (1) *is called the* compatibility condition.

Proof For each $\xi\in T^*N$ and each $X\in\mathcal{G}$

$$\big\langle J(\xi),X\big\rangle=\varphi(X)\big(\pi_N(\xi)\big)\,,$$

therefore

$$\left\langle\frac{\mathrm{d}}{\mathrm{d}t}\big(J\circ\psi(t)\big),X\right\rangle=\frac{\mathrm{d}}{\mathrm{d}t}\Big\langle\psi(t),\varphi(X)\big(\pi_N\circ\psi(t)\big)\Big\rangle.$$

Let $(x^1,\dots,x^n)$ be local coordinates on N, and $(x^1,\dots,x^n,p_1,\dots,p_n)$ be the associated local coordinates on T^*N. The smooth curves ψ and $\pi_N\circ\psi$ can be expressed as

$$t\mapsto\big(y^i(t),\varpi_i(t)\big)\quad\text{and}\quad t\mapsto\big(y^i(t)\big)\,,\quad(1\le i\le n)\,,$$

so we can write

$$\left\langle\frac{\mathrm{d}}{\mathrm{d}t}\big(J\circ\psi(t)\big),X\right\rangle=\frac{\mathrm{d}}{\mathrm{d}t}\left(\sum_{i=1}^n\varpi_i(t)\big(\varphi(X)\big)^i\big(y^1(t),\dots,y^n(t)\big)\right).$$

We have denoted by $\big(\varphi(X)\big)^i(x^1,\dots,x^n)$ the value of the i-th component of the vector field $\varphi(X)$, expressed as a function of the local coordinates x^i $(1\le i\le n)$.

The compatibility condition (1) becomes

$$\frac{\mathrm{d}y^k(t)}{\mathrm{d}t}=\Big(\varphi\big(V(t)\big)\Big)^k\big(y^1(t),\dots,y^n(t)\big)\,.$$

In what follows we write y^i for $y^i(t)$, ϖ_i for $\varpi_i(t)$, (y) for $\big(y^1(t),\dots,y^n(t)\big)$ and (y,ϖ) for $\big(y^1(t),\dots,y^n(t),\varpi_1(t),\varpi_n(t)\big)$. We have

$$\left\langle \frac{\mathrm{d}}{\mathrm{d}t}\bigl(J\circ\psi(t)\bigr),X\right\rangle = \sum_{i=1}^{n}\frac{\mathrm{d}\varpi_i}{\mathrm{d}t}\bigl(\varphi(X)\bigr)^i(y) + \sum_{(i,k)=(1,1)}^{(n,n)}\varpi_i\frac{\partial\bigl(\varphi(X)\bigr)^i(y)}{\partial x^k}\Bigl(\varphi\bigl(V(t)\bigr)\Bigr)^k(y)\,.$$

By using the local expression of the bracket of vector fields

$$\Bigl[\varphi\bigl(V(t)\bigr),\varphi(X)\Bigr]^i(x) = \sum_{k=1}^{n}\Bigl(\varphi\bigl(V(t)\bigr)\Bigr)^k(x)\frac{\partial\bigl(\varphi(X)\bigr)^i(x)}{\partial x^k} - \sum_{k=1}^{n}\bigl(\varphi(X)\bigr)^k(x)\frac{\partial\Bigl(\varphi\bigl(V(t)\bigr)\Bigr)^i(x)}{\partial x^k}$$

and taking into account the fact that, φ being a Lie algebras homomorphism,

$$\Bigl[\varphi\bigl(V(t)\bigr),\varphi(X)\Bigr] = \varphi\Bigl(\bigl[V(t),X\bigr]\Bigr)\,,$$

we get

$$\left\langle \frac{\mathrm{d}}{\mathrm{d}t}\bigl(J\circ\psi(t)\bigr),X\right\rangle = \Bigl\langle\psi(t),\varphi\Bigl(\bigl[V(t),X\bigr]\Bigr)\Bigr\rangle + \sum_{i=1}^{n}\bigl(\varphi(X)\bigr)^i(y)\left(\frac{\mathrm{d}\varpi_i}{\mathrm{d}t}+\sum_{k=1}^{n}\varpi_k\frac{\partial\Bigl(\varphi\bigl(V(t)\bigr)\Bigr)^k(y)}{\partial x^i}\right).$$

The first term in the right hand side can be written

$$\Bigl\langle\psi(t),\varphi\Bigl(\bigl[V(t),X\bigr]\Bigr)\Bigr\rangle = \Bigl\langle J\circ\psi(t),\bigl[V(t),X\bigr]\Bigr\rangle = \Bigl\langle \mathrm{ad}^*_{V(t)}\bigl(J\circ\psi(t)\bigr),X\Bigr\rangle.$$

For all $(x,X)\in N\times\mathcal{G}$ we have

$$\overline{L}(x,X) = L\bigl(\varphi(X)(x)\bigr)\,.$$

For any point $x\in N$ and any vector $w\in T_xN$, there exists a smooth curve $s\mapsto z(s)$ in N such that $z(0)=x$ and $\dfrac{\mathrm{d}z(s)}{\mathrm{d}s}\Big|_{s=0}=w$. We easily obtain $\Bigl\langle \mathrm{d}_1\overline{L}\bigl(x,V(t)\bigr),w\Bigr\rangle$ by taking the derivative of $\overline{L}\bigl(x(s),V(t)\bigr)$ with respect to s (t remaining fixed), then making $s=0$. We obtain

$$\Big\langle \mathrm{d}_1\overline{L}\big(x, V(t)\big), w\Big\rangle = \sum_{i=1}^{n} w^i \left(\frac{\partial L(x,v)}{\partial x^i} + \sum_{k=1}^{n} \frac{\partial L(x,v)}{\partial v^k} \frac{\partial \Big(\varphi\big(V(t)\big)\Big)^k (x)}{\partial x^i} \right).$$

Let us set $x = \pi_N \circ \psi(t)$, $w = \varphi(X)\big(\pi_N \circ \psi(t)\big)$. We observe that

$$\Big\langle \mathrm{d}_1\overline{L}\big(\pi_N \circ \psi(t), V(t)\big), \varphi(X)\big(\pi_N \circ \psi(t)\big)\Big\rangle = \Big\langle J\Big(d_1\overline{L}\big(\pi_N \circ \psi(t), V(t)\big)\Big), X\Big\rangle.$$

Now we take into account the well known relations which exist between the partial derivatives of the Lagrangian and of the Hamiltonian expressed in local coordinates

$$\frac{\partial L(x,v)}{\partial x^i} = -\frac{\partial H(x,p)}{\partial x^i} = \frac{d\varpi_i}{dt}\,, \quad \frac{\partial L(x,v)}{\partial v^k} = \varpi_k\,,$$

and we obtain

$$\Big\langle J\Big(d_1\overline{L}\big(\pi_N \circ \psi(t), V(t)\big)\Big), X\Big\rangle = \sum_{i=1}^{n} \big(\varphi(X)\big)^i (y) \left(\frac{\mathrm{d}\varpi_i}{\mathrm{d}t} + \sum_{k=1}^{n} \varpi_k \frac{\partial \Big(\varphi\big(V(t)\big)\Big)^k (y)}{\partial x^i} \right).$$

Since X can be any element in $\mathcal{G}$, the Euler–Poincaré equation follows. □

Remark 5.2.2 The assumptions made by Poincaré in [36] are less restrictive than those made in 5.2.1: he uses the Lagrangian formalism for a smooth Lagrangian $L : TN \to \mathbb{R}$ which is not assumed to be hyper-regular. The associated Legendre map $\mathcal{L} : TN \to T^*N$ still exists as a smooth map (it is the vertical differential of L, see for example [44]), but may not be a diffeomorphism. Of course the momentum map $J : T^*N \to \mathcal{G}^*$ still exists and can be used, together with the Legendre map, to express Poincaré's results intrinsically [45], independently of any choice of local coordinates. Poincaré proves that if a smooth parametrized curve $\gamma : [t_0, t_1] \to N$ is an extremal of the action functional

$$I(\gamma) = \int_{t_0}^{t_1} L\left(\frac{\mathrm{d}\gamma(t)}{\mathrm{d}t}\right) \mathrm{d}t$$

for infinitesimal variations of γ with fixed end points, and if $V : [t_0, t_1] \to \mathcal{G}$ is a smooth parametrized curve which satisfies, for each $t \in [t_0, t_1]$, the compatibility condition

$$\varphi\big(V(t)\big)\big(\gamma(t)\big) = \frac{d\big(\gamma(t)\big)}{\mathrm{d}t}\,, \tag{1}$$

the parametrized curve $t \mapsto J \circ \mathcal{L} \circ \dfrac{\mathrm{d}\gamma(t)}{\mathrm{d}t}$ satisfies the Euler–Poincaré equation

$$\left(\frac{\mathrm{d}}{\mathrm{d}t} - \mathrm{ad}^*_{V(t)}\right)\left(J \circ \mathcal{L} \circ \frac{\mathrm{d}\gamma(t)}{\mathrm{d}t}\right) = J\Bigl(d_1\overline{L}\bigl(\gamma(t), V(t)\bigr)\Bigr). \tag{2}$$

The Euler–Poincaré equation can be written under a slightly different form in which, instead of the Legendre map $\mathcal{L} : TN \to T^*N$, the partial differential $\mathrm{d}_2\overline{L} : N \times \mathcal{G} \to \mathcal{G}^*$ of the map $\overline{L} : N \times \mathcal{G} \to \mathbb{R}$ with respect to its second variable is used. We have indeed, for all $x \in N$ and $X \in \mathcal{G}$,

$$\mathrm{d}_2\overline{L}(x, X) = J \circ \mathcal{L}\bigl(\varphi(X)(x)\bigr),$$

which allows to write the Euler–Poincaré equation under the form

$$\left(\frac{\mathrm{d}}{\mathrm{d}t} - \mathrm{ad}^*_{V(t)}\right)\Bigl(\mathrm{d}_2\overline{L}\bigl(\gamma(t), V(t)\bigr)\Bigr) = J\Bigl(d_1\overline{L}\bigl(\gamma(t), V(t)\bigr)\Bigr). \tag{3}$$

5.2.1 Use of the Euler–Poincaré Equation for Reduction

Poincaré observes in his Note [36] that the Euler–Poincaré equation can be useful mainly when its right hand side vanishes and when it reduces to an autonomous differential equation on $\mathcal{G}^*$ for the parametrized curve $t \mapsto J \circ \psi(t)$. We will see in Sect. 6.4 that the first condition is satisfied when the Hamiltonian system under consideration describes the motion of a rigid body around a fixed point in the absence of external forces (Euler–Poinsot problem). The second condition generally is not satisfied, since the Euler–Poincaré equation involves the parametrized curve $t \mapsto V(t)$ in $\mathcal{G}$, whose dependence on $J \circ \psi(t)$ is complicated.

This simplification occurs when there exists a smooth function $h : \mathcal{G}^* \to \mathbb{R}$ such that

$$H = h \circ J,$$

which implies that H is constant on each level set of J. Then it can be shown that the Euler–Poincaré equation becomes the Hamilton equation on $\mathcal{G}^*$ for the Hamiltonian h and its canonical Poisson structure.

If we assume that the manifold N is a Lie group G and that the action $\varphi : \mathcal{G} \to A^1(G)$ of its Lie algebra is the action associated to the action of G on itself by translations on the left (respectively, on the right), $\widehat{\varphi}$ is the Lie algebra action associated to the canonical lift to T^*G of the canonical action of G on itself by translations on the left (respectively, on the right). The conditions under which the Euler–Poincaré equation can be used for reduction are exactly the same as those under which the Marsden–Weinstein reduction method can be applied, but for the canonical lift to T^*G of the action of G on itself by translations *on the right* (respectively, on the left). Moreover, applications of these two reduction methods lead to essentially the same equations: the only difference is that the Euler–Poincaré reduction method leads to a differential equation on $\mathcal{G}^*$, while the Marsden–Weinstein reduction method leads, for each value of the momentum map, to the same differential equation restricted to a coadjoint orbit of $\mathcal{G}^*$. The reader will find the proof of these assertions in [45, 46].

6 Examples of Hamiltonian Dynamical Systems

We present in this section three classical examples of Hamiltonian dynamical systems in which the previously discussed concepts (symmetry groups, momentum maps and first integrals, reduction methods) are illustrated. The configuration space of the first system (the spherical pendulum) is a sphere embedded in physical space; each point of the sphere is a possible position of a material point which moves on that sphere. The third example (the Kepler problem) deals with the motion of a material point in the acceleration field created by an attracting centre; the configuration space is the physical space minus one point (the attractive centre). In the second example (the motion of a rigid body around a fixed point) the configuration space is a little more complicated: it is the set of all maps which send the material body onto one of its possible positions in space.

6.1 The Mathematical Description of Space and Time

The framework in which the motions of material bodies occur is the physical space-time. It will be mathematically described here as it is usually done in classical (non-relativistic) Mechanics. In a Galilean reference frame, once units of length and of time are chosen, the physical space and the physical time are mathematically described by affine Euclidean spaces E and T, respectively three-dimensional and one-dimensional. We will consider E and T as *oriented*: T has a natural orientation (towards the future), while by convention, an arbitrary orientation of E is chosen. The choice of a particular element of T as origin will allow us to identify T with the real line $\mathbb{R}$.

In the three examples treated below there exists a privileged element of E (the centre of the sphere, the fixed point and the attractive centre, respectively in the first, second and third examples) which will be taken as origin. The space E will therefore be considered as an Euclidean three-dimensional *vector* space. For the same reason the abstract space S of material points used in the second example will be considered too as an Euclidean three-dimensional *vector* space.

In our three examples, the configuration space of the system will be denoted by N: therefore in the first example N is the sphere embedded in E centered on the origin on which the material point is moving; in the third example $N = E\backslash\{O\}$, where O is the attractive centre; and we will see that in the second example, $N = \mathrm{Isom}(S, E)$ is the space of orientation preserving linear isometries of an abstract three-dimensional Euclidean vector space S (the space of material points) onto the physical space E.

6.2 Vector Calculus in a Three-dimensional Oriented Euclidean Vector Space

The group SO(E) of orientation preserving linear isometries of E, isomorphic to SO(3), acts on the space E, and so does its Lie algebra $\mathfrak{so}(E)$, isomorphic to $\mathfrak{so}(3)$, by the associated action. The Euclidean vector space E being three-dimensional and oriented, there exists an isomorphism of $\mathfrak{so}(E)$ onto the space E itself widely used in elementary vector calculus, in which an element $X \in \mathfrak{so}(E)$, which is a linear map $E \to E$ represented, in some orthonormal positively oriented basis $(\overrightarrow{e}_1, \overrightarrow{e}_2, \overrightarrow{e}_3)$ of E, by the skew-symmetric 3×3-matrix

$$\begin{pmatrix} 0 & -c & b \\ c & 0 & -a \\ -b & a & 0 \end{pmatrix}$$

is identified with the vector $\overrightarrow{X} = a\overrightarrow{e}_1 + b\overrightarrow{e}_2 + c\overrightarrow{e}_3$. With this identification, the bracket in $\mathfrak{so}(E)$, in other words the map $(X, Y) \mapsto [X, Y] = X \circ Y - Y \circ X$, corresponds to the *vector product* $(\overrightarrow{X}, \overrightarrow{Y}) \mapsto \overrightarrow{X} \times \overrightarrow{Y}$. Expressed in terms of the vector product, the Jacobi identity becomes

$$\overrightarrow{X} \times (\overrightarrow{Y} \times \overrightarrow{Z}) + \overrightarrow{Y} \times (\overrightarrow{Z} \times \overrightarrow{X}) + \overrightarrow{Z} \times (\overrightarrow{X} \times \overrightarrow{Y}) = 0\,. \tag{1}$$

Let us recall another very useful formula which expresses, in terms of the scalar and vector products, the ad-invariance of the pairing between $\mathfrak{so}(E)$ and its dual by means of the scalar product. For any triple $(\overrightarrow{u}, \overrightarrow{v}, \overrightarrow{w}) \in E \times E \times E$, we have

$$\overrightarrow{u} \,.\, (\overrightarrow{v} \times \overrightarrow{w}) = \overrightarrow{v} \,.\, (\overrightarrow{w} \times \overrightarrow{u}) = \overrightarrow{w} \,.\, (\overrightarrow{u} \times \overrightarrow{v})\,. \tag{2}$$

The map $(\overrightarrow{u}, \overrightarrow{v}, \overrightarrow{w}) \mapsto \overrightarrow{u} \,.\, (\overrightarrow{v} \times \overrightarrow{w})$ is therefore a skew-symmetric trilinear form on E sometimes called the *mixed product*.

The dual E^* of E will be identified with E, with the scalar product $(\overrightarrow{u}, \overrightarrow{v}) \mapsto \overrightarrow{u} \,.\, \overrightarrow{v}$ as pairing by duality. The tangent and cotangent bundles TE and T^*E will therefore both be identified with $E \times E$, the canonical projections $\tau_E : TE \to E$ and $\pi_E : T^*E \to E$ both being the projection of $E \times E$ onto its first factor. The Lie algebra action of $\mathfrak{so}(E)$ on E associates, to each $\overrightarrow{X} \in \mathfrak{so}(E) \equiv E$, the vector field $\overrightarrow{X_E}$ on E whose value at an element $\overrightarrow{x} \in E$ is

$$\overrightarrow{X_E}(\overrightarrow{x}) = (\overrightarrow{x}, \overrightarrow{X} \times \overrightarrow{x})\,. \tag{3}$$

Since we have identified $\mathfrak{so}(E)$ with E, its dual space $\mathfrak{so}(E)^*$ is identified with E^*, which we have identified with E by means of the scalar product. Therefore $\mathfrak{so}(E)^*$ too will be identified with E.

The canonical lift to the cotangent bundle of the action of SO(E) on E is a Hamiltonian action (4.3.5) whose momentum map $J_E : T^*E \equiv E \times E \to \mathfrak{so}(E)^* \equiv E$ can easily be expressed in terms of the vector product. Indeed the map J_E must satisfy, for each $\overrightarrow{X} \in \mathfrak{so}(E) \equiv E$ and each $(\overrightarrow{x}, \overrightarrow{p}) \in T^*E \equiv E \times E$,

$$\big\langle J_E(\overrightarrow{x}, \overrightarrow{p}), X\big\rangle = \big\langle (\overrightarrow{x}, \overrightarrow{p}), \overrightarrow{X_E}(\overrightarrow{x})\big\rangle = \overrightarrow{p}.(\overrightarrow{X} \times \overrightarrow{x}) = \overrightarrow{X}.(\overrightarrow{x} \times \overrightarrow{p}),$$

the last equality being obtained by using the above formula ((2)). We therefore see that

$$J_E(\overrightarrow{x}, \overrightarrow{p}) = \overrightarrow{x} \times \overrightarrow{p}\,. \tag{4}$$

Expressed in terms of the vector product, the adjoint and coadjoint actions become

$$\mathrm{ad}_{\overrightarrow{X}} \overrightarrow{Y} = \overrightarrow{X} \times \overrightarrow{Y}\,, \quad \mathrm{ad}^*_{\overrightarrow{X}} \overrightarrow{\xi} = -\overrightarrow{X} \times \overrightarrow{\xi} = \overrightarrow{\xi} \times \overrightarrow{X}\,, \tag{5}$$

where $\overrightarrow{X}$ and $\overrightarrow{Y} \in \mathfrak{so}(E) \equiv E$ and $\overrightarrow{\xi} \in \mathfrak{so}(E)^* \equiv E$.

Of course all the above properties hold for the three-dimensional Euclidean oriented vector space S of material points which is used in the second example, for the group SO(S) of its linear orientation preserving isometries and for its Lie algebra $\mathfrak{so}(S)$.

6.3 The Spherical Pendulum

6.3.1 Mathematical Description of the Problem

Let us consider a heavy material point of mass m constrained, by an ideal constraint, on the surface of a sphere N of centre O and radius R embedded in the physical space E. Since the action of SO(E) on E maps N onto itself, SO(E) acts on N on the left, and so does its Lie algebra $\mathfrak{so}(E)$ by the associated action, which is locally (and globally) transitive. The configuration space N is the set of vectors $\overrightarrow{x} \in E$ which satisfy $\overrightarrow{x}.\overrightarrow{x} = R^2$ and its tangent bundle TN is the subset of $TE \equiv E \times E$ of pairs $(\overrightarrow{x}, \overrightarrow{v})$ of vectors which satisfy

$$\overrightarrow{x}.\overrightarrow{x} = R^2\,, \quad \overrightarrow{x}.\overrightarrow{v} = 0\,.$$

We assume that the material point is submitted to a constant acceleration field $\overrightarrow{g}$ (which, in most applications, will be the vertical gravity field directed downwards). The Lagrangian of the system is

$$L(\overrightarrow{x}, \overrightarrow{v}) = \frac{m\|\overrightarrow{v}\|^2}{2} + m\,\overrightarrow{g}.\overrightarrow{x}\,.$$

The Legendre map $\mathcal{L} : TN \to T^*N$ is expressed as

$$\mathcal{L}(\overrightarrow{x}, \overrightarrow{v}) = (\overrightarrow{x}, \overrightarrow{p}) \quad \text{with } \overrightarrow{p} = m\overrightarrow{v} .$$

The Hamiltonian of the system is therefore

$$H(\overrightarrow{x}, \overrightarrow{p}) = \frac{\| \overrightarrow{p} \|^2}{2m} - m\overrightarrow{g} . \overrightarrow{x} .$$

The momentum map J_E of the canonical lift to the cotangent bundle of the Lie algebra action φ, expressed in terms of the vector product, is given by Formula (4) in Sect. 6.2.

6.3.2 The Euler–Poincaré Equation

The map $\widetilde{\varphi} : N \times \mathfrak{so}(E) \to TN$ defined by $\widetilde{\varphi}(\overrightarrow{x}, \overrightarrow{X}) = \overrightarrow{X_N}(\overrightarrow{x})$, expressed, in terms of the vector product, is $\widetilde{\varphi}(\overrightarrow{x}, \overrightarrow{X}) = (\overrightarrow{x}, \overrightarrow{X} \times \overrightarrow{x})$. Using Formula (2) of Sect. 6.2, we easily obtain the expression of $\overline{L} = L \circ \widetilde{\varphi} : N \times \mathfrak{so}(E) \to \mathbb{R}$:

$$\overline{L}(\overrightarrow{x}, \overrightarrow{X}) = \frac{mR^2}{2}\left(\| \overrightarrow{X} \|^2 - \frac{(\overrightarrow{X} . \overrightarrow{x})^2}{R^2} \right) + m\overrightarrow{g} . \overrightarrow{x} .$$

The partial differentials of $\overline{L}$ with respect to its first and second variables are

$$\begin{aligned} \mathrm{d}_1\overline{L}(\overrightarrow{x}, \overrightarrow{X}) &= \left(\overrightarrow{x}, m\Big(\overrightarrow{g} - (\overrightarrow{X} . \overrightarrow{x})\overrightarrow{X} + \frac{(\overrightarrow{x} . \overrightarrow{X})^2 - \overrightarrow{g} . \overrightarrow{x}}{R^2} \overrightarrow{x} \Big) \right), \\ \mathrm{d}_2\overline{L}(\overrightarrow{x}, \overrightarrow{X}) &= m\big(R^2\overrightarrow{X} - (\overrightarrow{X} . \overrightarrow{x})\overrightarrow{x} \big) . \end{aligned}$$

Let $t \mapsto \overrightarrow{x}(t)$ be a smooth curve in N, parametrized by the time t, solution of the Euler–Lagrange equation for the Lagrangian L. The compatibility condition (1) of 5.2.2 becomes, for a smooth map $t \mapsto V(t)$ in $\mathfrak{so}(E)$,

$$\frac{\mathrm{d}\overrightarrow{x}(t)}{\mathrm{d}t} = \overrightarrow{V}(t) \times \overrightarrow{x}(t) ,$$

and the Euler–Poincaré equation (3) of 5.2.2 is

$$\frac{\mathrm{d}}{\mathrm{d}t}\Big(mR^2\overrightarrow{V}(t) - m\big(\overrightarrow{x}(t) . \overrightarrow{V}(t) \big)\overrightarrow{x}(t) \Big) = m\overrightarrow{x}(t) \times \overrightarrow{g} .$$

This equation can easily be obtained by much more elementary methods: it expresses the fact that the time derivative of the angular momentum at the origin is equal to the moment at that point of the gravity force (since the moment at the origin of the constraint force which binds the material point to the surface of the sphere vanishes).

The Euler–Poincaré equation allows a reduction of the problem if and only if its right hand side vanishes, which occurs if and only if $\overrightarrow{g} = 0$. When that condition is satisfied, it can be written as

$$m\frac{\mathrm{d}}{\mathrm{d}t}\left(\overrightarrow{x}(t) \times \frac{\mathrm{d}\overrightarrow{x}(t)}{\mathrm{d}t}\right) = 0\,,$$

which implies that the material point moves on a great circle of the sphere N, in the plane through its centre orthogonal to the constant vector $\overrightarrow{x}(t) \times \dfrac{\mathrm{d}\overrightarrow{x}(t)}{\mathrm{d}t}$. Using the conservation of energy H, we see that $\|\overrightarrow{v})\|$ remains constant during the motion.

6.3.3 Reduction by the Use of First Integrals

Equation (4) of Sect. 6.2 shows that the map

$$J(\overrightarrow{x}, \overrightarrow{p}) = \overrightarrow{x} \times \overrightarrow{p}\,, \quad \text{with } (\overrightarrow{x}, \overrightarrow{p}) \in T^*N \equiv N \times E\,,$$

is a momentum map for the canonical lift to T^*N of the action of $\mathrm{SO}(E)$ on N. When $\overrightarrow{g} \neq 0$ that action does not leave invariant the Hamiltonian H, but its restriction to the subgroup G_1 of rotations around the vertical line through the centre of the sphere N does leave H invariant. The Lie algebra of G_1 and its dual being identified with $\mathbb{R}$, the momentum map of this restricted action is

$$J_1(\overrightarrow{x}, \overrightarrow{p}) = \overrightarrow{e}_g.(\overrightarrow{x} \times \overrightarrow{p})$$

where $\overrightarrow{e_g}$ is the unit vector such that $\overrightarrow{g} = ge_g$, with $g > 0$. The only singular value of J_1 is 0. It is reached when the three vectors $\overrightarrow{x}$, $\overrightarrow{p}$ and $\overrightarrow{e_g}$ lie in the same vertical plane. Therefore, for any $\zeta \neq 0$, $J_1^{-1}(\zeta)$ is a three-dimensional submanifold of T^*N which does not contain $T^*_{R\overrightarrow{e}_g}N \cup T^*_{-R\overrightarrow{e}_g}N$ and which remains invariant under the action of G_1. The set of orbits of this action is the Marsden–Weinstein reduced symplectic manifold for the value ζ of the momentum map. On this two-dimensional reduced symplectic manifold all the integral curves of the Hamiltonian vector field associated to the reduced Hamiltonian H_ζ are periodic.

6.4 The Motion of a Rigid Body Around a Fixed Point

6.4.1 Mathematical Description of the Problem

We consider the motion of a rigid body containing at least three non-aligned material points. A configuration of the body in space is mathematically represented by an affine, isometric and orientation preserving map defined on an abstract Euclidean three-dimensional oriented affine space S (called the *space of material points*), with values in E, the three-dimensional Euclidean oriented affine space which mathematically describes the physical space. When the configuration of the body is represented by the map $x : S \to E$, the position in space of the material point of the body represented by $z \in S$ is $x(z)$.

We assume that one geometric point of the rigid body is constrained, by an ideal constraint, to keep a fixed position in the physical space. By using this fixed point as origin, both for S and for E, we can now consider these spaces as *vector* spaces. Each configuration of the body in space is therefore represented by a *linear* isometry. The set N of all possible configurations of the material body in space is therefore $\text{Isom}(S, E)$, the set of linear orientation-preserving isometries of S onto E.

The Lie groups $\text{SO}(S)$ and $\text{SO}(E)$ of linear orientation-preserving isometries, respectively of S and of E, both isomorphic to $\text{SO}(3)$, act on N, respectively on the left and on the right, by the two commuting actions Φ_S and Φ_E

$$\Phi_S(x, g_S) = x \circ g_S\,,\quad \Phi_E(g_E, x) = g_E \circ x\,,\quad g_E \in \text{SO}(E)\,,\ g_S \in \text{SO}(S)\,,\ x \in N\,.$$

The values at $x \in N$ of the fundamental vector fields on N associated to $X \in \mathfrak{so}(S)$ and $Y \in \mathfrak{so}(E)$ are

$$X_N(x) = \frac{\mathrm{d}\big(x \circ \exp(sX)\big)}{\mathrm{d}s}\Big|_{s=0}\,,\quad Y_N(x) = \frac{\mathrm{d}(\exp(sY) \circ x)}{\mathrm{d}s}\Big|_{s=0}\,.$$

The Lie algebra actions $\varphi_S : \mathfrak{so}(S) \to A^1(N)$ and $\varphi_E : \mathfrak{so}(E) \to A^1(N)$ associated to the Lie group actions Φ_S and Φ_E are, respectively, the maps

$$\varphi_S(X) = X_N\,,\quad \varphi_E(Y) = Y_N\,,\quad X \in \mathfrak{so}(S)\,,\quad Y \in \mathfrak{so}(E)\,.$$

One should be careful with signs: since Φ_S is an action of $\text{SO}(S)$ on the right, the bracket of elements in the Lie algebra $\mathfrak{so}(S)$ for which φ_S is a Lie algebras homomorphism is the bracket of *left-invariant* vector fields on the Lie group $\text{SO}(S)$; similarly, since Φ_E is an action of $\text{SO}(E)$ on the left, the bracket of elements in the Lie algebra $\mathfrak{so}(E)$ for which φ_E is a Lie algebras homomorphism is the bracket of *right-invariant* vector fields on the Lie group $\text{SO}(E)$.

Let $\widetilde{\varphi}_S : N \times \mathfrak{so}(S) \to TN$ and $\widetilde{\varphi}_E : N \times \mathfrak{so}(E) \to TN$ be the vector bundles isomorphisms

$$\widetilde{\varphi}_S(x, X) = \varphi_S(X)(x), \quad \widetilde{\varphi}_E(x, Y) = \varphi_E(Y)(x).$$

Let $\Omega_S : TN \to \mathfrak{so}(S)$ and $\Omega_E : TN \to \mathfrak{so}(E)$ be the vector bundles maps ($\mathfrak{so}(S)$ and $\mathfrak{so}(E)$ being considered as trivial vector bundles over a base reduced to a singleton)

$$\Omega_S(v) = \pi_{\mathfrak{so}(S)} \circ \widetilde{\varphi}_S^{-1}(v), \quad \Omega_E(v) = \pi_{\mathfrak{so}(E)} \circ \widetilde{\varphi}_E^{-1}(v), \quad v \in TN,$$

where $\pi_{\mathfrak{so}(S)} : N \times \mathfrak{so}(S) \to \mathfrak{so}(S)$ and $\pi_{\mathfrak{so}(E)} : N \times \mathfrak{so}(E) \to \mathfrak{so}(E)$ are the projections of these two products on their respective second factor.

A *motion* of the rigid body during a time interval $[t_0, t_1]$ is mathematically described by a smooth parametrized curve $\gamma : [t_0, t_1] \to N$. In his beautiful paper [47], Vladimir Arnold clearly explained the physical meaning, for each $t \in [t_0, t_1]$, of $\dfrac{d\gamma(t)}{dt}$, $\Omega_S\left(\dfrac{d\gamma(t)}{dt}\right)$ and $\Omega_E\left(\dfrac{d\gamma(t)}{dt}\right)$:

- $\dfrac{d\gamma(t)}{dt} \in T_{\gamma(t)}N$ is the value, at time t, of the *true angular velocity* of the body,
- $\Omega_S\left(\dfrac{d\gamma(t)}{dt}\right)$ is the value, at time t, of the *angular velocity of the body seen by an observer bound to the moving body* and moving with it,
- and $\Omega_E\left(\dfrac{d\gamma(t)}{dt}\right)$ is the value, at time t, of the *angular velocity of the body seen by an observer bound to the Galilean reference frame in which the motion is studied* and at rest with respect to that reference frame.

The following comments may be useful to explain Arnold's assertions. To shorten the notations, let us state, for some time $t \in [t_0, t_1]$, $x = \gamma(t) \in N$, $v = \dfrac{d\gamma(t)}{dt} \in T_xN$, $X = \Omega_S(v) \in \mathfrak{so}(S)$ and $Y = \Omega_E(v) \in \mathfrak{so}(E)$. We have

$$\widetilde{\varphi}_S(x, X) = \widetilde{\varphi}_E(s, Y) = v.$$

Let $z \in S$ be some material point of the moving body. Its position at time t is $x(z) \in E$ and its velocity is $\dfrac{d}{dt}\big(\gamma(t)(z)\big) \in T_{x(z)}E$. It depends only of $v = \dfrac{d\gamma(t)}{dt} \in T_xN$, not of the whole curve γ. We can therefore replace γ by the parametrized curve $s \mapsto \exp(sY) \circ x$, since we have

$$\frac{d\big(\exp(sY) \circ x\big)}{ds}\bigg|_{s=0} = Y_N(x) = v.$$

Therefore the velocity at time t of the material point $z \in S$ is

$$\frac{d\big(\exp(sY) \circ x(z)\big)}{ds}\bigg|_{s=0} = Y_E\big(x(z)\big) = \big(\overrightarrow{x}(z), \overrightarrow{Y} \times \overrightarrow{x}(z)\big) \in TE \equiv E \times E,$$

where we have denoted by Y_E the fundamental vector field on E associated to $Y \in \mathfrak{so}(E)$for the action Φ_E, and used Formula (3) of Sect. 6.2. This proves that the fundamental vector field Y_E is the velocity field of the rigid body as it appears in space E at time t, and explains why Arnold calls Y the angular velocity of the body seen by an observer bound to the Galilean frame in which the motion is studied.

Since $\dfrac{\mathrm{d}\gamma(t)}{\mathrm{d}t} = \dfrac{\mathrm{d}\bigl(x \circ \exp(sX)\bigr)}{\mathrm{d}s}\Big|_{s=0}$, the value at $x(z) \in E$ of the fundamental vector field Y_E is also given by $\dfrac{\mathrm{d}\bigl(x \circ \exp(sX)(z)\bigr)}{\mathrm{d}s}\Big|_{s=0}$. The pull-back $x^*(Y_E)$ by the isomorphism $x : S \to E$ of the fundamental vector field Y_E, i.e. of the velocity field of the moving body in space E at time t, is the vector field on S whose value at $z \in S$ is

$$x^*(Y_E)(z) = \frac{\mathrm{d}\bigl(\exp(sX)(z)\bigr)}{\mathrm{d}s}\Big|_{s=0} = X_S(z) = (\overrightarrow{z}, \overrightarrow{X} \times \overrightarrow{z}) \in TS \equiv S \times S\,,$$

where we have again used Formula (3) of Sect. 6.2. The pull-back $x^*(Y_E)$ of the velocity field of the moving body in space E at time t, by the isomorphism $x : S \to E$, is therefore the fundamental vector field X_S associated to $X \in \mathfrak{so}(S)$, for the action Φ_S. That explains why Arnold calls X the angular velocity of the body seen by an observer bound to the moving body.

The above observations prove that for any $x \in N$ and $v \in T_xN$,

$$x\bigl(\overrightarrow{\Omega_S}(v)\bigr) = \overrightarrow{\Omega_E}(v)\,, \tag{1}$$

the arrows over $\Omega_S(v) \in \mathfrak{so}(S)$ and $\Omega_E(v) \in \mathfrak{so}(E)$ indicating that they are here considered as vectors in S and in E, respectively.

When the true angular velocity of the body is $v \in TN$, its *kinetic energy* is

$$\mathbb{T}(v) = \frac{1}{2}I\bigl(\Omega_S(v), \Omega_S(v)\bigr)\,,$$

where $I : \mathfrak{so}(S) \times \mathfrak{so}(S) \to \mathbb{R}$ is a symmetric, positive definite bilinear form which describes the *inertia properties* of the body. The assumed rigidity of the body is mathematically described by the fact that the bilinear form I does not depend on time, nor on the configuration $\tau_N(v)$ of the body. Let us set, for each pair (v, w) of vectors in TN such that $\tau_N(v) = \tau_N(w)$,

$$\widetilde{\mathbb{T}}(v, w) = \frac{1}{2}I\bigl(\Omega_S(v), \Omega_S(w)\bigr)\,, \quad \text{so we can write } \mathbb{T}(v) = \widetilde{\mathbb{T}}(v, v)\,.$$

The symmetric bilinear form $\widetilde{\mathbb{T}}$ is a Riemannian metric on the manifold N. Let us consider the effects on $\widetilde{\mathbb{T}}$ of the canonical lifts to TN of the actions Φ_E and Φ_S on the manifold N. For each g_E in $\mathrm{SO}(E)$, $g_S \in \mathrm{SO}(S)$, we denote by $\Phi_{E\,g_E} : N \to N$ and by $\Phi_{S\,g_S} : N \to N$ the diffeomorphisms

$$\Phi_{E\,g_E}(x) = \Phi_E(g_E, x) = g_E \circ x\,, \quad \Phi_{S\,g_S}(x) = \Phi_S(x, g_S) = x \circ g_S\,, \quad x \in N\,.$$

For each $v \in TN$, with $\tau_N(v) = x \in N$, we have of course

$$\widetilde{\varphi}_S\bigl(x, \Omega_S(v)\bigr) = v\,.$$

Since the actions Φ_E and Φ_S commute we have, for any $g_E \in \mathrm{SO}(E)$, $v \in TN$, $t \in \mathbb{R}$ and $x = \tau_N(v) \in N$,

$$\Phi_E\Bigl(g_E, x \circ \exp\bigl(t\Omega_S(v)\bigr)\Bigr) = g_E \circ x \circ \exp\bigl(t\Omega_S(v)\bigr) = \Phi_S\Bigl(g_E \circ x, \exp\bigl(t\Omega_S(v)\bigr)\Bigr)\,.$$

By taking the derivative with respect to t, then setting $t = 0$, we get

$$T\Phi_{E\,g_E}(v) = \widetilde{\varphi}_S\bigl(g_E \circ x, \Omega_S(v)\bigr)\,,$$

which means that

$$\Omega_S\bigl(T\Phi_{E\,g_E}(v)\bigr) = \Omega_S(v)\,.$$

The Riemannian metric $\widetilde{\mathbb{T}}$ therefore satisfies, for each $g_E \in \mathrm{SO}(E)$ and each pair (v, w) of vectors in TN which satisfy $\tau_N(v) = \tau_N(w)$,

$$\widetilde{\mathbb{T}}\bigl(T\Phi_{E\,g_E}(v), T\Phi_{E\,g_E}(w)\bigr) = \widetilde{\mathbb{T}}(v, w)\,.$$

This result means that the Riemannian metric $\widetilde{\mathbb{T}}$ remains invariant under the canonical lift to TN of the action Φ_E.

A similar calculation, in which $g_E \in \mathrm{SO}(E)$ is replaced by $g_S \in \mathrm{SO}(S)$, proves that, for each $v \in TN$,

$$\Omega_S\bigl(T\Phi_{S\,g_S}(v)\bigr) = \mathrm{Ad}_{g_S^{-1}}\bigl(\Omega_S(v)\bigr)\,,$$

so we have, for v and $w \in TN$ satisfying $\tau_N(v) = \tau_N(w)$,

$$\widetilde{\mathbb{T}}\bigl(T\Phi_{Sg_S}(v), T\Phi_{Sg_S}(w)\bigr) = \frac{1}{2} I\bigl(\mathrm{Ad}_{g_S^{-1}} \circ \Omega_S(v), \mathrm{Ad}_{g_S^{-1}} \circ \Omega_S(w)\bigr)\,.$$

For a general rigid body, the kinetic energy $\mathbb{T}$ and the Riemannian metric $\widetilde{\mathbb{T}}$ do not remain invariant under the canonical lift to TN of the action Φ_S. However, let us define an action on the left of G_S on the vector space of bilinear forms forms on $\mathfrak{so}(S)$ by setting, for each such bilinear form B and each $g_S \in G_S$

$$(g_S.B)(X_S, Y_S) = B\bigl(\mathrm{Ad}_{g_S^{-1}}(X_S), \mathrm{Ad}_{g_S^{-1}}(Y_S)\bigr)\,, \quad X_S \text{ and } Y_S \in \mathfrak{so}(S)\,.$$

We see that the kinetic energy $\mathbb{T}$ and the Riemannian metric $\widetilde{\mathbb{T}}$ remain invariant under the action of an element $g_S \in G_S$ if and only if $g_S.I = I$, i.e. if and only if g_S is an element of the isotropy subgroup if I for the above defined action of G_S on the space of bilinear forms on $\mathfrak{so}(S)$. This happens, for example, when the body has a symmetry axis, the isotropy subgroup of I being the group of rotations around that axis.

When the configuration of the body is $x \in N$, its *potential energy* is

$$U(x) = -\bigl\langle P, x(\overrightarrow{a})\bigr\rangle,$$

where $\overrightarrow{a} \in S$ is the vector whose origin is the fixed point O_S and extremity the centre of mass of the body, and $P \in E^*$ is the gravity force. Since E is identified with its dual E^*, the pairing by duality being the scalar product, P can be seen as a fixed vertical vector $\overrightarrow{P} \in E$ directed downwards, equal to the weight of the body (product of its mass with the gravity acceleration), and the potential energy can be written

$$U(x) = -\overrightarrow{P}.x(\overrightarrow{a}) = -x^{-1}(\overrightarrow{P}).\overrightarrow{a}\,.$$

We used the fact that the transpose $x^T : E^* \to S^*$ of the orthogonal linear map $x : S \to E$ is expressed, when S and E are identified with their dual spaces by means of the scalar product, as $x^{-1} : E \to S$.

When either $\overrightarrow{a} = 0$ or $\overrightarrow{P} = 0$ the potential energy vanishes, therefore remains invariant under the actions Φ_E of G_E and Φ_S of G_S on the manifold N. When both $\overrightarrow{a} \neq 0$ and $\overrightarrow{P} \neq 0$, the above formulae show that the potential energy remains invariant by the action of an element $g_E \in \mathrm{SO}(E)$ if and only if $g_E(\overrightarrow{P}) = \overrightarrow{P}$, which means if and only if g_E is an element of the isotropy group of $\overrightarrow{P}$ for the natural action of G_E on E. This isotropy subgroup is the group of rotations of E around the vertical straight line through the fixed point. Similarly, the potential energy remains invariant by the action of an element $g_S \in \mathrm{SO}(S)$ if and only if $g_S(\overrightarrow{a}) = \overrightarrow{a}$, which means if and only if g_S is an element of the isotropy group of $\overrightarrow{a}$ for the natural action of G_S on S. This isotropy subgroup is the group of rotations of S around the straight line which joins the fixed point and the centre of mass of the body.

The motion of the rigid body can be mathematically described by a Lagrangian system whose Lagrangian $L : TN \to \mathbb{R}$ is given, for $v \in TN$, by

$$L(v) = \widetilde{\mathbb{T}}(v,v) + \overrightarrow{P}.\tau_N(v)(\overrightarrow{a}) = \widetilde{\mathbb{T}}(v,v) + \bigl(\tau_N(v)\bigr)^{-1}(\overrightarrow{P}).\overrightarrow{a}\,.$$

We denote by $\widetilde{\mathbb{T}}^\flat : TN \to T^*N$ the map determined by the equality, in which v and $w \in TN$ satisfy $\tau_N(v) = \tau_N(w)$,

$$\bigl\langle \widetilde{\mathbb{T}}^\flat(v), w\bigr\rangle = \widetilde{\mathbb{T}}(v,w)\,.$$

The Legendre map $\mathcal{L} : TN \to T^*N$ determined by the Lagrangian L is

$$\mathcal{L} = 2\widetilde{\mathbb{T}}^\flat\,.$$

Its linearity and the positive definiteness of I ensure that it is a vector bundles isomorphism. The motion of the rigid body can therefore be described by a Hamiltonian system whose Hamiltonian $H : T^*N \to \mathbb{R}$ is given, for $p \in T^*N$, by

$$H(p) = \frac{1}{4}\langle p, (\widetilde{\mathbb{T}}^\flat)^{-1}(p)\rangle - \overrightarrow{P}.\pi_N(p)(\overrightarrow{a}) = \frac{1}{4}\langle p, (\widetilde{\mathbb{T}}^\flat)^{-1}(p)\rangle - \bigl(\pi_N(p)\bigr)^{-1}(\overrightarrow{P}).\overrightarrow{a}\,.$$

6.4.2 The Hamiltonian in Terms of Momentum Maps

Let $x \in N$ be fixed. The maps

$$\Omega_{Sx} = \Omega_S \bigm|_{T_xN} : T_xN \to \mathfrak{so}(S) \quad \text{and} \quad \Omega_{Ex} = \Omega_E \bigm|_{T_xN} : T_xN \to \mathfrak{so}(E)$$

are vector spaces isomorphisms. Their transpose

$$\Omega_{Sx}^T : \mathfrak{so}(S)^* \to T_x^*N \quad \text{and} \quad \Omega_{Ex}^T : \mathfrak{so}(E)^* \to T_x^*N$$

are too vector spaces isomorphisms. Their inverses are closely linked to the momentum maps $J_S : T^*N \to \mathfrak{so}(S)^*$ and $J_E : T^*N \to \mathfrak{so}(E)^*$ of the canonical lifts to T^*N of the actions Φ_S of G_S and Φ_E of G_E, respectively, on the manifold N. We have indeed, for any $x \in N$,

$$J_S \bigm|_{T_x^*N} = (\Omega_{Sx}^T)^{-1}\,, \quad J_E \bigm|_{T_x^*N} = (\Omega_{Ex}^T)^{-1}\,.$$

As above, let $x \in E$ be fixed and let v and $w \in T_xN$. The Legendre map $\mathcal{L} : TN \to T^*N$ satisfies

$$\begin{aligned}
\bigl\langle \mathcal{L}(v), w\bigr\rangle &= \frac{1}{2}\frac{\mathrm{d}}{\mathrm{d}s} I\bigl(\Omega_S(v+sw), \Omega_S(v+sw)\bigr) \bigm|_{s=0} \\
&= I\bigl(\Omega_S(v), \Omega_S(w)\bigr) \\
&= \bigl\langle I^\flat \circ \Omega_S(v), \Omega_S(w)\bigr\rangle \\
&= \bigl\langle \Omega_{Sx}^T \circ I^\flat \circ \Omega_S(v), w\bigr\rangle\,,
\end{aligned}$$

where $I^\flat : \mathfrak{so}(S) \to \mathfrak{so}(S)^*$ is the map defined by

$$\bigl\langle I^\flat(X_S), Y_S\bigr\rangle = I(X_S, Y_S)\,, \quad X_S \text{ and } Y_S \in \mathfrak{so}(S)\,.$$

So we can write

$$\mathcal{L} \bigm|_{T_xN} = \Omega_{Sx}^T \circ I^\flat \circ \Omega_S \bigm|_{T_xN}\,,$$

which shows that the momentum map J_S composed with the Legendre map $\mathcal{L}$ has the very simple expression

$$J_S \circ \mathcal{L} = I^\flat \circ \Omega_S\,.$$

The momentum map J_E composed with $\mathcal{L}$ has a slightly more complicated expression, valid for each $x \in N$,

$$J_E \circ \mathcal{L} \mid_{T_xN} = (\Omega_{Ex}^T)^{-1} \circ \Omega_{Sx}^T \circ I^\flat \circ \Omega_S \mid_{T_xM} .$$

Let $I^* : \mathfrak{so}(S)^* \times \mathfrak{so}(S)^* \to \mathbb{R}$ be the symmetric, positive definite bilinear form on $\mathcal{G}^*$

$$I^*(\xi, \eta) = I\big((I^\flat)^{-1}(\xi), (I^\flat)^{-1}(\eta)\big) = \big\langle \xi, (I^\flat)^{-1}(\eta)\big\rangle = \big\langle \eta, (I^\flat)^{-1}(\xi)\big\rangle .$$

The above expression of $J_S \circ \mathcal{L}$ and the bilinear form I^* allow us to write the Hamiltonian H as

$$H(p) = \frac{1}{2} I^*\big(J_S(p), J_S(p)\big) - \pi_N(p)^{-1}(\overrightarrow{P}).\overrightarrow{a} , \quad p \in T^*N .$$

Although the kinetic energy remains invariant under the canonical lift to T^*N of the action Φ_E, the expression of H in terms of the other momentum map J_E is too complicated to be useful.

6.4.3 The Euler–Poincaré Equation

We use the vector bundles isomorphism $\widetilde{\varphi}_S : N \times \mathfrak{so}(S) \to TN$ to derive the Euler–Poincaré equation. The map $\overline{L} = L \circ \widetilde{\varphi}_S : N \times \mathfrak{so}(S) \to \mathbb{R}$ is

$$\overline{L}(x, X) = \frac{1}{2} I(X, X) + \overrightarrow{P}.x(\overrightarrow{a}) , \quad X \in \mathfrak{so}(S) , \ x \in N .$$

Its partial differential $\mathrm{d}_2\overline{L}$ with respect to its second variable is

$$\mathrm{d}_2\overline{L}(x, X) = I^\flat(X) \in \mathfrak{so}(S)^* \equiv S .$$

A calculation similar to those of Sect. 6.4 leads to the following expression of J_S composed with the partial differential of $\overline{L}$ with respect to its first variable:

$$J_S \circ \mathrm{d}_1\overline{L}(x, X) = \overrightarrow{a} \times x^{-1}(\overrightarrow{P}) \in \mathfrak{so}(S)^* \equiv S .$$

Let $t \mapsto x(t)$ be a smooth curve in N solution of the Euler–Lagrange equation for the Lagrangian L, and $t \mapsto X(t)$ a smooth curve in $\mathfrak{so}(S)$ which satisfies the compatibility condition (1) of 5.2.2

$$\frac{\mathrm{d}x(t)}{\mathrm{d}t} = \widetilde{\varphi}_S\big(x(t), X(t)\big) . \qquad (1)$$

The Euler–Poincaré equation (3) of 5.2.2, satisfied by the smooth curve $t \mapsto \bigl(x(t), X(t)\bigr)$ in $N \times \mathfrak{so}(S)$, is

$$\left(\frac{\mathrm{d}}{\mathrm{d}t} - \mathrm{ad}^*_{X(t)}\right)\Bigl(I^\flat\bigl(X(t)\bigr)\Bigr) = \overrightarrow{a} \times x^{-1}(\overrightarrow{P})\,.$$

Using the expression of ad^* given by Formula (4) of Sect. 6.2, we can write the Euler–Poincaré equation as

$$\frac{\mathrm{d}}{\mathrm{d}t}I^\flat\bigl(\overrightarrow{X}(t)\bigr) - I^\flat\bigl(\overrightarrow{X}(t)\bigr) \times \overrightarrow{X}(t) = \overrightarrow{a} \times \overrightarrow{P_S}(t)\,, \tag{2}$$

where we have set $\overrightarrow{P_S}(t) = x(t)^{-1}(\overrightarrow{P})$. The physical meaning of the quantities which appear in this equation is the following: $\overrightarrow{X}(t)$ is the angular velocity, $I^\flat\bigl(\overrightarrow{X}(t)\bigr)$ the angular momentum and $\overrightarrow{P_S}(t)$ the weight of the moving body, all three a time t and seen by an observer bound to the body, therefore considered as vectors in S. We recognize the classical *Euler equation* for the motion of a rigid body around a fixed point.

Of course $x(t)\bigl(\overrightarrow{P_S}(t)\bigr) = \overrightarrow{P}$ is a constant vector in E, therefore

$$\frac{\mathrm{d}\Bigl(x(t)\bigl(\overrightarrow{P_S}(t)\bigr)\Bigr)}{\mathrm{d}t} = \frac{\mathrm{d}x(t)}{\mathrm{d}t}\bigl(\overrightarrow{P_S}(t)\bigr) + x(t)\left(\frac{\mathrm{d}\overrightarrow{P_S}(t)}{\mathrm{d}t}\right) = 0\,.$$

The first term in the right hand side, $\dfrac{\mathrm{d}x(t)}{\mathrm{d}t}\bigl(\overrightarrow{P_S}(t)\bigr)$, is the value at $\overrightarrow{P} \in E$ of the velocity field in E of the moving body. Therefore

$$\frac{\mathrm{d}x(t)}{\mathrm{d}t}\bigl(\overrightarrow{P_S}(t)\bigr) = \overrightarrow{\Omega_E}(v) \times \overrightarrow{P}\,, \quad \text{with } v = \frac{\mathrm{d}x(t)}{\mathrm{d}t} \in T_{x(t)}N\,.$$

Therefore we have

$$\frac{\mathrm{d}\overrightarrow{P_S}(t)}{\mathrm{d}t} = -x(t)^{-1}\Bigl(\overrightarrow{\Omega_E}(v) \times \overrightarrow{P}\Bigr) = -\overrightarrow{X}(t) \times \overrightarrow{P_S}(t)\,,$$

since, by Formula (1) of Sect. 6.4, $x(t)^{-1}\bigl(\overrightarrow{\Omega_E}(v)\bigr) = \overrightarrow{\Omega_S}(v) = \overrightarrow{X}(t)$. The compatibility condition and the Euler–Poincaré equation (Eqs. (1) and (2) of this Section) have lead us to the differential equation on $S \times S$, for the unknown parametrized curve $t \mapsto \bigl(\overrightarrow{X}(t), \overrightarrow{P_S}(t)\bigr)$,

$$\begin{cases} \dfrac{\mathrm{d}I^\flat\bigl(\overrightarrow{X}(t)\bigr)}{\mathrm{d}t} = I^\flat\bigl(\overrightarrow{X}(t)\bigr) \times \overrightarrow{X}(t) + \overrightarrow{a} \times \overrightarrow{P_S}(t)\,, \\ \dfrac{\mathrm{d}\overrightarrow{P_S}(t)}{\mathrm{d}t} = -\overrightarrow{X}(t) \times \overrightarrow{P_S}(t)\,. \end{cases} \tag{3}$$

When the right hand side of Eq. (2) vanishes, which occurs when the fixed point is the centre of mass of the body ($\overrightarrow{a} = 0$) or when there is no gravity field ($\overrightarrow{P} = 0$), the Euler–Poincaré equation yields an important reduction, since the first equation of (3) becomes an autonomous differential equation on the three-dimensional vector space S for the smooth curve $t \mapsto I^\flat\bigl(\overrightarrow{X_S}(t)\bigr)$, while the Euler–Lagrange equation or the Hamilton equation live on the six-dimensional manifolds TN or T^*N, respectively. Under these assumptions, the study of all possible motions of the rigid body is known in Mechanics as the *Euler–Poinsot problem*. The reader will find in [48] a very nice and thorough geometric presentation of the phase portrait of this problem.

When $\overrightarrow{a} \neq 0$ and $\overrightarrow{P} \neq 0$, the first equation of (3) is no more autonomous: one has to solve (3) on the six-dimensional vector space $S \times S$. The use of the Euler–Poincaré equation does not allow a reduction of the dimension of the phase space, but (3) may be easier to solve than the Euler–Lagrange equation or the Hamilton equation, because it lives on a vector space instead of on the tangent or cotangent bundle to a manifold.

6.4.4 Use of the Lie Algebra of Euclidean Displacements

As explained for example in Theorem 4.1 of [46] or in Proposition 13 and Example 14 of [45], there exists a Hamiltonian action on T^*N of the semi-direct product $G_S \times S$, (the group of Euclidean displacements, generated by rotations and translations, of the Euclidean affine space S) which extends the canonical lift to T^*N of the action Φ_S, such that the Hamiltonian H can be expressed as composed of the momentum map of that action with a smooth function $h : \mathfrak{so}(S)^* \times S^* \to \mathbb{R}$. We briefly explain below the construction of that action.

For each $\overrightarrow{b} \in S$, let $f_{\overrightarrow{b}} : N \to \mathbb{R}$ be the smooth function

$$f_{\overrightarrow{b}}(x) = \bigl\langle x^T(P), \overrightarrow{b}\bigr\rangle = \overrightarrow{b}.x^{-1}(\overrightarrow{P}), \quad x \in N.$$

The map $\Psi : T^*N \times S \to T^*N$ defined by

$$\Psi(p, \overrightarrow{b}) = p - \mathrm{d}f_{\overrightarrow{b}} \circ \pi_N(p), \quad p \in T^*N, \ \overrightarrow{b} \in S,$$

is a Hamiltonian action of S on the symplectic manifold $(T^*N, \mathrm{d}\eta_N)$: the Lie algebra of S can indeed be identified with S, the exponential map becoming the identity of S, and for each $\overrightarrow{b} \in S$, the vector field on T^*N whose flow is the one-parameter group of transformations of T^*N

$$\Bigl\{p \mapsto p - t\mathrm{d}f_{\overrightarrow{b}}\bigl(\pi_N(p)\bigr) \,;\, t \in \mathbb{R}\Bigr\}$$

is Hamiltonian an admits as Hamiltonian the function

$$p \mapsto f_{\overrightarrow{b}} \circ \pi_N(p) == \overrightarrow{b}.\pi_N(p)^{-1}(\overrightarrow{P}), \quad p \in T^*N.$$

This formula proves that Ψ is a Hamiltonian action which admits

$$J_\Psi : T^*N \to S^* \equiv S\,, \quad J_\Psi(p) = \pi_N(p)^{-1}(\overrightarrow{P})$$

as a momentum map. Gluing together Ψ with the canonical lift $\widehat{\Phi}_S$ of Φ_S to the cotangent bundle, we obtain a Hamiltonian action on the right Ξ of the semi-direct product $G_S \times S$ on the symplectic manifold $(T^*N, \mathrm{d}\eta_N)$:

$$\Xi\big(p, (g_S, \overrightarrow{b})\big) = \Psi(\widehat{\Phi}_S(p, g_S), \overrightarrow{b})$$

with $(J_S, J_\psi) : T^*N \to \mathfrak{so}(S)^* \times S^* \equiv S \times S$ as a momentum map. The function $h : \mathfrak{so}(S)^* \times S^* \equiv S \times S \to \mathbb{R}$

$$h(\overrightarrow{\eta}, \overrightarrow{\zeta}) = \frac{1}{2} I^*(\overrightarrow{\eta}, \overrightarrow{\eta}) - \overrightarrow{\zeta}\,\overrightarrow{\eta}$$

is such that the Hamiltonian $H : T^*N \to \mathbb{R}$ can be written as $H = h \circ (J_S, J_\Psi)$, and Eq. (3) of Sect. 6.4 is the Hamilton equation on $\mathfrak{so}(S)^* \times S^* \equiv S \times S$ (endowed with its canonical Poisson structure) for the Hamiltonian h. This result is in agreement with the fact that (J_S, J_Ψ) is an ad*-invariant Poisson map (4.4.11).

6.4.5 Reduction by the Use of First Integrals

The effects on the kinetic and potential energies of the Hamiltonian actions $\widehat{\Phi}_E$ and $\widehat{\Phi}_S$ were discussed in Sect. 6.4. When $\overrightarrow{a} \neq 0$ and $\overrightarrow{P} \neq 0$, the Hamiltonian H remains invariant under the restriction of the action $\widehat{\Phi}_E$ to the subgroup of rotations around the vertical axis through the fixed point. The corresponding momentum map, which is the orthogonal projection of the momentum map J_E on the vertical direction, is therefore a first integral. Another first integral is the total energy, i.e. the Hamiltonian H itself. For a general rigid body, no other independent first integrals are known. However, in two special cases of particular rigid bodies, there exists another independent first integral.

The first case, known as the *Euler–Lagrange problem* in Mechanics, is when the straight line which joins the fixed point and the centre of mass of the body is an axis of symmetry for the inertia properties of the body. The Hamiltonian H remains then invariant under the restriction of the action $\widehat{\Phi}_S$ to the subgroup of $\mathrm{SO}(S)$ of rotations around this straight line. The corresponding momentum map is the orthogonal projection of the momentum map J_S on the direction of the symmetry axis.

The second case, discovered by the Russian mathematician Sonya Kovalevskaya (1850–1891) [49] is when two of the principal moments of inertia of the body are equal to twice the third and when the centre of mass of the body lies in the plane of the two equal moments of inertia. The explanation of the existence, in this very special case, of an additional integral is much more complicated than that of the existence of an additional integral for the Euler–Lagrange problem, and involves mathematical tools which are not discussed in the present paper. The reader is referred to the book by Audin [50] for a discussion of these tools and to the beautiful other book by the same author [51] for a very moving presentation of the life of Sonya Kovalevskaya.

When $\overrightarrow{a} = 0$ or $\overrightarrow{P} = 0$ (the Euler–Poinsot problem) the Hamiltonian H remains invariant under the action $\widehat{\Phi}_E$ of the full group $SO(E)$, so the corresponding momentum map J_E is (as already seen in Sect. 6.4) a (vector valued) first integral.

6.5 The Kepler Problem

6.5.1 Mathematical Description of the Problem

We consider the motion in space of a material point of mass m submitted to the gravitational field created by an attractive centre O. Taking O as origin allows us to consider E as a *vector* Euclidean three-dimensional oriented space. The configuration space, i.e. the set of all possible positions of the material point, is $N = E\backslash\{O\}$. The tangent bundle TN and the cotangent bundle T^*N will both be identified with $N \times E$. An element of TN is therefore a pair $(\overrightarrow{x}, \overrightarrow{v}) \in E \times E$ satisfying $\overrightarrow{x} \neq 0$. Similarly an element of T^*N is a pair $(\overrightarrow{x}, \overrightarrow{p}) \in E \times E$ satisfying $\overrightarrow{x} \neq 0$.

The kinetic energy $\mathbb{T} : TN \equiv N \times E \to \mathbb{R}$ and the potential energy $U : N \to \mathbb{R}$ are

$$\mathbb{T}(\overrightarrow{x}, \overrightarrow{v}) = \frac{1}{2}m\|\overrightarrow{v}\|^2, \quad U(\overrightarrow{x}) = \frac{mk}{\|\overrightarrow{x}\|}.$$

The Lagrangian $L : TN \equiv N \times E \to \mathbb{R}$ of the Kepler problem is therefore

$$L(\overrightarrow{x}, \overrightarrow{v}) = \frac{1}{2}m\|\overrightarrow{v}\|^2 + \frac{mk}{\|\overrightarrow{x}\|}.$$

The Legendre map $\mathcal{L} : TN \equiv N \times E \to T^*N \equiv N \times E$ is

$$\mathcal{L}(\overrightarrow{x}, \overrightarrow{v}) = (\overrightarrow{x}, \overrightarrow{p}), \quad \text{with } \overrightarrow{p} = m\overrightarrow{v}.$$

The Kepler problem can therefore be mathematically formulated as a Hamiltonian dynamical system on $T^*N \equiv N \times E$, with the Hamiltonian

$$H(\overrightarrow{x}, \overrightarrow{p}) = \frac{1}{2m}\|\overrightarrow{p}\|^2 - \frac{mk}{\|\overrightarrow{x}\|}.$$

The natural action Φ_E of SO(E) on E leaves invariant $N = E\backslash\{O\}$, therefore is an action of SO(E) on N. With the identifications of TN and T^*N with $N \times E$ which we have made, the canonical lifts $\overline{\Phi}_E$ and $\widehat{\Phi}_E$ of that action to the tangent and cotangent bundles, respectively, are expressed as

$$\overline{\Phi}_E\big(g_E, (\overrightarrow{x}, \overrightarrow{v})\big) = \big(g_E(\overrightarrow{x}), g_E(\overrightarrow{v})\big), \quad \widehat{\Phi}_E\big(g_E, (\overrightarrow{x}, \overrightarrow{p})\big) = \big(g_E(\overrightarrow{x}), g_E(\overrightarrow{p})\big).$$

Since the norm of a vector in E remains invariant under the action Φ_E, the Lagrangian L and the Hamiltonian H remain invariant under the actions $\overline{\Phi}_E$ and $\widehat{\Phi}_E$, respectively. The action $\widehat{\Phi}_E$ is Hamiltonian, and we know (Formula (4) of Sect. 6.2) that its momentum map $J_E : T^*N \equiv N \times E \to \mathfrak{so}(E)^* \equiv E$ is

$$J_E(\overrightarrow{x}, \overrightarrow{p}) = \overrightarrow{x} \times \overrightarrow{p}\,.$$

The map J_E is the *angular momentum* of the moving material point with respect to the attractive centre. Noether's theorem (4.4.5) shows that it is a first integral of the Kepler problem. Another first integral of the Kepler problem is the total energy H, as shown by 3.3.4.

6.5.2 The Euler–Poincaré Equation

The Lie group SO(E) does not act transitively on N by the action Φ_E, since the orbits of this action are spheres centered on O. However, extending this action by homotheties of strictly positive ratio, we obtain a transitive action on N of the direct product SO(E)$\times\,]0, +\infty[$

$$\Psi_E\big((g_E, r), \overrightarrow{x}\big) = r g_E(\overrightarrow{x}), \quad g_E \in \mathrm{SO}(E), \ r \in]0, +\infty[, \ \overrightarrow{x} \in N.$$

Let $\psi_E : \mathfrak{so}(E) \times \mathbb{R} \to A^1(N)$ be the associated action of the Lie algebra $\mathfrak{so}(E) \times \mathbb{R}$. The map $\widetilde{\psi}_E : N \times \big(\mathfrak{so}(E) \times \mathbb{R}\big) \to TN$, $\widetilde{\psi}_E\big(\overrightarrow{x}, (\overrightarrow{X}, \lambda)\big) = \psi_E(\overrightarrow{X}, \lambda)(\overrightarrow{x})$, can be written, with the identifications of $\mathfrak{so}(E)$ with E and of TN with $N \times E$,

$$\widetilde{\psi}_E\big(\overrightarrow{x}, (\overrightarrow{X}, \lambda)\big) = \big(\overrightarrow{x}, \overrightarrow{X} \times \overrightarrow{x} + \lambda\overrightarrow{x}\big).$$

The function $\overline{L} = L \circ \widetilde{\psi}_E$ is therefore

$$\overline{L}\big(\overrightarrow{x}, (\overrightarrow{X}, \lambda)\big) = \frac{1}{2}m\|\overrightarrow{x}\|^2(\|\overrightarrow{X}\|^2 + \lambda^2) - \frac{1}{2}m(\overrightarrow{X}.\overrightarrow{x})^2 + \frac{mk}{\|\overrightarrow{x}\|}\,.$$

Its partial differentials $\mathrm{d}_1\overline{L}$ and $\mathrm{d}_2\overline{L}$ with respect to its first variable $\overrightarrow{x}$ and to its second variable $(\overrightarrow{X}, \lambda)$ are, with the identifications of E^* and $\big(\mathfrak{so}(E) \times \mathbb{R}\big)^*$ with, respectively, E and $E \times \mathbb{R}$,

$$\mathrm{d}_1\overline{L}(\overrightarrow{x}, (\overrightarrow{X}, \lambda)) = \left(m(\|\overrightarrow{X}\|^2 + \lambda^2) - \frac{mk}{\|\overrightarrow{x}\|^3}\right)\overrightarrow{x} - m(\overrightarrow{X}.\overrightarrow{x})\overrightarrow{X},$$
$$\mathrm{d}_2\overline{L}(\overrightarrow{x}, (\overrightarrow{X}, \lambda)) = \left(m\|\overrightarrow{x}\|^2\overrightarrow{X} - m(\overrightarrow{X}.\overrightarrow{x})\overrightarrow{x}, m\|\overrightarrow{x}\|^2\lambda\right).$$

The canonical lift $\widehat{\Psi}_E$ of Ψ_E to the cotangent bundle is a Hamiltonian action, whose momentum map $(J_E, K_E) : T^*N \to \mathfrak{so}(E)^* \times \mathbb{R}$ has J_E as first component. Its second component is

$$K_E(\overrightarrow{x}, \overrightarrow{p}) = \overrightarrow{x}.\overrightarrow{p}.$$

Let $t \mapsto \overrightarrow{x}(t)$ be a smooth curve in N, parametrized by the time t, solution of the Euler–Lagrange equation for the Lagrangian L. The compatibility condition (1) of 5.2.2, for a smooth map $t \mapsto \left(\overrightarrow{X}(t), \lambda(t)\right)$ in $\mathfrak{so}(E) \times \mathbb{R}$, can be written as

$$\frac{\mathrm{d}\overrightarrow{x}(t)}{\mathrm{d}t} = \overrightarrow{X}(t) \times \overrightarrow{x}(t) + \lambda(t)\overrightarrow{x}(t). \tag{1}$$

This equation does not involve the component of $\overrightarrow{X}(t)$ parallel to $\overrightarrow{x}(t)$, since the vector product of this component with $\overrightarrow{x}(t)$ vanishes.

The Euler–Poincaré equation (3) of 5.2.2 has now two components, on $\mathfrak{so}(E)^*$ and on $\mathbb{R}^*$ identified, respectively, with E and with $\mathbb{R}$. With the above expressions of $\mathrm{d}_1\overline{L}$, $\mathrm{d}_2\overline{L}$, J_E and K_E, we obtain for its first component

$$\begin{aligned}
&\frac{\mathrm{d}}{\mathrm{d}t}\left(m\|\overrightarrow{x}(t)\|^2\left(\overrightarrow{x}(t) - \frac{\overrightarrow{x}(t).\overrightarrow{x}(t)}{\|\overrightarrow{x}\|^2}\overrightarrow{x}(t)\right)\right)\\
&\quad = m\left(\overrightarrow{x}(t).\overrightarrow{x}(t)\right)\overrightarrow{x}(t) \times \overrightarrow{x}(t) - m\left(\overrightarrow{x}(t).\overrightarrow{x}(t)\right)\mathrm{ad}^*_{\overrightarrow{x}(t)}\overrightarrow{x}(t) = 0,
\end{aligned}$$

where we have used Formula (4) of 6.2. Its second component is

$$\frac{\mathrm{d}}{\mathrm{d}t}\left(m\|\overrightarrow{x}(t)\|^2\lambda\right) = m\left(\|\overrightarrow{x}(t)\|^2 + \lambda^2 - \frac{\left(\overrightarrow{x}(t).\overrightarrow{x}(t)\right)^2}{\|\overrightarrow{x}\|^2}\right)\|\overrightarrow{x}(t)\|^2 - \frac{mk}{\|\overrightarrow{x}(t)\|}.$$

The vector $\overrightarrow{X}(t)$ is the sum of two components $\overrightarrow{X_1}(t)$ orthogonal to $\overrightarrow{x}(t)$ and $\overrightarrow{X_2}(t)$ parallel to $\overrightarrow{x}(t)$. Since

$$\overrightarrow{X_1}(t) = \overrightarrow{X}(t) - \frac{\overrightarrow{X}(t).\overrightarrow{x}(t)}{\|\overrightarrow{x}(t)\|^2}\overrightarrow{x}(t),$$

the two components of the Euler–Poincaré equation become

$$\begin{cases} \dfrac{\mathrm{d}}{\mathrm{d}t}\Big(m\|\overrightarrow{x}(t)\|^2\overrightarrow{X_1}(t)\Big) = 0\,, \\ \dfrac{\mathrm{d}}{\mathrm{d}t}\Big(m\|\overrightarrow{x}(t)\|^2\lambda\Big) = m\left(\|\overrightarrow{X_1}(t)\|^2 + \lambda^2\right)\|\overrightarrow{x}(t)\|^2 - \dfrac{mk}{\|\overrightarrow{x}(t)\|}\,. \end{cases}$$

The first equation expresses the fact that J_E is a first integral of the Kepler problem, since we have

$$m\|\overrightarrow{x}(t)\|^2\overrightarrow{X_1}(t) = \overrightarrow{x}(t) \times \overrightarrow{p}(t) = J_E\big(\overrightarrow{x}(t), \overrightarrow{p}(t)\big)\,.$$

Similarly, the second equation can be written

$$\frac{\mathrm{d}}{\mathrm{d}t}\big(\overrightarrow{p}(t).\overrightarrow{x}(t)\big) = \frac{\|\overrightarrow{p}(t)\|^2}{m} - \frac{km}{\|\overrightarrow{x}\|}\,,$$

which is a direct consequence of Hamilton's equations for the Hamiltonian H of the Kepler problem.

Neither the Euler–Poincaré equation nor the compatibility condition involve the component $\overrightarrow{X_2}(t)$ of $\overrightarrow{X}(t)$ parallel to $\overrightarrow{x}(t)$. This illustrates the fact that the system made by these equations is underdetermined when the dimension of the Lie algebra which acts on the configuration space is strictly larger than the dimension of this space.

6.5.3 Hamilton's Method of Solving the Kepler Problem

The Hamiltonian H of the Kepler problem remains invariant under the canonical lift to T^*N of the action of $\mathrm{SO}(E)$. Noether's theorem (4.4.5) shows that the corresponding momentum map J_E is a first integral. Of course the total energy, i.e. the Hamiltonian H, is too a first integral (3.3.4). Following the method due to Hamilton [52], we explain below how the three Kepler laws can easily be deduced from the first integrals J_E and H.

Let us assume that at a particular time t_0, $\overrightarrow{x}(t_0)$ and $\overrightarrow{p}(t_0)$ are not collinear. The vector $\overrightarrow{\Omega} = J_E\big(\overrightarrow{x}(t), \overrightarrow{p}(t)\big) = \overrightarrow{x}(t) \times \overrightarrow{p}(t)$ does not depend on t since J_E is a first integral, and is $\neq 0$ since for $t = t_0$, $\overrightarrow{x}(t)$ and $\overrightarrow{p}(t)$ are not collinear. We choose an orthonormal positively oriented basis $(\overrightarrow{e_x}, \overrightarrow{e_y}, \overrightarrow{e_z})$ of E such that $\overrightarrow{\Omega} = \Omega\,\overrightarrow{e_z}$, with $\Omega > 0$. The vectors $\overrightarrow{x}(t)$ and $\overrightarrow{p}(t)$ remain for all times in the two-dimensional vector subspace F spanned by $(\overrightarrow{e_x}, \overrightarrow{e_y})$. Let $\theta(t)$ be the polar angle made by $\overrightarrow{x}(t)$ with $\overrightarrow{e_x}$. We have

$$\begin{aligned}
\overrightarrow{x}(t) &= r(t)\cos\theta(t)\overrightarrow{e_x} + r(t)\sin\theta(t)\overrightarrow{e_y}\,,\\
\overrightarrow{p}(t) &= m\left(\frac{\mathrm{d}r(t)}{\mathrm{d}t}\cos\theta(t) - r(t)\frac{\mathrm{d}\theta(t)}{\mathrm{d}t}\sin\theta(t)\right)\overrightarrow{e_x}\\
&\quad + m\left(\frac{\mathrm{d}r(t)}{\mathrm{d}t}\sin\theta(t) + r(t)\frac{\mathrm{d}\theta(t)}{\mathrm{d}t}\cos\theta(t)\right)\overrightarrow{e_y}\\
\overrightarrow{\Omega} &= mr^2\frac{\mathrm{d}\theta(t)}{\mathrm{d}t}\overrightarrow{e_z}\,.
\end{aligned}$$

Therefore

$$mr^2\frac{d\theta}{dt} = \Omega = \text{Constant}\,.$$

This is the *second Kepler law*, also called *law of areas*, since $\dfrac{\Omega}{2m}$ is the area swept by the straight line segment joining the moving material point to the attractive centre during an unit time. Since $t \mapsto \theta(t)$ is a strictly increasing function whose derivative never vanishes, we can take θ instead of time t as independent variable. Using Hamilton's equation (or Newton's equation), we can write

$$\frac{\mathrm{d}\overrightarrow{p}(\theta)}{\mathrm{d}\theta} = \frac{\mathrm{d}\overrightarrow{p}(t)}{dt}\frac{\mathrm{d}t}{d\theta} = \frac{mr(\theta)^2}{\Omega}\left(-\frac{mk}{r(\theta)^3}\overrightarrow{x}(\theta)\right) = -\frac{m^2k}{\Omega}(\cos\theta\overrightarrow{e_x} + \sin\theta\overrightarrow{e_y})\,.$$

This ordinary differential equation for the unknown $\overrightarrow{p}(\theta)$, which no more involves $\overrightarrow{x}(\theta)$, can be readily integrated:

$$\overrightarrow{p}(\theta) = \frac{m^2k}{\Omega}(-\sin\theta\overrightarrow{e_x} + \cos\theta\overrightarrow{e_y}) + \overrightarrow{c}\,,$$

where $\overrightarrow{c}$ is a (vector) integrating constant. We will choose $\overrightarrow{e_y}$ such that $\overrightarrow{c} = c\overrightarrow{e_y}$, where c is a numeric constant which satisfy $c \geq 0$.

With O as origin let us draw two vectors in the plane xOy, the first one (constant) being equal to $\overrightarrow{c}$, and the second one (which varies with θ) equal to $\overrightarrow{p}$. The end point of that second vector moves on a circle whose centre is the end point of the vector equal to $\overrightarrow{c}$, and whose radius is $\mathcal{R} = \dfrac{m^2k}{\Omega}$. The part of this circle swept by the end point of this second vector is (up to multiplication by m) the *hodograph* of the Kepler problem. A short calculation leads to the following very simple relation between the energy H of a motion, the radius $\mathcal{R}$ of its hodograph and the distance c from the attracting centre O to the centre of the hodograph:

$$2mH = c^2 - \mathcal{R}^2\,.$$

The right-hand side $c^2 - \mathcal{R}^2$ is the *power*[1] of O with respect to the hodograph.

We also obtain $r = \| \overrightarrow{x} \|$ as a function of θ

$$r(\theta) = \frac{\Omega^2}{m^2k + \Omega c \cos\theta} = \frac{\Lambda}{1 + \varepsilon\cos\theta}, \quad \text{with } \Lambda = \frac{\Omega^2}{m^2k}, \; \varepsilon = \frac{\Omega c}{m^2k}.$$

It is the polar equation of a conic section with O as focus point and ε as eccentricity. This conic section (or, when $\varepsilon > 1$, the arc of this conic swept by the moving material point) is the orbit in E of the moving material point. This result is the *first Kepler law*.

The modulus Ω of the angular momentum, the total energy H and the eccentricity ε satisfy

$$\varepsilon^2 - 1 = \frac{2\Omega^2 H}{m^3k^2}.$$

This formula shows that the orbit in E of the moving material point is an ellipse $(0 \leq \varepsilon < 1)$ if $H < 0$, a parabola $(\varepsilon = 1)$ if $E = 0$ and a connected component of a hyperbola $(\varepsilon > 1)$ if $H > 0$.

When $H < 0$, the orbit in E of the moving point is an ellipse and its motion is periodic. The period T is easily obtained by writing that the area swept in a time T by the straight line segment which joins the moving point to the attractive centre is the area A delimited by the orbit:

$$T = \frac{2mA}{\Omega} = \frac{2\pi m a^2\sqrt{1-\varepsilon^2}}{\Omega}$$

where a is the length of the half major axis of the orbit. By using the formula

$$\Omega^2 = m^2ka\sqrt{1-\varepsilon^2}$$

we obtain

$$T^2 = \frac{4\pi^2}{k}a^3.$$

We conclude that the square of the period is proportional to the third power of the length of the half major axis. This result is the *third Kepler law*.

Hamilton's method of solving the Kepler problem is much easier than the Marsden–Weinstein reduction procedure, to which it is only very loosely related. A non-zero vector $\overrightarrow{\Omega}$ is a regular value of J_E, so $J_E^{-1}(\overrightarrow{\Omega})$ is a smooth three-dimensional submanifold of T^*N: it is the set of pairs of vectors $(\overrightarrow{x}, \overrightarrow{p}) \in F \times F$ such that

[1] In plane Euclidean geometry, the *power* of a point O with respect to a circle $\mathcal{C}$ is the real number $\overrightarrow{OA}.\overrightarrow{OB}$, where A and B are the two intersection points of $\mathcal{C}$ with a straight line $\mathcal{D}$ through O. That number does not depend on $\mathcal{D}$ and is equal to $\|\overrightarrow{OC}\|^2 - \mathcal{R}^2$, where C is the centre and $\mathcal{R}$ the radius of $\mathcal{C}$.

$\overrightarrow{x} \times \overrightarrow{p} = \overrightarrow{\Omega}$, where F is the two-dimensional vector subspace of E orthogonal to $\overrightarrow{\Omega}$. This submanifold remains invariant under the action on T^*N of the one-dimensional subgroup of SO(E), isomorphic to the circle S^1, of rotations around the straight line through O parallel to $\overrightarrow{\Omega}$. The reduced Marsden–Weinstein symplectic manifold is the set of orbits of this action. It is isomorphic to the open half-plane $\{(r,\lambda) \in \mathbb{R}^2; r > 0\}$, and the projection of $J^{-1}(\overrightarrow{\Omega})$ onto the reduced symplectic manifold is the map $(\overrightarrow{x}, \overrightarrow{p}) \mapsto (r,\lambda)$, with $r = \| \overrightarrow{x} \|$, $\lambda = \overrightarrow{x} \,.\, \overrightarrow{p}$. The reduced symplectic form and Hamiltonian are, respectively,

$$\omega_{\overrightarrow{\Omega}} = \frac{1}{r} \mathrm{d}\lambda \wedge \mathrm{d}r\,, \quad H_{\overrightarrow{\Omega}} = \frac{m(\Omega^2 + \lambda^2)}{2r^2} - \frac{mk}{r}\,.$$

Instead of using this reduced symplectic manifold and this reduced Hamiltonian, Hamilton's method uses a clever choice of independent and dependent variables on $J_E^{-1}(\overrightarrow{\Omega})$ which leads to an easy to solve autonomous differential equation for $\overrightarrow{p}$ as a function of the polar angle θ of $\overrightarrow{x}$. It is successful essentially because the hodograph of the Kepler problem is a circle (or, when $H \geq 0$, a part of a circle).

6.5.4 The Eccentricity Vector

There exists still another vector valued first integral $\overrightarrow{\varepsilon}$ of the Kepler problem called the *eccentricity vector*, discovered by Jakob Hermann (1678–1753) three centuries ago [53, 54], often improperly called the *Laplace vector* or the *Ruge-Lenz vector*, whose expression is

$$\overrightarrow{\varepsilon} = -\frac{\overrightarrow{x}}{\| \overrightarrow{x} \|} + \frac{\overrightarrow{p} \times (\overrightarrow{x} \times \overrightarrow{p})}{m^2 k} = \left(\frac{\| \overrightarrow{p} \|^2}{m^2 k} - \frac{1}{\| \overrightarrow{x} \|} \right) \overrightarrow{x} - \frac{\overrightarrow{p} \,.\, \overrightarrow{x}}{m^2 k} \overrightarrow{p}\,.$$

For each motion of the moving material point, the eccentricity vector $\overrightarrow{\varepsilon}$ is a dimensionless vector parallel to the straight line segment which joins the attractive centre O to the perihelion of the orbit (i.e. the point of the orbit which is the nearest to the attractive centre), of length numerically equal to the eccentricity ε of the orbit. When the orbit is a circle, the perihelion is undetermined and $\overrightarrow{\varepsilon} = 0$. We briefly explain below the group theoretical origin of the eccentricity vector. A more detailed explanation can be found for example in [55]. Many other interesting results about the Kepler problem can be found in the excellent books [48, 56–59].

Motions $t \mapsto (\overrightarrow{x}(t), \overrightarrow{p}(t))$ of the Kepler problem in which $\overrightarrow{x}(t)$ and $\overrightarrow{p}(t)$ are parallel are not defined for all values of the time t: the curves drawn in E by the vectors $\overrightarrow{x}(t)$ and $\overrightarrow{p}(t)$ both are supported by the same straigh line through the attractive centre O, so the motion finishes, or begins, at a finite time, when $\overrightarrow{x}(t)$ reaches 0, i.e. when the moving point collides with the attractive centre or is expelled by it. When

t tends towards that final (or initial) instant, $\|\overrightarrow{p}(t)\|$ tends towards $+\infty$. This fact complicates the study of the global topological properties of the set of all possible motions of the Kepler problem.

For any motion $t \mapsto (\overrightarrow{x}(t), \overrightarrow{p}(t))$, the curves drawn in E by the vectors $\overrightarrow{x}(t)$ and $\overrightarrow{p}(t)$ are, respectively, the orbit and the hodograph of the motion. The exchange $(\overrightarrow{x}, \overrightarrow{p}) \mapsto (\overrightarrow{p}, \overrightarrow{x})$ is an anti-symplectic map, which allows us, at the price of a change of sign of the symplectic form, to consider the curve drawn by $\overrightarrow{p}(t)$ as the orbit of some Hamiltonian dynamical system and the curve drawn by $\overrightarrow{x}(t)$ as the corresponding hodograph. This remark offers a way of studying the global properties of the set of all possible motions: for a motion $t \mapsto (\overrightarrow{x}(t), \overrightarrow{p}(t))$ which starts or ends at finite instant by a collision with the attractive centre or an ejection by that point, the curve drawn by $\overrightarrow{p}(t)$, now considered as an orbit rather than a hodograph, goes to infininy when t tends towards this limit instant. By a inverse stereographic projection, E can be mapped on a three-dimensional sphere Q minus a point (the pole of the stereographic projection), and the curve drawn by $\overrightarrow{p}(t)$ is mapped onto a curve which tends towards the pole P of the stereographic projection. The canonical prolongation of the inverse stereographic projection to the cotangent bundles allows us to map the phase space of the Kepler problem onto the open subset of T^*Q complementary to the fibre T^*_PQ over the pole of the stereographic projection. On T^*Q, motions which reach T^*_PQ can be prolongated and no more appear as starting, or ending, at a finite instant of time. This idea, due to Fock [60] who applied it to the study of the hydrogen atom in quantum mechanics, was used by Moser [61] for the regularization of the Kepler problem for negative values of the Hamioltonian H. Györgyi [62, 63] used a similar idea. Since the inverse stereographic projection maps circles onto circles, the image of $\overrightarrow{p}(t)$ draws a circle on the three-dimensional sphere Q and, for a particular value of the total energy H, this circle is a great circle, i.e. a geodesic of Q. Györgyi [62, 63] proved that the cylindrical projection onto E of these great circles are ellipses centered on O whose eccentricity is the same as those of the orbits drawn on E by the corresponding vector $\overrightarrow{x}(t)$. The group SO(4) acts on the three-dimensional sphere Q and, by the canonical lift to the cotangent bundle, on T^*Q by a Hamiltonian action. The transformed Hamiltonian is not really invariant under that action and some more work (a reparametrization of time) is still needed, but finally Noether's theorem can be used. The eccentricity vector $\overrightarrow{\varepsilon}$ is (modulo the identification of the phase space of the Kepler problem with an open subset of T^*Q) the momentum map of that action, which explains why it is a first integral.

For motions with a positive value of the total energy, there exists a similar construction in which instead of a three-dimensional sphere, Q is a two-sheeted revolution three-dimensional hyperboloid. The symmetry group is the Lorentz group SO(3, 1); the eccentricity vector $\overrightarrow{\varepsilon}$ still is the momentum map of its action on T^*Q [64, 65]. For a motion with a zero value of H, the circle drawn in E by the vector $\overrightarrow{p}(t)$ contains the attractive centre O, so an inversion with O as pole transforms this circle into a straight line, i.e. a geodesic of E. The symmetry group is then the group of Euclidean displacements in E (generated by rotations and translations); the eccentricity vector $\overrightarrow{\varepsilon}$ still is the momentum map of its action on T^*E.

Ligon and Schaaf [66] used these results to construct a global symplectic diffeomorphism of the phase space of the Kepler problem (for negative values of H) onto an open subset of the cotangent bundle to a three-dimensional sphere. Györgyi had done that earlier [62, 63] but it seems that his work was not known by mathematicians. Later several other authors pursued these studies [67, 68].

Souriau [69] used a totally different approach. He built the regularized manifold of motions of the Kepler problem in a single step, for all values of the energy, by successive derivations of the equations of motion and analytic prolongation, calculated its symplectic form and directly determined its symmetry groups. The eccentricity vector appears again as a momentum map for the Hamiltonian actions of these groups.

Acknowledgments I address my thanks to Ivaïlo Mladenov for his kind invitation to present lectures at the International Conference on Geometry, Integrability and Quantization held in Varna in June 2013. For their moral support and many stimulating scientific discussions, I thank my colleagues and friends Alain Albouy, Marc Chaperon, Alain Chenciner, Maylis Irigoyen, Jean-Pierre Marco, Laurent Lazzarini, Fani Petalidou, Géry de Saxcé, Wlodzimierz Tulczyjew, Paweł Urbański, Claude Vallée and Alan Weinstein.

References

1. Abraham, R., Marsden, J.E.: Foundations of Mechanics, 2nd edn. Addison-Wesley, Reading (1978)
2. Arnold, V.I.: Mathematical Methods of Classical Mechanics, 2nd edn. Springer, New York (1978)
3. Arnold, V.I., Khesin, B.A.: Topological Methods in Hydrodynamics. Springer, New York (1998)
4. Iglesias, P.: Symétries et Moment. Hermann, Paris (2000)
5. Laurent-Gengoux, C., Pichereau, A., Vanhaecke, P.: Poisson Structures. Springer, Berlin (2013)
6. Kosmann-Schwarzbach, Y.: (éditrice), Siméon-Denis Poisson, les mathématiques au service de la science. Éditions de l'École polytechnique, Palaiseau (2013)
7. Vaisman, I.: Lectures on the Geometry of Poisson manifolds. Birkhäuser, Basel Boston Berlin (1994)
8. Gromov, M.: Pseudo holomorphic curves in symplectic manifolds. Invent. Math. **82**, 307–347 (1985)
9. Tulczyjew, W.M.: Geometric Formulations of Physical Theories. Monographs and Textbooks in Physical Science. Bibliopolis, Napoli (1989)
10. Dirac, P.A.M.: Generalized Hamiltonian dynamics. Can. J. Math. **2**, 129–148 (1950)
11. Dirac, P.A.M.: Lectures on Quantum Mechanics. Belfer graduate School of Science, Yeshiva University, New York (1964)
12. Lichnerowicz, A.: Les variétés de Poisson et leurs algèbres de Lie associées. J. Differ. Geom. **12**, 253–300 (1977)
13. Lichnerowicz, A.: Les variétés de Jacobi et leurs algèbres de Lie associées. Journal de Mathématiques Pures et Appliquées **57**, 453–488 (1979)
14. Kirillov, A.: Local Lie algebras. Russ. Math. Surv. **31**, 55–75 (1976)
15. Weinstein, A.: The local structure of Poisson manifolds. J. Differ. Geom. **18**, 523–557 (1983)

16. Weinstein, A.: Addendum to the local structure of Poisson manifolds. J. Differ. Geom. **22**, 255 (1985)
17. Libermann, P., Marle, C.-M.: Symplectic Geometry and Analytical Mechanics. D. Reidel Publishing Company, Dordrecht (1987)
18. Koszul, J.L.: Crochet de Schouten-Nijenhuis et cohomologie, in É. Cartan et les mathématiques d'aujourd'hui, Astérisque numéro hors série, pp. 257–271 (1985)
19. Marle C.-M., Calculus on Lie algebroids, Lie groupoids and Poisson manifolds, Dissertationes Mathematicae 457 Warszawa (2008) 1–57.
20. Libermann P.: Sous-variétés et feuilletages symplectiquement réguliers. In: Crumeyrolle, A., Grifone, J. (eds.) Symplectic Geometry, pp. 81–106. Pitman, London (1983)
21. Kirillov, A.: Éléments de la théorie des représentations. Éditions Mir, Moscou (1974)
22. Kostant, B.: Quantization and Unitary Representations, part 1. Prequantization, Lecture Notes in Mathematics, vol. 170, pp. 87–208 (1970)
23. Souriau, J.-M.: Structure des systèmes dynamiques. Dunod, Paris (1969)
24. Smale S.: Topology and mechanics, part I. Invent. Math. **10**, 305–331 (1970)
25. Smale S.: Topology and mechanics, part II. Invent. Math. **11**, 45–64 (1970)
26. Hilton, P.J., Stammbach, U.: A Course in Homological Algebra. Springer, New York (1994)
27. Kosmann-Schwarzbach, Y.: The Noether Theorems. Springer, New York (2011)
28. Atiyah, M.: Convexity and commuting Hamiltonians. Bull. Lond. Math. Soc. **14**, 1–15 (1982)
29. Guillemin, V., Sternberg, S.: Convexity properties of the moment mapping. Invent. Math. **67**, 491–513 (1982)
30. Guillemin, V., Sternberg, S.: Convexity properties of the moment mapping II. Invent. Math. **77**, 533–546 (1984)
31. Kirwan, F.: Convexity properties of the moment map III. Invent. Math. **77**, 547–552 (1984)
32. Delzant, T.: Hamiltoniens périodiques et images convexes de l'application moment. Bull. Soc. Math. Fr. **116**, 315–339 (1988)
33. Sniatycki, J., Tulczyjew, W.: Generating forms of Lagrangian submanifolds. Indiana Univ. Math. J. **22**, 267–275 (1972)
34. Meyer K., Symmetries and integrals in mechanics. In: Peixoto, M. (ed.) Dynamical Systems , pp. 259–273. Academic Press, New York (1973)
35. Marsden, J.E., Weinstein, A.: Reduction of symplectic manifolds with symmetry. Rep. Math. Phys. **5**, 121–130 (1974)
36. Poincaré H., Sur une forme nouvelle des équations de la Mécanique. C. R. Acad. Sci. Paris **CXXXII** (7), 369–371 (1901)
37. Cendra, H., Holm, D., Marsden, J.E., Ratiu, T.: Lagrangian reduction, the Euler–Poincaré equations and semidirect products. Am. Math. Soc. Transl. **186**, 1–25 (1998)
38. Cendra H., Marsden J.E., Pekarsky S., Ratiu, T.: Variational principles for Lie-Poisson and Euler-Poincaré equations. Moscow Math. J. **3**(3), 833–867 (2003)
39. Bott, R.: Nondegenerate critical manifolds. Ann. Math. **60**, 248–261 (1954)
40. Ortega, J.-P., Ratiu, T.S.: Momentum Maps and Hamiltonian Reduction. Birkhäuser, Boston (2004)
41. Guillemin, V., Sternberg, S.: Symplectic Techniques in Physics. Cambridge University Press, Cambridge (1984)
42. Sternberg, S.: Minimal coupling and the symplectic mechanics of a classical particle in the presence of a Yang-Mills field. Proc. Natl. Acad. Sci. USA **74**, 5453–5254 (1977)
43. Weinstein, A.: A universal phase space for particles in a Yang–Mills field. Lett. Math. Phys. **2**, 417–420 (1978)
44. Godbillon, C.: Géométrie différentielle et Mécanique analytique. Hermann, Paris (1969)
45. Marle, C.-M.: On Henri Poincaré's note "Sur une forme nouvelle des équations de la Mécanique". J. Geom. Symmetry Phys. **29**, 1–38 (2013)
46. Marle C.-M.: On mechanical systems with a Lie group as configuration space. In: de Gosson, M. (ed.) Jean Leray 99 Conference Proceedings: The Karlskrona Conference in the Honor of Jean Leray, pp. 183–203. Kluwer, Dordrecht (2003)

47. Arnold, V.I.: Sur la géométrie différentielle des groupes de Lie de dimension infinie et ses applications à l'hydrodynamique des fluides parfaits. Ann. Inst. Fourier Grenoble **16**, 319–361 (1966)
48. Cushman, R.H., Bates, L.M.: Global Aspects of Classical Integrable Systems. Birkhäuser, Basel-Boston-Berlin (1997)
49. Kowalevski, S.: Sur le problème de la rotation d'un corps solide autour d'un point fixe. Acta Mathematica **12**, 177–232 (1899)
50. Audin, M.: Spinning Tops, A Course on Integrable Systems. Cambridge University Press, Cambridge (1996)
51. Audin, M.: Souvenirs sur Sofia Kovalevskaya. Calvage et Mounet, Paris (2008)
52. Hamilton, W.R.: The hodograph or a new method of expressing in symbolic language the Newtonian law of attraction. Proc. Roy. Irish Acad. **3**, 287–294 (1846)
53. Bernoulli J.: Extrait de la réponse de M. Bernoulli à M. Herman, datée de Basle le 7 octobre 1710, Histoire de l'Académie Royale des Sciences, année M.DCC.X, avec les Memoires de Mathématiques et de Physique pour la même année, (1710), pp. 521–533. http://gallica.bnf.fr/ark:/12148/bpt6k34901/f707.image.pagination
54. Herman J.: Extrait d'une lettre de M. Herman à M. Bernoulli, datée de Padoüe le 12 juillet 1710, Histoire de l'Académie Royale des Sciences, année M.DCC.X, avec les Memoires de Mathématiques et de Physique pour la même année, pp. 519–521, (1710). http://gallica.bnf.fr/ark:/12148/bpt6k34901/f709.image.pagination
55. Marle, C.-M.: A property of conformally Hamiltonian vector fields; application to the Kepler problem. J. Geom. Mech. **4**, 181–206 (2012)
56. Cordani, B.: The Kepler Problem. Birkhäuser, Basel Boston Berlin (2003)
57. Goodstein D., Goodstein J., Feynman R.: Feynman's Lost Lecture. The Motion of Planets Around the Sun. W. W. Norton and Company Inc., New York (1996) (French translation: Cassini, Paris 2009)
58. Guichardet, A.: Le problème de Kepler; histoire et théorie. Éditions de l'École Polytechnique, Paris (2012)
59. Guillemin V., Sternberg, S.: Variations on a Theme by Kepler, vol. 42. American mathematical society colloquium publications, Providence, RI (1990)
60. Fock V. A.: Zur Theorie des Wasserstoffatoms. Zeitschrift für Physik **98**, 145–154 (1935)
61. Moser, J.: Regularization of Kepler's problem and the averaging method on a manifold. Commun. Pure Appl. Math. **23**, 609–636 (1970)
62. Györgyi G.: Kepler's equation, Fock variables, Bacry's generators and Dirac brackets, part I. Il Nuovo Cimento **53**, 717–736 (1968)
63. Györgyi G.: Kepler's equation, Fock variables, Bacry's generators and Dirac brackets, part II. Il Nuovo Cimento **62**, 449–474 (1969)
64. Anosov, D.V.: A note on the Kepler problem. J. Dyn. Control Syst. **8**, 413–442 (2002)
65. Milnor, J.: On the geometry of the Kepler problem. Am. Math. Monthly **90**, 353–365 (1983)
66. Ligon, T., Schaaf, M.: On the global symmetry of the classical Kepler problem. Rep. Math. Phys. **9**, 281–300 (1976)
67. Cushman, R., Duistermaat, J.J.: A characterization of the Ligon-Schaaf regularization map. Commun. Pure Appl. Math. **50**, 773–787 (1997)
68. Heckman G., de Laat T.: On the regularization of the Kepler problem, preprint. arXiv: 1007.3695
69. Souriau, J.-M.: Géométrie globale du problème à deux corps. Atti Accad. Sci. Torino Cl. Sci. Fis. Mat. Natur. **117**, 369–418 (1983)

Symmetries in Mechanics: From Field Theories to Master Responses in the Constitutive Modeling of Materials

Jean-François Ganghoffer

Abstract This chapter is concerned with the role of continuous symmetries in field theories in a broad sense and in particular in continuum mechanics. In the first part, we present in a synthetic and self-contained manner the formalism of classical and quantum field theory, focusing on the essential role of symmetries in connection with the Lagrangian and Hamitonian formalism. The second part highlights the importance of symmetries in continuum mechanics and mechanics of materials. Especially, a novel rational methology for constructing constitutive models of viscous materials combining Lie symmetries with experimental data is presented.

1 Introduction

This chapter is concerned with the role of continuous symmetries in field theories in a broad sense and in particular in continuum mechanics. In classical mechanics, interactions between particles are supposed to occur instantaneously, and the space-time, conceived as the frame of evolution of all physical phenomena, is Newtonian, so that the time is absolute. Those interactions are described by adding to the Lagrangian, only function of the particles position (and not of velocity), a supplementary term coined interaction potential. The picture is completely different in field theory, due to the finite speed of propagation of interactions evidenced by Michelson and Morley experiment, leading to special relativity: the forces acting at a given moment on any particle are not defined by the position of these particles at the same instant. *A* finite duration of the propagation is thus required so that the change of position of a given particle produces some effect on the other particles. In order to account for this modification, one introduces the concept of field: to the idea of an action of a

J.-F. Ganghoffer (✉)
LEMTA, Université de Lorraine, 2, Avenue de la Forêt de Haye. TSA 60604, 54518 Vandoeuvre-les-Nancy, France
e-mail: jean-francois.Ganghoffer@univ-lorraine.fr

J.-F. Ganghoffer and I. Mladenov (eds.), *Similarity and Symmetry Methods*, Lecture Notes in Applied and Computational Mechanics 73, DOI: 10.1007/978-3-319-08296-7_5,

particle on another particle, one substitutes the idea that it generates a field. Any particle influenced by the field will be submitted to a force: the particle thereby interacts with the field, which becomes an intrinsic reality having its own degrees of freedom. Field theory is constructed—in both the classical and quantum cases—from the symmetry transformations acting either on external degrees of freedom (space-time coordinates), or on internal d.o.f., referring in this last situation to gauge symmetries. Those symmetries reflect in the form taken by the total Lagrangian, sum of the Lagrangian of matter fields and of the Lagrangian of the interaction fields. The existence and properties of the interaction particles- called bosons -are deduced from symmetry requirements, and particularly the localization of the global gauge symmetry condition. The experimental validation of the existence of gauge bosons (which sometimes comes after the funding theory) explains the competition to develop high energy collisionners.

The principal goal of this chapter is to present in a synthetic and self-contained manner the formalism of field theory, focusing on the essential role of symmetries. The first part will be devoted to a synthetic exposition of classical and quantum field theories, and it will take the form of a course; the second part will highlight the importance of symmetries in continuum mechanics and mechanics of materials. Especially, we shall expose a novel approach of constructing constitutive models for various materials combining Lie symmetries with experimental data.

Regarding notations, vectors and tensors are represented either using the arrow notation (for the 3-vector part of four-vectors), or and more often using boldface symbols. The partial derivative is either denoted explicitly, as for instance $\dfrac{\partial f}{\partial a}$, or with the short hand notation $\partial_a f$. The nabla operator ∇ is associated to the spatial gradient of a function of the spatial coordinates and time. The shortcuts l.h.s. and r.h.s. will be used for left-hand side and right-hand side respectively.

2 Lagrangian Formulation of Classical Mechanics

The Lagrangian formulation of the laws of physics trace back to about 1790, and is frequently used in classical mechanics to write the equations of motion from a least action principle. As a generalization, a large number of physical laws have been derived from a Lagrangian formulation, allowing a non mechanistic reading of classical mechanics, and highlighting the symmetry properties. It allows furthermore the description of elementary phenomena (interacting particles) and it provides the linkage with quantum mechanics thanks to Hamiltonian formalism. The notion of symmetry reveals as a fundamental scientific concept that has invaded many scientific disciplines, from biology up to mathematics. Independently from its aesthetic attraction, the symmetry intrinsically contained in physical systems allows a concise description of both their invariance properties (geometrical, temporal, or under a given transformation of a set of relevant variables) and a systematic prediction of the various phenomena susceptible to occur in these systems. This in turn allows a classification of physical systems in terms of their symmetries, which clearly analogies and contributes to a unifying view of their behavior. The symmetry properties of new physical quantities have been investigated historically, in order to account for all invariance properties: scale transformations, time, parity and charge inversions, gauge symmetries, dynamical symmetries. Generalizing the discrete symmetries in use for the classification of crystals, the notion of continuous symmetries coined Lie symmetries to remind their main promotor (Sophus Lie), were developed by [1], followed by Noether [2], who evidenced the relationship between conservation laws (of energy, momentum,...) and associated Lie symmetries. The various field theories built in the twentieth century abundantly involve Lie symmetries in their construct, being especially reflected in the form taken by the Lagrangian function.

2.1 Dynamical Equations of Motion

The basic idea is to represent a system depending on N d.o.f. (a short cut for degrees of freedom that will be employed here and in the sequel) by a point or a vector made of N generalized coordinates $\{q_\alpha\}$ in a N dimensional space called the configuration space. The consideration of the velocities, quantities $\{\dot{q}_\alpha\}$, leads to the $2N$ dimensional phase space. The Hamilton principle of least action consists in determining the trajectory of a given system in phase space from a variational formulation (the initial state and forces being known). The system is characterized by a functional called Lagrangian function

$$L\left[q_\alpha, \dot{q}_\alpha, t\right]$$

depending on the generalized coordinates and velocities, and possibly on time (these arguments are indicated with a bracket notation) wich is a scalar valued function condensing all information on the dynamic evolution of the considered system. When classical mechanics is of concern, $L\left[q_\alpha, \dot{q}_\alpha, t\right]$ is a temporal density, which when integrated over time (from an initial time t_1 up to a final time t_2) leads to Hamilton-Jacobi action

$$S\left[q_\alpha\right] = \int_{t_1}^{t_2} L\left[q_\alpha, \dot{q}_\alpha, t\right] \mathrm{d}t.$$

The isochronal variation (at fixed time) of the action writes

$$\begin{aligned}
\delta\left[q_\alpha\right] &= \int_{t_1}^{t_2} \left\{ \frac{\partial L}{\partial q_\alpha} \delta q_\alpha + \frac{\partial L}{\partial \dot{q}_\alpha} \delta \dot{q}_\alpha \right\} \mathrm{d}t \\
&\equiv \int_{t_1}^{t_2} \left\{ \frac{\partial L}{\partial q_\alpha} - \frac{\mathrm{d}}{\mathrm{d}t}\left(\frac{\partial L}{\partial \dot{q}_\alpha} \right) \right\} \delta q_\alpha \mathrm{d}t + \left[\frac{\partial L}{\partial \dot{q}_\alpha} \delta q_\alpha \right]_{t_1}^{t_2}
\end{aligned}$$

and leads as a necessary condition to the Euler-Lagrange equations

$$\delta S\left[\mathrm{q}_\alpha\right]=0, \qquad \alpha\in\{1,2,\ldots,N\} \Rightarrow \frac{\partial L}{\partial q_\alpha}-\frac{\mathrm{d}}{dt}\left(\frac{\partial L}{\partial \dot{q}_\alpha}\right)=0$$

consisting of a set of N second order ODE's providing the dynamical path of evolution of the system of N particles provided a set of $2N$ independent initial conditions for the generalized coordinates and velocities have been prescribed. The virtual (virtual in the sense of mathematical variations from the actual trajectory; those variations lead to new virtual trajectories that only have to respect the end conditions) variations $\delta q_\alpha, \delta\dot{q}_\alpha$ are considered as mutually independent and performed at fixed time, so one should in order to be accurate write those variations $\delta_t q_\alpha, \delta_t\dot{q}_\alpha$, the subscript indicating that time is held fixed during the variation, a similar notion for fields will be encountered later on. Previous equations are the Newton equations of classical dynamics.

An equivalent formulation is obtained by introducing the Hamiltonian, another scalar valued function of an alternative set of variables, the generalized coordinates and the N canonical momenta, defined as

$$p_\alpha := \frac{\partial L}{\partial \dot{q}_\alpha} \to H\left[q_\alpha, p_\alpha, t\right] := p_\alpha \dot{q}_\alpha - L\left[q_\alpha, \dot{q}_\alpha, t\right].$$

In this new system of coordinates $\{q_\alpha, p_\alpha\}$, Jacobi action writes

$$S\left[q_\alpha\right] := \int_{t_1}^{t_2} \{p_\alpha \dot{q}_\alpha - L\}\mathrm{d}t.$$

The Hamiltonian is elaborated as a Legendre transformation of the Lagrangian density; it can be expressed in terms of the coordinates $\left(q_j, p_j, t\right)$ in the phase space as

$$H = H\left(q_j, p_j, t\right) = p_j\dot{q}_j\left(q_j, p_j, t\right) - L\left(q_j, \dot{q}_j\left(q_j, p_j, t\right), t\right).$$

Hamilton equations then easily follow from the expression of the differential of $H\left(q_j, p_j, t\right)$, as

$$q_j = \frac{\partial H}{\partial p_j}, \qquad p_j = -\frac{\partial H}{\partial q_j}, \qquad \frac{\partial L}{\partial t} = -\frac{\partial H}{\partial t}$$

which is interpreted as the equality between the variation of momentum and a force, quantity $\partial L/\partial q_\alpha$. Since the Hamiltonian is obtained by a substitution of the generalized velocities in the Lagrangian by the canonical momenta, both functions contain the same information but in different sets of variables.

Similarly, the dynamical evolution of the system is obtained as the unique solution of this set of $2N$ dynamical equations together with initial conditions for the generalized coordinates and momenta.

The interest of Hamilton formalism is the search for first integrals of motion using Poisson's bracket: for two functions $f(q_\alpha, p_\alpha, t)$ and $g(q_\alpha, p_\alpha, t)$, their Poisson's bracket is defined as (implicit summation over repeated indices is done)

$$[f, g] := \frac{\partial f}{\partial p_\alpha} \cdot \frac{\partial g}{\partial q_\alpha} - \frac{\partial f}{\partial q_\alpha} \cdot \frac{\partial g}{\partial p_\alpha}.$$

This expression leads to the time derivative of any function $f(q_\alpha = q_\alpha(t),\ p_\alpha = p_\alpha(t), t)$ as

$$\frac{\mathrm{d}f}{\mathrm{d}t} = \frac{\partial f}{\partial t} + [H, f].$$

As a corollary, if f does not explicitly depend on time, one can characterize f as a first integral of motion in terms of the necessary and sufficient condition

$$[H, f] = 0.$$

Observe further that the Poisson's bracket of two first integrals is also a first integral; this property may be used recursively to produce new first integrals. Let mention the particular cases

$$[p_\alpha, q_\beta] = \delta_{\alpha\beta}, \qquad [q_\alpha, q_\beta] = 0 = [p_\alpha, p_\beta].$$

2.2 *Noether's Theorem for Discrete Systems*

Conserved quantities play an important role in the analysis of dynamical systems, since they allow solving dynamical problems more easily and they highlight invariance properties of the system. For an isolated Newtonian system, there exists ten basic external conserved quantities, resulting from the invariance of the laws of physics w.r. to the Galilean symmetry transformations. There is a systematic procedure to find such symmetries for both discrete and continuous systems; we shall here consider this procedure for discrete dynamical systems. Noting that the dynamics of a system is independent of the choice of the generalized coordinates, general transformations of the generalized coordinates (called canonical transformations) are of the form

$$t \to \bar{t} = \bar{t}\,(t)\,, \qquad q \to \bar{q} = \bar{q}\,(q(t), t)$$

called point transformations. Examples include translations, rotations, boosts, and reflections which define the whole set of Galilean transformations. Here, q denotes a vector with N components as in previous subsection. Similar notions prevail for continuous (Lie symmetries) symmetries, as exposed in the contributions of N. Ibragimov and G. Bluman in this volume.

Requiring that the action built from the Lagrangian density $L(q, \dot{q}, t)$, the integral

$$S[q] = \int_{t_i}^{t_f} L(q, \dot{q}, t)\,\mathrm{d}t$$

remains invariant under previous transformations, one can easily show [3] that the transformed Lagrangian density is related to the initial density by

$$\overline{L}\left(\overline{q}, \dot{\overline{q}}, \overline{t}\right) = \frac{\partial t}{\partial \overline{t}} L(q, \dot{q}, t)$$

with $\partial t / \partial \overline{t}$ the inverse of the Jacobian of the transformation of times. Since the Jacobian equals unity for Galilean transformations, the Lagrangian is then a Galilean invariant (for isolated systems). The Euler-Lagrange equations take a new form after the transformation, so they describe the dynamics in the new generalized coordinates, traducing the canonical invariance of the Euler-Lagrange equations. In order for the transformation to be a symmetry, one must require some form invariance. For invariant Lagrangians, satisfying the condition $\overline{L}(q, \dot{q}, t) = L(q, \dot{q}, t)$, equations of motion will have the same form, so the transformation is a symmetry. This entails that the infinitesimal change of L under infinitesimal changes Δt, Δq shall satisfy the condition [3]

$$\frac{\partial(\Delta L)}{\partial t} + \Delta L = 0,\ \Delta L = \frac{\mathrm{d}L}{\mathrm{d}t}\Delta t = \left\{\frac{\partial L}{\partial t} + \frac{\partial L}{\partial q}\dot{q} + \frac{\mathrm{d}}{\mathrm{d}t}\frac{\partial L}{\partial \dot{q}}\ddot{q}\right\}\Delta t.$$

In a more general situation, a covariant Lagrangian is one that differs from the original one by a divergence, that is the total time derivative of a scalar function $F(q, t)$, namely

$$\overline{L}(q, \dot{q}, t) = L(q, \dot{q}, t) + \frac{\partial F(q, t)}{\partial t}.$$

For such Lagrangians, the equations of motion (Euler-Lagrange equations) will take the same form. Let then express the general variation of the action integral under a transformation acting on both the independent (time) and dependent variables,

$$t \rightarrow \overline{t}(t) = t + \delta t(t), \qquad q \rightarrow \overline{q}\left(\overline{t}\right) = q(t) + \delta q(t).$$

We shall distinguish the variation denoted by δ for a symmetry transformation from the more general variaion Δ (not necessarily associated to a symmetry). Note that in this case, it holds that $\delta \dot{q}(t) \neq \frac{\mathrm{d}}{\mathrm{d}t}(\delta q(t))$. The variation of the action integral, quantity $\delta S[q] := \int_{\overline{t}_i}^{\overline{t}_f} L\left(\overline{q}, \dot{\overline{q}}, \overline{t}\right) d\overline{t} - \int_{t_i}^{t_f} L(q, \dot{q}, t)\,\mathrm{d}t$ can be evaluated as the sum of a surface term and a volume term [3], viz

$$\delta S\left[q\right] := \left[p^{j}\delta q_{j} - H\delta t\right]_{t_i}^{t_f} + \int\limits_{\bar{t}_i}^{\bar{t}_f} EL^{j}\left(\delta q_{j} - \dot{q}_{j}\delta t\right)\mathrm{d}t$$

introducing therein the Euler derivative (or Euler operator acting on the Lagrangian)

$$EL^{j} := \frac{\partial L}{\partial q_{j}} - \frac{\mathrm{d}}{\mathrm{d}t}\left(\frac{\partial L}{\partial \dot{q}_{j}}\right).$$

This leads to Hamilton's modified principle of varied action. Obviously, when the independent variable is not varied, one recovers the Euler equations (and there is no boundary term).

We are now in a position to examine the conditions to be satisfied by the Lagrangian so that the transformation is a symmetry of the Euler-Lagrange equations, meaning that they leave unchanged those equations; as written before, such symmetries must satisfy the condition

$$\overline{L}\left(\overline{q}, \dot{\overline{q}}, \overline{t}\right) = \frac{\partial t}{\partial \overline{t}} L\left(q, \dot{q}, t\right)$$

or the alternative condition

$$\overline{L}\left(q, \dot{q}, t\right) = L\left(q, \dot{q}, t\right) + \frac{\mathrm{d}F\left(q, t\right)}{\mathrm{d}t}.$$

One can show that an equivalent writing of the covariance condition for the Lagrangian would read [3]

$$\frac{\partial\left(\Delta t\right)}{\partial t} + \Delta L = -\frac{\mathrm{d}\left(\Delta F\right)}{\mathrm{d}t}.$$

Writing this last condition with the arguments $\left(\overline{q}, \dot{\overline{q}}, \overline{t}\right)$, and inserting the result into the variation of the action gives

$$\Delta S = -\int\limits_{t_i}^{t_f} dt \frac{\mathrm{d}F\left(q, t\right)}{\mathrm{d}t}.$$

Using next the previously calculated general variation $\delta S\left[q\right]$, but with more specific variations $(\Delta t,\ \Delta q\left(t\right))$ leads after subtraction from previous expression of ΔS to the equation

$$\frac{\mathrm{d}}{\mathrm{d}t}\left[p^{j}\Delta q_{j} - H\Delta t + \Delta F\right]_{t_i}^{t_f} + \left(\frac{\partial L}{\partial q} - \frac{\mathrm{d}}{\mathrm{d}t}\left(\frac{\partial L}{\partial \dot{q}}\right)\right)\left(\Delta q - \dot{q}\Delta t\right) = 0.$$

Note that we have not yet used the equations of motion, so that this condition also holds on virtual paths of the dynamical system (those paths have a crucial importance in quantum mechanics). The previous equation results from demanding that the action being covariant, namely it is built from a covariant Lagrangian; note that an invariant Lagrangian corresponds to the special case $\Delta F = 0$.

We next suppose that the system has a Lie group of transformations, depending on a finite number of p parameters μ_i, $i = 1 \ldots p$, not depending on time; this entails the variation of the generalized coordinates and time as

$$\Delta t = \frac{\partial \Delta t\,(t)}{\partial \Delta \mu_i} \Delta \mu_i, \qquad \Delta q_j = \frac{\partial q_j\,(t)}{\partial \Delta \mu_i} \Delta \mu_i, \qquad j = 1 \ldots p.$$

Introducing those variations into the previous condition leads to

$$\left\{ \frac{\mathrm{d}Q_i}{\mathrm{d}t} + \frac{\partial L}{\partial q_j} - \frac{\mathrm{d}}{\mathrm{d}t}\left(\frac{\partial L}{\partial \dot{q}_j}\right)\left(\frac{\partial q_j\,(t)}{\partial \Delta \mu_i} - \dot{q}_j \frac{\partial \Delta t\,(t)}{\partial \Delta \mu_i}\right)\right\} \Delta \mu_i = 0$$

with the Noether charge Q_i defined as

$$Q_i = p^j \frac{\partial q_j\,(t)}{\partial \Delta \mu_i} - H \frac{\partial \Delta t\,(t)}{\partial \Delta \mu_i} + \frac{\partial \Delta F}{\partial \Delta \mu_i}.$$

Due to the independence of parameters μ_i, we deduce the expression of the total derivative

$$\begin{aligned} \frac{\mathrm{d}Q_i}{\mathrm{d}t} &= -\left(\frac{\partial L}{\partial q_j} - \frac{\mathrm{d}}{\mathrm{d}t}\left(\frac{\partial L}{\partial \dot{q}_j}\right)\right)\left(\frac{\partial q_j\,(t)}{\partial \Delta \mu_i} - \dot{q}_j \frac{\partial \Delta t\,(t)}{\partial \Delta \mu_i}\right) \\ &= -EL^j \left(\frac{\partial q_j\,(t)}{\partial \Delta \mu_i} - \dot{q}_j \frac{\partial \Delta t\,(t)}{\partial \Delta \mu_i}\right). \end{aligned}$$

Thus, for any symmetry of a dynamical system, there corresponds a quantity Q_i, called the charge, the total derivative of which is a linear combination of the Euler derivatives; note that this result is true for any path $q\,(t)$ in configuration space. Previous equations were published by Hamel (1904) and Hergoltz (1911) in the context of relativity theory, and were obtained as conservation theorems by E. Noether and Klein in 1908, and by Bessel-Hagen in 1921; they are generally referred to as the Noether relations. Noether$\tilde{s}$ theorem then states that on the path of motion (the Euler operator vanishes), the charge is conserved,

$$\frac{\mathrm{d}Q_i}{\mathrm{d}t} = 0 \qquad \Leftrightarrow \qquad Q_i = Cte.$$

Thereby, p integral of motions have been obtained, corresponding to the p symmetry transformations. This theorem is not valid in the quantum case, thus it deserves the name of weak conservation law, in contrast to local gauge invariance (to be developed later), a symmetry valid off the path. The Noether theorem will be extended to fields in forthcoming sections of this chapter.

2.3 Construction of the Lagrangian: Role of Symmetries

The form taken by the Lagrangian can be restricted by symmetry considerations; we call symmetry in a broad sense in field theory, any operation leaving the action invariant; those transformations are called variational symmetries, [4, 5]. Let consider as an example a scaling transformation acting on the spatial coordinates

$$x \to \lambda x$$

with λ a constant parameter. In order to find the transformation law of the field ψ, of the form $\psi \to T(\psi)$, let express the fact that ψ and $T(\psi)$ measure the same information at the same position $T(\psi(\lambda x)) = \psi(x)$, leading to the transformation law for the field $T(\psi)(x) = \psi(x/\lambda)$. When equations of motion have the same form, their solution are also identical; hence, one searches for transformations of the Lagrangian leaving the Euler equations unchanged. Two viewpoints can be considered a priori, called the passive and the active viewpoints, as explained next.

In the passive point of view, two observers O and O' use different rules to measure positions in their own referentials: O attributes coordinates x to the system, while O' attributes another set of coordinates x^1. Each observer evaluates the action in its own referential as

$$S[q_\alpha] := \int_{t_1}^{t_2} L(q_\alpha, \dot{q}_\alpha, t)\mathrm{d}t.$$

A sufficient condition for the transformation

$$O \to O'$$

to achieve a symmetry transformation is the Lagrangian being defined up to the addition of the derivative of an arbitrary function with respect to time, viz

$$L(\boldsymbol{x}, \dot{\boldsymbol{x}}) = L(\boldsymbol{x}, \dot{\boldsymbol{x}}) + \frac{\mathrm{d}K(\boldsymbol{x}')}{\mathrm{d}t}.$$

In the active viewpoint, there is only one observer, and the transformation $\boldsymbol{x} \to \boldsymbol{x}'$ is done in the referential linked to the observer, this second point of view is better adapted to continuous symmetries involving (continuous) parameters. A sufficient condition of $\boldsymbol{x} \to \boldsymbol{x}'$ being a symmetry is then that the action be equal in both coordinate systems, so that an extremum of the first action is also extremum of the second action, this is the case when $S\left[\boldsymbol{x}'\right] = S\left[\boldsymbol{x}\right] + \frac{\mathrm{d}K}{\mathrm{d}t}$.

In a Galilean referential (this defines a class of referentials in relative motion at uniform velocity), considering a uniform time (Newtonian absolute time) and a homogeneous (properties are the same whatever the position x in physical space, considered as a specific configuration space) isotropic (properties are the same in all directions) space, it holds as a consequence the conditions

$$\frac{\partial L}{\partial t} = 0 = \frac{\partial L}{\partial x} \Rightarrow L\left(x, \dot{x}, t\right) = \alpha \dot{x}^2 + \beta.$$

This holds true in fact for any generalized coordinate, thus one can write more generally $L\left(q, \dot{q}, t\right) = \alpha \dot{q}^2 + \beta$.

Application of the Euler equations to this Lagrangian independent of q (hence the absence of forces) then leads to

$$\frac{\partial L}{\partial q} = 0 \Rightarrow \frac{\mathrm{d}}{\mathrm{d}t}\left(\frac{\partial L}{\partial \dot{q}}\right) = 0 \Rightarrow \frac{\partial L}{\partial \dot{q}} = Cte \Rightarrow \dot{q} = Cte$$

leading to the law of inertia: a free particle moves with a constant velocity in any Galilean referential (note that the rest state is nothing but a particular case of a motion at nil velocity). Since now the Euler equations do not change if one rescales the Lagrangian by a multiplicative factor, one may adopt the kinetic energy for the free particle as a Lagrangian $L\left(q, \dot{q}, t\right) = \frac{1}{2} m \dot{q}^2$.

Interactions of the system of material points will be described by adding a potential energy contribution $V(q)$ for each particle, thus the Lagrangian

$$L\left(q, \dot{q}, t\right) = \frac{1}{2} m \dot{q}^2 - V\left(q\right).$$

As an illustration, for the specific choice $q \equiv x$, the stationnarity condition of Jacobi action

$$S\left[x\right] := \int_{t_1}^{t_2} \left\{\frac{1}{2} m \dot{x}^2 - V\left(x\right)\right\} \mathrm{d}t$$

leads to Newton's equations of motion

$$\delta S\left[q_\alpha\right] = 0 \Rightarrow m\ddot{x} = -\frac{\partial V}{\partial x} \equiv f.$$

In a general setting, the least action principle in classical mechanics is fully equivalent to the laws of mechanics, which are rooted in three conservation laws: conservation of linear and angular momentum, and conservation of energy. We will prove that these laws result from symmetries of the physical system incorporated into the Lagrangian

Table 1 Condensed form of Noether's theorem in classical mechanics

Non observable	Symmetry	Conservation law
Absolute origin of time	Temporal translation	Energy
Absolute origin of space	Spatial translation	Linear Momentum
Privileged direction	Rotation	Angular momentum

formulation, and associated to the non observability of certain quantities. The articulation between conservation laws, symmetries, and relativity principles (the absolute values of certain quantities are not measurable or observable) is encapsulated into

Noether's theorem: on the real trajectory of a dynamical system, a quantity is conserved for each symmetry (discrete or continuous). The measurement of observable physical quantities implies their invariance by a change of experimental conditions: this relativity principle entails conservation laws; the non observable quantities are then not measurable.

The following Table 1 summarizes the quantity being conserved in classical mechanics for each symmetry, and the associated non observable quantity. The proof of Noether's theorem is given for instance for the case of conservation of angular momentum: let consider the transformation $\boldsymbol{x} \mapsto \boldsymbol{x}+\delta\omega\boldsymbol{x}\wedge\boldsymbol{a}$ generating the following variation of the Lagrangian

$$\delta L = \frac{\partial L}{\partial \boldsymbol{x}} \cdot \delta \boldsymbol{x} + \frac{\partial L}{\partial \dot{\boldsymbol{x}}} \cdot \partial \dot{\boldsymbol{x}} \equiv \delta \boldsymbol{x} \cdot \left[\frac{\partial L}{\partial \boldsymbol{x}} - \frac{\mathrm{d}}{\mathrm{d}t} \frac{\partial L}{\partial \dot{\boldsymbol{x}}} \right] + \frac{\mathrm{d}}{\mathrm{d}t} \left(\delta \boldsymbol{x} \cdot \frac{\partial L}{\partial \dot{\boldsymbol{x}}} \right)$$

wherein the last equality follows after elementary calculations. Assuming rotation is a symmetry, the function in the total derivative leads to a variation of the action, hence the total derivative has to vanish: this leads to

$$\frac{\mathrm{d}}{\mathrm{d}t} \left(\delta \boldsymbol{x} \cdot \frac{\partial L}{\partial \dot{\boldsymbol{x}}} \right) = 0$$

with $\delta \boldsymbol{x} = \delta\omega \boldsymbol{x} \wedge \boldsymbol{a}$, $\frac{\partial L}{\partial \dot{x}} = m\dot{\boldsymbol{x}}$, $[\boldsymbol{x} \wedge \boldsymbol{a}] \cdot \dot{\boldsymbol{x}} = -\boldsymbol{a} \cdot \boldsymbol{x} \wedge \dot{\boldsymbol{x}}$, thus

$$\frac{\partial}{\partial t} (\boldsymbol{x} \wedge \dot{\boldsymbol{x}}) = 0$$

which is the conservation of angular momentum.

A corollary is the incompatibility between different physical quantities associated to each line of previous Table, for instance conservation of energy is associated with the non observable nature of absolute time; this can be formulated as the classical limit when Planck constant vanishes within a continuum viewpoint

$$\Delta E \cdot \Delta t = h \to 0.$$

Similar relations written as the well-known Heisenberg inequalities hold in quantum mechanics (Planck constant defines the lowest attainable scale of energy, which is quantified).

The laws of motion now follow from the Lagrangian formulation; conservation laws follow from symmetries, as one shall prove in the case of energy: this results from the absence of an absolute origin of time, thus

$$\frac{\partial (q_i, \dot{q}_i, t)}{\partial t} = 0 \Rightarrow \frac{\mathrm{d}L}{\mathrm{d}t} = \left(\frac{\partial L}{\partial q_i}\frac{\partial q_i}{\partial t} + \frac{\partial L}{\partial \dot{q}_i}\frac{\partial \dot{q}_i}{\partial t}\right) = \dot{p}_i\dot{q}_i + p_i\ddot{q}_i = \frac{\mathrm{d}}{\mathrm{d}t}(p_i\dot{q}_i)$$

using the definition of momentum and the Euler equations; hence, one obtains

$$\frac{\mathrm{d}}{\mathrm{d}t}(L - p_i\dot{q}_i) = 0 \Rightarrow \frac{\mathrm{d}H}{\mathrm{d}t} = 0 \Rightarrow E := L - p_i\dot{q}_i = Cte$$

which is nothing else than the conservation of energy. We have thus shown that energy conservation is equivalent to absolute time being non observable.

The construction of the Lagrangian (classical case) relies on symmetries that do restrict its form; we shall consider the following postulate [6]:

- Homogeneity of time: $\frac{\partial L}{\partial t} = 0$;
- Absence of privileged direction: $L(|v|) \to L(v^2)$. This specifies only the dependency of the Lagrangian upon velocity, but does not say anything as to other dependencies;
- Invariance under the transformations of Galilee group: $\frac{\partial L}{\partial q} = 0$. This means that equations have to be independent of the selected referential, leading to the form $L(v^2) = Cv^2$;
- Invariance of the equations of motion when the Lagrangian is scaled by an arbitrary constant: $C = m/2 \Rightarrow L = mv^2/2$.
  Note that a physical theory can be formulated either in causal form (force is the cause of the motion, as expressed by the differential equations of motion), or in weak form, as encapsulated in the Lagrangian (and Hamiltonian) formulations, both based on extremum principles.

2.4 Lagrangian Formulation in Special and General Relativity

In special relativity, the Lagrangian is required to be invariant with respect to the transformations of Poincaré group, which conserve the square of the distance in Minkoswki space-the metric has signature $(-1, -1, -1, +1)$

$$ds^2 = -(dx_1^2 + dx_2^2 + dx_3^2) + c^2 dt^2.$$

Such transformations are for this reason called isometries (they define the isometry group of Minkowski space), and they include the set of rotations, translations and so-called boosts (the specific form of those last transformations will be written later on).

Greek indices that shall appear in the sequel vary from 0 to 3; for any four-vector, the first index 0 denotes the time component, while the next indices {1,2,3 } indicate the spatial components.

The Jacobi action formally writes in terms of the velocity field v

$$S = \int_{t_1}^{t_2} L(v) dt.$$

The relativistic Lagrangian density $L(v)$ therein has been given previously from symmetry arguments. Let consider the proper time t_0, related to time by $dt_0 = (1 - v^2/c^2)^{1/2}$ dt $\equiv \gamma dt$ inserted this relation into the action gives

$$S = \int_{t_1}^{t_2} \frac{L(v)}{(1 - v^2/c^2)^{1/2}} dt_0.$$

Since dt_0 is a scalar invariant, it means that the integrand therein, quantity $L(v)/(1 - v^2/c^2)^{1/2}$, has to be a true scalar, say K, hence $S = \int_{t_1}^{t_2} K(1 - v^2/c^2)^{1/2} dt_0$ The classical limit (obtained when $c \to \infty$) delivers $L(v) = \frac{1}{2} m v^2$. One can also expand the relativistic Lagrangian for velocities much smaller than the speed of light as

$$L \cong K - Kv2/c^2 = \frac{1}{2} mv2 = K = -mc2 = S = -mc2 \int_{t_1}^{t_2} (1 - v^2/c^2)^{1/2} dt$$

The infinitesimal distance element can further be written

$$ds^2 = c^2 dt^2 - v^2 dt^2 = c(1 - v^2/c^2)^{1/2} dt.$$

which entails the following action having a quite simple form

$$S = -mc \int_{t_1}^{t_2} ds.$$

This expression shows that the action is simply the length of the path between the initial and final times (up to a scaling by a constant quantity). A simple calculation shows that the stationnarity of S leads to the law of inertia, that is a free particle moves at constant velocity

$$\frac{\mathrm{d}u^{\nu}}{\mathrm{d}s} = 0.$$

In order to be complete, let mention that the Lorentz transformation of two referentials in uniform translation along the first axis x_1 is obtained by the postulate (coming from an experimental fact) that the speed of light is absolute, and given by the boost along x_1

$$x_0^1 = x_0 \mathrm{ch}\psi + x^1 \mathrm{sh}\psi, \qquad x_1^1 = x_0 \mathrm{sh}\psi + x^1 \mathrm{ch}\psi$$
$$x_0 = ct, \qquad \beta = v/c, \qquad \mathrm{ch}\psi = (1-\beta^2)^{-1/2}, \qquad \mathrm{th}\psi = \beta.$$

It is worthwhile mentioning that those relations are precisely obtained by the requirement that the distance element ds be the same for two observers located in two referentials in uniform translation with respect to each other. Special relativity is associated to an invariance of the action under the transformations of Lorentz group; general relativity applies a similar covariance principle, but requiring now that the previous action built from the path length is invariant under an arbitrary change of coordinates

$$x^{\mu} \to x^{'\mu} = x^{\mu} + \xi^{\mu}(x).$$

This general gauge invariance of GR (short cut for general relativity to be used in the sequel) is associated to the idea that all reference systems are equivalent for the description of the laws of nature; the equations obtained from this principle are said to be *covariant*. The core idea in general relativity is that the metric field $g_{\mu v}$ characterizing the geometry of space plays the role of a gravitation field (the square of the infinitesimal element writes $\mathrm{d}s^2 = g_{\mu v}\mathrm{d}x^{\mu}\mathrm{d}x^{v}$); thus it is the energy and mass content of space (both incorporated into the energy momentum tensor) that in turn determine the metric, and thus the geometrical structure of space-time.

The gauge invariance in GR is associated to the idea that a gravitation field is locally equivalent (thus it can be compensated, and there is no way for the observer in a falling lift to decide whether he is submitted to a gravitational field or to an acceleration) to an accelerated referential; this follows from the identity of two masses in the relations

$$\boldsymbol{f} = m_i \boldsymbol{\gamma}, \quad \boldsymbol{f} = m_p \boldsymbol{g} \Rightarrow \boldsymbol{\gamma} = \frac{m_p}{m_i}\boldsymbol{g}$$

a result of the observation made by the Greek scientist Eötvös around 1885.

As shall appear later on, the fundamental difference between GR and (relativistic) electromagnetism is that the equations of motion are deduced from the field equation in the former situation, whereas one has to add a field-particle interaction term to get the equations of motion in the latter situation (Lorentz force is not included in

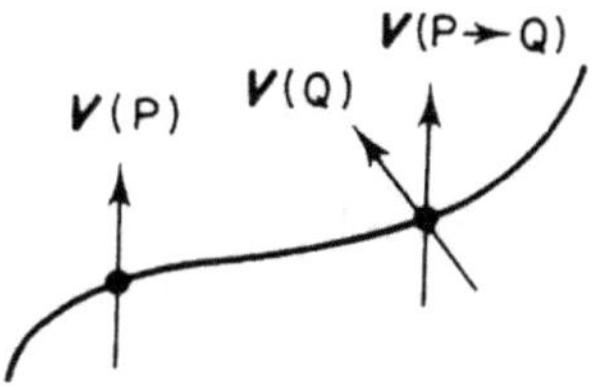

Fig. 1 Parallel transport of a vector field along a path, cf. [7]

Maxwell equations). For this reason, GR can be considered a pure field theory, with the physical field identical to the metric of space.

The equations of motion for a particle in the framework of GR is obtained by considering a general metric $g_{\mu v}$, such that the infinitesimal line element writes

$$\mathrm{d}s^2 = g_{\mu v}\mathrm{d}x^{\mu}\mathrm{d}x^{v}.$$

This entails the following form of the action of the single particle in which the Lagrangian density is the kinetic energy of the particle evolving in a space with metric $g_{\mu v}$, viz

$$S = \frac{1}{2}m\int_{t_1}^{t_2} \mathrm{d}\tau g_{\mu v}(x(\tau))\frac{\mathrm{d}x^{\mu}}{\mathrm{d}\tau}\frac{\mathrm{d}x^{v}}{\mathrm{d}\tau}.$$

The Euler equations for a free particle evolving on a geodesic are the equations of motion

$$\frac{Du^j}{Ds} = 0 = \frac{\mathrm{d}^2x^j}{\mathrm{d}s^2} + \Gamma^j_{rt}\frac{\mathrm{d}x^r}{\mathrm{d}s}\frac{\mathrm{d}x^t}{\mathrm{d}s} = 0$$

with the Christoffel symbols Γ^j_{rt} expressing an affine connection (notion of parallelism on a non euclidean manifold), and depending upon the first order derivatives of the metric coefficients

$$\Gamma^m_{ik} = g^{mj}\Gamma_{ik,j}, \qquad \Gamma_{ij,1} = \frac{1}{2}(\frac{\partial g_{1i}}{\partial x^j} + \frac{\partial g_{1j}}{\partial x^i} - \frac{\partial g_{ij}}{\partial x^1}).$$

The covariant derivative Du^j/Ds built from Christoffel symbols replaces the ordinary derivative; the covariant derivative of a vector remains a vector, allowing a comparison of vectors attached to different vector spaces. The parallel transport of a vector field along a path is shown on Fig. 1, with $V(P \to Q)$ the vector resulting from the parallel transport of $V(P)$ along the curve to point O, see Fig. 1.

Since vectors $V(Q)$ and $V(P \to Q)$ are expressed at the same point, and thus belong to the same vector space, it makes sense to define the covariant derivative along a path parameterized by scalar λ as

$$\frac{DV}{D\lambda} := \lim_{\delta\lambda\to 0} \frac{V(Q) - V(P \to Q)}{\delta\lambda} \equiv \Gamma^{\mu}_{\upsilon o}(P)V^{\upsilon}(P)\frac{\mathrm{d}x^{o}}{\mathrm{d}\lambda}.$$

Thus, the role played by the Christoffel symbols $\Gamma^{\mu}_{\upsilon o}(P)$ defining the affine connection gives a sense to the notion of parallelism of two vectors at different points of a non euclidean manifold (it means to define parallel transport of a vector along a curve).

In the previous geodesic, the contribution due to the affine connection, term $\Gamma^{j}_{rt}\frac{\mathrm{d}x^{r}}{\mathrm{d}s}\frac{\mathrm{d}x^{t}}{\mathrm{d}s}$, has the meaning of a force.

It is important to note that the structure of space-time changes radically in special and general relativity in comparison to classical mechanics: space and time are uncoupled in classical mechanics (Newtonian absolute time prevails), so that the Galilean spacetime is a simple direct product of space and time; time and space are linked in relativity theories, so that the length of any path becomes measurable thanks to Lorentz metric. The mathematical structure needed to accommodate such imbrications of time and space is a fiber bundle, with the base being time and $3D$ space defining the bundle over each instant. In GR, space-time itself has no intrinsic reality, and its structure adapts to the energy content so that the laws of physics keep the same form.

3 Field Theory

Before dwelling into the Lagrangian formulation of field theories in relation to symmetries and in order to set the stage, it is worthwhile recalling the concept of field. The simplest definition of a field is that of a region of space in which any point has certain properties depending upon its position and time. By extension, a field theory is any theory involving properties of space in order to describe interactions between material points. While analytical mechanics deals with the motion of discrete entities (the so-called particles), space and time become continuous in field theories, which describe forces having a continuous and smooth variation in both space and time. Note that field theory is incompatible with the existence of vacuum, and space, the arena of events in field theory, is endowed with physical attributes; the concept of field illustrate the rupture between a discontinuous (matter is punctual) and a continuous viewpoint (continuous space-time).

3.1 A Short Historical Vignette

From a historical perspective, Aristotle may be considered as the first philosopher of nature who imagined a field theory describing the impossibility of remote actions. According to his views, the characteristics of the locus dictate the natural motion of objects (theory of natural motion): the four basic elements (earth water, fire, air) migrate towards their natural place in his model of imbricated spherical shells centered on the earth. This faculty of space to attract objects on those spherical shells conceived by Aristotle differs from the modern classical theories of electric potentials, which only appear provided a charged body exists: space is passive in the Greek time, whereas it is active in modern physics. Much later, Descartes lays down the foundations of the first modern theory of continuous media, considering that space itself if a substance (matter) with a certain extent made of incompressible corpuscles of different sizes, in relative motion. All physical actions due to the impacts and pressures exerted by those corpuscles generate whirlwinds, inducing in turn an instantaneous displacement of light from point to point; this viewpoint will be frowned on by the subsequent theories. Nevertheless, Descartes provided a model of a mathematical theory of motion in fluids and in elastic solids (first developments were made later on by Euler and Bernoulli) which is a basis of the future field theories. Newton conceived in the Principia (1687) the gravitation theory, based on the idea of a remote interaction of masses; he was a source of inspiration for Faraday (1852), and gravitation became the paradigm for remote action. Faraday took inspiration in the works of Ruggiero Giuseppe Boscovich (Theorica Philosophiae Naturalis 1758): matter is composed of punctual masses surrounded by zones of attractive and repulsive forces. He was thereby trying to unify the explanations of mechanical, electrical, magnetic and chemical phenomena; mass exists in discrete form whereas forces are continuous. Faraday reversed this interpretation, considering that what is called matter is actually punctual atoms and forces, based on the belief that matter fills entire space where gravitation is acting. The line forces of the electrical and magnetic fields correspond to real modifications of the empty space located inbetween the charged bodies. The medium in which the field is acting is real (endowed with physical properties) without resembling ordinary matter.

Later on, the theories of ether shall rely on the mathematical foundations of continuous elastic media, whereby light propagates similar to a wave. James Clerc Maxwell developed the mathematical representation of Faraday line forces, and extends those ideas to the electromagnetic field (1860). He postulated the existence of ether, a quite complex medium, equipped with the mechanical properties of matter. At this stage, the matter conceived by Descartes and Newton (mass and motion) is replaced by energy, viewed as a specific substance of the field. This conception was then reinforced by Heinrich Hertz, who discovered the propagation of electromagnetic waves at the speed of light, thereby providing a confirmation of the unification of optics with electromagnetism done earlier on by Maxwell: Hertz proved the presence of energy in the field, propagating between the source and the receptor. The model of ether shall be abandoned with the advent of relativistic theories, and Michelson and Morley experiment.

The simplest way to translate the idea that physical systems mutually interact is to imagine that each system emits some kind of messenger which acts on other systems when reaching them; let then call field such a messenger, which then shall exist and be specific for each type of interaction (electromagnetic, nuclear, gravitational,...). *A* field is then the vehicle of an interaction, propagating at a finite speed; this implies that the forces acting at a given instant on a system are not defined by the state of sources (the other systems) at the same instant, since there is some delay before the effect of the forces or interactions manifest themselves. The concept of field is quite operative in the situation of electromagnetism, which requires special relativity. To the contrary, gravitation involves different concepts based on a modification of the structure of space-time due to the presence of objects. This is the fundamental idea at the origin of general relativity, which is the first occurrence of a geometrization of forces in modern physics: an inhabited space has a different geometrical structure (curvature) compared to empty space, and the local geometrical structure of space guides the evolution of the system instead of distant interactions. Note that this is consistent with Newtonian gravitation theory being non relativistic due to the instantaneous effect of test masses on other masses (hence, no field propagates here).

3.2 *Field Function and Field Equations: Analogies with Classical Mechanics*

The field is given all the attributes that usually pertain to a classical physical system, such as velocity, energy, linear and angular momentum, spin, eventually mass and charge. One difference in the vocabulary but also in the concept in comparison with classical mechanics is the fundamental equation satisfied by the field called the field equation, and not the equation of motion: the field occupies indeed all space at each instant, so that one does not have to bother about the position of the field, but rather what its value is at any point and any instant. The generalized coordinates of analytical mechanics (in finite number) are now replaced by a continuous function of space and time, called the field function $\psi(\boldsymbol{r}, t)$, with $\boldsymbol{r}$ the 3-vector of spatial coordinates, and t the time. Spatial and time coordinates are usually condensed into the four-vector $x = \{x^\mu\}$. One considers in general such functions which together with their first order derivatives attenuate with the distance (they vanish at infinity). To each interaction, one associates a specific field function, of appropriate mathematical nature (scalar, vectorial, tensorial, spinorial, ...).

Similar to the quantification of material systems, one may also quantify the fields themselves to highlight pseudo-particles, also called quanta of the field, which have similar attributes as the particles (mass, charge, spin, ...). The following rule can be proven: if there are N independent components of the field function, and if s is the spin of the associated quantum, then $N = 2s + 1$. For instance, a scalar field corresponds to zero spin; a Dirac spinor with two components corresponds to particles having spin $1/2$, and a quantum with spin unity (the photon) requires a vectorial field with 3 components.

The field equation is conveniently given from a Lagrangian density (per unit volume and time), a scalar valued function of the field, vector $\psi(r, t) \equiv \{\psi_k(r, t)\}$ and its first order spatial partial derivatives, the set of functions $\{\frac{\partial \psi_k}{\partial x^\mu}\}$, viz

$$L = L(\psi_k, \partial_i \psi_k) \to I = \int\limits_v \int\limits_{t_1}^{t_2} L\mathrm{d}^4 X \equiv \int\limits_x L\mathrm{d}^4 X.$$

The action integral is here the integral of the Lagrangian density over space and time; the integral over space and time has been conveniently expressed over the space-time set $X = V \times [t_1, t_2]$.

The transition from analytical mechanics to field theory involves going from the description of discrete system to that of a continuum, which is next described [8]. For a chain of N equidistant material points aligned along the axis Ox, and each moving parallel to the vertical direction Oy, the kinetic energy of the set of particles writes

$$T_\mathrm{n} = \frac{1}{2} m v_\mathrm{n}^2 \;\to\; K = \sum_{\mathrm{n}=1}^{N} T_\mathrm{n}$$

so that the overall kinetic energy K can be defined. The differential of K is easily evaluated as

$$\mathrm{d}K = \sum_{\mathrm{n}=1}^{N} m v_\mathrm{n} \mathrm{d} v_\mathrm{n}.$$

We now increase the particle number and their mutual distance, keeping a constant density of particles per unit length, ratio $\mu = m/a$, and total length $1 = Na$. Previous formula become

$$T(x) = \frac{1}{2} \mu v^2(x) \;\to\; K = \int\limits_0^1 T(x) \mathrm{d}x.$$

The discrete index has been replaced by the continuous variable x. The differential of the kinetic energy can now be expressed as

$$\delta K = \int\limits_0^1 \mathrm{d}x \frac{\overline{\partial} K}{\overline{\partial} v(x)} \delta v(x) = \int\limits_0^1 \mathrm{d}x \frac{\partial K}{\partial v(x)} \delta v(x) = \int\limits_0^1 \mathrm{d}x \mu v(x) \delta v(x)$$

involving the functional derivative $\overline{\partial} K / \overline{\partial} v(x)$ of the kinetic energy with respect to $v(x)$: this derivative indicates how the functional K changes when $v(x)$ varies by $\delta v(x)$ in the small interval $[x, x + \mathrm{d}x]$. The notion of functional derivative is useful each time a law of physics is formulated by a variational principle: for an action of the form

$$S = \int_{t_1}^{t_2} \mathrm{d}t L\left(x_j(t), \dot{x}_j(t), t\right) \mathrm{d}t$$

the Euler equations can be expressed into the form

$$\delta S = \int_{t_1}^{t_2} \frac{\overline{\partial} S}{\overline{\partial} x_j(t)} \delta x_j(t) \equiv 0, \qquad \delta x_j(t) \Rightarrow \frac{\overline{\partial} S}{\overline{\partial} x_j(t)} = 0.$$

The last equation is nothing else than the Lagrange equation relative to x_j, according to previous developments, we have replaced the generalized coordinates q_α by the x_j.

$$\begin{aligned}\delta S &= \int_{t_1}^{t_2} \left\{ \frac{\partial L}{\partial x_j} - \frac{\mathrm{d}}{\mathrm{d}t}\left(\frac{\partial L}{\partial \dot{x}_j}\right)\right\} \delta x_j \mathrm{d}t + \left[\frac{\partial L}{\partial \dot{x}_j}\delta x_j\right]_{t_1}^{t_2} \\ &\equiv \int_{t_1}^{t_2} \left\{ \frac{\partial L}{\partial x_j} - \frac{\mathrm{d}}{\mathrm{d}t}\left(\frac{\partial L}{\partial \dot{x}_j}\right)\right\} \delta x_j \mathrm{d}t.\end{aligned}$$

Identifying the two variations leads to the functional derivative of the action

$$\frac{\overline{\partial} S}{\overline{\partial} x_j(t)} = \left\{ \frac{\partial L}{\partial x_j(t)} - \frac{\mathrm{d}}{\mathrm{d}t}\left(\frac{\partial L}{\partial \dot{x}_j}\right)\right\} = 0.$$

We now turn to a dynamical system for which the Lagrangian becomes itself a functional of the form

$$L = \int \mathrm{d}^3 X \tilde{L}\left(A_j, \dot{A}_j, \partial_i A_j\right).$$

Note that the arguments $A_j, \dot{A}_j$ therein are independent, but not A_j and $\partial_i A_j$, since the knowledge of the first determines the second one. The differential of the Lagrangian is evaluated using the functional derivative as

$$\delta L = \int \mathrm{d}^3 X \left\{ \frac{\overline{\partial} L}{\overline{\partial} A_j} \delta A_j + \frac{\overline{\partial} L}{\overline{\partial} A_j} \delta \dot{A}_j \right\}$$

with the identification of the functional derivatives therein that easily follows as

$$\frac{\overline{\partial} L}{\overline{\partial} A_j} \equiv \frac{\partial \tilde{L}}{\partial A_j} - \partial_i \left(\frac{\partial \tilde{L}}{\partial\left(\partial_i A_j\right)}\right), \qquad \frac{\overline{\partial} L}{\overline{\partial} \dot{A}_j} \equiv \frac{\overline{\partial} \tilde{L}}{\partial \dot{A}_j}.$$

For a continuous system, the action is a functional of the A_j, elaborated as

$$S = \int_{t_1}^{t_2} \mathrm{d}t L(t) = \int_{t_1}^{t_2} \mathrm{d}t \int \mathrm{d}^3X\, \tilde{L}\left(A_j, \dot{A}_j, \partial_i A_j\right)$$

$$\to \delta S = \int_{t_1}^{t_2} \mathrm{d}t \int d^3X \frac{\overline{\partial} S}{\overline{\partial} A_j} \delta A_j.$$

The least action principle, equation $\frac{\overline{\partial} S}{\overline{\partial} A_j} = 0$, can be made more explicit

$$\delta S = \int_{t_1}^{t_2} \mathrm{d}t \delta L(t) = \int_{t_1}^{t_2} \mathrm{d}t \int \mathrm{d}^3X \left\{ \frac{\overline{\partial} L}{\overline{\partial} A_j} - \frac{\mathrm{d}}{\mathrm{d}t} \frac{\overline{\partial} L}{\overline{\partial} \dot{A}_j} \right\} \delta A_j$$

$$\Rightarrow \frac{\overline{\partial} S}{\overline{\partial} A_j} = \frac{\overline{\partial} L}{\overline{\partial} A_j} - \frac{\mathrm{d}}{\mathrm{d}t} \frac{\overline{\partial} L}{\overline{\partial} \dot{A}_j} = 0.$$

The use of the functional derivatives leads to field equations that are formally identical to the discrete case. The Hamilton equations of motion for a field shall also take a similar form: let first define the momentum of the field and the Hamiltonian

$$\Pi_j := \frac{\partial L}{\partial \dot{A}_j} \quad \to \quad H := \int \mathrm{d}^3X \left(\Pi_j \dot{A}_j - L\right).$$

An easy calculation of the variation of H using the functional derivative and the definition of H leads to the Hamilton equations

$$\dot{\Pi}_j = -\frac{\overline{\partial} H}{\overline{\partial} A_j}, \quad \dot{A}_j := \frac{\overline{\partial} H}{\overline{\partial} \Pi_j}.$$

The field equations are obtained from the stationnarity of S, considering that L vanishes at infinity and at initial and final times (hence the boundary terms vanish): a simple calculation leads to the variation

$$\delta I = \int_x \left(\frac{\partial L}{\partial \Psi_k} - \sum_i \partial^\mu \left(\frac{\partial L}{\partial^\mu \left(\partial \Psi_k\right)} \right) \right) \partial \Psi_k \mathrm{d}^4X + \int_\Sigma \frac{\partial L}{\partial \left(\partial^\mu \Psi_k\right)} \delta \Psi_k \mathrm{d}\Sigma$$

involving a volume integral (the spacetime volume is denoted X) and a surface integral written over the boundary $\Sigma = \partial X$. Due to the conditions chosen for the variation of the field (the field itself is supposed to vanish at infinity), the boundary integral vanishes, thus it remains as a necessary stationnarity condition

$$\delta I = 0 \quad \Rightarrow \quad \frac{\partial L}{\partial \Psi_k} - \sum_i \frac{\partial L}{\partial\left(\partial \Psi_k\right)} = 0.$$

This defines a set of N second order differential equations (as many equations as field components), called the field equation for the free field.

3.3 Field Equations and Noether Currents

The fields shall here in and the sequel be denoted collectively as the vector $\psi = \psi(\mathbf{x}) = \left\{\psi_j(\mathbf{x})\right\}_{j=1..d}$, with the components therein being the continuous analogues of the generalized coordinates $\left\{q_j\right\}$ in a discrete description, see Section 2. The action functional is here also built from a Lagrangian density in space, $L\left(\psi, \partial_\mu \psi\right)$, a function of the field and their 4D gradient (in time and space), so that

$$S = \int_D \mathrm{d}^4 x L\left(\psi, \partial_\mu \psi\right)$$

with D a compact region of space-time. The Lagrangian density is non singular provided $\det\left(\frac{\partial^2 L}{\partial_0 \psi_j \partial_0 \psi_k}\right) \neq 0$. Noether theorem is established next in a manner similar to classical mechanics, as previously exposed: the independent and dependent variables are transformed according to

$$x \mapsto x + \delta x(x), \quad \psi \mapsto \psi + \delta \psi(x).$$

One deduces

$$\delta \psi = \delta_t \psi + \delta \mathbf{x}.\partial \psi, \quad \delta \partial_\mu \psi = \delta_t \partial_\mu \psi + \delta \mathbf{x}.\partial \partial_\mu \psi$$

with $\delta_t \psi$ the variation taken at fixed time, and where the notation $\partial \psi$ stands for the 4D gradient. The volume of integration varies as

$$\delta\left(\mathrm{d}^4 x\right) = \mathrm{d}^4 x.\delta \mathbf{x}(\mathbf{x}).$$

This entails the following general variation of the action integral

$$\delta S = \int_{A_i}^{A_f} \mathrm{d}^4 x \left[\left(\delta^\mu_{.\nu} L - \frac{\partial L}{\partial\left(\partial_\mu \psi\right)} \partial_\nu \psi\right) \delta x^\nu + \frac{\partial L}{\partial\left(\partial_\mu \psi\right)} \delta \psi \right.$$
$$\left. + \left(\frac{\partial L}{\partial \psi} - \partial_\mu\left(\frac{\partial L}{\partial\left(\partial_\mu \psi\right)}\right)\right)(\delta \psi - \delta x.\partial \psi)\right].$$

From this, application of Hamilton principle leads to the Euler-Lagrange equations

$$\frac{\partial L}{\partial \psi} - \partial_\mu \left(\frac{\partial L}{\partial \left(\partial_\mu \psi\right)} \right) = 0.$$

For a Lie symmetry transformation $(\Delta x^\mu, \Delta\psi)$, with Lie group parameters $\{\mu_i\}_{i=1\ldots p}$, we obtain the identity (in a manner similar as for the previous discrete situation)

$$\partial_\mu j_i^\mu + \left(\frac{\partial L}{\partial \psi} - \partial_\mu \left(\frac{\partial L}{\partial \left(\partial_\mu \psi\right)} \right) \right) \left(\frac{\partial \Delta \psi}{\partial \Delta \mu_i} - \frac{\partial \Delta x^\mu}{\partial \Delta \mu_i} \partial_\mu \psi \right) = 0, \quad i = 1 \ldots p$$

The Noether currents j_i^μ therein are elaborated as

$$j_i^\mu = \left(\frac{\partial L}{\partial \left(\partial_\mu \psi\right)} \partial_\nu \psi - \delta^\mu_{.\nu} L \right) \frac{\partial \Delta x^\mu}{\partial \Delta \mu_i} - \frac{\partial L}{\partial \left(\partial_\mu \psi\right)} \frac{\partial \Delta \psi}{\partial \Delta \mu_i} - \frac{\partial \left(\Delta \Lambda^\mu\right)}{\partial \Delta \mu_i}, \quad i = 1 \ldots p$$

The variations of the arbitrary function $\Lambda^\mu = \Lambda^\mu(x^\mu)$ intervene in the covariant Lagrangians that are not invariant, so that

$$\partial.\left(\Delta \mathbf{x}\right) + \Delta L = -\partial.\left(\Delta \Lambda\right)$$

The symbol ∂ therein denotes the nabla operator, so $\partial.\Delta\mathbf{x}$ is the divergence of the quadri-vector $\mathbf{x}$. On the stationary path on which Euler equations are satisfied, the Noether currents satisfy the continuity equation

$$\partial_\mu j_i^\mu, \qquad i = 1 \ldots p.$$

Integrating over a spacelike surface A gives the conserved Noether charge

$$Q_i := \int\limits_A \mathrm{d}A \, j_i^\mu.$$

One can check that in the specific case of translations, $x^\mu \mapsto x^\mu + a^\mu$, with a^μ constants, so that $\Delta x^\mu = a^\mu$ and for fixed fields, thus $\Delta\psi = 0$, Noether theorem leads in the case of invariant Lagrangians

$$\Delta \Lambda \equiv 0$$

to the conservation of the canonical energy momentum tensor

$$T^{\mu\nu} := -c \left(\frac{\partial L}{\partial \left(\partial_\mu \psi\right)} \partial_\nu \psi - \delta^\mu_{.\nu} L \right) \rightarrow \partial_\mu T^{\mu\nu} = 0.$$

The conserved 4-momenutm is defined as the integral of the energy-momentum tensor over any space-like surface (cross section of a 4D volume at fixed time), so that

$$p^{\mu} = \frac{1}{c}\int\limits_{A} \mathrm{d}A T^{\mu\nu}.$$

The covariant momentum densities are further defined as

$$\pi^{\mu} = \left(\pi^{0}\ \pi^{k}\right) \ \rightarrow\ \pi^{\mu} := \frac{\partial L}{\partial\left(\partial_{\mu}\psi\right)}.$$

It is to be noticed that the Hamiltonian density of the field is the pure time-like component of the energy momentum tensor, so that

$$H := T^{00}.$$

We shall next extend those notions from a slightly different perspective in order to fully characterize a field and give it all attributes a discrete mechanical system has, such as 4-momentum, energy, kinetic moment, spin.

3.4 Noether's Theorem and Conservation Laws

Recall that for a material system, the symmetries imposed to the system (spatial translation, temporal translation and rotation) are reflected in the form taken by its Lagrangian; those symmetries lead to conserved quantities (respectively conservation of linear momentum, energy and angular momentum). Similarly, requiring the invariance of physics of the field, in terms of the action integral, with respect to the same continuous transformations, leads to the identification of conserved quantities such as energy, linear and angular momentum for the field. Since we deal with continuous transformations (corresponding to Lie groups), we can equivalently restrict ourselves to infinitesimal transformations. Although it may be represented as a column vector, the spatial coordinate will be denoted as a scalar (without boldface). Let then consider an infinitesimal change of referential

$$X_i \rightarrow X_i + \delta X_i, \quad \Psi\left(X\right) \rightarrow \Psi'\left(X'\right) = \Psi\left(X\right) + \delta\Psi\left(X\right).$$

The local variation $\delta\psi(X)$ therein compares the fields $\psi(X)$, $\psi^{'}(\mathrm{X}')$ at two different points; one next elaborates the *proper variation of the field* (local variation of the field) as the difference of ψ evaluated at the same point

$$\bar{\delta}\psi(X) := \psi'(X) - \psi(X).$$

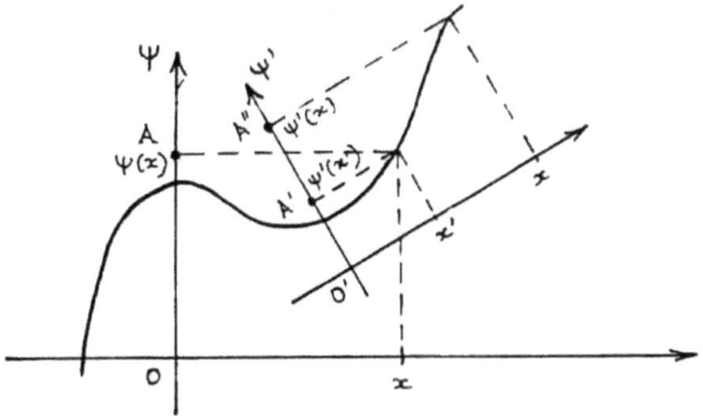

Fig. 2 Proper variation of a field, cf. [9]

A first order Taylor expansion then leads to

$$\psi'(X') = \psi'(X) + \frac{\partial \psi'(X)}{\partial X_i}\delta X_i.$$

This results in the expression

$$\Psi'\left(X'\right) = \Psi\left(X\right) + \bar{\delta}\Psi + \frac{\partial \Psi\left(X\right)}{\partial X_i}\delta X_i + \frac{\partial\left(\bar{\delta}\Psi\right)\left(X\right)}{\partial X_i}\delta X_i \equiv \Psi\left(X\right) + \delta\Psi\left(X\right).$$

Observe that the term $\frac{\partial(\bar{\delta}\Psi)(X)}{\partial X_i}\delta X_i$ is a second order contribution. The last identity allows identifying the proper variation of the field as

$$\bar{\delta}\Psi = \delta\Psi - \frac{\partial\Psi\left(X\right)}{\partial X_i}\delta X_i$$

see Fig. 2.

The invariance of the action integral under a change of referential leads to

$$0 = \delta I = \int\limits_V \int\limits_{t_1}^{t_2} L\left(\Psi + \bar{\delta}\Psi, \partial_i\Psi + \bar{\delta}\partial_i\Psi\right)\mathrm{d}^4X' - \int\limits_V \int\limits_{t_1}^{t_2} L\left(\Psi, \partial_i\Psi\right)\mathrm{d}^4X$$

involving the proper variations of the field and its first order spatial derivatives. The Lagrangian function has to be invariant under space and time translations, thus it does not explicitly depend on the coordinates.

The Jacobean in the first integral of last equality is evaluated at first order as $J = 1 + \partial_i\delta X_i + \mathrm{o}(\delta X_i)$. We can further expand the first Lagrangian density as

$$L\left(\Psi+\bar{\delta}\Psi,\partial_i\Psi+\bar{\delta}\partial_i\Psi\right)\cong L\left(\Psi,\partial_i\Psi\right)+\frac{\partial L}{\partial\Psi}\cdot\bar{\delta}\Psi+\frac{\partial L}{\partial\left(\partial_i\Psi\right)}\cdot\bar{\delta}\partial_i\Psi.$$

Accounting for the relations

$$\bar{\delta}\partial_i\Psi=\partial_i\bar{\delta}\Psi$$

and

$$\frac{\partial L}{\partial\left(\partial_i\Psi\right)}\cdot\partial_i\left(\bar{\delta}\Psi\right)=\partial_i\left(\frac{\partial L}{\partial\left(\partial_i\Psi\right)}\cdot\bar{\delta}\Psi\right)-\partial_i\frac{\partial L}{\partial\left(\partial_i\Psi\right)}\cdot\bar{\delta}\Psi$$

then leads to the variation

$$\delta I=\int_V\int_{t_1}^{t_2}\sum_i\partial_i\left\{\left(L\delta_{ij}-\frac{\partial L}{\partial\left(\partial_i\Psi\right)}\partial_j\Psi\right)\delta X_j+\frac{\partial L}{\partial\left(\partial_i\Psi\right)}\cdot\delta\Psi\right\}\mathrm{d}^4X=0.$$

The last equality can be written more compactly into the form (the symbol $\Box$ denotes the quadrivergence or d'Alembertian)

$$\partial_i f_i=0\leftrightarrow\Box f=0$$

with

$$f_i:=\sum_j\left(L\delta_{ij}-\frac{\partial L}{\partial\left(\partial_i\Psi\right)}\partial_j\Psi\right)\delta X_j+\frac{\partial L}{\partial\left(\partial_i\Psi\right)}\cdot\delta\Psi\equiv\sum_j T_{ij}\delta X_j+\frac{\partial L}{\partial\left(\partial_i\Psi\right)}\cdot\delta\Psi.$$

Previous conservation equation (the quadrivergence of the force like quadrivector f vanishes) highlights the contribution of the celebrated energy-momentum tensor of the field, the second order tensor with components

$$T_{ij}=L\delta_{ij}-\frac{\partial L}{\partial\left(\partial_i\Psi\right)}\partial_j\Psi.$$

The previous variation of the action includes a contribution due to the variation of the field and a variation of the independent variable X, acted upon by tensor T. Integrating previous conservation law onto the infinite $3D$-volume and isolating the spatial and time-like force components therein leads to

$$\mathbf{f}=\left(\bar{f},f^4\right),\quad\partial_i f_i=0\rightarrow\operatorname{div}\bar{f}+\frac{\partial f_4}{\partial t}=0\Rightarrow\int_{V_3}\operatorname{div}\bar{f}\,\mathrm{d}V+\frac{\mathrm{d}}{\mathrm{d}t}\int_{V_3}f_4\,\mathrm{d}V.$$

Green's formula implies the vanishing of the first integral, hence one arrives at the conservation law

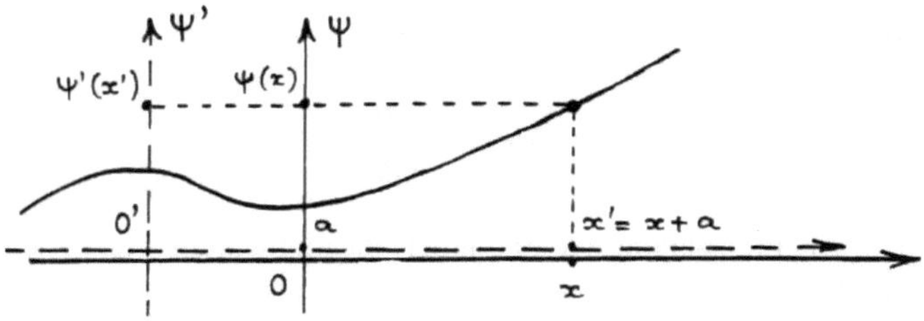

Fig. 3 Invariance by translation, cf. [9]

$$F := \int_{V_3} f_4 \, \mathrm{d}V \equiv \left\{ \sum_j T_{4j} \delta X_j + \frac{\partial L}{\partial (\partial_4 \Psi)} \cdot \delta \Psi \right\} \mathrm{d}V = Cte$$

traducing Noether's theorem [10]: any invariance of physics by a continuous transformation leads to the conservation of a physical quantity.

Note that one is led to generalize Noether's theorem in the quantum domain by including non continuous transformations (discrete symmetries such as inversions).

The postulate of homogeneity of the four-space leads to the invariance of physics under spatio-temporal translation by the infinitesimal vector: the transformation in Fig. 3, i.e.,

$$X \to X' = X + a \Rightarrow \delta X_j = a_j$$

results in

$$\Psi' \left(X' \right) = \Psi \left(X \right) \Rightarrow \delta \Psi = 0.$$

An immediate application of previous general conservation law and the consideration of the arbitrariness of the constants a_j lead to

$$P_j := \int_{v3} T_{4j} \mathrm{d}V = Cte$$

traducing the conservation of the 4-momentum. One can accordingly rewrite the conservation law in the form

$$\sum_i \partial_i f_i = 0 \to \sum_j \partial_j T_{4j} = 0$$

One identifies in previous identity the 3-momentum with components

$$P_\mu := \int_{V_\infty = V} T_{4\mu} \mathrm{d}V = - \int_{V_\infty = V} \frac{\partial L}{\partial (\partial_4 \Psi)} \partial_\mu \Psi \mathrm{d}V \to \boldsymbol{P} = - \int_{V_\infty = V} \frac{\partial L}{\partial (\partial_4 \Psi)} \nabla \Psi \mathrm{d}V$$

involving the quantity

$$\Pi(X) := \frac{\partial L}{\partial(\partial_4 \Psi)}$$

called the field function, playing a role similar to the momentum in analytical mechanics. Accordingly, the momentum of a field is defined as the 3-vector built from the volumetric density

$$\boldsymbol{P}(X) := -\Pi(X)\nabla\Psi(X) \rightarrow \boldsymbol{P} = -\int_{V_\infty = V_3} \Pi(X)\nabla\Psi(X)\,\mathrm{d}V.$$

The energy of the field is defined as the remaining time-like component

$$P_4 = W := \int_{V_\infty = V_3} T_{44}\mathrm{d}V = \int_{V_\infty = V_3} \{-L + \Pi(X)\,\partial_4\Psi\}\,\mathrm{d}V.$$

The density of energy, the scalar

$$H := \Pi(X)\,\partial_4\Psi - L$$

plays a role similar to the Hamiltonian in classical mechanics.

We next define the *kinetic moment* of a field, vector $\boldsymbol{L}$ and its *spin* $\boldsymbol{I}$: this last vector does not depend on the choice of an origin of space, and it represents an intrinsic property of the field. Contrary to this, the *total moment of the field*, vector I, is a conserved quantity depending upon the selected origin of space; it vanishes for a central field. For a scalar field, the operators I_μ vanish, hence the spin is nil (only one component for the field). The field ψ may be complex valued, but the Lagrangian density L is real, thus it contains only combinations of the type $\psi^*\psi$ (with ψ^* the complex conjugate of ψ), which means that L is invariant under the transformation $\psi \rightarrow \psi' = \mathrm{e}^{\mathrm{i}\alpha}\psi$, called a *gauge transformation of the first kind*. When parameter α is constant, one can approximate previous finite transformation by the expansion

$$\psi' \cong (1 + \mathrm{i}\alpha)\psi \rightarrow \delta\psi = \mathrm{i}\alpha\psi,\ \delta\psi^* = -\mathrm{i}\alpha\psi^*$$

Noether's theorem then leads to the force

$$f_i = \frac{\partial L}{\partial(\partial_i\psi)}.\psi - \frac{\partial L}{\partial(\partial_i\psi^*)}.\psi^*$$

The associated four-vector is characterized by the current density

$$j_i = \frac{\partial L}{\partial(\partial_i\psi)}.\psi - \psi^*.\frac{\partial L}{\partial(\partial_i\psi^*)}$$

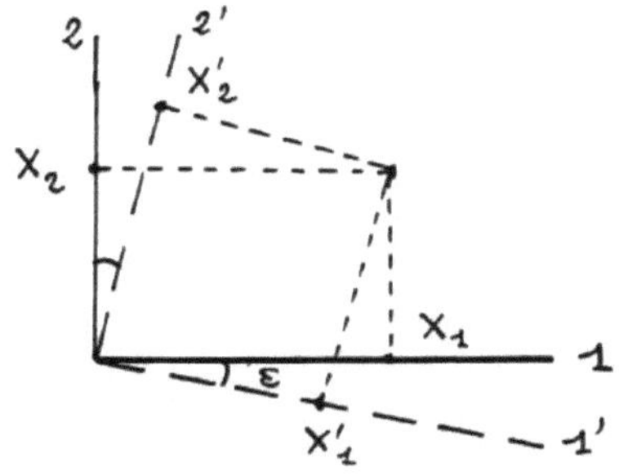

Fig. 4 Rotation of the referential, cf. [9]

and it satisfies the conservation of the field charge defined as the integral of the temporal component of the current

$$Q := \int_{v_3} j_0(X)\mathrm{d}V = \int_{v_3} \{\Pi(X).\psi(X) - \psi^*(X).\Pi^*(X)\}\mathrm{d}V.$$

In order to highlight the conservation of the total moment of the field and considering a specific rotation, we perform an infinitesimal rotation of axis X_3 and angle ε, a small parameter (see Fig. 4).

hence the variation of coordinates

$$\begin{aligned} &X_1' = X_1 \cos\varepsilon - X_2 \sin\varepsilon \cong X_1 - \varepsilon X_2 \Rightarrow \delta X_1 = -\varepsilon X_2 \\ &X_2' = X_1 \sin\varepsilon - X_2 \cos\varepsilon \cong X_2 - \varepsilon X_1 \Rightarrow \delta X_2 = -\varepsilon X_1 \\ &X_3' = X_3 \Rightarrow \delta X_3 = 0. \end{aligned}$$

This transformation leads to $\delta\Psi = I_3$ with I_3 the infinitesimal generator of the rotation, which depends on the field ψ. For an isotropic space, we then obtain the conserved integral quantity

$$\int_{v_3} \{\epsilon\Pi(X)I_3\psi(X) + T_{41}\delta X_1 + T_{41}\delta X_1 + T_{42}\delta X_2\}\mathrm{d}V = Cte.$$

We can rewrite previous conservation law as

$$\int_{v_3} \Pi(X)I_3\psi(X)\mathrm{d}V - \int_{v_3} \{\Pi(X).(X_1\partial_2\psi - X_2\partial_1\psi)\}\mathrm{d}V = Cte$$

in which the quantity $(X_1\partial_2\psi - X_2\partial_1\psi)$ is recognized as the third component of the vectorial product $\boldsymbol{r} \times \nabla\boldsymbol{\Psi}$.

This highlights the angular moment of the field identified as the vector

$$L := \int_{V_3} \boldsymbol{r} \times \boldsymbol{d}\mathrm{d}V$$

with

$$\boldsymbol{p} := -\Pi\,(X)\,.\nabla\boldsymbol{\Psi}\,.$$

The invariance by rotation around the third axis then entails the conservation of the third component

$$J_3 := \int_{V_3} I_3 \mathrm{d}V + \int_{V_3} \Pi\,(X)\,\boldsymbol{I}.\Psi\,(X)\,\mathrm{d}V$$

with $\boldsymbol{I} := (\boldsymbol{I}_1\ \boldsymbol{I}_2\ \boldsymbol{I}_3)$ the set of three infinitesimal generators of rotations in 3D space. In a general case, the invariance by rotation thus leads to the conservation of the total momentum, sum of the angular momentum of the field, vector $\boldsymbol{L}$, and of the spin, vector $\boldsymbol{S}$, with

$$\boldsymbol{S} := \int_{V_3} \Pi\,(X)\boldsymbol{I}.\Psi\,(X)\,\mathrm{d}V.$$

We shall further distinguish external symmetries, namely symmetry transformations acting on the space time coordinates from internal (or gauge) symmetries, acting on internal degrees of freedom. The space-time coordinate vector $\mathbf{x} = \{x^{\mu}\} = \{ct, x, y, z\}$ may be transformed by a linear mapping L, thus defining the new coordinates $\mathbf{x}' = \left\{x'^{\mu}\right\}$, such that $\mathbf{x}' = L.\mathbf{x}+\mathbf{a}$, $\mathrm{L}^T\eta L = \eta$, where η is Minkowski metric. Transformations of this kind define the full Poincaré group of transformations; the full Lorentz group is obtained as a specific case for $a = 0$ and L satisfying previous equation.

4 Lagrangian Formulation of Electromagnetism and Relativity Theories

We herewith adopt the viewpoint of symmetries inherently contained in the form of the Lagrangian density: the fields and the Lagrangian will be introduced for several field theories without so much physical justification, as it is not the goal and viewpoint of this chapter to explain the physical content of the theories, but rather to illustrate the powerfulness of symmetries in the construction of field theories [11]. As previously mentioned, those theories can be classified according to the spin, which determines the number of independent components of the field.

• A massive scalar field (zero spin) can be defined by a real function of the space-time coordinates, $\varphi\left(x_{\mu}\right)$, and a Lagrangian density

$$L = \frac{1}{2}(\partial_{\mu}\varphi\partial^{\mu}\varphi - m^2\varphi)$$

The Euler-Lagrange equations lead to Klein-Gordon equation for a free massive scalar field

$$\left(\Box + m^2\right)\varphi = 0$$

with $\Box$ the d'Alembertien. Considering an interaction with an external potential $J(\boldsymbol{x},t)\,\varphi(\boldsymbol{x},t) = V(\boldsymbol{x},t)$ leads to the new Lagrangian density

$$L = \frac{1}{2}\left(\partial_\mu \partial^\mu \varphi - m^2\varphi\right) - J\varphi$$

It results in the modified Klein-Gordon equation evidencing the current J as a source term:

$$\left(\Box + m^2\right)\varphi = J$$

• The free behavior of fermions (spin is $1/2$) such as electrons is described by a four-spinor ψ, with the Lagrangian (involving ψ and its complex conjugate ψ)

$$L_f = \bar{\Psi}\left(\mathrm{i}\gamma^\mu \partial_\mu \Psi - m\right)\Psi$$

with γ^μ the 4×4 Dirac matrix built from Pauli matrices; it entails the Dirac equation

$$(\mathrm{i}\gamma^\mu \partial_\mu - m)\psi = 0$$

• The electromagnetic field is a massless vectorial field characterized by the vector potential $A = (\varphi, \boldsymbol{A})$, and the Faraday tensor

$$F_{\mu v} := \partial_\mu A_v - \partial_v A_\mu \;\rightarrow\; F^{\mu\nu} = \begin{pmatrix} 0 & -E_1 & -E_2 & -E_3 \\ E_1 & 0 & -B_3 & B_2 \\ E_2 & B_3 & 0 & -B_1 \\ E_3 & -B_2 & B_1 & 0 \end{pmatrix}.$$

The metric tensor relates the covariant to the contravariant components of any tensor; for instance, the components of Faraday tensor are related by $F_{\alpha\beta} = \eta_{T\alpha}\eta_{0\beta}F^{T0}$ in Minkowski space-time.

The electric and magnetic fields, vectors $\boldsymbol{E}$, $\boldsymbol{B}$ are not determined in a unique way, since any gauge transformation of the form

$$\boldsymbol{A}' = \boldsymbol{A} + \nabla\chi, \quad \phi' = \phi + \partial_t \chi$$

with χ an arbitrary scalar function of coordinates, leaves the fields $\boldsymbol{E}$, $\boldsymbol{B}$ invariant, due to the relations

$$\boldsymbol{E} = -\nabla\phi + -\partial_t \boldsymbol{A}, \qquad \boldsymbol{B} = \nabla \wedge \boldsymbol{A}.$$

In order to remove the gauge freedom, a certain number of *gauge conditions* have been proposed, amongst which the Lorentz gauge and the Coulomb gauge (a subcase of Lorentz gauge) are the most popular. The Lorentz gauge, equation

$$\partial_t \phi + \nabla . \boldsymbol{A} = 0 \Leftrightarrow \partial_\mu A^\mu = 0$$

is further compatible with the conservation of the electric current, identity $\partial_t \rho + \nabla . \boldsymbol{j} = 0$ involving the density of charge ρ and the current $\boldsymbol{j} = \sigma \boldsymbol{E}$ (σ is the conductivity). Previous definition of Faraday tensor leads to the first group of Maxwell equations

$$\partial_\mu F_{\lambda\gamma} + \partial_\lambda F_{\gamma\mu} + \partial_\gamma F_{\mu\lambda} = 0 \Leftrightarrow \left| \begin{array}{l} \partial_t \boldsymbol{B} + \mathrm{rot} \boldsymbol{E} = 0 \\ \mathrm{div} \boldsymbol{B} = 0. \end{array} \right.$$

The Lagrangian density of the electromagnetic field in presence of sources described by the current $J^\mu \equiv e\bar{\psi}\gamma^\mu\psi$ write together with the action

$$\begin{aligned} L_{\mathrm{e}}\left(A_\mu, \partial_\mu A_\upsilon\right) &= -\frac{1}{16\pi} F_{\mu\upsilon} F^{\mu\upsilon} - J^\mu A_\mu \\ &\equiv L_{em} + L_{\mathrm{int}} \rightarrow S_{\mathrm{e}}[A_\mu] := \int_\upsilon L_{\mathrm{e}}(A_\mu, \partial_\mu A_\upsilon) \mathrm{d}^4 X. \end{aligned}$$

The first term on the r.h.s. has the physical significance of the density of electromagnetic energy, with

$$L_{em} = -\frac{1}{16\pi} F_{\mu\upsilon} F^{\mu\upsilon} \equiv \frac{1}{8\pi\mu} \left(\frac{E^2}{c^2} - B^2 \right)$$

while the second Lagrangian L_{mt} describes the interaction between the field and the sources (particles). This Lagrangian leads to the Euler equations (note that the Lagrangian of matter does not play a role, since the variation is taken versus the potentials A_μ)

$$\delta S_e \left[A_\mu \right] = 0 \Rightarrow \partial_\gamma F^{\mu\gamma} + 4\pi J^\mu = 0$$

identified to the second group od Maxwell equations when Lorentz gauge is imposed

$$\Box A_\mu = J_\mu \Leftrightarrow \left| \begin{array}{l} \nabla . \boldsymbol{D} = 4\pi\rho \\ \mathrm{rot} \boldsymbol{H} = 4\pi \boldsymbol{j} + \partial_t \boldsymbol{D}. \end{array} \right.$$

One has to add to the field equations the expression of the Lorentz force exerted on a charged particle (with electrical charge e), given in vectorial and tensorial format as

$$\boldsymbol{F} = e\left(\boldsymbol{E} + \boldsymbol{v} \wedge \boldsymbol{B} \right) \Leftrightarrow f_\nu = F_{\mu\nu} J^\mu.$$

The equation of motion of a charged and massive particle is then obtained as the Euler equation traducing the equality of the Lorentz force with the derivative of the 4-momentum with respect to the proper time,

$$\frac{\mathrm{d}p^{v}}{\mathrm{d}\tau} := m\frac{\mathrm{d}u^{v}}{\mathrm{d}\tau} = f^{v}$$

involving the 4-velocity vector

$$u^{\nu} = \left(u^{0}, u^{i}\right) = \left(\gamma, \gamma v^{i}\right), \qquad \gamma = \left(1 - v^{2}/c^{2}\right)^{1/2}.$$

The energy momentum tensor of the electromagnetic field in the absence of charge (tensor of the pure field) is conserved

$$T_{\mu}^{v} = \frac{\partial L_{em}}{\partial(\partial_{v}A_{\beta})}\partial_{v}A_{\beta} - \delta_{\mu}^{v}L_{em} \equiv \frac{1}{4\pi}(F_{o}^{\mu}F^{vo} - \frac{1}{4}\eta^{\mu v}F_{0\lambda}F^{0\lambda}) \rightarrow \partial_{\gamma}T_{v}^{\gamma} = 0.$$

The Hamiltonian is elaborated from the Lagrangian density, introducing the conjugated moments, and leading to Hamilton equations for the field as

$$\Pi_{\mu} := \frac{\partial L_{e}}{\partial \dot{A}_{\mu}} \rightarrow H_{e}\left(A_{\mu}, \Pi_{\mu}\right) := \Pi_{\mu}\dot{A}_{\mu} - L_{e} \rightarrow \dot{A}_{\mu} = \frac{\partial H_{e}}{\partial \Pi_{\mu}}$$

$$\dot{\Pi}_{\mu} = \frac{\partial H_{e}}{\partial A_{\mu}} - \partial_{\alpha}\left(\frac{\partial H_{e}}{\partial\left(\partial_{\alpha}A_{\mu}\right)}\right).$$

Coming back to the comparison between electromagnetism and relativity, the determination of the field equations follows similar principles:

• The equations of motion are obtained considering that the field is given, hence

$$\delta S(\text{field}) \equiv 0 \rightarrow \delta S(\text{particle}) + \delta S(\text{interaction}) \equiv 0.$$

The variation of the Lagrangian densities is here done with respect to the coordinates.

• The field equations are obtained considering that the motion of charges (particles) is given, thus

$$\delta S(\text{matter}) \equiv 0 \rightarrow \delta S(\text{field}) + \delta S(\text{interaction}) \equiv 0.$$

Now, the variation of the Lagrangian densities is performed with respect to the potentials (internal d.o.f.) viewed as generalized coordinates.

This can be illustrated in the case of GR by the derivation of Einstein equation from the action integral, sum of the action integral of energy and that of geometry, respectively integrals

$$S_e = \frac{1}{c}\int \sqrt{-g}L_e \mathrm{d}^4x, \qquad S_g = \gamma \int \sqrt{-g}R\mathrm{d}^4x.$$

The Lagrangian density for energy (including matter and the electromagnetic field), quantity $L_\mathrm{e} = L_\mathrm{e}(g_{\mu v}, \partial_\lambda g_{\mu v})$, is a scalar built from the metric tensor and its first order derivatives. The geometrical Lagrangian density is selected as the scalar curvature $R = g^{ik}R_{ik}$, built from the second order partial derivatives of the metric tensor, and $:= -\dfrac{1}{2\chi c}$, with $\chi = \dfrac{2\pi G}{c^4}$ Einstein coefficient. The geometrical action is then

$$S_g := \int \sqrt{-g}R\mathrm{d}^4x.$$

One then obtains the Euler equations

$$\delta\left(S_\mathrm{e} + S_g\right) = 0 = R_{ij} - \frac{1}{2}g_{ij}R = \chi T_{ij}$$

involving the Ricci second order curvature tensor R and the energy-momentum tensor $\mathbf{T}$ with components

$$T_{ij} = \frac{1}{2\sqrt{-g}}\left\{\frac{\partial\left(\sqrt{-g}L_e\right)}{\partial g^{ij}} - \partial_k\left(\frac{\partial\left(\sqrt{-g}L_e\right)}{\partial\left(\partial_k g^{ij}\right)}\right)\right\}.$$

The obtained field equations link the energy momentum tensor to curvature, reflecting the idea that mass and energy induce a modification of the geometrical structure of space-time.

Note that when deriving the Euler equations, we have assumed that the boundary term

$$\int_{\partial V = \Sigma} \frac{\partial L}{\partial\left(\partial^\mu\phi\right)}\delta\phi\mathrm{d}\Sigma$$

resulting from integration by parts vanishes, considering nil variations of the field on the boundary Σ of the 4-volume V.

To summarize, gravitational forces essentially arise from the communication between different points of space-time; this communication is done from a technical and operative viewpoint by the affine connection, which defines parallel transport. In a metric space, the coefficients of the connection are expressible versus the metric coefficients, and they lead to gravitational forces. It will appear in subsequent developments that all forces arise essentially in this way; anticipating next section, consider a particle described by the complex wave function $\phi(x) = \phi_1(x) + \mathrm{i}\phi_2(x)$. It is clear that the phase of $\phi(x)$ is not observable; this implies the absence of privileged orientation in the (ϕ_1, ϕ_2) plane (considered as an internal space), and it requires a rule for comparing the orientation of the wave function at different points in this plane. Similar to the concept of covariant derivative, we need a rule to construct the wave function $\phi(x_1 \to x_2)$ resulting from the transport of $\phi(x)$ from x_1 to x_2, in order to give a physical meaning to the phase variation $\delta\phi = \phi(x_2) - \phi(x_1 \to x_2)$, we introduce connection coefficients $\Gamma_{ij\mu}$, such that

$$\phi_i(x \to x + \Delta x) = \phi_i(x) - \Gamma_{ij\mu}(x)\phi_j(x)\Delta x^\mu$$

and further require that the amplitude of $\phi(x)$ remains unchanged during the transport. A connection satisfying those conditions then must be antisymmetric, of the form

$$\Gamma_{ij\mu}(x) = -\lambda\epsilon_{ij}A_\mu(x)$$

wherein $A_\mu(x)$ is the electromagnetic 4-potential vector, and ϵ_{ij} the $2D$ Levi-Civita symbol.

4.1 Symmetry Principles in Continuum Mechanics: Material Frame Indifference, Material Symmetries and Eshelbian Mechanics

Symmetry requirements form a cornerstone in continuum mechanics. One may classify such symmetries as material symmetries (acting in the reference configuration, they determine what is called the material symmetry group of a given material), and spatial symmetries, which have to be universally satisfied by any theory of finite deformation, and are usually referred to as material frame indifference. Following an early work [12], [13] was one of the few researchers to brought further the idea of general point transformations of both the Lagrangian coordinates and of the fields, thereby extending the concept of material symmetries described by a mere change of the reference configuration. Those generalized coordinate transformations are invariance properties in the sense that they are compatible with the Galilean invariance of the strain energy density function, namely, the material frame indifference is preserved. Going one step further, one may suspect in addition to material frame indifference (a postulate valid independently of the physical problem) the existence of hidden symmetries in the space of control and internal variables, which can be computed from the set of governing equations for a specific problem. In the context of continuum solids mechanics, Lie groups have been applied to solve the Navier and the Lame equations ([14–16]), or, in a similar spirit and extending this view to dissipation, to partially solve the ideal plasticity equation ([17–19]). The concept of nonlocal symmetries allows to construct novel BVP in continuum mechanics (and group invariant solutions), involving potential variables, thereby extending the classical picture relying on the traditional Lagrangian and Eulerian viewpoints, [4, 20–24].

Symmetry methods can also be successfully applied to media presenting internal length effects expressed by generalized continuum models; for instance, the symmetry analysis of nonlocal elasticity model has already been done in the literature, in [25] and [26]. When the nonlocality is decribed by higher order gradients of the internal variables, such as in the recent contribution of [27], the symmetry generators shall include the prolongation of the vector field associated to higher order gradients of the variables. The Lie symmetries of finite strain perfectly plastic equations were computed in [28] and numerical schemes preserving group properties were accordingly developed, having the advantage of satisfying the consistency condition exactly. The issue of symmetries is not only important from a theoretical and constitutive point of view, but it also has an impact as to computational issues. Numerical schemes which preserve symmetry and utilize some induced conservation laws have long term stability and are endowed with improved efficiency and stability.

As shown previously, the field of Eshelbian Mechanics (in the honour of the works of Eshelby [29]), otherwise coined *Configurational Mechanics*, relies on translational symmetries in the so-called material space, for the writing of the field equations in terms of Eshelby stresses [30, 31]. Those symmetries extended to rotations and dilatations have been intensively used to construct the well known J-integrals [32]; further conservation laws in 2D nonlinear elastodynamics have been proposed in [18], also leading to path-independent integrals when written in integral form (due to their divergential form). The fundamental postulates in continuum mechanics valid irrespective of the form of the constitutive law of the considered material (usually based on experimental data, thus of an approximate nature) can be summarized as follows [6]:

- Principle of determinism for the stress: the stress response within a continuum body is completely determined by the history of its motion;
- Principle of local action: the stress at a given material point only depends on the state of motion in a small neighborhood of that point;
- Principle of material objectivity: this principle also coined material frame-indifference enunciates that constitutive equations must be independent of the observer. It leads to the same stress measure for two observers in relative motion with respect to each other; for a tensor valued function of a tensor argument such as the Cauchy stress in elasticity, $\sigma(\mathbf{F})$, this principle expresses as the transformation rule for the stress response under an arbitrary rotation of the deformed configuration, in the form
  $\sigma(Q.\mathbf{F}) = Q.\sigma(\mathbf{F}).Q^T$.
- Principle of material symmetry: the symmetry properties of a given material can be traduced by a material symmetry group acting in the reference configuration (those transformations act in referential coordinates), and they influence the stress response according to certain rules.

The combination of material symmetries with the principle of material objectivity highlights the role of isotropic tensor functions when expressing constitutive laws. For a symmetry transformation g that belongs to the material symmetry group of an hyperelastic material with strain energy function $W(\mathbf{F})$, with $\mathbf{F}$ the transformation gradient, it holds that

$$W(\mathbf{F}.g) = W(\mathbf{F}).$$

When the constitutive law is non dissipative, one may formulate the equilibrium as an extremum principle, and Noether's theorem then articulates the functional invariance (under symmetry transformations) called variational symmetries, conservation laws and the non observability of certain quantities. Let consider an action integral $S := \int_{\Omega} L(\mathbf{X}, \varphi, \nabla\varphi)\, \mathrm{d}\Omega$ built from a Lagrangian density per unit volume

$$L(\mathbf{X}, \varphi, \nabla\varphi) = W(\mathbf{X}, \varphi, \nabla\varphi)$$

denoting by $\mathbf{X}$ the independent variables associated to the parameterization, and $\varphi = \varphi(\mathbf{X}, t)$ the dependent variables identified to the fields. The Lagrangian identified in continuum mechanics to a strain energy density function is here chosen to depend at most on the first order gradient of the field, and will explicitly not depend on $\mathbf{X}$, so that

$$L(\mathbf{X}, \varphi, \nabla\varphi) = W(\varphi, \nabla\varphi).$$

The general variation of the action integral under an arbitrary Lie group of transformations acting on both the dependent and independent variables expresses as

$$\begin{aligned}\delta S &= \mu \int_{\Omega} \left(\frac{\partial W}{\partial \varphi_k} - D_i \frac{\partial W}{\partial \varphi_{k,i}} \right) \left(\phi_k - \xi_j \varphi_{k,j} \right) \mathrm{d}\Omega \\ &+ \mu \int_{\partial\Omega} \left(W\xi_i + \left(\phi_k - \varphi_{k,j}\xi_j \right) \frac{\partial W}{\partial \varphi_{k,i}} \right) n_i \mathrm{d}(\partial\Omega)\end{aligned}$$

wherein the components of the vector field generator of the Lie group are denoted $(\xi, \phi) \equiv \mu(\delta\mathbf{X}, \delta\varphi)$, with μ the group (small) parameter. One may observe that the contribution

$$\int_{\partial\Omega} \left(W\xi_i - \varphi_{k,j}\xi_j \frac{\partial W}{\partial \varphi_{k,i}} \right) n_i \mathrm{d}(\partial\Omega) \equiv \mu \int_{\partial\Omega} \left(W\mathbf{I} - \nabla^T \varphi . \frac{\partial W}{\nabla'} \right) . \delta\mathbf{X} \mathrm{d}(\partial\Omega)$$

of the previous boundary integral traduces the domain variation (this term only involves the variation $\delta\mathbf{X}$), and it highlights the Eshelby energy momentum tensor

$$\Sigma := L\mathbf{I} - \nabla^T \varphi \cdot \frac{\partial L}{\nabla\varphi}.$$

The first integral on the right hand side has an integrant involving the characteristic $Q_k \equiv \phi_k - \xi_j u_{k,j}$, as a factor of the Euler operator applied to the Lagrangian density.

Previous variation can be rewritten in a more compact differential form a ła Cartan $L_X\omega = i_X \mathrm{d}\omega + \mathrm{d}(i_X\omega)$ allowing a more compact writing of Noether's theorem: under the condition of vanishing of the Lie derivative of the one form $\omega = Ldt$, that is $L_X\omega = 0$ (invariance of $\int_\Omega \omega \equiv \int_\Omega W\mathrm{d}X\mathrm{d}t$by the group generated by $\mathbf{X}$) and $i_X \mathrm{d}\omega = 0$ (Euler equations are satisfied as a necessary optimality condition $\delta S = 0$), the following conservation law (since it appears in divergence form) is satisfied

$$\mathrm{Div}\left(W\xi_i + (\phi_k - \varphi_{k,j}\xi_j)\frac{\partial W}{\partial \varphi_{k,i}}\right) = 0.$$

As a specific case, for a purely horizontal variation (the fields are fixed, only parameterization is varied by translation, $\delta\varphi \equiv 0 \to \phi = 0$, one recovers the conservation law of Eshelby stress in the absence of dynamical effects

$$\mathrm{Div}\left(W\delta_{ij} - \varphi_{k,j}\frac{\partial W}{\partial \varphi_{k,i}}\right)\xi_j = 0$$
$$\to \quad D_i\left(W\delta_{ij} - \varphi_{k,j}\frac{\partial W}{\partial \varphi_{k,i}}\right)\delta X_j = 0, \qquad \delta X_j \;\to\; \mathrm{Div}\Sigma = 0$$

where D_i is the total derivative with respect to X_i. This conservation law is associated to the non existence of an absolute origin of space for a homogeneous body; it is a generalization to the continuum theory of the conservation of momentum previously obtained in analytical mechanics. The integration of previous conservation law around the tip of a crack leads by construction to a path-independent integral representing the energy release rate.

The translational invariance with respect to the field itself, more precisely the invariance of the strain energy function under the transformation $\varphi \to \varphi + \mathbf{c}$ with $\mathbf{c} \in \mathbb{R}^3$ an arbitrary constant (φ is here a vector in 3D euclidean space), leads to $D_i\left(\frac{\partial W}{\partial \varphi_{k,i}}\right) = 0, k = 1\ldots 3$. Those equations are nothing else than the Euler-Lagrange equations of the functional built from W. Further conservation laws result from the postulate of invariance with respect to (continuous) rotations and expansions of the material coordinates: for a isotropic material, the invariance of L under the material symmetry group $\mathbf{X} \mapsto Q.\mathbf{X}$, $Q \in SO(3)$, the group of finite rotations acting on the material coordinates, the condition $W(\nabla\varphi.Q) = W(\nabla\varphi)$ leads to the 3 conservation laws

$$D_i\left((X_j\varphi_{p,k} - X_k\varphi_{p,j})\frac{\partial W}{\partial \varphi_{p,i}} + (\delta_{ij}X_k - \delta_{ik}X_j)\,W\right) = 0, \quad p = 1\ldots 3$$

associated to the infinitesimal generator

$$X_k\frac{\partial}{\partial X_j} - X_j\frac{\partial}{\partial X_k}$$

for rotations. The principle of material indifference specialized to rotations acting now in the space of the field, associated to the infinitesimal generator

$$\varphi_k \frac{\partial}{\partial \varphi_j} - \varphi_j \frac{\partial}{\partial \varphi_k}$$

leads to the conservation law of angular momentum

$$D_i \left(\varphi_p \frac{\partial W}{\partial \varphi_{q,i}} - \varphi_q \frac{\partial W}{\partial \varphi_{p,i}} \right) = 0, \quad p, q = 1 \ldots 3.$$

The scaling group $(\mathbf{X}, \varphi) \mapsto \left(\lambda \mathbf{X}, \lambda^{(n-3)/n} \varphi\right)$, $\lambda > 0$, is a variational symmetry with associated infinitesimal generator

$$X_k \frac{\partial}{\partial X_j} + \frac{n-p}{n} \varphi_j \frac{\partial}{\partial \varphi_k}$$

leading to the conservation law

$$D_i \left(\frac{(n-p)}{n} \varphi_p \frac{\partial W}{\partial \varphi_{p,i}} + X_i W - X_j \varphi_{p,j} \frac{\partial W}{\partial \varphi_{p,i}} \right) = 0.$$

Note that an individual scaling of $\mathbf{X}$ or φ leads to a symmetry of the Euler-Lagrange equations, but not to a variational symmetry. Recall that a variational symmetry group of a functional is also a symmetry group of the Euler-Lagrange equations, but the converse is false, as previous situation shows.

The previous conservation laws when written in integral form lead to the well-known path-integral integrals in fracture mechanics, used to characterize the stress singularity close to the crack tip by performing integration far from the singularity.

5 Lagrangian Formulation and Internal Symmetries: Gauge Theories

The expression of the Lagrangian reflects the laws of physics, amongst which symmetries play a prominent role, as previous section has shown. Since a physical system evolves in space-time but also includes internal d.o.f., such symmetries can be classified into two categories [11, 33]:

- External symmetries acting on the space-time coordinates of the scene of events.
- Internal symmetries acting on internal parameters, such as potentials, charges, the wave function; such internal symmetries are also coined gauge symmetries.

Both external and internal symmetries leave invariant the laws of physics, and as such constitute an extension of the geometrical interpretation of general rela-

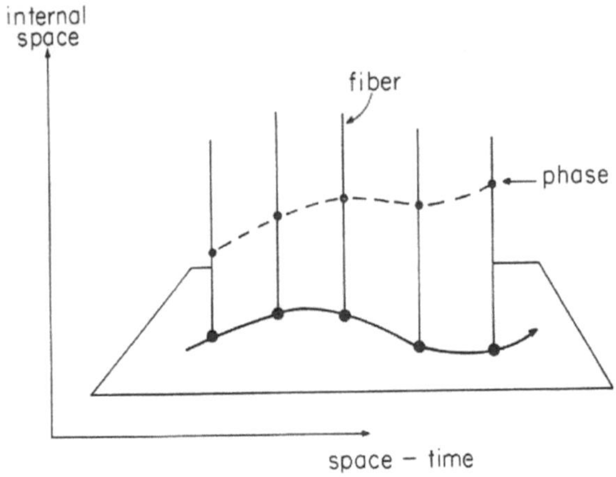

Fig. 5 Parallel transport of a vector field along a path, cf. [33]

tivity to all fundamental interactions, of a quantum nature. Recall that a concept common to special and general relativity is the absence of absolute referential. In special relativity, the class of equivalent referentials is defined by Poincaré group of transformations; all referentials in this class can be located at arbitrary distances from each other, so that the symmetry transformations induced by Poincaré group are global. The picture is different in RG, since the postulated equivalence between the gravitation field and the inertial frame is valid only locally (the orientation of the gravitation field varies from point to point), thus the inertial referential equipped with the metric field g is only locally defined (where g is uniform): RG is thus a local theory. This comparison hints at the key idea of Weyl's gauge theory (1919), which was the first historical attempt to generalize the idea that the effect of the gravitation field can be described by a connection giving the relative orientation of frames in space-time. Weyl then advanced the idea that the norm of any vector associated to a physical quantity should not have an absolute magnitude, but instead would depend on its position in space-time.

The set of all transformations associated to internal symmetries form a Lie group, acting on internal d.o.f.; those internal d.o.f. f belong to the fiber at each point of the base, as pictured on Fig. 5.

In the sequel, the wave function will be represented without the boldface notation pertaining to vector-like quantities, since it may have different representations (vectorial, spinorial, matricial, …). In electrodynamics, the fibers consist of the phases of the wave function, and each fiber can be conceived as a copy of the unit circle in the complex plane. The angle $\theta(x) \in [O, 2\pi]$ specifies a transformation of the phase $\phi(x)$ into

$$\phi(x) := e^{i\theta(x)} |\phi(x)|$$

The set of all transformations parameterized by the angle $\Theta(x)$ constitutes a Lie group; it is further clear that a global change of the phase of $\phi(x)$ by a uniform factor $e^{i\Theta}$, with $\Theta = \text{Cte}$, will not change the field equations. Such an invariance of the field equations (or of the action integra) by an internal symmetry is called *gauge invariance*.

Gauge theories rely on a construction of the Lagrangian in the space of internal d.o.f. being gauge invariant (in addition to being invariant with respect to external symmetries); the gauge transformation will be specified by the global action of a Lie group of transformations, and the gauge invariance shall specify the form taken by the Lagrangian function. The wording global means that the Lie group parameters are initially not depending on space or time. For a given interaction (weak, electromagnetic, ...), the dimension of the Lie group determines the number of interaction particles mediating the interaction; those particles are called gauge bosons (recall that matter particles are defined as fermions). The geometrical interpretation of local gauge invariance is the freedom given to any observer located in the base space to choose the base of the internal space (the fiber above each point). Note the recent applications of gauge invariance in the field of elastodynamics [23]. We shall next show how gauge theories are built, considering as a specific case quantum electrodynamics (QED in short), describing the interactions of fermions with the electromagnetic field.

5.1 Gauge Invariance and Noether's Theorem: Case of QED

We adopt a sequel a system of units in which the speed of light is unity. We focus on the Dirac equation satisfied by a fermion (the spin is $1/2$), the origin of which is recalled: the linearized relativistic energy writes $E = \mathbf{p}.\mathbf{v} - L$, with $L = -m(1 - v^2)^{1/2}$, hence

$$E = p.v + (1 - v^2)m.$$

One then admits the existence of a Hamiltonian having the same form

$$H = \alpha.\mathbf{p} + \beta m$$

with (α, β) matrices. The eigenvalue problem for the linearized Hamiltonian writes

$$H\psi = E\psi \rightarrow (\alpha.\mathbf{p} + \beta m)\psi = E\psi.$$

According to the correspondence principle, one replaces energy and momentum by the differential operators

$$E \rightarrow i\hbar\partial/\partial t, \qquad p_i \rightarrow -i\hbar\partial/\partial x^i.$$

Previous equation then rewrites

$$i\partial_t\psi = (\alpha.\mathbf{p} + \beta m)\psi.$$

This becomes with the matrices $\alpha = (\alpha_\mu)_{\mu=0,1,2,3}$ the equation

$$i(\partial_0 + \alpha_k\partial_k)\psi - m\beta\psi = 0.$$

Multiplying previous equation from the left by β and introducing the quantities $\gamma^0 = \beta,\ \gamma^i = \beta\alpha_i$ results in previously written Dirac equation. The coefficients (α, β) are obtained by applying the energy operator E to the equation

$$(\alpha.p + \beta m)\psi = E\psi$$

expanding the left-hand side and identifying the right hand side with $(p^2 + m^2)\psi$, from which it is found that matrices $(\alpha_\mu)_{\mu=0,1,2,3}$ and β are traceless, and have real eigenvalues equal to ± 1, with an equal number of $+\ 1$ an -1.

In Dirac equation, matrices $(^\mu)$ satisfy the condition

$$\{\gamma^\mu, \gamma^\upsilon\} := \gamma^\mu\gamma^\upsilon + \gamma^\upsilon\gamma^\mu = 2\eta^{\mu\upsilon}I,\quad \gamma^0 = \beta = \begin{pmatrix} I & 0 \\ 0 & -I \end{pmatrix},\quad \gamma^k = \begin{pmatrix} 0 & \sigma_k \\ -\sigma_k & 0 \end{pmatrix}$$

with I the second order identity tensor, and σ_k Pauli matrices. The column four-vector $\psi = \{\psi_1, \psi_2, \psi_3, \psi_4\}^t$ has an hermitic conjugate given by the row vector $\psi^+ = \left(\psi_1^*\ \psi_2^*\ \psi_3^*\ \psi_4^*\right)$.

Let then introduce the complex conjugate

$$\bar{\Psi} = \psi^+\gamma^0 = \left(\psi_1^*\ \psi_2^*\ \psi_3^*\ \psi_4^*\right)\begin{pmatrix} I & 0 \\ 0 & -I \end{pmatrix} = \psi^+ = \left(\psi_1^*\ \psi_2^*\ \psi_3^*\ \psi_4^*\right).$$

The adjoint Dirac equation is obtained as follows: let start by taking the hermitic conjugate of

$$i\partial_t\psi = \alpha(-i\nabla\psi) + m\beta\psi$$

Hence (with $\beta^+ = \beta$)

$$-i\partial_t\psi^+ = i\nabla\psi^+\alpha + m\psi^+\beta.$$

Multiplying previous equation to the right by $\gamma^0 = \beta$, with $\left(\gamma^0\right)^2 = I$ delivers

$$-i\partial_t\bar{\psi} = i\nabla\bar{\psi}\gamma + m\bar{\psi}.$$

Finally, the adjoint Dirac equation writes

$$i\partial_\mu\overline{\psi}^\mu + m\overline{\psi} = 0.$$

The Lagrangian is then constructed based on both the Dirac equation and its adjoint as

$$L = \mathrm{i}\bar{\psi}\gamma^{\mu}\partial_{\mu}\bar{\psi} - m\bar{\psi}\psi = \mathrm{i}\bar{\psi}\bar{\partial}\psi - m\bar{\psi}\psi = -\bar{\psi}\left(\mathrm{i}\bar{\partial} - m\right)\psi = -\bar{\psi}\left(\mathrm{i}\bar{\partial} + m\right)\psi$$

introducing therein the new derivative $\bar{\partial} := \gamma^{\mu}\partial_{\mu}$, and the left and right arrows indicating on which side the operator is acting. It is easy to verify that the stationarity of L with respect to $\bar{\psi}$ and ψ leads respectively to Dirac equation and to its adjoint. The probability density is defined as the scalar

$$\dot{j}_k(x) = \bar{\psi}(x)\gamma^k\psi(x)$$

wherein the last relation follows from the identity $(\gamma^0)^2 = I$. Similarly, the density of current probability is elaborated as

$$j_k(x) = \overline{\psi}(x)^k\psi(x).$$

Multiplying Dirac equation from the left by $\bar{\psi}$ and its adjoint from the right by ψ, and summing up both relations rescaled by the charge of electron, we arrive at the conservation law of the electric current

$$\partial_{\mu}(ej^{\mu}(x)) = 0$$

The density of electric charge therein is the temporal component $e\psi^{+}(x)\psi(x)$.

The conservation of the electric current can be deduced from a global gauge invariance of the Lagrangian, following the transformation

$$\psi_i(x) \rightarrow \psi_i' - \exp\left(-\mathrm{i}\Lambda T_{ij}\right)\psi_j(x)$$

with Λ a constant parameter, and T_{ij} a constant matrix. A Taylor expansion of previous transformation at first order gives the variation

$$\psi_i'(x) - \psi_i(x) = \delta\psi_i(x) = -\mathrm{i}\Lambda T_{ij}\psi_j(x).$$

This leads to the variation of the Lagrangian

$$\delta L = \frac{\partial L}{\delta\psi_i}\delta\psi_i + \frac{\partial L}{\partial(\partial_{\mu}\psi_i)}\delta(\partial_{\mu}\psi_i) = \partial_{\mu}(\frac{\partial L}{\partial(\partial_{\mu}\psi_i)}\delta\psi_i) - (\partial_{\mu}(\frac{\partial L}{\partial(\partial_{\mu}\psi_i)}) - \frac{\partial L}{\partial(\partial_{\mu}\psi_i)})\delta\psi_i.$$

The last contribution vanishes for the optimal trajectory (when Dirac equation is satisfied), thus we get the conservation law

$$\delta L = 0 \Rightarrow \partial_\mu(-\mathrm{i}\frac{\partial L}{\partial(\partial_\mu \psi_i)} T_{ij}\psi_j) = \partial_\mu j^\mu = 0$$

of the electric 4-current, illustrating Noether's theorem.

The Dirac Lagrangian is invariant under the global gauge transformation (when Λ is uniform); since previous result is valid for any constant matrix of coefficients T_{ij}, the specific choice $T_{ij} = \delta_{ij}$ leads to the expression of the current $j^\mu(x) = \overline{\psi}(x)^\mu \psi(x)$ obtained previously.

A stronger condition of local gauge invariance is next required, letting the group parameter depend upon the coordinates, viz $\Lambda = \Lambda(x)$, selecting $T_{ij} = q\delta_{ij}$, with q an electric charge, hence resulting in the transformations

$$\psi_i(x) \rightarrow \psi_i^1(x) = \exp(-\mathrm{i}q\Lambda(x)T_{ij})\psi_j(x)$$

A straightforward calculation then leads to the modified Lagrangian

$$L' = L + q\overline{\psi}\gamma^\mu\psi\partial_\mu\Lambda = L + q^\mu\partial_\mu\Lambda$$

One next introduces the covariant derivative

$$D_\mu = \partial_\mu + iqA_\mu$$

with A_μ the gauge field responsible for the interaction between fermions and the electromagnetic field. The Lagrangian is then invariant under the previous Lie group transformation

$$L_F = L - qj^\mu A_\mu \equiv \bar{\Psi}\left(\mathrm{i}\overline{D} - m\right)\psi \rightarrow L_F{}' = L_F$$

involving the new covariant derivative

$$\overline{D} := \gamma^\mu D_\mu$$

The Lagrangian of fermions L_F is accordingly invariant under a local gauge transformations, provided the partial derivative ∂_μ is replaced by the covariant derivative D_μ; the field A_μ is coined a *compensating field* or a *gauge field*, which allows localizing the initially global gauge transformation. The Lagrangian of the developed theory, called quantum electrodynamics (it is a relativistic theory) then writes as the sum of the Lagrangian of matter (fermions) and of the free electromagnetic field

$$L_{QED} = L_F + L_\mathrm{e} = L_F - \frac{1}{4}F_{\mu\upsilon}F^{\mu\upsilon}.$$

Note that since the Lagrangian L_F does not include a kinetic term quadratic in the field A_μ, we deduce that the photon—the vehicle of electromagnetic interactions—is massless.

5.2 Generalization to all Fundamental Interactions

Fundamental interactions are described by the action of Lie groups with p continuous parameters $(\Theta_a)_{a=1\ldots p}$, thereby defining transformations of the form $V = e^{i\Theta_a T_a}$, with the Hermitian matrices T_a satisfying the following commutation relations of the Lie algebra of the group

$$[T_a, T_b] = iC_{abc}T_c$$

for certain group structures set of constants C_{abc}. The matrices $(T_a)_{a=1\ldots p}$ constitute a N dimensional representation of the Lie algebra of the group.

The column vector $\phi = (\phi_1, \phi_2, \ldots, \phi_N)^t$ with N components, transforms according to

$$\phi(x) \to \phi'(x) = V[\phi(x)] \to \phi_r{}'(x) = V_{rs}\phi_s(x).$$

Under an infinitesimal transformation, one has the transformation relation of the field

$$\phi(x) \to \phi_{r'}(x) = \phi_r(x) + i\Theta_a(T_a)_{rs}\phi_s(x)$$

inducing the following transformation of the Lagrangian density

$$\delta L = \partial_\mu\left(\frac{\partial L}{\partial(\partial_\mu\psi_i)}\delta\phi_s\right) = -\Theta_a\partial_\mu\left(\frac{\partial L}{\partial(\partial_\mu\psi_i)}(T_a)_{rs}\phi_s\right) \equiv -\Theta_a\partial_\mu j_a^\mu(x)$$

with the quadricurrent $j_a^\mu(x)$ being conserved

$$j_a^\mu(x) = -i\frac{\partial L}{\partial(\partial_\mu\psi_i)}(T_a)_{rs}(|)s \to \partial_\mu j_a^\mu = 0.$$

Accordingly, the invariance of the Lagrangian imposes the conservation of the current and of the charges, quantities

$$Q_a := \int j_a^\mu(\mathbf{x}, t)\,d^3x.$$

The charges can be rewritten in terms of the conjugated field momentum

$$\Pi_r := \frac{\partial L}{\partial\phi_r}.$$

The existing interactions in field theories can be classified according to the structure of their associated Lie groups: U(1), the group of unimodular matrices (complex numbers of modulus unity) for quantum electrodynamics, with one generator; the interaction particle (gauge boson) is the photon, the existence of which is deduced from the localization of the gauge invariance;

SU(2), the special unitary group of order 2, group of 2×2 matrices with unit determinant, describing weak interactions, having 3 generators $W^+,\ W^-,\ Z^0$;

SU(3), the special unitary group of order 3, for quantum chromodynamics (QED in short), describing interactions between quarks, and having 8 generators called gluons.

QED is a quantum relativistic theory, Abelian (the group generators commute with each other, since there is only one generator); this is not the case of weak interactions (the Lie group is non Abelian), the Lie algebra of which is characterized by Pauli matrices τ satisfying the commutation relations

$$[\tau^A, \tau^B] = 2\mathrm{i}C^{ABC}\tau_c$$

with A,$B,\ C \in \{1, 2, 3\},\ \tau^1 = \begin{pmatrix} 0 & 1 \\ -1 & 0 \end{pmatrix},\ \tau^2 = \begin{pmatrix} 0 & \mathrm{i} \\ -\mathrm{i} & 0 \end{pmatrix},\ \tau^3 = \begin{pmatrix} 1 & 0 \\ 0 & -1 \end{pmatrix}$, and C^{ABC} the group structure constants. The gauge transformation acts on a doublet of scalar fields $\varphi =$ according to $\varphi \to \varphi^1 = \exp(-\frac{i}{2}\tau_A\alpha^A)\varphi$, with α^A the group parameters.

The gauge transformation is non Abelian: due to the approximation

$$\exp(-\frac{i}{2}\tau_A\alpha^A) \cong 1 - \frac{\mathrm{i}}{2}\tau_A\alpha^A$$

the product of two successive transformations,

$$\varphi_{\alpha\beta} = \exp(-\frac{\mathrm{i}}{2}\tau_A\alpha^A)\exp(-\frac{\mathrm{i}}{2}\tau_B\varphi^B)\varphi$$

leads to

$$\varphi_{\alpha\beta} - \varphi_{\beta\alpha} = -\frac{1}{4}[\tau_A, \tau_B]\alpha^A\alpha^B\varphi = -\mathrm{i}\alpha^A\alpha^B C_{ABC}\frac{\tau^C}{2}\varphi.$$

When the group parameters depend on coordinates, viz $\alpha^A = \alpha^A(x)$, the gauge transformation is local. Let then introduce the three gauge fields A_μ^N for $N = 1, 2, 3$, and define $A_\mu := \frac{1}{2}\tau_N A_\mu^N$. The Lagrangian of the field is then constructed as

$$L_M = (D_\mu\varphi^+)(D_\mu\varphi) - V\varphi^+\varphi$$

with $D_\mu = \partial_\mu + \mathrm{i}gA_\mu$ the covariant derivative, such that the covariant derivative transforms as $D_\mu^\dagger\varphi^1 = \mathrm{U}(D_\mu\varphi)$, with U a transformation of the group and g a constant, called the gyromagnetic factor. The gauge field tensor is then defined as

$$F_{\mu\upsilon} = \frac{1}{2}\tau_N F_{\mu\upsilon}^N;\ F_{\mu\upsilon}^N = \partial_\mu A_\upsilon^N - \partial_\upsilon A_\mu^N - g\epsilon^{NPQ}A_\mu^P A_\upsilon^Q$$

and further rewritten as

$$F_{\mu v} = \partial_\mu A_\mu - \partial_v A_v + ig[A_\mu, A_v] = -\frac{i}{g}[D_\mu, D_v]$$

The Lagrangian of the field is then written in a form similar to electromagnetism, that is

$$L_{Field} = -\frac{1}{4} F_{\mu v} F^{\mu v}$$

with $F_{\mu v}$ satisfying the conservation law $D_\mu F^{\mu v} = 0$.

The gauge fields appearing due to the requirement of localization of the (global) gauge invariance correspond to forces and gauge bosons, the particles mediating the interaction. The phase change of the wave function is treated as gauge invariance; a similar procedure holds for other interactions: the change of color in QCD (quantum chromodynamics), and the change of a charged lepton in its neutrino in weak interactions. The nature of the interactions between different fields is uniquely determined by postulating the local gauge invariance. This amounts to the freedom of any observer to choose at any point of the configuration space (the base in the introduced fiber bundle) the basis of the internal space containing the internal d.o.f. This can be translated in mathematical terms as the decomposition of the Hilbert space for the system H into a direct product of the type $H = H_{\text{ext}} \otimes H_{\text{int}}$, with H_{ext} the set of square integrable functions and H_{int} a finite dimensional vector space of the internal d.o.f's. The wave function takes locally the form

$$\psi(x) = \psi_k(x) X_k(x), \qquad \psi_k(x) \in H_{\text{ext}}, \qquad \psi_k(x) \in H_{\text{int}}$$

Two observers located at different points X_1 and X_2 must be able to compare their observations: one way to do this is to make a parallel transport from X_1 to X_2, such that

$$D_\mu \chi = 0$$

The covariant derivative D_μ is built as $D_\mu = \partial_\mu - \mathrm{i} A_\mu^A T_A$, involving the generators of the Lie algebra, quantites T_A, and the constant A_μ^A.

The articulation of symmetry groups, associated conservation laws and existence of non observable quantities is summarized in Table 2, including both classical and quantum mechanics as well as statistical mechanics. The dimension of the Lie group determines the number of interaction particles.

6 Scale Invariance in Physics

Spatial and temporal fluctuations appear on a wide spectrum of scales and scale invariance and concern numerous areas, not only the physics of critical points in magnetic systems and simple fluids; let mention soft matter (polymers, liquid crystals),

Table 2 Noether's theorem in classical and quantum mechanics

Non observable	Symmetry	Conservation law
Absolute spatial position	Space translation	Linear momentum
Absolute time	Time translation	Energy
Absolute spatial direction	Rotation	Angular momentum
Absolute velocity	Lorentz Transformation	Generators of Lorentz group
Difference between identical particles	Permutation of identical particles	Fermi-Dirac or Bose-Einstein statistics
Absolute right or left	Inversion $X \to -X$	Parity
Absolute sign of the charge	Particles transformed into their antiparticles	Charge conjugation
Absolute phase of a charge matter field	Change of phase	Electrical charge, generators in U(1)
Difference between coherent mixtures of colored quarks	Change of color	Color generator, belong to group SU (3)
Difference between coherent mixtures of charged leptons and neutrinos	Transformation of a lepton in its neutrino	Weak isospin generators, belong to group SU (2)

systems outside equilibrium (growth, aggregation, fracture, coalescence), critical self-organized systems, turbulence, fractals, complex dynamical systems (economy, biology, astrophysics). The development of the concept of renormalization group in quantum field theory and its applications to statistical physics from the early 1970 has profoundly modified modern physics [34]. It allows to understand that quantum theories depend upon the energy scale of the phenomena they try to describe (or the spatial and temporal resolution), and the concept of renormalization emerged as a criterion to judge the consistency of the proposed quantum theories.

6.1 Renormalization Group

We briefly summarize the formulation of the renormalization group conceived initially by K. Wilson [34, 35]. We consider in order to simplify the presentation a system with a continuous order parameter (a field) ϕ and linear scale transformations of the fields. The model is defined on a square lattice Λ in dimension D, the d.o.f. are the classical fields ϕ_i attached to the sites i of the network, with the ϕ_i being continuous real variables with a measure $D[\phi] = \prod_i \mathrm{d}\phi_i$. The possible constraints on those variables (for instance $|\phi_i| = 1$ for Ising model of spins) are incorporated into the microscopic Hamiltonian H. The dynamics of the system is given by a local microscopic Hamiltonian $H[\phi]$ -having the sense of energy –expressed as a sum of local observables O_α of the ϕ_i and their derivatives evaluated at site i (the gradients

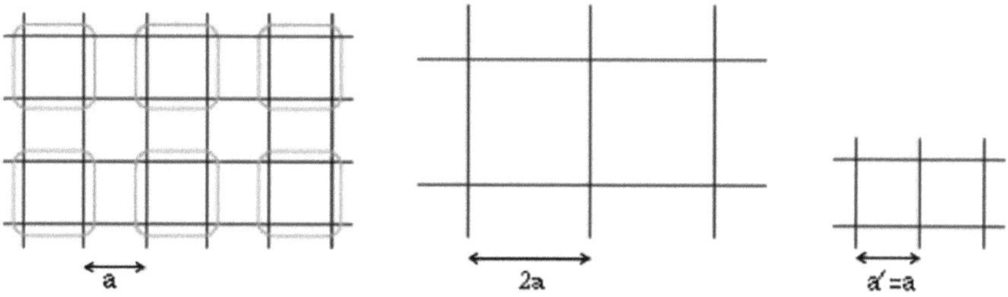

Fig. 6 The three steps of an elementary transformation of the renormalization group, cf. [34]

are replaced by finite differences)

$$H[\phi] = \int d^D x \sum_{\alpha} \kappa^{\alpha} O_{\alpha} = \alpha^D \sum_{i \in \Lambda} \kappa^{\alpha} O_{\alpha} [\phi_i, \nabla \phi_i, ...].$$

The coefficients κ^{α} are the coupling constants of the model, including parameters such as temperature. The partition function of the network model writes

$$Z = \sum e^{-H[\phi]} = \int \prod_{i \in \Lambda} d\phi_i e^{-H[\phi]}.$$

We assume to be sufficiently close to a critical point so that the correlation length ξ is large

$$\xi \gg a.$$

The transformations of the renormalization group then allow to analyze the system at intermediate scales

$$a \ll 1 \ll \xi.$$

This is done according to the three following steps pictured in Fig. 6:

1. Decimation and bloc variables: in order to analyze the average dynamics of the spins in a region of finite size, the initial network is divided into blocks b of size B (square of B^D sites). The average field on block b is defined by the average of the spins on the same block

$$\phi_b := \frac{1}{B^D} \sum_{i \in b} \phi_i$$

and B is the scale factor in this decimation procedure.

2. Effective Hamiltonian for the block variables: one performs an averaging over the fluctuations of spins inside the blocks in order to arrive at an effective Hamiltonian $H_{\text{eff}}[\tilde{\phi}_b]$ for the block variables $\tilde{\phi}_b$ attached to the size of network $\tilde{\Lambda}$ with elementary mesh size $\tilde{a} = Ba$

$$\sum \mathrm{e}^{-H_{\mathrm{eff}}\left[\tilde{\phi}_b\right]} = \int \prod_{i \in \Lambda} \mathrm{d}\phi_i \delta\left(\tilde{\phi}_b - \frac{1}{B^D}\sum_{i \in b}\phi_i\right)\sum e^{-H\left[\tilde{\phi}_b\right]}$$

and the partition function is defined as

$$Z = \int \prod_{b \in \Lambda} \mathrm{d}\tilde{\phi}_b \mathrm{e}^{-H_{\mathrm{eff}}\left[\tilde{\phi}_b\right]}.$$

3. The third step consists in comparing previous effective Hamiltonian describing the dynamics of blocks at scale

$$\tilde{a} = Ba$$

with the initial Hamiltonian (that describes the dynamics at the scale a of the network), by a scale change in order to obtain renormalized quantities. One first rescales distances

$$x \to x' = x/B$$

so that the new network $\tilde{\Lambda}$ of mesh size $\tilde{a} = Ba$ becomes again the initial network Λ with mesh size a:

$$\tilde{\Lambda} \to \Lambda' = \Lambda, \qquad b \in \tilde{\Lambda} \to i' \in \Lambda' = \Lambda.$$

One then performs an adapted scale change on the field of the form

$$\phi_b = -B^{-\Delta}\phi_i'$$

with a factor Δ adjusted so that the renormalized Hamiltonian (after this scaling) for the ϕ_i', scalar $H\left[\phi'_i\right]$, defined by the equality

$$\mathrm{d}\tilde{\phi}_b e^{-H_{\mathrm{eff}}\left[\tilde{\phi}_b\right]} = \mathrm{d}\phi_i' \mathrm{e}^{-H'\left[\tilde{\phi}_i'\right]}$$

resembles as much as possible the initial Hamiltonian $H[\phi_i]$. Since the physics did not change during renormalization, relations exist between the observables of the initial system and the system described by the renormalized Hamiltonian.

The partition functions of both systems are identical

$$Z = \int \mathrm{d}\phi e^{-H[\phi]} = \int \mathrm{d}\phi' \mathrm{e}^{-H'[\phi]'}.$$

The correlation functions of the initial and renormalized systems are related by simple relations

$$\langle \phi(x) \rangle_H = \langle \phi_i \rangle_H = \left\langle \tilde{\phi}_b \right\rangle_{H_{eff}} = B^{-\Delta} \langle \phi(x/B) \rangle_{H'}.$$

For the two-points functions, as long as

$$|i_1 - i_2| \gg B, \qquad i_1 \in b_1,\ i_2 \in b_2$$

it holds that

$$\langle \phi(x_1)\, \phi(x_2) \rangle_H = B^{-2\Delta} \langle \phi(x_1/B)\, \phi(x_2/B) \rangle_{H'}.$$

Similar equalities hold for the K points functions.

The long distance properties of a system (typically the correlation length) are determined through the iteration of the renormalization procedure: in order to reach a length l starting from the microscopic length of the lattice a, one has to perform a scale transformation by a factor $S = 1/a$, thus it has to be iterated $N = \log S/\log B$ time to reach a renormalized Hamiltonian H_s. The properties of the initial system at large scales $x \sim l$ are then equivalent to the short distance properties

$$x' \sim a = 1/S$$

of the renormalized system, which is a priori more simple to analyze.

The iterated transformations give a flow in the space of Hamiltonians, of the form

$$H \to H' \to H'' \to H''' \to \ \ldots \to H^{(N-1)} \to H^{(N)} = H_S.$$

from the initial Hamiltonian H to the renormalized one H_s, versus the scaling factor $S = B^N$. One performs in parallel a series of scaling transformations of the fields

$$\phi(x) \to B^{-\Delta}\phi^{\dagger}(x^{\text{l}}) \to B^{-\Delta-\Delta^{\dagger}}\phi^{\dagger\dagger}(x^{\text{ll}}) \to \ \ldots \to B^{-(\Delta+\Delta^{\dagger}+\ldots+\Delta^{(N-1)})}\phi^{(N)}$$

leading to a renormalized field ϕ_s linked to the initial field ϕ by a multiplicative renormalization factor $Z(S)$

$$\phi_s(x_s) = \phi^{(N)}(x_s) = Z_s\phi(x), \quad x_s = x/S, \quad Z(S) = Z_s = B^{\Delta+\Delta^{\dagger}+\ldots+\Delta^{(N-1)}}.$$

Let then note R_s the transformation of the Hamiltonian corresponding to the scale factor S

$$H \overset{R_S}{\to} H_S.$$

These transformations form a multiplicative semi-group

$$R_{S_2} \circ R_{S_1} = R_{S_2} R_{S_1}$$

which is thus logarithmic in the scale factor S (playing the role of a discrete time in the scaling transformations), thus

$$s = \log S = N\Delta s$$

with time increment $\Delta s = \log B$. The set of all such transformations in the space of Hamiltonians forms the *renormalization group*. The action of the group on observable is the same as in the decimation step: denoting $Z_H(L)$ the partition function for the initial system of size L, one has

$$Z_H\ (SL) = Z_{H_S}(L).$$

For the correlation functions, it holds that

$$\langle \phi(Sx_1)\ldots\phi(Sx_K)\rangle_H = Z(S)^{-K}\langle \phi_s(x_1)\ldots\phi_s(x_K)\rangle_{H_S}.$$

This results in the important relation between the correlation lengths ξ of the initial and renormalized systems

$$\xi_H = S\xi_{H_s}.$$

Let consider to be more specific Hamiltonian of the form

$$H[\phi] = \sum_{i\in\Lambda}\sum_{\alpha} K^{\alpha} O_{\alpha}[\phi_i]$$

with O_α local operators in the fields, of the general form

$$O_\alpha[\phi_i] = \nabla^{\mathrm{n}_1}\phi_i\nabla^{\mathrm{n}_1}\phi_i\ldots\nabla^{\mathrm{n}_k}\phi_i$$

and with ∇ the finite difference operator on the network

$$\nabla_\mu\phi_i = \phi_{i+\mathrm{e}} - \phi_i.$$

Examples of such operators include the following terms

$$O_\alpha[\phi_i] = 1, \phi, \phi^2, \phi^4, \phi^6, (\nabla\phi)^2, (\nabla\phi)^4, (\Delta\phi)^2.$$

The κ^α are the coupling constants, also called external fields associated to the operators; they represent coordinates in the space of Hamiltonians. The renormalization group then acts in the space of couplings according to

$$\kappa \xrightarrow{R_S} \kappa\,(S) = K\,(\kappa, s = \log S)\,.$$

Such transformations are continuous in the couplings κ; since they form an additive semi-group in the variable $s = \log S$, it holds that

$$K(K(\kappa, s_1), s_2) = K(\kappa, s_1 + s_2).$$

In practical situations, one has to reduce the space of Hamiltonians to a subspace involving a finite number of couplings (coupling parameters). When the number of iterations become very large, $N \gg 1$, one may consider the scale parameter S as a continuous parameter, this is justified in the vicinity of fixed points of the group, where the speed of the flow is small. In this case, iterations of the renormalization group take the form of flow equations generated by a vector in the space of couplings. Indeed, the transformations of the renormalization group write in differential form as

$$S\frac{\partial}{\partial S}K^{\alpha}(S) = W^{\alpha}(\kappa(S))$$

with Wilson functions given by

$$W^{\alpha}(\kappa) = \frac{\kappa^{\alpha} - \kappa^{\alpha}}{\log B} = \frac{K^{\alpha}(\kappa, \Delta s) - \kappa}{\Delta s} = \lim_{s \to 1} \frac{\kappa_S^{\alpha} - \kappa^{\alpha}}{\log S}.$$

The flow equations then take the form

$$\frac{\partial}{\partial s}\kappa(S) = S\frac{\partial}{\partial S}\kappa(S) = W(\kappa(S)).$$

Wilson functions W^{α} are the components of a vector field W in the space of couplings, which generates the flow of the renormalization group. Similarly, the renormalization factor $Z(S)$ of the fields present in the correlation functions obeys the equation

$$S\frac{\partial}{\partial S}\log[Z(S)] = \Delta(\kappa(S)).$$

The function $\Delta(\kappa)$ is called the scale dimension of the field ϕ, denoted Δ_{ϕ}; it is in general dependent on the couplings κ.

The fact that the transformations of the renormalization group derive from a flow implies that Wilson functions are transformed as contravariant vectors under a change of the coordinate system $\kappa \to \tilde{\kappa}$ in the space of Hamiltonians, Fig. 7.

$$\kappa^{\alpha} \to \tilde{\kappa}^{\alpha}, \qquad \tilde{W}^{\alpha}(\tilde{\kappa}) = \frac{\partial \tilde{\kappa}^{\alpha}}{\partial \tilde{\kappa}^{\beta}} W^{\beta}(\kappa).$$

The dimension of the field transforms like a scalar

$$\Delta_{\varphi} \to \tilde{\Delta}_{\varphi} = \Delta_{\varphi}.$$

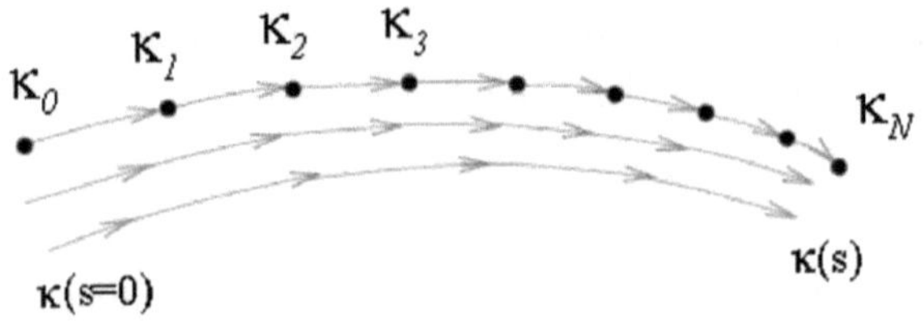

Fig. 7 Continuous flow in the space of couplings, cf. [34]

The field itself ϕ is an operator associated to the coupling h (external field) added to the microscopic Hamiltonian

$$H[\phi] \to H[\phi] - \delta H[\phi], \qquad \delta H[\phi] = h \int \mathrm{d}^D x \phi(x) = a^D \sum_{i \in \Lambda} h\phi_i.$$

In an elementary step of the renormalization procedure, such a term transforms as

$$\sum_{i \in \Lambda} h\phi_i = \sum_{b \in \Lambda} hB^D \tilde{\phi}_b = \sum_{i' \in \Lambda} hB^{D-\Delta}\phi_i.$$

Hence, the coupling is renormalized as

$$h \to h' = B^{D-\Delta}h.$$

Iterating to obtain a rescaling by a scale factor S, the coupling h thus transforms as

$$h \to h(S) = S^D Z(S)^{-1} h$$

and the Wilson function associated to h is simply

$$W^\alpha(\kappa) = S\frac{\partial}{\partial S}h(S) = (D - \Delta(\kappa))h.$$

The dimension of the field Δ_ϕ is thus simply related to the flow of the coupling constant h (the external field) by

$$W^h = (D - \Delta_\phi)h.$$

6.2 Example: Ginsburg-Landau-Wilson Theory

Landau approach (1937) was developed in the context of phase transitions between different states of matter (magnetic to paramagnetic, liquid-gas, supraconductor to

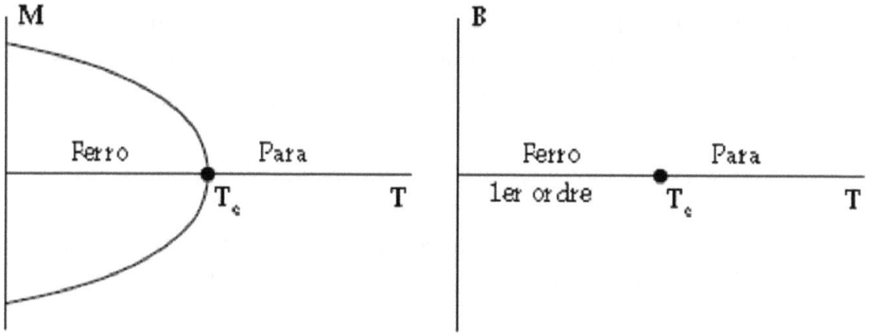

Fig. 8 Phase diagram temperature-magnetization (*left*) and temperature-field (*right*), cf. [34]

conductor, metal-insulating, ...). The states on either side of the transition line are characterized by their symmetry. For instance, water is a more symmetrical phase in comparison to ice, since it has rotational and translation symmetry with arbitrary parameters, whereas ice is a network with a six fold symmetry axis. We exemplify the description of critical phenomena by the classical example of the ferromagnetic-paramagnetic transition in magnetic materials close to the Curie point. There exists in these materials a spontaneous global magnetization at low temperature under the action of an external magnetic field, Fig. 8.

The critical point separates two thermodynamic phases having different symmetries: above the critical temperature ($T > T_c$), the Curie temperature, the material is invariant under a sign change of the magnetization $M \to -M$ (paramagnetic phase); below T_c, this symmetry is spontaneously broken (ferromagnetic phase).

Magnetization is here the order parameter of the transition; it is defined from the partial derivative of the free energy F

$$-M = \frac{1}{V}\frac{\partial F}{\partial B}$$

with V the system volume, and B the applied magnetic field. In order to describe the state of a system (e.g. magnetization), one introduces an order parameter M considered as a field, for instance a vector depending on position, so that $M \equiv M(x)$. In the magnetic state, one component of M is non zero, whereas all components vanish in the paramagnetic state; let denote M_z this component. Restricting to continuous transitions (second order phase transitions), the order parameter is continuous during the transition; below a critical temperature T_c, M_z is non zero, and it becomes zero above T_c

$$\lim_{T \to T_C} M_z(T) = 0.$$

An important feature of a continuous transition is the existence of important thermodynamic fluctuations occurring on large spatial and temporal scales, which entails singularities in the thermodynamic quantities (they verify scaling laws versus the difference $|T - T_c|$). The correlations are measured by the spatial correlations of the variation of the local magnetization m around the average magnetization $M := \frac{1}{V}\langle \int \mathrm{d}x m(x) \rangle = \langle m \rangle$, quantified by the two points correlation function

$$G(x, \mathrm{y}) = \langle \Delta m(x) \Delta m(\mathrm{y}) \rangle = \langle m(x) m(\mathrm{y}) \} - \langle m(x) \} \langle m(\mathrm{y}) \rangle.$$

This function measures the correlations between the fluctuations of m at points x and y. For large distances $G(x, \mathrm{y})$ decreases exponentially with distance

$$G(x, \mathrm{y}) \propto \exp(-|x - \mathrm{y}|/\xi), \qquad |x - \mathrm{y}| \to \infty$$

with ξ the correlation length for the fluctuation of magnetization. Far from the critical point, ξ is small, but it becomes very large close to the critical point where it diverges; this divergence reflects the presence of critical fluctuations: the system becomes scale invariant at the critical point. This divergence is translated into the relation

$$\xi \propto |T - T_c|^{-\nu}$$

with ν the critical exponent of the correlation length. The mean field theory predicts the value $\nu = 1/2$.

Thermodynamic fluctuations are present at all spatial and time scales close to the critical point; the correlation functions take universal forms depending only upon the ratio of distances to correlation lengths; for instance, the correlation length behaves as

$$G(x, \mathrm{y}, ; T) \propto H(|x - \mathrm{y}|/\xi(T))$$

with H a universal scaling function (independent of the considered physical system).

Landau made the assumption that the thermodynamic potentials are analytic functions of the order parameter; if temperature and pressure are the controlled parameters, M is determined from the condition of vanishing functional derivative

$$\frac{\delta G}{\delta M\,(x)} = 0$$

since enthalpy $G = G[M(x)]$ is a functional of the order parameter.

Landau assumes for small values of M a power expansion of G. In a first scheme, one may assume such an expansion involving only even powers

$$G = G_0 + a(T, p)M^2 + u(T, p)M^4$$

so that enthalpy is invariant under a change of sign of M. The state minimizing enthalpy is thus degenerated and the symmetry of funtion $G = G(M)$ reflects the symmetry of the system. If now $G = G(M)$ includes an odd power of M, viz

$$G = G_0 + a(T, p)M^2 + b(T, p)M^3 + u(T, p)M^4$$

so that $G = G(M)$ no more has the symmetry $M \to -M$. This second situation is characteristic of first order transitions for which the order parameter experiences a jump during the transition. Considering now a second order (continuous) transition, $G(M)$ has two minima according to the sign of coefficient $a(T, p)$

- If $a > 0$, then $M = 0$ is the only minimum and the state is paramagnetic.
- If $a < 0$, $G(M)$ has two minima separated by a maximum at $M = 0$, given by $M_\pm = \pm(-\frac{a}{2u})^{1/2}$, which has to vanish at $T = T_c$, thus one may select

$$a(T, p) = a_2(T - T_c), \qquad a_2 > 0$$

leading in turn to

$$M(T) = 0 \quad \text{if} \quad T > T_c \quad and \quad M(T) = \left(\frac{a_2}{2u}\right)^{1/2}(T_c - T)^{1/2} \quad if \quad T < T_c.$$

This Landau type energy can be generalized to account for spatial variations of the order parameter, thus defining enthalpy through a spatial density depending upon gradient terms of M, as

$$G = \int_\Omega \mathrm{d}x\tilde{G} = \int_\Omega \mathrm{d}x\left[\frac{c}{2}(\nabla M)^2 + a_2(T - T_c)^2 M(x)^2 + uM(x)^4\right].$$

Coming back to an Hamiltonian depending on the field labeled ϕ in initial developments of this section,

$$H[\phi] = \int_\Omega \mathrm{d}^D x\left[\frac{1}{2}(\nabla\phi)^2 + \frac{r_0}{2}\phi^2 + \frac{u_0}{4}\phi^4 - h_0\phi\right]$$

with r_0 playing the role of temperature, the two first steps (decimation and determination of the effective Hamiltonian) are trivial, since the field does not fluctuate in a block, thus

$$\tilde{\phi}(x) = \phi(x).$$

Next, the change of scale

$$x = Bx', \qquad \tilde{\phi} = B^{-\Delta}\phi$$

gives the renormalized Hamitonian

$$H'[\phi] = \int_\Omega \mathrm{d}^D x^1[\frac{B^{D-2-2\Delta}}{2}(\nabla\phi')^2 + \frac{B^{D-2\Delta}r_0}{2}\phi' + \frac{B^{D-4\Delta}u_0}{4}\phi' - B^{D-\Delta}h_0\phi'].$$

In order to find a Hamiltonian of the same form as the initial Hamiltonian, the field dimension Δ must accordingly be

$$\Delta = \frac{D-2}{2}.$$

The flow of the renormalization group in the space of couplings writes

$$r_0 \to r_s = S^2 r_0, \qquad u_0 \to u_s = S^{4-D} u_0, \qquad h_0 \to h_s = S^{\frac{D+2}{2}} r_0$$

corresponding to the dimensions of the field and of the couplings

$$[\phi] = \frac{D-2}{2}, \qquad [r] = 2, \qquad [u] = 4 - D, \qquad [h] = \frac{D+2}{2}.$$

6.3 Symmetry Breaking

Symmetry breaking can be categorized into either explicit symmetry breaking, whereby terms included into the Lagrangian do not respect the symmetry, or spontaneous symmetry breaking: it is a mode of symmetry breaking in physical systems, where the underlying laws are invariant under a symmetry transformation, but the system as a whole changes under such transformations. A system in an initially symmetrical state ends up in an asymmetrical state. It thus describes systems where the equations of motion or the Lagrangian obey certain symmetries, but the lowest energy solutions do not exhibit that symmetry. Most phases of matter can be understood through the prism of spontaneous symmetry- breaking. For example, crystals are periodic arrays of atoms that are not invariant under all translations (only under a small sub-set of translations by a lattice vector). Magnets have north and south poles that are oriented in a specific direction, breaking rotational symmetry. Other examples of symmetry-breaking phases of matter are nematic phases of liquid crystals, charge- and spin-density waves, superfluids. The strong, weak, and electromagnetic forces can all be understood as arising from gauge symmetries. The Higgs mechanism, the spontaneous-symmetry breaking of gauge symmetries, is an important component in understanding the superconductivity of metals and the origin of particle masses in the standard model of particle-physics. Spontaneous symmetry breaking can be best illustrated relying on Goldstone model: we consider a scalar massless field Φ (a complex field) and the Lagrangian

$$L = \partial_\mu \Phi^+ \partial^\mu \Phi - V(\Phi^+ \Phi)$$

which is clearly invariant under the global gauge transformation

$$\Phi \to \Phi' = \exp(\mathrm{i}\Lambda)\Phi.$$

We further choose the following form of the potential

$$\rho V(\Phi^+\Phi) \equiv V(\rho) = \mu^2\rho + h\rho^2, \qquad \rho := \Phi^+\Phi, \qquad h > 0.$$

For $\mu^2 > 0,\ V(p)$ is a symmetrical potential having a minimum at the origin. For $\mu^2 < 0,\ V(\rho)$ has a minimum for a non zero value of ρ given by $\rho_m = \dfrac{-\mu^2}{2h} > 0$; this corresponds to a circle in the complex plane. The Euler equation writes

$$(\mu^2 + 2h\Phi^+\Phi)\Phi = 0.$$

The value $\Phi = 0$ is a trivial solution corresponding to an unstable point for $\mu^2 < 0$, each value of Φ such that $\Phi^+\Phi = \dfrac{-\mu^2}{2h} > 0 \rightarrow \Phi = (\dfrac{-\mu^2}{2h})^{1/2}\mathrm{e}^{\mathrm{i}\beta}$ gives a minimum of the potential. Last relation implies that the Hamiltonian density

$$H = \mu^2\rho + h\rho^2 = \mu^2|\Phi|^2 + h|\Phi|^4$$

has a minimum for $\dfrac{\partial H}{\partial \rho} = 0 \Rightarrow \rho = (\dfrac{-\mu^2}{2h})(1/2) \equiv \rho_{min}$.

Previous solution thus corresponds to a minimum of H, given by

$$\Phi_{min} = \langle 0|\Phi|0\rangle = (\frac{-\mu^2}{2h})^{1/2}\mathrm{e}^{\mathrm{i}\beta} \equiv \frac{v}{\sqrt{2}}\mathrm{e}^{\mathrm{i}\beta}.$$

After the global gauge transformation, this minimum becomes

$$\langle 0\,|\Phi'\,|0\,\rangle = \mathrm{e}^{\mathrm{i}\Lambda}\,\langle 0\,|\Phi\,|0\,\rangle \neq \langle 0\,|\Phi\,|0\,\rangle\,.$$

The minimum has thus broken the initial symmetry.

Since the phase β is arbitrary, let fix it to zero, thus $\Phi_{Imn} = \langle 0|\Phi|0\rangle = \dfrac{v}{\sqrt{2}}$, and redefine the scalar complex field $\Phi = \Phi_{\mathrm{mln}} + (\Phi_1 + \mathrm{i}\Phi_2)/\Psi 2$, with Φ_1, Φ_2 real fields. Inserting this decomposition into the equation of motion delivers the two coupled equations

$$(\partial_\mu\partial^\mu - 2\mu^2)\Phi_1 = -h(3\Phi_1^2 + \Phi_2^2 + \Phi_1^3 + \Phi_1\Phi_2^2)$$
$$\partial_\mu\partial^\mu\Phi_2 = -h(2\Phi_1\Phi_2 + \Phi_1^2\Phi_2 + \Phi_2^3).$$

It thus appear that field Φ_1 has acquired a mass $m_1^2 = -2\mu^2$, whereas the field Φ_2 has no mass. The spontaneous breaking of the global symmetry has evidenced the existence of massless particles, called Goldstone bosons, and has generated a massive field. For the electroweak model, a component of the Higgs field provides the order parameter breaking the electroweak gauge symmetry to the electromagnetic gauge symmetry. Let consider to illustrate this the Abelian electromagnetic field; the

Lagrangian writes

$$L = -F_{\mu\nu}F^{\mu\nu} + (D_\mu\varphi^+)(D_\mu\varphi) - V(\varphi^+\varphi)$$

with the covariant derivative therein $D_\mu = \partial_\mu - ieA_\mu$, and Higgs potential

$$V(\varphi^+\varphi) = \mu^2\varphi^+\varphi + h(\varphi^+\varphi)^2, \qquad h > 0, \qquad \mu^2 < 0.$$

The local gauge transformation $\varphi \to \varphi' = \exp(i\alpha(x))\varphi$ leaves previous Lagrangian invariant. Let search as previously the solutions of the equation of motion for fields φ corresponding to the minimum of energy.

$$\varphi^+\varphi = \frac{-\mu^2}{2h} > 0$$

so that the minimum of H is obtained for $\varphi_{mm} = \langle 0|\varphi|0\rangle = \left(\frac{-\mu^2}{2h}\right)^{1/2}$.

As before, we choose the fundamental state such that

$$\langle 0|\varphi_1|0\rangle = v, \qquad \langle 0|\varphi_2|0\rangle = 0, \qquad \langle 0|\varphi|0\rangle = v/\sqrt{2}$$

and the gauge $\alpha(x)$ such that

$$\varphi_1 = \frac{1}{\sqrt{2}}(v + \varphi_1), \qquad \varphi_2 = 0$$
$$A'_\mu = B_\mu, \qquad B_{\mu v} = \partial_\mu B_v - \partial_v B_\mu$$

The Lagrangian density can then be decomposed using these new terms into

$$\begin{aligned}
L = &-\frac{1}{4}B_{\mu v}B^{\mu v} + \frac{1}{2}[\partial^\mu\varphi_1\partial_\mu\varphi_1 + e^2 B_\mu B^\mu(v+\varphi_1)^2] - \frac{\mu^2}{2}(v+\varphi_1)^2 - \frac{h}{4}(v+\varphi_1)^4\\
\equiv &-\frac{1}{4}B_{\mu v}B^{\mu v} + \frac{1}{2}e^2v^2 B_\mu B^\mu + ve^2 B_\mu B^\mu\varphi_1 + \frac{1}{2}e^2 B_\mu B^\mu\varphi_1^2\\
&+\frac{1}{2}[\partial^\mu\varphi_1\partial_\mu\varphi_1 + 2\mu^2\varphi_1^2] + \frac{\mu^2}{v}\varphi_1^3 + \frac{\mu^2}{4v^2}\varphi_1^4 - \frac{1}{4}v^2\mu^2
\end{aligned}$$

The Lagrangian is decomposed into contributions (from the second line) which have the following meaning:

- second line: interaction of the massive vectorial field B_μ, with mass $|ev|$;
- third line: interaction of the massive vectorial field B_μ with the scalar field φ_1 with the coupling constants ve^2 and $e^2/2$;
- fourth line: massive scalar field transported by Higgs bosons of mass $\sqrt{-2\mu^2}$;
- fifth line: self interaction of the scalar field φ_1, up to the constant term $-\frac{1}{4}v^2\mu^2$.

The Goldstone boson has here been absorbed by the vectorial massive boson B_μ. The system has initially four d.o.f., two for the scalar fields ϕ_1, ϕ_2, and two for the massless vector field A_μ, which become after symmetry breaking three d.o.f. linked to B_μ, and one for φ_1.

In the standard model of particle physics, spontaneous symmetry breaking of the SU(2) $\times$ U(1) gauge-symmetry associated with the electro-weak force generates masses for several particles, and separates the electromagnetic and weak forces. Bosons W and Z are the elementary particles that mediate the weak interaction, while the photon mediates the electromagnetic interaction. At energies much greater than 100 GeV, all these particles behave in a similar manner. The Weinberg-Salam theory predicts that, at lower energies, this symmetry is broken so that the photon and the massive W and Z bosons emerge.

Other examples of symmetry breaking in diverse fields such as physics, biology, mechanics, include:

- the laws of motion for a solid, which are invariant under the full Euclidean group, but the solid itself spontaneously breaks this group down to a space group. Here, the displacement and orientation are the order parameters.
- General relativity has a Lorentz symmetry, but in Friedman-Lemaitre-Robertson-Walker cosmological models, the mean four-velocity field defined by averaging over the velocities of the galaxies acts as an order parameter breaking this symmetry.
- In superconductors, there is a condensed-matter collective field Ψ, which acts as the order parameter breaking the electromagnetic gauge symmetry.
- Take a thin cylindrical plastic rod and push both ends together. Before buckling, the system is symmetric under rotation, and so visibly cylindrically symmetric. But after buckling, it looks asymmetric, so buckling has broken the symmetry by forcing the system to select one post- buckled state.
- In biology, there exists at least one asymmetric carbon atom in the molecules being the 'bricks of life' (nucleotides, amino acids), leading to the homochirality of biomolecules: amino acids are left-handed (levorotary compound), whereas nucleotides are right-handed (dextrorotary compound). Such chiral molecules are non-superposable with their mirror image. Two possible explanations have been advanced for this asymmetry: an amplification of random fluctuations by some self-catalytic process, or a more fundamental dissymmetry of universe. In line with this second scheme, let mention the violation of the parity of weak interactions: electrons (matter) who dominate over positrons are left-handed, whereas positrons are right-handed.
- Irreversible processes have a dual role, since they destroy order close to equilibrium, but generate order far from equilibrium. This transition towards organized states in non equilibrium situations is due to the generation of order by an amplification of fluctuations and percolation phenomena, generating the so-called dissipative structures, as exemplified by crystals [36].

7 Symmetry Methods in Continuum Mechanics of Materials

The main problematic of continuum mechanics applied to material modeling is to find objective tensorial functionals (traducing the invariance of the material's response under the group of rotations) and invariant under the action of the material symmetry group (discrete symmetries), as mentioned in [37] or [38]. For dissipative materials, the non-equilibrium thermodynamics based on thermodynamic potentials, essentially the free energy for writing the state law and dissipation potentials to express the evolution laws of internal variables, is the natural framework for the writing of the constitutive laws, [39]. It is accordingly natural that those potentials reflect the symmetry properties of any material. The analysis of the Lie symmetries of constitutive laws for dissipative materials written within a thermoydnamical framework of relaxation have been proposed in [40]; especially, a Lagrangian formulation of the state laws and the kinetic equations for the internal variables has been achieved. Both the local and variational symmetries of the constitutive laws have been determined, and have been shown to deliver the same Lie groups; they further lead to the construction of master responses for viscoelastic or viscoplastic materials, [41, 42].

The potentialities of Lie symmetries have not been so much explored in the field of mechanics of materials. Lie groups appear in fact as a predictive and systematic methodology to obtain general invariance properties of materials: from the knowledge of the constitutive law of a given material, Lie symmetries prove able to predict its response under various control conditions, that is when parameters of the constitutive law such as temperature, strain rate do change. Lie groups have been considered as a tool to analyze the experimental behaviour of materials and to model their mechanical response in [43], a pioneering contribution at the crossing of material constitutive modelling and symmetry analysis. The authors develop more specifically a methodology to model the mechanical behaviour of an acrylic stick at high strain rates, which however has a wide range of potential applications for different classes of materials.

7.1 Classification of Approaches: The Direct and Inverse Methods

The methodology basing on Lie groups can follow three interrelated routes [44]:

- The Direct Method (abbreviation DM) aims at finding the symmetries associated to the constitutive law of a given material. This methodology has been followed historically by several authors in the literature, see e.g. [45–47]. In continuum solid mechanics, the method is especially promising for dissipative materials. The invariants associated with those symmetries may then be computed and used in a practical way to synthesize the material's response into master curves traducing invariance relations in a graphical manner, and revealing the variation of the material's response when some control parameters (such as temperature or strain rate) vary;

- The Extended Direct Method (abbreviated as EDM) is an extension of the direct problem, and consists of finding and classifying the symmetries in a given BVP involving some initially unspecified constitutive functions (and also possibly loading functions). Invariants can be constructed from the obtained symmetries; consideration of nonlocal variables leads further to additional symmetries and invariants, and to more general formulations in Continuum Mechanics, as exemplified in nonlinear elasticity [20];
- The Inverse Method (abbreviated as IM) relies on experimental data of mechanical tests to construct the constitutive law of a given material in terms of initially unknown constitutive functions, the form of which is determined by Lie symmetries. One may here view those symmetries as an interpolation method to link experimental data, but our belief is that they have a more profound meaning highlighting the symmetry structure of the constitutive equations. This is especially true when the constitutive behavior is expressed in terms of thermodynamic potentials, under the umbrella of irreversible thermodynamics. A feedback from the inverse method to the direct method clearly exists: once a possible material constitutive law has been constructed, the material's response can be predicted outside the range of variation of the control variables involved in the measurements and for varying experimental conditions, thereby using the Lie symmetries inherent to the obtained constitutive law as an extrapolation technique.

This methodology exploiting Lie symmetries can be potentially applied to a wide class of materials—including polymers, metals, ceramics, metallic and polymeric foams—and constitutive laws, such as nonlinear elasticity, viscoelasticity and viscoplasticity. The DM algorithm is summarized in the diagram of Fig. 9.

The algorithm of the DM can be extended to a *class* of constitutive equations: if a general class of behaviors is considered, written in terms of generic functions, the search of the associated Lie groups allows a classification of symmetries depending on the mathematical structure of these functions. It may result in (i) a synthetic calculation of master curves belonging to the same "family", and (ii) the prediction of master curves in a broader constitutive framework (e.g. non-linear elasticity, viscoplasticity, etc...). The algorithm of this Extended Direct Method has been recently applied to the case of non-linear elastodynamics in [20, 48] and, in the same manner, to the case of non-local elasticity in [49].

A novel and rational approach based on Lie analysis is next exposed, to investigate the mechanical behaviour of materials presenting experimental master curves, relying on several works of the author of this chapter with collaborators, [40–44, 50]. This approach provides *a priori* ways of formulating constitutive laws from data and the possibility of predicting new master curves and material charts. Invariance relations—together with master curves, which are their graphical counterpart—are currently used in experimental rheology to synthesize the constitutive response of various materials submitted to mechanical or thermal solicitations. By "master curve", we mean a graphical superposition of different experimental curves based

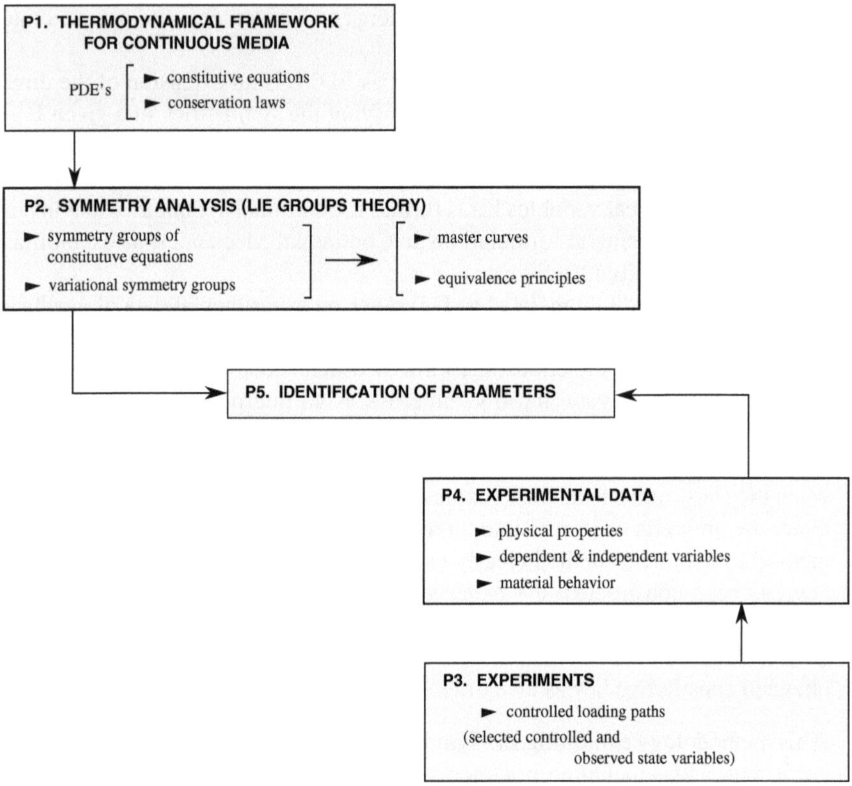

Fig. 9 Diagram associated with the Direct Method, cf. [44]

on some geometrical mapping. For instance, the so-called time-temperature equivalence principle stating an equivalence between the effect of time and of temperature, as proposed originally by Williams, Landel and Ferry (giving rise to the so-called WLF model), reveals the possibility of plotting a unique master curve from several isothermal mechanical responses. Despite the high practical interest of these relationships, the mathematical background for the prediction of the invariance relations and the associated master curves has not yet been fully developed in the literature to our knowledge.

More precisely, for different plots representing the logarithm of the isothermal creep compliance $J(t, T)$—ratio of the Henky strain $\varepsilon(t)$ to the Cauchy stress $\sigma(t)$ at a given time t and temperature T versus the logarithm of time $\log t$, a unique curve C can be obtained by shifting the different isothermal responses (curves) along the $\log t$ axis, as pictured in Fig. 10.

The resulting curve C is called the master curve, but the same expression can also refer to the geometrical mapping allowing the construction of C (a horizontal translation in the present case). It is relevant to associate a Lie symmetry to any

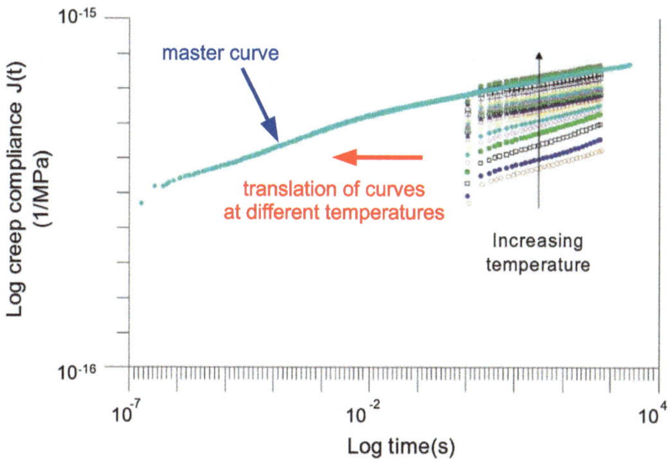

Fig. 10 Example of master curve obtained from creep tests on polymers (nylon 6/6, slightly adapted from [51])

master curve (and conversely), since the master curve is (beyond the geometrical nature of the mapping) by construction governed by three principles closely related to the three basic axioms of the Lie groups theory. Denoting by C_T the isothermal plot of $\log\ J(t, T)$ versus $\log t$ at a given temperature T, we have indeed:

- any creep compliance curve C_T is obtained by a nil translation along the $\log t$ axis; this particular translation is nothing but the identity application, the **neutral element** of translations (in the sense of composition).
- if $C_{T'}$ is obtained from C_T by a translation of μ along the $\log t$ axis, then C_T is obtained by shifting $C_{T'}$ of—μ along the same axis. This property is a **reflexivity property**.
- if $C_{T''}$ is obtained from $C_{T'}$ by a translation of μ' along the $\log t$ axis, and if $C_{T'}$ is obtained from C_T by a translation of μ, then $C_{T''}$ can be obtained by translating C_T of $\mu' + \mu$ (**transitivity property**).

The scalar μ represents the Lie group parameter, allowing a continuous mapping from one curve to another; it represents the length of the translation vector. In a more general case, the geometrical mapping leading to a master curve may be more elaborated than a mere translation: the group parameter μ may represent a rotation angle, or e^{μ} may represent the ratio of a homothetic transformation.

The general notion of invariant (and their combination) associated to Lie groups provides a relevant fundation to investigate constitutive equations in accordance with the observed symmetries. The general methodology splits into three main steps exposed in a synthetic form in the sequel, with more details provided in [44, 50]:

- Formulation of Lie symmetries from measurements. The construction of constitutive equations linking a set of observable variables

$\boldsymbol{u} = \{u_1, u_2, \cdots, u_n\}$ to a set of parameters $\boldsymbol{p} = \{p_1, p_2, \cdots, p_m\}$ relies on the analysis of experimental data and the finding of suitable functions or functionals linking those data. In terms of vocabulary, observable variables and parameters differ in the sense that parameters are quantities being controlled during the test, contrary to observable variables. From a physical point of view, parameters define the loading, while observable variables characterize the mechanical response. For instance, for a uniaxial compression test, we can select $u_1 = \sigma$ as the uniaxial Cauchy stress, while the controlled strain rate $p_1 = \dot{\varepsilon}$ or the controlled room temperature $p_2 = T$ are parameters, as well as time $p_3 = t$, since there is no aboslute origin of time. We will focus on the existence of parametric applications $G_i^{\text{obs}}(\mu, \boldsymbol{u}, \boldsymbol{p})$ (the superscript "obs" stands for "observed"), $i = 1, \cdots, q$, defined as follows

$$\mathcal{P} \times \mathcal{U} \times \mathbb{R} \xrightarrow{G_i^{\text{obs}}} \mathcal{P} \times \mathcal{U}, \qquad [\boldsymbol{p}, \boldsymbol{u}, \mu] \to [\bar{\boldsymbol{p}}, \bar{\boldsymbol{u}}] = [\boldsymbol{\Phi}^{\boldsymbol{p}}(\boldsymbol{p}, \boldsymbol{u}, \mu), \boldsymbol{\Phi}^{\boldsymbol{u}}(\boldsymbol{p}, \boldsymbol{u}, \mu)] \tag{1}$$

with $\boldsymbol{\Phi}^{\boldsymbol{p}} = \{\Phi^{p_1}, \ldots, \Phi^{p_m}\}$, $\boldsymbol{\Phi}^{\boldsymbol{u}} = \{\Phi^{u_1}, \ldots, \Phi^{u_n}\}$, μ the parameter, and such that there exists a subset of the experimental data which is "approximatively transformed" into another subset of experimental data. We assume that the applications $G_i^{\text{obs}}(\mu, \boldsymbol{u}, \boldsymbol{p})$ satisfy the three axioms of a Lie group. A generator $\boldsymbol{v}_i^{\text{obs}}$ is then associated with the Lie group G_i^{obs}, defined by

$$\boldsymbol{v}_i^{\text{obs}} = \sum_{k=1}^{m} \left.\frac{\partial \Phi^{p_k}}{\partial \mu}\right|_{\mu=0} \frac{\partial}{\partial p_k} + \sum_{k=1}^{n} \left.\frac{\partial \Phi^{u_k}}{\partial \mu}\right|_{\mu=0} \frac{\partial}{\partial u_k}, \qquad i = 1, \cdots, q. \tag{2}$$

At this stage, we have considered q observed symmetries $\boldsymbol{v}_i^{\text{obs}}$, $i = 1, \cdots, q$. The previously-built observed symmetries are next applied to the (currently unknown) postulated constitutive equations, written as a set of partial differential equations (PDE) written in the general form

$$\boldsymbol{\Delta} = \{\Delta_i = 0\ ;\ i = 1 \ldots n\} \tag{3}$$

in which the constitutive model Δ_i may depend on $\boldsymbol{p}$ and all the derivatives of $\boldsymbol{u}$ (including quantities $\boldsymbol{u}$ themselves). We assume that there are as many equations as observables, hence the index i varies from 1 to n in (3). Equation (3) hides unknown constitutive functions of the variables, which shall be identified by applying the symmetry conditions (see e.g. [43]), and making some assumption relative to the mathematical structure of equation (3) according to the considered constitutive framework (e.g. hyperelasticity, viscoelasticity, viscoplasticity). The symmetry conditions write

$$\begin{aligned}
\mathrm{pr}^{()}\boldsymbol{v}_k^{\mathrm{obs}}\Delta_1 &= 0 \quad \text{whenever } \Delta_1 = 0, \qquad k \in K_1 \\
\mathrm{pr}^{()}\boldsymbol{v}_k^{\mathrm{obs}}\Delta_2 &= 0 \quad \text{whenever } \Delta_2 = 0, \qquad k \in K_2 \\
&\cdots \\
\mathrm{pr}^{()}\boldsymbol{v}_k^{\mathrm{obs}}\Delta_n &= 0 \quad \text{whenever } \Delta_n = 0, \qquad k \in K_n
\end{aligned} \tag{4}$$

in which $\mathrm{pr}^{(\kappa)}\boldsymbol{v}_k^{\mathrm{obs}}$ or $\mathrm{pr}^{()}\boldsymbol{v}_k^{\mathrm{obs}}$ stand for the κ-th order prolongation — $\mathrm{pr}^{()}\boldsymbol{v}_k^{\mathrm{obs}}$ is used if the order κ is not prescribed—of the vector field $\boldsymbol{v}_k^{\mathrm{obs}}$.

- Constitutive equations in terms of invariants For the purpose of clarifying the mathematical expression of the (unknown) constitutive equations, one next calculates the invariants of the generators $\boldsymbol{v}_k^{\mathrm{obs}}$ and find suitable combinations of them that fit the experimental data. Let accordingly consider $\boldsymbol{v}^{\mathrm{obs}}$ as any vector field given by (2) and rewritten as:

$$\boldsymbol{v}^{\mathrm{obs}} = \sum_{k=1}^{m} \phi^{p_k}\frac{\partial}{\partial p_k} + \sum_{k=1}^{n} \phi^{u_k}\frac{\partial}{\partial u_k} \tag{5}$$

with

$$\phi^{p_k} = \left.\frac{\partial \Phi^{p_k}}{\partial \mu}\right|_{\mu=0}, \qquad \phi^{u_k} = \left.\frac{\partial \Phi^{u_k}}{\partial \mu}\right|_{\mu=0}. \tag{6}$$

If $\boldsymbol{v}^{\mathrm{obs}}$ has s non vanishing components amongst $\boldsymbol{\phi^p}$ and $\boldsymbol{\phi^u}$, it then has $s-1$ invariants $I_1, I_2, \cdots, I_{s-1}$ given by the solution of the characteristic system

$$\frac{\mathrm{d}p_j}{\phi^{p_j}} = \frac{\mathrm{d}u_i}{\phi^{u_i}}. \tag{7}$$

It is straightforward to show that any function $f(I_1, I_2, \cdots, I_{s-1})$ of the invariants is also invariant under $\boldsymbol{v}^{\mathrm{obs}}$, since

$$\begin{aligned}
\boldsymbol{v}^{\mathrm{obs}}[f(I_1, .., I_{s-1})] &= \sum_{k=1}^{m} \phi^{p_k}\frac{\partial f}{\partial p_k} + \sum_{k=1}^{n} \phi^{u_k}\frac{\partial f}{\partial u_k} \\
&= \sum_{k=1}^{m} \phi^{p_k}\sum_{l=1}^{s-1}\frac{\partial f}{\partial I_l}\frac{\partial I_l}{\partial p_k} + \sum_{k=1}^{n} \phi^{u_k}\sum_{l=1}^{s-1}\frac{\partial f}{\partial I_l}\frac{\partial I_l}{\partial u_k} \\
&= \sum_{l=1}^{s-1}\frac{\partial f}{\partial I_l}\underbrace{\left[\sum_{k=1}^{m} \phi^{p_k}\frac{\partial I_l}{\partial p_k} + \sum_{k=1}^{n} \phi^{u_k}\frac{\partial I_l}{\partial u_k}\right]}_{\boldsymbol{v}^{\mathrm{obs}}(I_l)=0} = 0.
\end{aligned} \tag{8}$$

Hence, the constitutive equation Δ_i can be expressed by combining the invariants $I_1, I_2, ...I_{s-1}$, provided it fits the experimental data.

- Computation of the Lie algebra and predictions
  This third step consists in a complete Lie analysis of $\boldsymbol{\Delta}$: for each Δ_i, one computes the Lie algebra

$$A_i = \{\boldsymbol{v}_i^1, \boldsymbol{v}_i^2, \cdots, \boldsymbol{v}_i^{n_i}\} \tag{9}$$

containing the n_i vector fields that can generate (by linear combination) any Lie symmetry of Δ_i. Obviously, the vector fields $\boldsymbol{v}_k^{\text{obs}}$, $k \in K_i$, have to be generated by the vector fields (9), that is

$$\boldsymbol{v}_k^{\text{obs}}\ (k \in K_i), \qquad \boldsymbol{\alpha} \in \mathbb{R}^{n_i} \qquad \text{such that} \qquad \boldsymbol{v}_k^{\text{obs}} = \sum_{j=1}^{n_i} \alpha_j \boldsymbol{v}_i^j. \tag{10}$$

Among the set of all combinations of $\boldsymbol{v}_i^j$, we can focus on vector fields for which (i) the component of a given observable u_j vanishes and (ii) the other components do not depend on u_j

$$\boldsymbol{v} = \sum_{k=1}^{m} \phi^{p_k}(\boldsymbol{p}, \boldsymbol{u}^{-j}) \frac{\partial}{\partial p_k} + \sum_{k=1, k\neq j}^{n} \phi^{u_k}(\boldsymbol{p}, \boldsymbol{u}^{-j}) \frac{\partial}{\partial u_k} \tag{11}$$

with $\boldsymbol{u}^{-j} = \{u_1, \ldots, u_{j-1}, u_{j+1}, \ldots, u_n\}$. The flow of such a vector field takes the particular form

$$\bar{\boldsymbol{p}} = \bar{\boldsymbol{p}}(\boldsymbol{p}, \boldsymbol{u}^{-j}, \mu), \qquad \bar{\boldsymbol{u}}^{-j} = \bar{\boldsymbol{u}}^{-j}(\boldsymbol{p}, \boldsymbol{u}^{-j}, \mu), \qquad \bar{u}_j = u_j. \tag{12}$$

Hence, eliminating the parameter μ allows *a priori* to determine the equations of the "iso-u_j" curves in the space $\mathcal{P} \times \mathcal{U}^{-j}$, with

$$\mathcal{U}^{-j} = U_1 \times \cdots U_{j-1} \times U_{j+1} \cdots \times U_n. \tag{13}$$

This kind of chart allows to extrapolate experimental data when the value of some parameter p_k cannot be reached in the experimental setting. The three steps of the Inverse Method are condensed into the diagram of Fig. 11. Since the results induced by the Lie algebra have to be validated by experiments, this step is a phase of prediction and validation.

8 Invariance Relations for the Creep and breakRupture Behavior of 9Cr1Mo

High temperature creep of metallic alloys is an important field of application of invariance relations, especially due to industrial aspects. The Larson-Miller and the Dorn models are two well-knonw invariance relations [52] that allow to extrapolate

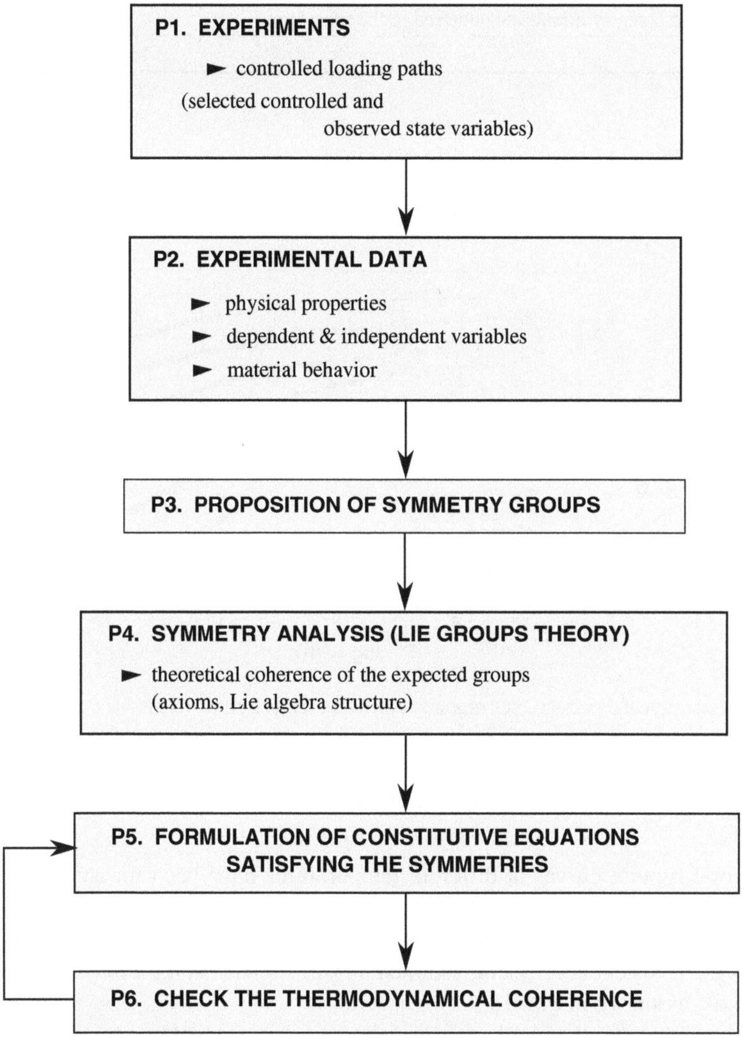

Fig. 11 Diagram associated with the Inverse Method, [44]

experimental data at different temperatures, assuming that the microstructure of the material remains stable during the creep test. The most widespread creep invariants are shown in Table 3.

The two main objectives of this kind of investigation are either the determination of the time needed to reach a given strain for fixed stress and temperature, or the estimation of the time leading to rupture, in the same controlled conditions. We focus on the creep behaviour of the 9Cr1Mo martensitic stainless steel, also known for its good thermal-fatigue strength and oxydation resistance [53]. This section condensates the methodology and results expressed in [50].

Table 3 Some creep invariants encountered in the literature

Authors	Invariant(s)
Larson & Miller	$T(\log t + \log B_2)$
Dorn	$\frac{B_1}{T} - \log t$
Monkman-Grant	$\log t_R + m \log \dot{\varepsilon}_{min}$
Modified Monkman-Grant	$\log\left(\frac{t_R}{\varepsilon_R}\right) + m' \log \dot{\varepsilon}_{min}$

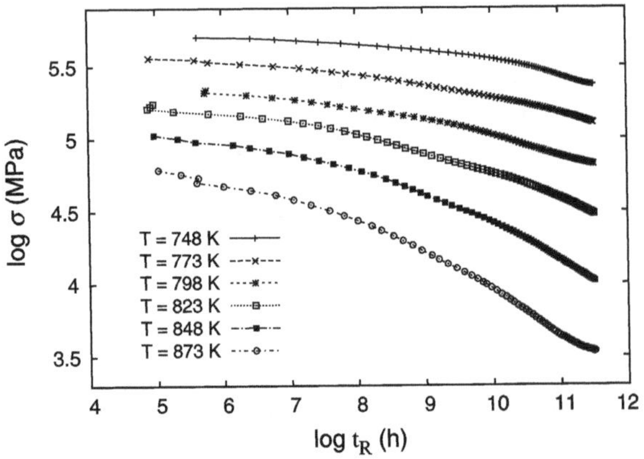

Fig. 12 Experimental stress versus rupture at different temperatrure, from [54]

8.1 Master Curves for Creep and Rupture

Isothermal rupture curves at different temperatures have been measured; they are represented in the logarithmic plane on Fig. 12.

Focusing on "simple" geometrical mappings such as translations or homothetic mappings, a master curve can be obtained by shifting the curves along the $\log t_R$ and $\log \sigma$ axis, as shown in Fig. 13.

It means that one then looks for the following geometrical mapping

$$\bar{T} = T + \mu, \qquad \log \bar{t}_R = a_1(\mu) + \log t_R, \qquad \log \bar{\sigma} = b_1(\mu) + \log \sigma \tag{14}$$

with $a_1(\mu)$ and $b_1(\mu)$ shift factors along the $\log t_R$ and $\log \sigma$ axis respectively. If the (arbitrary) chosen value of $\bar{T}$ is 748 K, then $a_1(\mu)$ and $b_1(\mu)$ correspond to the shift required to move the curve at T to make it coincide with those at $\bar{T}$, if μ is defined by $\bar{T} = T + \mu$. If we assume that Eq. (14) define a Lie group, then the fullfilment of the three Lie group axioms leads to the following conditions to be satisfied by functions $a_1(\mu)$ and $b_1(\mu)$:

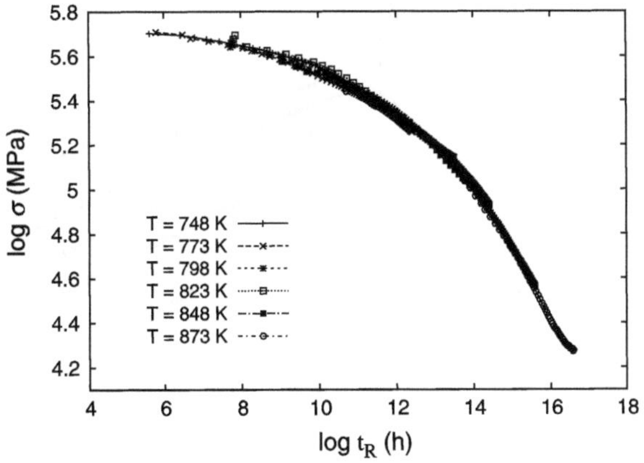

Fig. 13 Experimental master curve at the reference temperature $\bar{T} = 748\,\mathrm{K}$ obtained by shifting the curves of Fig. 12 along the $\log t_R$ and $\log \sigma$ axis, [50]

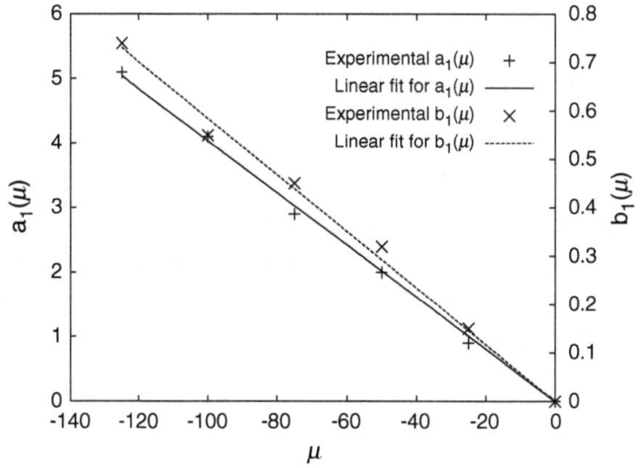

Fig. 14 Linear fits of the shift factors $a_1(\mu)$ and $b_1(\mu)$, [50]

$$\begin{aligned} &a_1(0) = 0, \quad &a_1(-\mu) = -a_1(\mu), \quad &a_1(\mu_1 + \mu_2) = a_1(\mu_1) + a_1(\mu_2) \\ &b_1(0) = 0, \quad &b_1(-\mu) = -b_1(\mu), \quad &b_1(\mu_1 + \mu_2) = b_1(\mu_1) + b_1(\mu_2) \end{aligned} \tag{15}$$

for all μ, μ_1, μ_2 in $\mathbb{R}$. It results from Eq. (15), the linearity of $a_1(\mu)$ and $b_1(\mu)$ with respect to μ the expressions

$$a_1(\mu) = \alpha\mu, \qquad b_1(\mu) = \beta\mu \tag{16}$$

where α and β are true constants.

These last relations fully agree with the experimental values of the shift factors, since the linear fits of $a_1(\mu)$ and $b_1(\mu)$ pictured in Fig. 14 exhibit good correlations.

The two fits give the following values of the coefficients

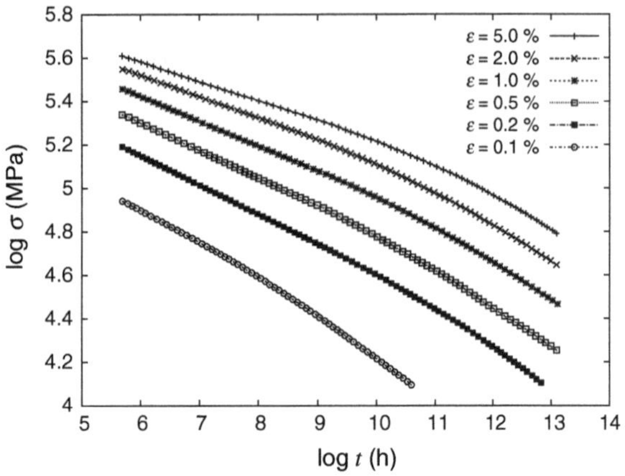

Fig. 15 Experimental stress versus time responses at different strain levels in the logarithmic plane ($T = 773$ K), from [54]

$$\alpha = -0.0403636, \qquad \beta = -0.00584727. \tag{17}$$

Inserting (16) into (14) leads to the following experimental Lie group

$$G_1^{\text{obs}} : \bar{T} = T + \mu, \qquad \bar{t}_R = \mathrm{e}^{\alpha\mu} t_R, \qquad \bar{\sigma} = \mathrm{e}^{\beta\mu}\sigma \tag{18}$$

mapping an experimental curve into another one in the range of the explored values of T, but its validity is presently assumed for all T between the extremal values 748 K and 923 K. The components of the generator $\boldsymbol{v}_1^{\text{obs}}$ associated with G_1^{obs} are given by the derivation of equations (18) with respect to μ at $\mu = 0$, viz

$$\boldsymbol{v}_1^{\text{obs}} = \alpha\, t_R \frac{\partial}{\partial t_R} + \beta\, \sigma \frac{\partial}{\partial \sigma} + \frac{\partial}{\partial T}. \tag{19}$$

The Lie group generated by this vector field allows to shift all rupture responses in the stress-rupture time plane to the response obtained for a chosen reference temperature; the curve associated with this response is called the master curve.

A similar strategy can be applied to the the "isostrain" creep curves in the logarithmic plane $\log \sigma$ versus $\log t$ presented in Fig. 15.

Proceeding in the same manner, a second Lie group G_2^{obs} is obtained

$$G_2^{\text{obs}} : \bar{\varepsilon} = \mathrm{e}^{\mu}\varepsilon, \qquad \bar{t} = \mathrm{e}^{\gamma\mu} t, \qquad \bar{\sigma} = \mathrm{e}^{\delta\mu}\sigma \tag{20}$$

with a generator $\boldsymbol{v}_2^{\text{obs}}$ of G_2^{obs} given by

$$\boldsymbol{v}_2^{\text{obs}} = \gamma\, t \frac{\partial}{\partial t} + \delta\sigma \frac{\partial}{\partial \sigma} + \varepsilon \frac{\partial}{\partial \varepsilon}. \tag{21}$$

The prolongations of the previously obtained vector fields $\boldsymbol{v}_1^{\text{obs}}$ and $\boldsymbol{v}_2^{\text{obs}}$ are computed with the prolongation formulae [5]

$$\text{pr}^{(1)}\boldsymbol{v}_1^{\text{obs}} = \alpha t_R \frac{\partial}{\partial t_R} + \beta\sigma\frac{\partial}{\partial\sigma} + \frac{\partial}{\partial T} \tag{22}$$

$$\text{pr}^{(1)}\boldsymbol{v}_2^{\text{obs}} = \gamma t\frac{\partial}{\partial t} + \delta\sigma\frac{\partial}{\partial\sigma} + \varepsilon\frac{\partial}{\partial\varepsilon} + \dot{\varepsilon}(1-\gamma)\frac{\partial}{\partial\dot{\varepsilon}}.$$

Based on the general methodology exposed previously, we search a set of constitutive equations able to capture the creep behaviour of the material. The set of constitutive equations is split into two relations revealing two different physical informations: the first relation expresses the time to rupture t_R versus the loading parameters (T and σ in the present case), while the second one gives the evolution of the strain rate $\dot{\varepsilon}$ as a function of t, σ, ε, and T

$$\Delta_1 = t_R - f(\sigma, T) = 0 \tag{23}$$

$$\Delta_2 = \dot{\varepsilon} - h(t, \sigma, \varepsilon, T) = 0. \tag{24}$$

with h and f therein unknown functions that will be further determined. It seems reasonable to apply the symmetry condition $\boldsymbol{v}_1^{\text{obs}}$ to Δ_1 and $\boldsymbol{v}_2^{\text{obs}}$ to Δ_2. Indeed, the generator $\boldsymbol{v}_1^{\text{obs}}$ ($\boldsymbol{v}_2^{\text{obs}}$ respectively) has been built by only considering rupture curves (creep curves respectively). The satisfaction of those symmetry conditions indicates that the constitutive law (creep and rupture) remains valid when the considered parameter changes. The set of symmetry conditions finally writes

$$\text{pr}^{(1)}\boldsymbol{v}_1^{\text{obs}}\Delta_1 = 0, \quad \text{whenever}\Delta_1 = 0 \tag{25}$$

$$\text{pr}^{(1)}\boldsymbol{v}_2^{\text{obs}}\Delta_2 = 0, \quad \text{whenever}\Delta_2 = 0 \tag{26}$$

and is associated with the sets $K_1 = \{1\}$ and $K_2 = \{2\}$. The next step consists in writing the constitutive models in terms of invariants.

8.2 Formulation of the Constitutive Equations for Creep and Rupture

Since $m_1 = 1$ and $m_2 = 1$, one can apply the two methodologies described previously, and consider more specifically a combination of invariants, in the spirit of extrapolation methods of the literature, [52]. One accordingly search two new functions $\Sigma_1(I_1, I_2)$ and $\Sigma_2(J_1, J_2, J_3)$ such that

$$\Sigma_1(I_1, I_2) = 0 \Leftrightarrow \Delta_1 = 0 \tag{27}$$

$$\Sigma_2(J_1, J_2, J_3) = 0 \Leftrightarrow \Delta_2 = 0 \tag{28}$$

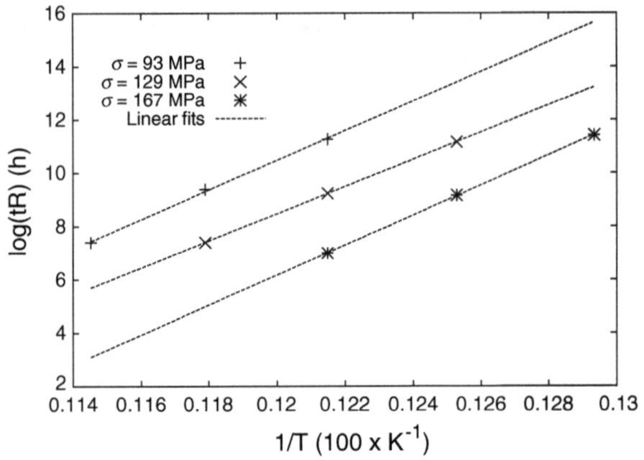

Fig. 16 Experimental responses $\log(t_R)$ versus $1/T$ at different stress levels (computed from data presented in [54]), [50]

with I_1, I_2 the invariants of $\boldsymbol{v}_1^{\text{obs}}$ given by the solution of the system (7)

$$\frac{\mathrm{d}t_R}{\alpha t_R} = \frac{\mathrm{d}\sigma}{\beta\sigma} = \mathrm{d}T \tag{29}$$

and J_1, J_2, J_3 the invariants of $\boldsymbol{v}_2^{\text{obs}}$ solution of the same characteristic system

$$\frac{\mathrm{d}t}{\gamma t} = \frac{\mathrm{d}\sigma}{\delta\sigma} = \frac{\mathrm{d}\varepsilon}{\varepsilon} = \frac{\mathrm{d}\dot{\varepsilon}}{(1-\gamma)\dot{\varepsilon}}. \tag{30}$$

Solving (29) and (30) delivers the following rupture invariants

$$I_1 = \log t_R - \alpha T, \qquad I_2 = \log\sigma - \beta T \tag{31}$$

and creep invariants

$$J_1 = \log t - \gamma\log\varepsilon, \quad J_2 = \log\sigma - \delta\log\varepsilon, \quad J_3 = \log\dot{\varepsilon} - (1-\gamma)\log\varepsilon. \tag{32}$$

Considering now rupture, the experimental representation of $\log(t_R)$ versus $1/T$ pictured in Fig. 16 suggests to propose a relation of the form

$$\Sigma_1(I_1, I_2) = \frac{K_1}{I_2} + K_2 I_1 + K_3 I_2 = 0 \tag{33}$$

with K_1, K_2, K_3 constants, which warrants a hyperbolic dependence with respect to T. Equation (33) leads to

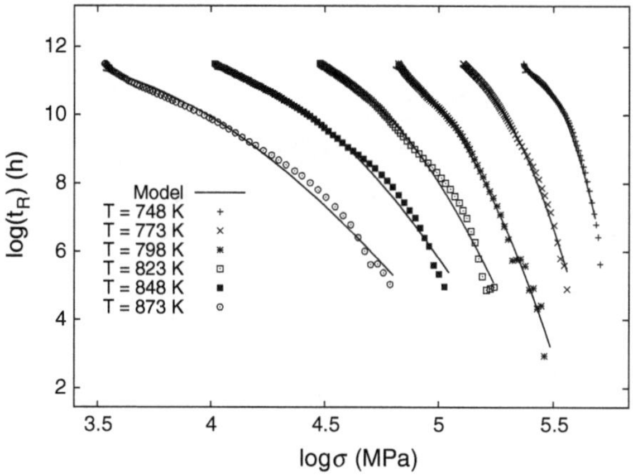

Fig. 17 Validation of Eq. (34) (from data presented in [54]), [50]

$$\log t_R = \frac{a}{\log\sigma - \beta T} + b\log\sigma + (\alpha - b\beta)T \tag{34}$$

with $a = -K_1/K_2$ and $b = -K_3/K_2$ new constants. The validation of equation (34) is pictured in Fig. 17, in which a good agreement with the experimental data presented in [54] is observed (the coefficients a and b are adjusted).

As to the creep strain rate, we introduce in accordance with the "convenient creep laws" of [54] the power function

$$\Sigma_2(J_1, J_2, J_3) = e^{J_3} - Ae^{cJ_1}e^{dJ_2} = \dot{\varepsilon}\varepsilon^{-(1-\gamma)} - At^c\varepsilon^{-c\gamma}\sigma^d\varepsilon^{-d\delta} = 0 \tag{35}$$

with A, c and d constants, leading to

$$\dot{\varepsilon} = At^c\sigma^d\varepsilon^{1-\gamma-c\gamma-d\delta}. \tag{36}$$

Denoting by $\varepsilon(t = 0, \sigma, T) = \varepsilon_0(\sigma, T)$ the initial strain ε and introducing the exponent $\nu = \gamma + c\gamma + d\delta$, the resolution of Eq. (36) allows writing

$$\varepsilon(t) = (1+c)^{-1/\nu}\left[(1+c)\varepsilon_0(\sigma, T)^\nu + A\nu t^{c+1}\sigma^{\mathrm{d}}\right]^{1/\nu}. \tag{37}$$

If the temperature T and the strain ε are assigned constant values, and assuming that the initial strain obeys a Hookean relation

$$\varepsilon_0(\sigma, T) = \frac{\sigma}{E(T)} \tag{38}$$

with $E(T)$ the Young modulus at T, then the "isostrain" curves may be obtained by expressing the time $t(\sigma)$ from equation (37), viz

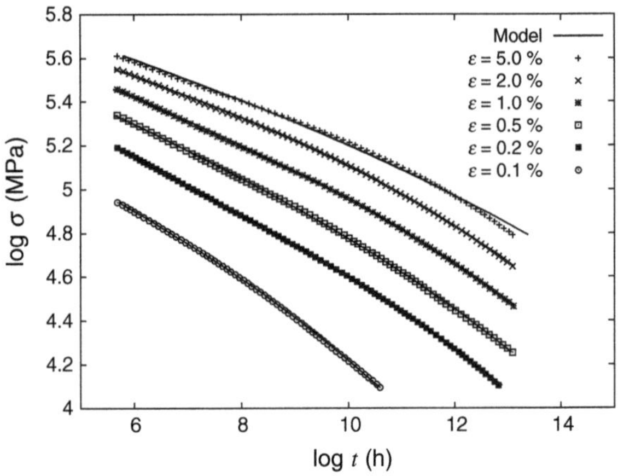

Fig. 18 Validation of Eq. (39) at $T = 773$ K (from data presented in [54]), [50]

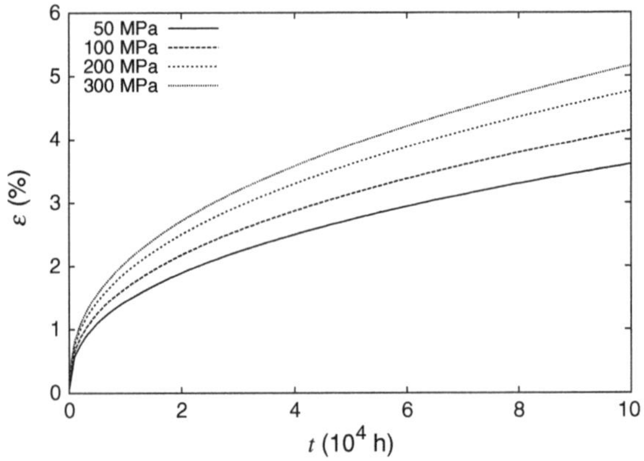

Fig. 19 Creep curves $\varepsilon(t)$ at different stress levels predicted by the present model, [50]

$$t(\sigma) = \left[\frac{(c+1)(\varepsilon^{\nu} - \sigma^{\nu}E(T)^{-\nu})}{A\nu\sigma^{\mathrm{d}}}\right]^{1/(1+c)} . \tag{39}$$

The model $\log\sigma$ versus $\log t$ is validated by the comparison of the inverse functional dependence $\sigma(t)$ numerically calculated from equation (39) with experimental data, both pictured in Fig. 18.

The coefficients $E(T)$, A and c, d are here adjusted so that a good agreement is obtained.

The Fig. 19 represents the creep curves at different stress levels as predicted by the present model. The following constitutive equations have been obtained as representative of the creep and rupture behaviours of the considered 9Cr1Mo stainless steel

Table 4 Commutator table of the Lie algebra of Eq. (40)

	$\boldsymbol{v}_2^1$	$\boldsymbol{v}_2^2$	$\boldsymbol{v}_2^3$	$\boldsymbol{v}_2^4$
$\boldsymbol{v}_2^1$	0	0	$(c+1)\boldsymbol{v}_2^1$	$2(c+1)\boldsymbol{v}_2^1$
$\boldsymbol{v}_2^2$	0	0	$(c+1)\boldsymbol{v}_2^2$	$(c+1)\boldsymbol{v}_2^2$
$\boldsymbol{v}_2^3$	$-(c+1)\boldsymbol{v}_2^1$	$-(c+1)\boldsymbol{v}_2^2$	0	0
$\boldsymbol{v}_2^4$	$-2(c+1)\boldsymbol{v}_2^1$	$-(c+1)\boldsymbol{v}_2^2$	0	0

$$\dot{\varepsilon} = At^c \sigma^d \varepsilon^{1-\gamma-c\gamma-d\delta} \quad \text{if} \quad t \leq t_R \tag{40}$$

$$t_R = \sigma^b \mathrm{e}^{\frac{a}{\log\sigma - \beta T} + (\alpha - b\beta)T} \tag{41}$$

Those forms of the constitutive laws are compatible with the observed master curves. The Lie algebra of the formulated creep and rupture constitutive models can lastly be derived.

8.2.1 Lie Algebra of the Creep and Rupture Models

The Lie algebra of the creep model is computed according to the methodology described in [5]; it contains 4 generators $\boldsymbol{v}_2^i$, $i = 1, \ldots, 4$, with prolongations given by

$$\mathrm{pr}^{(1)}\boldsymbol{v}_2^1 = \sigma^d t^{-c} \frac{\partial}{\partial t} + c\dot{\varepsilon}\sigma^d t^{-c-1} \frac{\partial}{\partial \dot{\varepsilon}} \tag{42}$$

$$\mathrm{pr}^{(1)}\boldsymbol{v}_2^2 = \sigma^d \varepsilon^{1-\nu} \frac{\partial}{\partial \varepsilon} + (1-\nu)\sigma^d \varepsilon^{-\nu}\dot{\varepsilon} \frac{\partial}{\partial \dot{\varepsilon}} \tag{43}$$

$$\mathrm{pr}^{(1)}\boldsymbol{v}_2^3 = t\frac{\partial}{\partial t} + \left(\frac{c+1}{\nu}\right)\varepsilon\frac{\partial}{\partial \varepsilon} + \left(\frac{c+1-\nu}{\nu}\right)\dot{\varepsilon}\frac{\partial}{\partial \dot{\varepsilon}} \tag{44}$$

$$\mathrm{pr}^{(1)}\boldsymbol{v}_2^4 = t\frac{\partial}{\partial t} - \left(\frac{c+1}{d}\right)\sigma\frac{\partial}{\partial \sigma} - \dot{\varepsilon}\frac{\partial}{\partial \dot{\varepsilon}} \tag{45}$$

with $\nu = \gamma + c\gamma + d\delta$. The commutation table of this Lie algebra is given in Table 4.

The Lie algebra of the rupture model equation (40) has two generators $\boldsymbol{v}_1^i$, $i = 1, 2$, given by

$$\boldsymbol{v}_1^1 = t_R \left(\frac{b\,(\log\sigma - \beta T)^2 - a}{\log\sigma - \beta T}\right)\frac{\partial}{\partial t_R} + \sigma(\log\sigma - \beta T)\frac{\partial}{\partial \sigma} \tag{46}$$

$$\boldsymbol{v}_1^2 = t_R \left(\frac{(\alpha - b\beta)\,(\log\sigma - \beta T)^2 + a\beta}{\log\sigma - \beta T}\right)\frac{\partial}{\partial t_R} + (\log\sigma - \beta T)\frac{\partial}{\partial T}\cdot \tag{47}$$

Its commutator table is given in Table 5.

Table 5 Commutator table of the Lie algebra of Eq. (40)

	$\boldsymbol{v}_1^1$	$\boldsymbol{v}_1^2$
$\boldsymbol{v}_1^1$	0	$\beta\boldsymbol{v}_1^1+\boldsymbol{v}_1^2$
$\boldsymbol{v}_1^2$	$-\beta\boldsymbol{v}_1^1-\boldsymbol{v}_1^2$	0

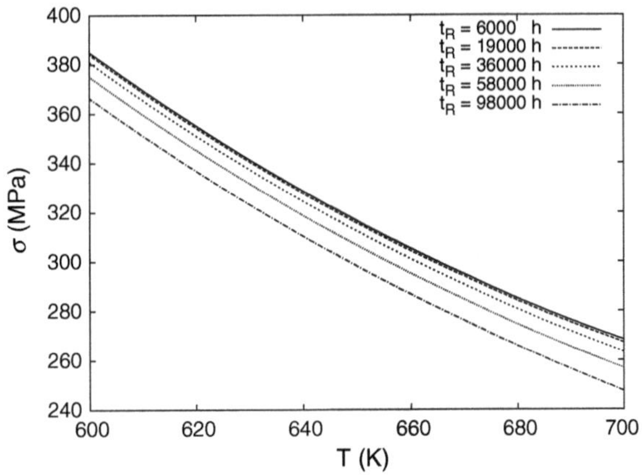

Fig. 20 Predicted "iso-t_R" curves given by the numerical solution of Eqs. (49) and (50), [50]

Denoting by ϕ^{1,t_R} and ϕ^{2,t_R} the component in t_R of $\boldsymbol{v}_1^1$ and $\boldsymbol{v}_1^2$ respectively, then the following local combination

$$\boldsymbol{v}=\frac{\phi^{2,t_R}\boldsymbol{v}_1^1-\phi^{1,t_R}\boldsymbol{v}_1^2}{t_R}=\sigma\left[a\beta+(\alpha-b\beta)(\log\sigma-\beta T)^2\right]\frac{\partial}{\partial\sigma}$$
$$+\left(a-b(\log\sigma-\beta T)^2\right)\frac{\partial}{\partial T} \tag{48}$$

is a symmetry of Eq. (40), with no component in t_R and without any dependency of the components σ and T on t_R. Consequently, $\boldsymbol{v}$ takes the form of Eq. (11), and the flow of this vector field is given by the solution of the system of differential equations

$$\frac{\mathrm{d}\bar{\sigma}}{\mathrm{d}\mu}=\bar{\sigma}\left[a\beta+(\alpha-b\beta)(\log\bar{\sigma}-\beta\bar{T})^2\right] \tag{49}$$

$$\frac{\mathrm{d}\bar{T}}{\mathrm{d}\mu}=a-b(\log\bar{\sigma}-\beta\bar{T})^2 \tag{50}$$

with initial conditions

$$\bar{\sigma}(0)=\sigma,\qquad \bar{T}(0)=T \tag{51}$$

and the equality $\bar{t}_R=t_R$. Hence, the numerical solution of (49) and (50) provides a parametric representation $(T(\mu),\sigma(\mu))$ of the "iso-t_R" curves in the (T,σ) plane, as shown in Fig. 20.

This kind of theoretical chart is of high practical interest, since it allows to extrapolate the rupture time of a specimen at any stress level or any temperature, provided the rupture time is known for a given value of σ and T.

8.3 Some Open Problems in Continuum Mechanics and Mechanics of Materials

There are many open problems and perspectives of applications of symmetry methods in continuum mechanics; these include the following:

- Find closed form solutions and invariance relations for wave propagation problems in hyperelastic solids. This is especially important in the context of soft biological tissues, for which one has additionnaly to incorporate the material anisotropy due to the presence of populations of fibers.
- Find invariance relations for BVP's involving materials having a complex rheological behavior, including time-dependence, and non linear stress-strain relations. This has been exemplified in the present contribution in the case of creep of mettalic alloys, and has to be extended to polymers. One then has to find efficient algorithms to construct invariance relations between the (many) variables and parameters involved in the constitutive law.
- Clarify the fundamental significance of the symmmetries computed for materials having a complex rheological behavior. Especially, the symmetries inherent to the thermodynamic potentials involved in the construction of the constitutive laws have to be evidenced, and their significance clarified. Such an interpretation will further nicely fit within the rational framework of continuum mechanics, which relies on symmetry arguments for its construction (including material frame indifference).
- Find approximate symmetries for materials and systems endowed with an uncertainty in their behavior, for instance a statistical variability in their material properties, requiring the use of stochastic models. This is especially important in modelling the response of soft biological tissues.

References

1. Hamel, G.: Die Lagrange-Eulerschen Gleichungen der Mechanik. Z. Mathematik und Physik, vol. 50 (1984)
2. Noether, E.: Invariante variationsprobleme. Nachr. Konig. Gesell. wissen. Gottingen, Math-Phys. Kl. 235–257 (1918)
3. Doughty, N.A.: Lagrangian interactions. An Introduction to Relativistic Symmetry in Electrodynamics and Gravitation. Addison-Wesley, Reading (1990)
4. Bluman, G.W., Cheviakov, A.F., Anco, S.C.: Applications of Symmetry Methods to Partial Differential Equations. Springer, New York (2010)
5. Olver, P.: Application of Lie Groups to Differential Equations. Springer, New York (1993)

6. Basar, Y., Weichert, D.: Nonlinear Continuum Mechanics of Solids. Fundamental Mathematical and Physical Concepts. Springer, New York (2000)
7. Moriyasu, K.: An Elementary Primer for Gauge Theory. World scientific, Singapore (1985)
8. Cohen-Tannoudji, C., Dupont-Roc, J., Grynberg, G.: Photons et Atomes. Introduction à l'électrodynamique quantique, InterEditions/Editions du CNRS (1987)
9. Laverriere, R.: Electrodynamique Classique, vol. 2 : Electromagnétisme. Bréal, (1992)
10. Noether, E.: Invariante variationsprobleme. Nachr. König. Gesell. Wissen. Göttingen, Math.-Phys. 235–257 (1918)
11. Elbaz, E.:De, : l'électromagnétisme à l'électrofaible. Monopôles Magnétiques, Ellipses (1989)
12. Adeleke, S.: On material symmetry in mechanics. Int. J. Solids Struct. **16**, 199–215 (1980)
13. Ericksen, J.: On invariance groups for equilibrium theories. J. Elasticity **59**(1):9 (2000)
14. Chien, N., Honein, T., Hermann, G.: Dissipative systems, conservation laws and symmetries. Int J. Solids Struct. **33**(20–22), 2959–2968 (1996)
15. Chien, N., Honein T., Hermann G.: Conservation laws for linear viscoelasticity. Z. Angew. Math. Phys. **44** (1993)
16. Ozer, T.: Symmetry group classification for one dimensional elastodynamics problems in non-local elasticity. Mech. Res. Comm. **30**, 539–546 (2003)
17. Annin, B.D., Bytev, V.O., Senashov, S.I.: Group Properties of Equations of Elasticity and Plasticity. Nauka, Novosibirsk (1985)
18. Cheviakov, A., Ganghoffer J.F., Rahouadj R.: Finite strain plasticity models revealed by symmetries and integrating factors: the case of dafalias spin model. Int. J. Plasticity. http://dx.doi.org/10.1016/j.ijplas.2012.12.001 (2013)
19. Senashov, S.I.: Invariant Spatial Solutions for Equations of Ideal Plasticity. Zhurnal Prikladnoi Mechaniki i Tekhnicheskoi Fiziki, 2 (1980)
20. Bluman, G.W., Cheviakov, A.F., Ganghoffer, J.-F.: Nonlocally related PDE systems for one-dimensional nonlinear elastodynamics. J. Eng. Math. **62**, 203–221 (2008)
21. Bluman, G.W., Cheviakov A.F., Ganghoffer, J.-F.: On the nonlocal symmetries, group invariant solutions and conservation laws of the equations of nonlinear dynamical compressible elasticity. In: Proceedings of IUTAM Symposium on Progress in the Theory and Numerics of Configurational Mechanics, Springer, New York, pp. 107–120 (2009)
22. Bluman, G.W., Cheviakov, A.F.: Nonlocally related systems, linearization and nonlocal symmetries for the nonlinear wave equation. JMAA **333**, 93–111 (2007)
23. Bluman, G.W., Ganghoffer, J.-F.: Connecting Euler and Lagrange systems as nonlocally related systems of dynamical nonlinear elasticity. Arch. Mech. **63**, 363–382 (2011)
24. Cheviako, A.F., "GeM": A Maple Module for Symmetry and Conservation Law Computation for PDEs/ODEs. http://www.math.usask.ca/ cheviakov/gem/ 2013
25. Ozer, T.: Symmetry group classification for one-dimensional elastodynamics problems in non-local elasticity. Mech. Res. Comm. **30**, 539–546 (2003)
26. Ozer, T.: Symmetry group classification for two-dimensional elastodynamics problems in non-local elasticity. Int. J. Engng Sci. **41**, 2193–2211 (2003)
27. Voyiadjis, G.Z., Pekmezi, G., Deliktas, B.: Nonlocal gradient-dependent modeling of plasticity with anisotropic hardening. Int. J. Plasticity **26**, 1335–1356 (2010)
28. Liu, C.S.: Lie symmetries of finite strain elastic-perfectly plastic models and exactly consistent schemes for numerical integrations. Int. J. Solids Struct. **41**(7), 1823–1853 (2004)
29. Eshelby, J.D., The Mechanics of Defects and Inhomogeneities - Collected works of J.D. Eshelby, Editor Markenscoff, Xanthippi, Gupta, Anurag. Solid Mechanics and Its Applications (2006) 133.
30. Maugin, G.A.: Material inhomogeneities in elasticity. Cambridge University Press, Cambridge (1993)
31. Steinmann, P., Maugin, G.A.: Mechanics of Material Forces vol. 11. Advances in Mechanics and Mathematics, New York (2005)
32. Budianski, B., Rice, J.R.: Conservation laws and energy-release rate. ASME J. Appl. Mech. **40**, 201–203 (1973)
33. Lawrie, D.: A Unified Grand Tour of Theoretical Physics. Adam Hilger, Bristol (1990)

34. David, F.: Introduction à la Théorie Statistique des Champs et au Groupe de Renormalisation. Cours (2006)
35. Le Bellac M.: Quantum and Statistical Field Theory. Oxford University Press, Oxford (1992)
36. Glansdorff, P., Prigogine, I.: Structure, Stabilité et Fluctuation. Wiley, New York (1971)
37. Weyl, H.: The Classical Group, their Invariants, and Representations. Princeton Landmarks in Mathematics. Princeton University Press, Princeton (1997)
38. Fulton, W., Harris, J.: Representation Theory: A First Course. Springer, New York (1991)
39. De Groot, S.R., Mazur, P.: Non-equilibrium thermodynamics. North-Holland, Amsterdam (1969)
40. Magnenet, V., Rahouadj, R., Ganghoffer, J.F., Cunat, C.: Continuous symmetries and constitutive laws of thermo-elasto(visco)plastic materials within a thermodynamical framework of relaxation. Part I: Formal Aspects. Int. J. Plasticity **23**(1) 87–113 (2007)
41. Magnenet, V., Rahouadj, R., Ganghoffer, J.F., Cunat, C.: Continuous symmetry analysis of a dissipative constitutive law: application to the time-temperature superposition. Eur. J. Mech. - A/Solids **28**(4), 744–751 (2009)
42. Magnenet, V., Rahouadj, R., Ganghoffer, J.F.: A new methodology for determining the mechanical behavior of polymers exploiting lie symmetries: application to a stick-like material. Mech. Mat. **41**(9), 1017–1024 (2009)
43. Magnenet, V., Rahouadj, R., Ganghoffer, J.F.: A new methodology for determining the mechanical behavior of polymers exploiting lie symmetries: application to a stick-like material. Mech. Mat. **41**, 1017–1024 (2009)
44. Ganghoffer, J.F., Magnenet, V., Rahouadj, R.: Relevance of symmetry methods in mechanics of materials. Int. J. Eng. Math. **66**, 103–119 (2010)
45. Ibragimov, N.H.: CRC Handbook of lie group analysis of differential equations. Symmetries, exact solutions, and conservation laws, 2nd Edn, **2**. CRC Press, Boca Raton (1994)
46. Ibragimov, N.H.: CRC Handbook of Lie Group Analysis of Differential Equations. Applications in Engineering and Physical Sciences, vol. 2. CRC Press, Boca Raton (1995)
47. Ibragimov, N.H., CRC Handbook of Lie Group Analysis of Differential Equations. New Trends in Theoretical Developments and Computational Methods, 2nd Edn, vol. 2. CRC Press, Boca Raton (1996)
48. Cheviakov, A., Ganghoffer, J.F.: Symmetry properties of two-dimensional ciarlet-mooney-rivlin constitutive models in nonlinear elastodynamics. J. Math. Anal. Appl. **396**, 625–639 (2012)
49. Ozer, T.: Symmetry Group Analysis and Similarity Solutions of Invariant Nonlinear Long-wave Equations. Solitons and Fractals, Chaos **38**, 722–730 (2008)
50. Magnenet, V., Rahouadj, R., Ganghoffer, J.F.: Symmetry analysis and invariance relations in creep. Math. Mech. Solids. (1081). doi: 286513498331 (2013)
51. Chevali, V.S., Dean, D.R., Janowski, G.M.: Flexural Creep behavior of discontinuous thermoplastic composites: non-linear viscoelastic modeling and time-temperature-stress superposition. Composites Part A **40**, 870–877 (2009)
52. Saint-Antonin, F.: Essais de Fluage. Techniques de l'ingénieur, M140, 1995.
53. Kim, W.G., Kim, S.H., Ryu, W.S.: Evaluation of monkman-grant parameters for type 316LN and modified 9Cr-Mo stainless steels. KSME Inter. J. **16**, 1420–1427 (2002)
54. Skrzypek, J.J., Hetnarski, R.B.: CRC Press, Plasticity and Creep (1993)

Symmetries and Some Special Solutions of the Helfrich Model

Vladimir I. Pulov, Mariana Ts. Hadzhilazova and Ivaïlo M. Mladenov

Abstract The goal of this chapter is to present the results of the Lie group analysis in application to the Helfrich spontaneous curvature model. Special attention is paid to the translationally invariant solutions and the corresponding cylindrical equilibrium shapes. Graphs of closed diretrices of the obtained cylindrical surfaces in fixed and moving reference frame are presented.

1 Introduction

The *Helfrich model* (also known as the *spontaneous curvature model*) describes equilibrium shapes of *fluid membranes*—elastic membranous systems formed in aqueous solutions, such as biological membranes, soap bubbles, etc. [1, 2]. The governing equation of the Helfrich model is the *Helfrich equation*. In Mongé representation the Helfrich equation is a fourth order nonlinear partial differential equation. By using conformal coordinates and new dependent variables the Helfrich equation is transformed to a system of four nonlinear partial differential equations of second order of the derivatives [3, 4] (Sect. 2). We name this system the *Helfrich system*. Our main objective in this chapter is to present the results of the Lie symmetry group analysis in application to the Helfrich system of differential equations. By

V. I. Pulov (✉)
Department of Physics, Technical University of Varna, Studentska Str. 1,
9010 Varna, Bulgaria
e-mail: vpulov@tu-varna.bg; vpulov@hotmail.com

M. Ts. Hadzhilazova · I. M. Mladenov
Institute of Biophysics and Biomedical Engineering, Bulgarian Academy of Sciences,
Acad. G. Bonchev Str. Bl. 21, 1113 Sofia, Bulgaria
e-mail: murryh@bio21.bas.bg

I. M. Mladenov
e-mail: mladenov@bio21.bas.bg

J.-F. Ganghoffer and I. Mladenov (eds.), *Similarity and Symmetry Methods*, Lecture Notes in Applied and Computational Mechanics 73, DOI: 10.1007/978-3-319-08296-7_6,

creating and solving the so called *determining system* of equations [5] we obtain the Lie symmetry algebra (respectively the Lie symmetry group) of the Helfrich system (Sect. 3). It is shown in [3] that all one-dimensional Lie subalgebras of the general Lie symmetry algebra of the Helfrich system are equivalent to each other under the group of inner automorphisms. Based on this knowledge we apply a Lie symmetry group reduction technique [5] to obtain a translationally invariant solution and the corresponding cylindrical equilibrium shapes of the Helfrich model (Sects. 4 and 5). The solution and the directrices of the cylindrical surfaces are expressed by the Weierstrassian functions. Plots of graphs of closed diretrices for some of the surfaces in fixed and moving reference frame are presented.

2 The Helfrich Model

In the Helfrich model the fluid membrane is described as a smooth surface $\mathcal{S}$ in the Euclidean space $\mathbb{R}^3$ with the mean and the Gaussian curvatures H and K, respectively. The equilibrium shapes of the membrane is determined by solving the *Helfrich equation* [6]

$$\Delta H + 2(H^2 + \mathbb{h}H - K)(H - \mathbb{h}) - \frac{2\lambda H}{k} + \frac{p}{k} = 0. \tag{1}$$

The above equation is (a coordinate free) Euler-Lagrange equation for the minimum of the free elastic energy

$$\mathcal{F} = k\int_{\mathcal{S}} (H - \mathbb{h})^2 \mathrm{d}S + \bar{k}\int_{\mathcal{S}} K\mathrm{d}S \tag{2}$$

obtained under the constraints of fixed enclosed volume and surface area of the membrane. The free energy $\mathcal{F}$ is the bending energy due to the curvatures H and K through the two elastic moduli of the membrane – the bending and the Gaussian rigidities k and $\bar{k}$, respectively and the parameter $\mathbb{h}$ is the so-called *spontaneous mean curvature*. Other physical characteristics involved are the tensile stress λ, the osmotic pressure p and Δ is the Laplace-Beltrami operator on $\mathcal{S}$.

We are interested in immersed surfaces in the Euclidean space that satisfy the Helfrich equation (1). The necessary and sufficient condition for the immersed surface to exist is that the coefficients of its first and second fundamental form satisfy the so called Gauss-Codazzi-Mainardi integrability equations [7]. For this reason, when looking for the equilibrium surfaces of the Helfrich model, the Eq. (1) has to be solved together with the Gauss-Codazzi-Mainardi equations. The system of differential equations formed thereby is named the *Helfrich system* and its solutions – the *Helfrich surfaces.*

Given some local coordinates on the surface $\mathcal{S}$, the Helfrich system takes the form of a system of nonlinear partial differential equations. The number of equations and the order of the derivatives being involved depend on the coordinates that have been specified. In our work we introduce the conformal coordinates (x, y), defined by the metric (*conformal metric*) on $\mathcal{S}$

$$ds^2 = 4q^2\varphi^2(dx^2 + dy^2) \tag{3}$$

and the matrix of the *second fundamental form*

$$b = \begin{pmatrix} \theta & \omega \\ \omega & 8q^2\varphi(1 + \mathbb{h}\varphi) - \theta \end{pmatrix} \tag{4}$$

depending on four unknown functions

$$q = q(x, y), \qquad \varphi = \varphi(x, y), \qquad \theta = \theta(x, y), \qquad \omega = \omega(x, y).$$

Under this choice of the variables the Helfrich system is recast into a system of four second order partial differential equations [3, 4, 8]

$$\begin{aligned}
&q^2(\varphi_{xx} + \varphi_{yy}) + 2q\varphi(q_{xx} + q_{yy}) \\
&\qquad -2\varphi(q_x^2 + q_y^2) + q^4(8\varphi + \alpha_2\varphi^2 + \alpha_3\varphi^3 + \alpha_4\varphi^4) = 0 \\
&\theta_y - \omega_x - (8 + \frac{\alpha_2}{3}\varphi)q(\varphi q_y + q\varphi_y) = 0 \\
&\omega_y + \theta_x - \frac{\alpha_2}{3}q\varphi(\varphi q_x + q\varphi_x) - 8q\varphi q_x = 0 \\
&4q^2\varphi(\varphi_{xx} + \varphi_{yy}) + 4q\varphi^2(q_{xx} + q_{yy}) \\
&\qquad -4\varphi^2(q_x^2 + q_y^2) - 4q^2(\varphi_x^2 + \varphi_y^2) - \omega^2 - \theta^2 + (8 + \frac{\alpha_2}{3}\varphi)q^2\varphi\theta = 0
\end{aligned} \tag{5}$$

where $\alpha_2 = 24\mathbb{h}$, $\alpha_3 = 8(2\mathbb{h}^2 - \frac{\lambda}{k})$ and $\alpha_4 = \frac{4p}{k} - \frac{8\lambda\mathbb{h}}{k}$ are phenomenological constants, and $\varphi_x = \partial\varphi/\partial x$, etc. denote partial derivatives. In comparison with the fourth order nonlinear partial differential equation in the Mongé representation of the Helfrich equation (see e.g. [9]), the system of differential equations (5) is a simpler version of the Helfrich model.

In the conformal metric coordinates, defined by (3) and (4), the mean H and the Gaussian K curvatures take the form

$$H = \frac{1}{\varphi} + \mathbb{h} \tag{6}$$

$$K = \frac{1}{4q^4\varphi^4}[\varphi^2(q_x^2 + q_y^2) + q^2(\varphi_x^2 + \varphi_y^2) - q\varphi^2(q_{xx} + q_{yy}) - q^2\varphi(\varphi_{xx} + \varphi_{yy})].$$

The latter is obtained by using the Brioschi formula

$$K = -\Delta \log(2q\varphi) \tag{7}$$

by making use of the respective Laplace-Beltrami operator

$$\Delta = \frac{1}{4q^2\varphi^2}\left(\frac{\partial^2}{\partial x^2} + \frac{\partial^2}{\partial y^2}\right).$$

In this setting the functional (2) reads

$$\mathcal{F} = 4k \iint q^2 \mathrm{d}x\mathrm{d}y + \bar{k} \iint \frac{8q^2\varphi\theta(1 + \mathrm{h}\varphi) - \theta^2 - \omega^2}{4q^2\varphi^2} \mathrm{d}x\mathrm{d}y$$

and this is an obvious manifestation of the fact that the squared function q plays the role of a free elastic energy density due to the mean curvature H of the membrane (cf. [3, 4]).

3 The Determining System and Symmetries

A *symmetry* of a given system of differential equations is a transformation of the independent and dependent variables that leaves the system invariant. The *determining system* is a system of partial differential equations used for finding the Lie group of symmetries (*Lie symmetry group*) of the considered system of differential equations [5].

In the case of the Helfrich system (5) the one-parameter Lie group of symmetry transformations has the form

$$\begin{aligned} \tilde{x}^i &= \Phi^i(\mathbf{x}, \mathbf{u}, \varepsilon), \qquad \Phi^i|_{\varepsilon=0} = x^i, \qquad i = 1, 2 \\ \tilde{u}^\alpha &= \Psi^\alpha(\mathbf{x}, \mathbf{u}, \varepsilon), \qquad \Psi^\alpha|_{\varepsilon=0} = u^\alpha, \qquad \alpha = 1, 2, 3, 4 \end{aligned}$$

where ε ($\varepsilon \in I \subset \mathbb{R}, 0 \in I$) is the group parameter. The vectors $\mathbf{x} = (x^1, x^2)$ and $\mathbf{u} = (u^1, u^2, u^3, u^4)$ denote the independent and the dependent variables, respectively: $x^1 = x, x^2 = y, u^1 = q, u^2 = \varphi, u^3 = \theta, u^4 = \omega$. The functions $\Phi^i(\mathbf{x}, \mathbf{u}, \varepsilon)$ and $\Psi^\alpha(\mathbf{x}, \mathbf{u}, \varepsilon)$ are found by solving the *Lie equations* [5]

$$\begin{aligned} \frac{\mathrm{d}\Phi^i}{\mathrm{d}\varepsilon} &= \xi^i(\boldsymbol{\Phi}, \boldsymbol{\Psi}), \qquad \Phi^i|_{\varepsilon=0} = x^i, \qquad i = 1, 2 \\ \frac{\mathrm{d}\Psi^\alpha}{\mathrm{d}\varepsilon} &= \eta^\alpha(\boldsymbol{\Phi}, \boldsymbol{\Psi}), \qquad \Psi^\alpha|_{\varepsilon=0} = u^\alpha, \qquad \alpha = 1, 2, 3, 4 \end{aligned}$$

where $\xi^i(\mathbf{x}, \mathbf{u})$ and $\eta^\alpha(\mathbf{x}, \mathbf{u})$ are the coefficients of the Lie group generator

$$X = \sum_{i=1}^{2} \xi^i(\mathbf{x}, \mathbf{u}) \frac{\partial}{\partial x^i} + \sum_{\alpha=1}^{4} \eta^\alpha(\mathbf{x}, \mathbf{u}) \frac{\partial}{\partial u^\alpha}.$$

The coordinates of the vectors $\boldsymbol{\Phi}$ and $\boldsymbol{\Psi}$ are Φ^i and Ψ^α, respectively. The set of all Lie group generators constitute a Lie algebra – the *Lie symmetry algebra* of the Lie symmetry group. In order to find the Lie symmetry algebra one has to solve the determining system of equations.

The determining system is a linear homogeneous system of partial differential equations for the coefficient functions $\xi^i(\mathbf{x}, \mathbf{u})$ and $\eta^\alpha(\mathbf{x}, \mathbf{u})$ of the Lie group generator. For most of the important physical applications the determining system consists of hundreds of equations. Creating and solving of such a large system of differential equations, though algorithmically straightforward, may cause serious technical difficulties. In order to cope with the great number of tedious calculations we take advantage of the specially developed Mathematica package *LieSymm-PDE* [10]. By applying the *LieSymm-PDE* package to the considered Helfrich system (5), we have obtained a determining system of 206 first and second order partial differential equations. All these equations have the general form

$$\sum_k \mu_k (u^1)^j (u^2)^l (u^3)^m (u^4)^n f_k(\mathbf{x}, \mathbf{u}) = 0, \qquad j, l, m, n = 0, 1, \ldots, 7$$

where μ_k are real constants and $f_k(\mathbf{x}, \mathbf{u})$ are either one of the functions $\xi^i(\mathbf{x}, \mathbf{u})$, $\eta^\alpha(\mathbf{x}, \mathbf{u})$ or their first, or second order derivatives. Thirty five equations are with more than 10 addends (expressions of the above form), six are with more than 20 addends. The largest are two equations with 43 and 44 addends. Many of these equations are equivalent to each other or functionally dependent, which means that the determining system is overdetermined. Nevertheless, manipulation of so many equations without making errors is quite boring and time consuming. With the aid of the *LieSymm-PDE* facilities for solving determining systems we managed to do all symbolic calculations automatically, eluding the tedious substitutions, transformations and other technicalities, which otherwise we should had made by hand.

We started up the solving process by invoking the *LieSymm-PDE* iterative function for solving some predetermined types of equations with known solutions. If *LieSymm-PDE* identifies such an equation, its solution is substituted for the respective variable in the remainder part of the equations. In this way the determining system of the Helfrich model has been reduced to 29 partial differential equations for six unknown functions of the form

$$\begin{aligned} \xi^1 &= h(x, y), & \eta^1 &= v(x, y, q, \varphi), & \eta^3 &= g(x, y, q, \varphi, \theta, \omega) \\ \xi^2 &= r(x, y), & \eta^2 &= w(x, y, q, \varphi), & \eta^4 &= \rho(x, y, q, \varphi, \theta, \omega) \end{aligned}$$

where $h(x, y)$ and $r(x, y)$ satisfy the Cauchy-Riemann conditions

$$\frac{\partial h}{\partial y} = -\frac{\partial r}{\partial x}, \qquad \frac{\partial h}{\partial x} = \frac{\partial r}{\partial y}. \tag{8}$$

We proceeded with applying the *LieSymm-PDE* package in an interactive mode feeding back the program with the solutions we had found. After seven interactive cycles two of the coefficient functions changed their form to

$$\eta^1 = q\sigma(x, y), \qquad \eta^2 = -C\varphi$$

($C \in \mathbb{R}$) and the determining system reduced to the 10 equations below

$$\begin{aligned}
q\varphi(24 + \alpha_2\varphi)\rho_\theta + 3\rho_q &= 0\\
24q^2 r_x + \alpha_2 q^2\varphi\rho_\theta + 3\rho_\varphi &= 0\\
q^2\varphi(24 + \alpha_2\varphi)\sigma_x - 3g_x - 3\rho_y &= 0\\
q^2\varphi(24 + \alpha_2\varphi)\sigma_y - 3g_y + 3\rho_x &= 0\\
24q^2 r_x - q^2(24 + \alpha_2\varphi)\rho_\theta - 3\rho_\varphi &= 0\\
3g_\varphi + \alpha_2 q^2\varphi\rho_\omega - 2\alpha_2 q^2\varphi\sigma + 2\alpha_2 C q^2\varphi &= 0\\
3g_\varphi + q^2(24 + \alpha_2\varphi)\rho_\omega - 2q^2(24 + \alpha_2\varphi)\sigma + 2Cq^2(12 + \alpha_2\varphi) &= 0\\
3g_q + q\varphi(24 + \alpha_2\varphi)\rho_\omega - 2q\varphi(24 + \alpha_2\varphi)\sigma + 2Cq\varphi(12 + \alpha_2\varphi) &= 0
\end{aligned}$$

$$\begin{aligned}
&2\sigma_{xx} + 2\sigma_{yy} + 2q^2(8 + \alpha_2\varphi + \alpha_3\varphi^2 + \alpha_4\varphi^3)r_y\\
&\quad + 2q^2(8 + \alpha_2\varphi + \alpha_3\varphi^2 + \alpha_4\varphi^3)\sigma - Cq^2\varphi(\alpha_2 + 2\alpha_3\varphi + 3\alpha_4\varphi^2) = 0
\end{aligned}$$

$$\begin{aligned}
&2(\alpha_2 q^2\varphi^2\theta - 6\alpha_2 q^4\varphi^3 - 6\alpha_3 q^4\varphi^4 - 6\alpha_4 q^4\varphi^5 - 48q^4\varphi^2\\
&\quad + 24q^2\varphi\theta - 3\theta^2 - 3\omega^2)r_y + (\alpha_2 q^2\varphi^2 + 24q^2\varphi - 6\theta)g - 6\omega\rho\\
&\quad - 6(2\alpha_2 q^4\varphi^3 + 2\alpha_3 q^4\varphi^4 + 2\alpha_4 q^4\varphi^5 + 16q^4\varphi^2 - \theta^2 - \omega^2)\sigma\\
&\quad + 6Cq^4\varphi^3(\alpha_2 + 2\alpha_3\varphi + 3\alpha_4\varphi^2) + 6C(4q^2\varphi\theta - \theta^2 - \omega^2) = 0.
\end{aligned}$$

Continuing in the same manner of solving we have obtained the solution of the determining system, arriving at the Lie symmetry algebra of the Helfrich system for the two distinguished cases (compare with [3])

Case 1. $|\alpha_2| + |\alpha_3| + |\alpha_4| \neq 0$

$$\begin{aligned}
X^{\mathrm{I}}(\xi^1, \xi^2) = \xi^1\partial_x + \xi^2\partial_y - q\xi^1_x\partial_q - 2(\theta\xi^1_x + \omega\xi^2_x)\partial_\theta\\
-2\left[\omega\xi^1_x - \left(\theta - 4q^2\varphi - \frac{\alpha_2 q^2\varphi^2}{6}\right)\xi^2_x\right]\partial_\omega
\end{aligned}$$

Case 2. $\alpha_2 = \alpha_3 = \alpha_4 = 0$

$$X^{\mathrm{II}}(\xi^1, \xi^2) = X_1(\xi^1, \xi^2) + cX_2, \qquad c \in \mathbb{R}$$

$$\begin{aligned} X_1(\xi^1, \xi^2) &= \xi^1 \partial_x + \xi^2 \partial_y - q\xi^1_x \partial_q - 2(\theta \xi^1_x + \omega \xi^2_x)\partial_\theta \\ &\quad - 2\left[\omega \xi^1_x - \left(\theta - 4q^2\varphi\right)\xi^2_x\right]\partial_\omega \\ X_2 &= \varphi\partial_\varphi + \theta\partial_\theta + \omega\partial_\omega \end{aligned}$$

where $\xi^1 = h(x, y)$, $\xi^2 = r(x, y)$ are arbitrary real-valued harmonic functions satisfying the Cauchy-Riemann conditions (8) and $\partial_x \equiv \partial/\partial x$, etc.

The full sets of group generators $X^{\mathrm{I}}(\xi^1, \xi^2)$ and $X^{\mathrm{II}}(\xi^1, \xi^2)$ constitute two symmetry Lie algebras L^{I} and L^{II} for each one of the considered cases. The Lie algebras L^{I} and L^{II} are infinite dimensional with the commutator operator defined by

$$[X(\xi^1, \xi^2), X(\widehat{\xi}^1, \widehat{\xi}^2)] = X(\Xi^1, \Xi^2)$$

where

$$\Xi^1 = \xi^1\widehat{\xi}^1_x - \xi^2\widehat{\xi}^2_x - \widehat{\xi}^1\xi^1_x + \widehat{\xi}^2\xi^2_x, \qquad \Xi^2 = \xi^2\widehat{\xi}^1_x + \xi^1\widehat{\xi}^2_x - \widehat{\xi}^2\xi^1_x - \widehat{\xi}^1\xi^2_x$$

(X equals X^{I} or X^{II} respectively).

4 A Group-Invariant Solution

Under the action of the symmetry group each solution of the considered system of differential equations is transformed also into a solution of the system. The *group-invariant solutions* are the fixed points of this action. Any group-invariant solution can be obtained by applying a group reduction technique based on constructing and solving of the so-called *reduced system* of differential equations [5]. It is of great practical importance that in comparison with the original system the reduced system of equations has fewer independent variables. Particularly, if the number of the group parameters is one less the number of the independent variables, then the initially given system of partial differential equations is reduced to a much simpler system of ordinary differential equations.

As it is shown in [3], all the group-invariant solutions of the Helfrich system (5), related to the one-parameter Lie symmetry subgroups of the general Lie symmetry group, are equivalent (conjugate) to each other in a sense that the related subalgebras are pairwise conjugate under the group of the adjoint representations [5]. Hence, it suffices to consider one representative of the conjugacy class of the one-parameter group-invariant solutions from which every other such solution can be derived by acting with a transformation of the symmetry group.

In this chapter we look for a group-invariant solution of the Helfrich system (5), related to the subalgebra of the general symmetry Lie algebra L^{I} (respectively L^{II})

spanned by the group generator $X^{\mathrm{I}}(1,0) \equiv \partial_x$, i.e., a solution invariant under the translations of the variable x. The symmetry reduction procedure leads to the invariant solution of the form [3]

$$q = q(y), \qquad \varphi = \varphi(y), \qquad \theta = \theta(y), \qquad \omega(y) = \alpha_5 \equiv \text{const}$$

and the reduced system of ordinary differential equations

$$\begin{aligned} q^2\varphi_{yy} + 2q\varphi q_{yy} - 2\varphi q_y^2 + q^4(8\varphi + \alpha_2\varphi^2 + \alpha_3\varphi^3 + \alpha_4\varphi^4) = 0 \\ \theta_y - q(8 + \frac{\alpha_2}{3}\varphi)(\varphi q_y + q\varphi_y) = 0 \qquad (9) \\ 4q\varphi^2 q_{yy} + 4\varphi q^2\varphi_{yy} - 4\varphi^2 q_y^2 - 4q^2\varphi_y^2 - \alpha_5^2 - \theta^2 + q^2\varphi\theta(8 + \frac{\alpha_2}{3}\varphi) = 0. \end{aligned}$$

By imposing the relation

$$q(y) = \frac{1}{\varphi(y)} \tag{10}$$

from equations (9) we obtain the system

$$\theta(y) = 0, \qquad \omega(y) = 0 \tag{11}$$

$$\int (C_1\varphi^4 - 2\alpha_4\varphi^3 - \alpha_3\varphi^2 - \frac{2}{3}\alpha_2\varphi - 4)^{-1/2}\mathrm{d}\varphi = y + C_2 \tag{12}$$

in which C_1, C_2 are integration constants. The latter integral can be expressed in terms of elliptic functions [11] (and in some special cases in elementary functions), which means that we have arrived at a group-invariant solution of the Helfrich system given by the Eqs. (10)–(12).

Now, we proceed with the introduction of new variables that will return us to the original geometrical considerations. As it follows from the Brioschi formula (7), the Gaussian curvature K of the Helfrich surfaces corresponding to the above solution vanishes

$$K = 0$$

which implies the existence of cylindrical Helfrich surfaces $\mathcal{S}$, having as directrices plane curves Γ (Fig. 1), whose curvature $\kappa(s) = 2H(s)$ is parametrized by the arclength $s = 2y$ according to the conformal metric (3). By passing to the new variables $(s, \kappa(s))$ in (12), via the expression (6) for H, the equation for the curvature $\kappa(s)$ takes the form

$$\kappa(s) = a + \frac{f'(a)}{4}\left(\wp(\frac{s}{2} + C_2) - \frac{f''(a)}{24}\right)^{-1} \tag{13}$$

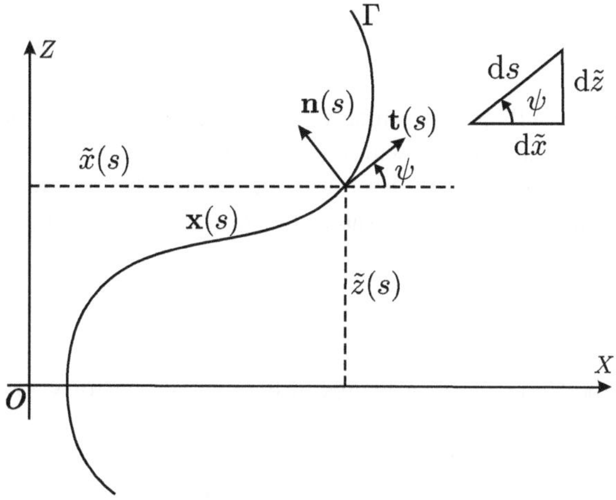

Fig. 1 Geometry of a *plane curve*

where $\wp(s) \equiv \wp(s; g_2, g_3)$ is the Weierstrassian elliptic function [11] and a is an arbitrary root of the polynomial

$$f(t) = -t^4 + 2\mu t^2 + 4\nu t + 8E.$$

The parameters

$$g_2 = \frac{\mu^2}{3} - 8E, \qquad g_3 = \nu^2 - \frac{\mu^3}{27} - \frac{8\mu E}{3}$$

are the invariants of $\wp(s)$ with $\mu = 4(\mathbb{h}^2 + \frac{\lambda}{k})$, $E = \frac{4\mathbb{h}}{k}(p - \lambda\mathbb{h}) - 2\mathbb{h}^4 + \frac{C_1}{2}$, $\nu = -\frac{4p}{k}$, and $f'(a) \equiv \mathrm{d}f(t)/\mathrm{d}t_{|_{t=a}}, \ldots$. As a result we obtain a special type of *cylindrical Helfrich surfaces* $\mathcal{S}$ with directrices Γ, defined by their *intrinsic equation* (13) in terms of the Weierstrassian $\wp$-function.

5 Cylindrical Helfrich Surfaces

Given the intrinsic equation $\kappa = \kappa(s)$ of a plane curve Γ, it is possible to recover (up to a rigid motion) the position vector of the curve $\mathbf{x}(s) = (\tilde{x}(s), \tilde{z}(s))$ in the plane $\mathbb{R}^2$ in a standard manner by calculating the quadratures [7]

$$\tilde{x}(s) = \int \cos\psi(s)\mathrm{d}s, \qquad \tilde{z}(s) = \int \sin\psi(s)\mathrm{d}s \tag{14}$$

where $\psi(s)$ is the slope angle of Γ (Fig. 1), expressed by the formula

$$\psi(s) = \int \kappa(s)\mathrm{d}s. \tag{15}$$

In the following we will confine ourselves to the case $\nu \neq 0$. For the directrices of the Helfrich cylindrical surfaces with intrinsic equation (13) the integration in (14), for $\nu \neq 0$, can be avoided (cf [12] and references cited therein), which is a specific consequence of the fact that the curvature function (13) satisfies the equation

$$2\frac{\mathrm{d}^2\kappa(s)}{\mathrm{d}s^2} + \kappa^3(s) - \mu\kappa(s) - \nu = 0.$$

As a result the Cartesian coordinates of the position vector $\mathbf{x}(s) = (\tilde{x}(s), \tilde{z}(s))$ take the form [12]

$$\begin{aligned} \tilde{x}(s) &= \frac{2}{\nu}\frac{\mathrm{d}\kappa(s)}{\mathrm{d}s}\cos\psi(s) + \frac{\kappa^2(s) - \mu}{\nu}\sin\psi(s) \\ \tilde{z}(s) &= \frac{2}{\nu}\frac{\mathrm{d}\kappa(s)}{\mathrm{d}s}\sin\psi(s) - \frac{\kappa^2(s) - \mu}{\nu}\cos\psi(s) \end{aligned} \tag{16}$$

where the slope angle $\psi(s)$ is expressed explicitly, via the quadrature (15), in terms of the Weierstrassian function $\wp(s)$, and its integrals $\sigma(s)$ and $\zeta(s)$ [11]

$$\psi(s) = as + \frac{f'(a)}{2\wp'(\mathring{s})}\left[s\zeta(\mathring{s}) + \ln\frac{\sigma(\frac{s}{2} - \mathring{s})}{\sigma(\frac{s}{2} + \mathring{s})}\right].$$

Here $\mathring{s}$ is defined by the equation $\wp(\mathring{s}) = \frac{f''(a)}{24}$ $(C_2 = 0)$.

Taken with respect to the moving reference frame $(\mathbf{t}(s), \mathbf{n}(s))$, built up by the tangent $\mathbf{t}(s) = (\mathrm{d}\tilde{x}/\mathrm{d}s, \mathrm{d}\tilde{z}/\mathrm{d}s)$ and the normal $\mathbf{n}(s) = (-\mathrm{d}\tilde{z}/\mathrm{d}s, \mathrm{d}\tilde{x}/\mathrm{d}s)$ vectors to the curve (Fig. 1), the coordinates $(\tilde{\xi}(s), \tilde{\eta}(s))$ of the position vector

$$\mathbf{x}(s) = \tilde{\xi}(s)\mathbf{t}(s) + \tilde{\eta}(s)\mathbf{n}(s)$$

have the form (compare with (16)) and are depicted via the curves in the second row in Figs. 2 and 3

$$\tilde{\xi}(s) = \frac{2}{\nu}\frac{\mathrm{d}\kappa(s)}{\mathrm{d}s}, \qquad \tilde{\eta}(s) = -\frac{\kappa^2(s) - \mu}{\nu}.$$

Several graphs of closed directrices of the so obtained cylindrical Helfrich surfaces, defined by $(\tilde{x}, \tilde{z})$ and $(\tilde{\xi}, \tilde{\eta})$ in the respective fixed and moving reference frame for different values of the parameters ν, μ, E and a, are presented in Figs. 2 and 3.

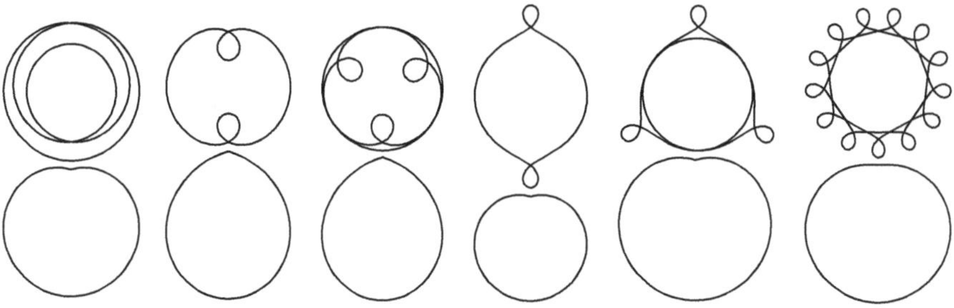

Fig. 2 Closed directrices with self-intersections in the fixed (*top*) and the moving (*bottom*) reference frame

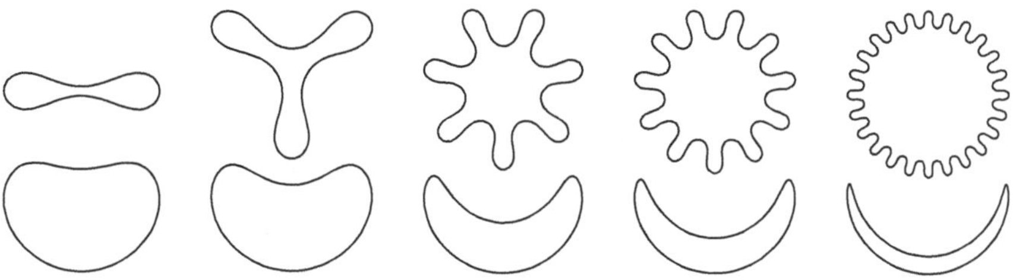

Fig. 3 Closed directrices without self-intersections in the fixed (*top*) and the moving (*bottom*) reference frame

6 Conclusion

The results presented in this chapter are obtained by an application of the Lie group analysis to the Helfrich model of the shapes of the biological membranes. The Helfrich model is considered in conformal metric representation. The Lie group analysis has been carried out by the help of the program *LieSymm-PDE* within the computer system `Mathematica`®. The determining system of equations for the admissible group of point symmetry transformations has been created. The determining system consists of 206 first and second order partial differential equations. With the help of the program *LieSymm-PDE* the determining system has been solved in explicit form. Its solution constitutes an infinite dimensional Lie algebra of symmetries of the Helfrich model. Based on this a group reduction method for finding group-invariant solutions has been applied and a solution of the Helfrich system has been obtained. The Helfrich surfaces corresponding to this solution are generalized cylinders whose directrices are plane curves with curvatures expressed by the Weierstrassian functions. Plots of several graphs of the directrices of some of these cylindrical surfaces in fixed and moving reference frame are presented.

References

1. Brown, F.: Elastic modeling of biomembranes and lipid bilayers. Annu. Rev. Phys. Chem. **59**, 685–712 (2008)
2. Faller, R. (ed.): Biomembrane frontiers: nanostructures, models, and the design of life, handbook of modern biophysics, vol. 2. Humana Press, New York (2009)
3. de Matteis, G., Manno, G.: Lie algebra symmetry analysis of the Helfrich and Willmore surface shape equations. Comm. Pure Appl. Anal. **13**, 453–481 (2014)
4. Konopelchenko, B.: On solutions of the shape equation for membranes and strings. Phys. Lett. B **414**, 58–64 (1997)
5. Olver, P.: Applications of Lie groups to differential equations. Springer, Berlin (1986)
6. Helfrich, W.: Elastic properties of lipid bilayers: theory and possible experiments. Z. Naturforsch C **28**, 693–703 (1973)
7. Oprea, J.: Differential geometry and its applications. Mathematical Association of America, Washington D.C. (2007)
8. Pulov, V., Chacarov, E., Hadzhilazova, M., Mladenov, I.: Symmetry properties of the membrane shape equation. Geom. Integr. Quant. **14**, 152–159 (2013)
9. Vassilev, V., Djondjorov, P., Mladenov, I.: Symmetry groups, conservation laws and group-invariant solutions of the membrane shape equation. Geom. Integr. Quant. **7**, 265–279 (2006)
10. Pulov, V., Chacarov, E., Uzunov, I.: A computer algebra application to determination of Lie symmetries of partial differential equations. Serdica J. Comput. **1**, 505–518 (2007)
11. Abramowitz, M., Stegun, I. (eds.): Handbook of mathematical functions. Dover, New York (1972)
12. Vassilev, V., Djondjorov P., Mladenov, I.: Cylindrical equilibrium shapes of fluid membranes. J. Phys. A Math. Theor. **41**, 435201(16pp) (2008)

Lie Group Analysis of the Willmore and Membrane Shape Equations

Vassil M. Vassilev, Petar A. Djondjorov and Ivaïlo M. Mladenov

Abstract The present paper is concerned with the geometric Lie symmetry groups of the Willmore and shape equations—the Euler-Lagrange equations associated with the Willmore and Helfrich functionals. The ten-parameter group of special conformal transformations in the three-dimensional Euclidean space, which in known to be the symmetry group of the Willmore functional, is recognized as the largest group of geometric transformations admitted by these equations in Monge representation. The conserved currents of ten linearly independent conservation laws, which correspond to the variational symmetries of the Willmore equation and hold on its smooth solutions, are derived. The shape equation is found to admit only a six-parameter subgroup of the aforementioned ten-parameter group. Each symmetry admitted by the shape equation is its variational symmetry as well and the corresponding conserved currents are obtained.

1 Introduction

A wide variety of objects exhibit elastic behaviour in ordinary operation, and "thin" or "thin-walled" ones are of special interest concerning various human activities. In many cases their equilibrium shapes are formed due to bending. The problem for determination of the equilibrium shapes of such objects (bars, rods, archs, rings, pipes, baloons, etc.) is usually reduced to analysis of geometric objects—curves and surfaces [6]. Such an analysis is based on two concepts—extrema of the curvature

V. M. Vassilev (✉) · P. A. Djondjorov
Institute of Mechanics, Bulgarian Academy of Sciences,
Acad. G. Bonchev Str., Block 4, 1113 Sofia, Bulgaria
e-mail: vasilvas@imbm.bas.bg

I. M. Mladenov
Institute of Biophysics and Biomedical Engineering, Bulgarian Academy of Sciences,
Acad. G. Bonchev Str., Block 21, 1113 Sofia, Bulgaria

J.-F. Ganghoffer and I. Mladenov (eds.), *Similarity and Symmetry Methods*, Lecture Notes in Applied and Computational Mechanics 73, DOI: 10.1007/978-3-319-08296-7_7,

and symmetry of the shape. Obtaining the curvature of such geometric objects it is possible to determine the stresses [13] and even the parametric equations of the curve/surface in some cases (see, e.g., [5, 27]).

To the best of our knowledge, the first model in this field was suggested by Daniel Bernoulli (see [1]) for plane curves and it states that the shapes of each such curve $\mathcal{L}$ are extrema of the functional

$$\mathcal{B} = \int_{\mathcal{L}} \kappa^2 \mathrm{d}s$$

where κ is the curvature of the curve $\mathcal{L}$. It took more that 50 years until Siméon Denis Poisson [20] and a bit later Marie-Sophie Germain proposed a similar idea concerning surfaces, namely the equilibrium shapes of an elastic surface are extrema of the total squared mean curvature of this surface (see, e.g. [8])

$$\mathcal{W} = \int_{\mathcal{S}} H^2 \mathrm{d}A \tag{1}$$

where H is the mean curvature of the surface $\mathcal{S}$. The Euler-Lagrange equation for this functional reads

$$\Delta H + 2(H^2 - K)H = 0 \tag{2}$$

in which Δ is the Laplace-Beltrami operator on the surface $\mathcal{S}$ and K is the Gaussian curvature of $\mathcal{S}$.

According to Thomsen [23], W. Schadow was the first who had derived Eq. (2) in 1922 as the Euler-Lagrange equation for the variational problem

$$\int_{\mathcal{S}} \left(\frac{1}{R_1} - \frac{1}{R_2} \right)^2 \mathrm{d}A \tag{3}$$

where $1/R_1$ and $1/R_2$ are the two principal curvatures of the surface $\mathcal{S}$. This variational problem is studied in Thomsen's thesis in 1923 (see [23]), where a reference to the aforementioned (and probably unpublished) result by Schadow was given. Actually, the Lagrangian densities of the functionals (1) and (3) are proportional up to the divergence term $2K$ and that is why they lead to the same Euler-Lagrange equation.

However, these results remain unnoticed and when Thomas Willmore [29] suggested again the functional (1) in 1965, it along with the Eq. (2) were named after him. Thus, surfaces providing extremum to the functional (1) are now referred to as Willmore surfaces.

The study of the extrema of functional (1), i.e. the Willmore surfaces, turned out to be of great importance not only for differential geometry (in connection with the Willmore problem and conformal geometry) but also for the 2D string theory and

2D gravity based on the Polyakov integral over surfaces (see [21]). In these theories, the functional (1) is known as the Polyakov's extrinsic action. The properties of the Polyakov's extrinsic action as well as various generalizations, such as the Polyakov-Kleinert rigid string action [11, 22]

$$\mathcal{A} = \int_{\mathcal{S}} \left(\alpha H^2 + \gamma\right) \mathrm{d}A, \qquad \alpha, \gamma - \text{constant} \tag{4}$$

for instance, have been studied in a number of papers (see the review article [7]).

The functional (1) has found application in biophysics too. In Helfrich theory [9], the bending energy of a homogeneous vesicle membrane is assumed to be given by the functional

$$\mathcal{F}_c = k \int_{\mathcal{S}} H^2 \mathrm{d}A + \bar{k} \int_{\mathcal{S}} K \mathrm{d}A$$

where k and $\bar{k}$ are real constants representing the bending and Gaussian rigidity of the membrane. The equilibrium shape of the vesicle is supposed to be determined by an extremum of the Helfrich curvature free energy (shape energy)

$$\mathcal{F} = k \int_{\mathcal{S}} (H - \mathrm{I\!h})^2 \, \mathrm{d}A + \bar{k} \int_{\mathcal{S}} K \mathrm{d}A + \lambda \int_{\mathcal{S}} \mathrm{d}A + \Delta p \int \mathrm{d}V \tag{5}$$

where $\mathrm{d}V$ is the volume element, $\mathrm{I\!h}$, λ and Δp are real constants and denote the so-called spontaneous curvature, tensile stress and osmotic pressure difference between the outer and inner media. The corresponding Euler-Lagrange equation

$$k \Delta H + 2k (H - \mathrm{I\!h}) (H^2 - K) - 2k (H - \mathrm{I\!h})^2 H - 2\lambda H + \Delta p = 0 \tag{6}$$

(derived in [15, 16]) is referred to as the Helfrich's membrane shape equation.

There is a vast amount of papers in which the extrema of functionals (1) and (5), i.e., the solutions of Eqs. (2) and (6) are studied (see, e.g., [12, 17, 18, 24, 25] and the references therein). However, the determined exact solutions of these equations are quite a few and all of them rely on an assumption of a certain symmetry of the equilibrium shape. This observation is a good motivation to study Eqs. (2) and (6) applying Lie group analysis (details on this approach can be found in [10, 14, 19]) in order to determine all symmetries of these equations. The aim of this paper is to summarize the achievements of the authors in Lie group analysis of Eqs. (2) and (6) in Monge representation and thus to provide a ground knowledge for determination of their group-invariant solutions.

2 Willmore and Shape Equations in Monge Representation

Let (x^1, x^2, x^3) be a fixed right-handed rectangular Cartesian coordinate system in the three-dimensional Euclidean space $\mathbb{R}^3$ in which a surface $\mathcal{S}$ is immersed, and let this surface be given by the equation

$$\mathcal{S} : x^3 = w(x^1, x^2), \qquad (x^1, x^2) \in \Omega \subset \mathbb{R}^2 \tag{7}$$

where $w : \mathbb{R}^2 \to \mathbb{R}$ is a single-valued and smooth function possessing as many derivatives as may be required on the domain Ω. Let us take x^1, x^2 to serve as Gaussian coordinates on the surface $\mathcal{S}$. Then, relative to this coordinate system, the components of the first fundamental tensor $g_{\alpha\beta}$, the second fundamental tensor $b_{\alpha\beta}$, and the alternating tensor $\varepsilon^{\alpha\beta}$ of $\mathcal{S}$ are given by the expressions

$$g_{\alpha\beta} = \delta_{\alpha\beta} + w_\alpha w_\beta, \qquad b_{\alpha\beta} = g^{-1/2} w_{\alpha\beta}, \qquad \varepsilon^{\alpha\beta} = g^{-1/2} e^{\alpha\beta} \tag{8}$$

where

$$g = \det(g_{\alpha\beta}) = 1 + (w_1)^2 + (w_2)^2 \tag{9}$$

$\delta_{\alpha\beta}$ is the Kronecker delta symbol and $e^{\alpha\beta}$ is the alternating symbol. The contravariant components $g^{\alpha\beta}$ of the first fundamental tensor read

$$g^{\alpha\beta} = g^{-1}\delta^{\alpha\beta} + \varepsilon^{\alpha\mu}\varepsilon^{\beta\nu} w_\mu w_\nu = g^{-1}\left(\delta^{\alpha\beta} + e^{\alpha\mu}e^{\beta\nu} w_\mu w_\nu\right). \tag{10}$$

Here and in what follows: Greek indices have the range 1, 2, and the usual summation convention over a repeated index is employed, $w_{\alpha_1 \ldots \alpha_k}$ $(k = 1, 2, \ldots)$ denote the kth order partial derivatives of the function w with respect to the variables x^1 and x^2, i.e.,

$$w_{\alpha_1\alpha_2\ldots\alpha_k} = \frac{\partial^k w}{\partial x^{\alpha_1} \ldots \partial x^{\alpha_k}}, \qquad k = 1, 2, \ldots$$

The mean curvature H of the surface $\mathcal{S}$ and its Gaussian curvature K are given as follows

$$H = \frac{1}{2} g^{\alpha\beta} b_{\alpha\beta}, \qquad K = \frac{1}{2} \varepsilon^{\alpha\mu}\varepsilon^{\beta\nu} b_{\alpha\beta} b_{\mu\nu} \tag{11}$$

that is

$$H = \frac{1}{2} g^{-3/2}\left(\delta^{\alpha\beta} w_{\alpha\beta} + e^{\alpha\mu}e^{\beta\nu} w_{\alpha\beta} w_\mu w_\nu\right), \qquad K = \frac{1}{2} g^{-2} e^{\alpha\mu}e^{\beta\nu} w_{\alpha\beta} w_{\mu\nu}. \tag{12}$$

In the above Monge representation the Willmore functional (1) reads

$$\mathcal{W} = \iint_{\Omega} \frac{1}{4} g^{-5/2} \left(\delta^{\alpha\beta} w_{\alpha\beta} + e^{\alpha\mu} e^{\beta\nu} w_{\alpha\beta} w_\mu w_\nu\right)^2 \mathrm{d}x^1 \mathrm{d}x^2. \tag{13}$$

The application of the Euler operator

$$E = \frac{\partial}{\partial w} - D_\mu \frac{\partial}{\partial w_\mu} + D_\mu D_\nu \frac{\partial}{\partial w_{\mu\nu}} - \cdots$$

where

$$D_\alpha = \frac{\partial}{\partial x^\alpha} + w_\alpha \frac{\partial}{\partial w} + w_{\alpha\mu} \frac{\partial}{\partial w_\mu} + w_{\alpha\mu\nu} \frac{\partial}{\partial w_{\mu\nu}} + w_{\alpha\mu\nu\sigma} \frac{\partial}{\partial w_{\mu\nu\sigma}} + \cdots$$

denote the total derivative operators, on the Lagrangian density

$$L = H^2 g^{1/2} = \frac{1}{4} g^{-5/2} \left(\delta^{\alpha\beta} w_{\alpha\beta} + e^{\alpha\mu} e^{\beta\nu} w_{\alpha\beta} w_\mu w_\nu\right)^2 \tag{14}$$

of the Willmore functional leads, after taking into account expressions (8), (9), (10) and (12), to the expression

$$E(L) = \Delta H + 2(H^2 - K)H.$$

Actually, in Monge representation, the Willmore equation $E(L) = 0$ is to be regarded as a fourth-order partial differential equation in two independent variables x^1, x^2 and one dependent variable w—the displacement field. This equation belongs to the class of equations of the form

$$\mathcal{E} \equiv \frac{1}{2} g^{-1/2} g^{\alpha\beta} g^{\mu\nu} w_{\alpha\beta\mu\nu} + \Phi\left(x_1, x_2, w, w_1, \ldots, w_{222}\right) = 0 \tag{15}$$

where $\Phi\left(x_1, x_2, w, w_1, \ldots, w_{222}\right)$ is a differential function depending on the independent and dependent variables and the derivatives of the dependent variable up to third order. Indeed, using expressions (8), (9), (10), (12) and the well-known formula

$$\Delta = g^{-1/2} \frac{\partial}{\partial x^\alpha} \left(g^{1/2} g^{\alpha\beta} \frac{\partial}{\partial x^\beta}\right) = g^{\alpha\beta} \frac{\partial^2}{\partial x^\alpha \partial x^\beta} + g^{-1/2} \frac{\partial}{\partial x^\alpha} \left(g^{1/2} g^{\alpha\beta}\right) \frac{\partial}{\partial x^\beta}$$

one can represent the Willmore equation (2) in the form (15). The same procedure can be applied to the functional (5) and Eq. (6). Omitting the lengthy but straightforward computations, this equation is found to belong also to the class (15) with a certain differential function $\Phi(x_1, x_2, w, w_1, \ldots, w_{222})$.

3 Lie Group Analysis of the Willmore Equation

3.1 Symmetry Groups

The main objective of the present Section is to establish, following [10, 14, 19], the invariance properties of the Willmore equation (2) relative to local one-parameter Lie groups of local point transformations acting on open subsets of the three-dimensional Euclidean space $\mathbb{R}^3$, with coordinates (x^1, x^2, w), representing the involved independent and dependent variables x^1, x^2 and w, respectively. For that purpose Lie infinitesimal technique is used and the results obtained are expressed in terms of the infinitesimal generators (operators) of the respective groups. In the present case, the latter are vector fields on $\mathbb{R}^3$ of the form

$$\mathbf{v} = \xi^{\mu}\frac{\partial}{\partial x^{\mu}} + \eta\frac{\partial}{\partial w} \tag{16}$$

where ξ^{μ} and η are functions of the variables x^1, x^2 and w.

The infinitesimal criterion of invariance

$$\mathrm{pr}^{(4)}\mathbf{v}\,(\mathcal{E}) = 0 \text{ whenever } \mathcal{E} = 0 \tag{17}$$

where $\mathrm{pr}^{(n)}\mathbf{v}$ denotes the nth prolongation of the vector field $\mathbf{v}$ (see [14]), leads, through the standard computational procedure (see, e.g. [14] or [19]), to the following result:

Proposition 1 *The ten-parameter Lie group G_{SCT} of special conformal transformations in $\mathbb{R}^3$ (whose basic generators $\mathbf{v}_j$, $j = 1, \ldots, 10$, their characteristics, commutators and the corresponding finite transformations are given in Tables* 1, 2, 3*) is the largest group of geometric transformations of the involved independent and dependent variables that a generic equation of form (15) could admit.*

Let us denote by L_{SCT} the Lie algebra corresponding to the group G_{SCT}, i.e. L_{SCT} is the ten-dimensional Lie algebra spanned by the vector fields $\mathbf{v}_j$, $j = 1 \ldots 10$. Actually, the group G_{SCT} is a representation of the Lie group $O\,(4, 1)$ in the vector space $\mathbb{R}^3$, which corresponds to the action of $O\,(4, 1)$ on $\mathbb{R}^3$ determined by the representation L_{SCT} of its Lie algebra $O\,(4, 1)$ in $\mathbb{R}^3$.

Proposition 2 *In Monge representation, the Willmore equation (2) admits all the transformations of the group G_{SCT}.*

3.2 Conservation Laws

A particular interest exists for the variational symmetries of Eq. (2)—the Lie groups generated by the so-called infinitesimal divergence symmetries (see Definition 4.33 in

Table 1 Generators and characteristics of the group of special conformal transformations in $\mathbb{R}^3$

Generators	Characteristics
Translations	
$\mathbf{v}_1 = \frac{\partial}{\partial x^1}$	$Q_1 = -w_1$
$\mathbf{v}_2 = \frac{\partial}{\partial x^2}$	$Q_2 = -w_2$
$\mathbf{v}_3 = \frac{\partial}{\partial w}$	$Q_3 = 1$
Rotations	
$\mathbf{v}_4 = -x^2\frac{\partial}{\partial x^1} + x^1\frac{\partial}{\partial x^2}$	$Q_4 = x^2 w_1 - x^1 w_2$
$\mathbf{v}_5 = -w\frac{\partial}{\partial x^1} + x^1\frac{\partial}{\partial w}$	$Q_5 = x^1 + w w_1$
$\mathbf{v}_6 = -w\frac{\partial}{\partial x^2} + x^2\frac{\partial}{\partial w}$	$Q_6 = x^2 + w w_2$
Dilatation	
$\mathbf{v}_7 = x^1\frac{\partial}{\partial x^1} + x^2\frac{\partial}{\partial x^2} + w\frac{\partial}{\partial w}$	$Q_7 = w - x^1 w_1 - x^2 w_2$
Inversions	
$\mathbf{v}_8 = \chi^1\frac{\partial}{\partial x^1} + 2x^1x^2\frac{\partial}{\partial x^2} + 2x^1 w\frac{\partial}{\partial w}$	$Q_8 = 2x^1 w - \chi^1 w_1 - 2x^1x^2 w_2$
$\mathbf{v}_9 = 2x^2x^1\frac{\partial}{\partial x^1} + \chi^2\frac{\partial}{\partial x^2} + 2x^2 w\frac{\partial}{\partial w}$	$Q_9 = 2x^2 w - 2x^1x^2 w_1 - \chi^2 w_2$
$\mathbf{v}_{10} = 2x^1 w\frac{\partial}{\partial x^1} + 2x^2 w\frac{\partial}{\partial x^2} + \chi^3\frac{\partial}{\partial w}$	$Q_{10} = \chi^3 - 2x^1 w w_1 - 2x^2 w w_2$

Here, the following notation is used: $\chi^1 = (x^1)^2 - (x^2)^2 - w^2$, $\chi^2 = (x^2)^2 - (x^1)^2 - w^2$ and $\chi^3 = w^2 - (x^2)^2 - (x^1)^2$

Table 2 Commutator table of the generators. Here, the entry in row i and column j represents the commutator $[\mathbf{v}_i, \mathbf{v}_j]$

	$\mathbf{v}_1$	$\mathbf{v}_2$	$\mathbf{v}_3$	$\mathbf{v}_4$	$\mathbf{v}_5$	$\mathbf{v}_6$	$\mathbf{v}_7$	$\mathbf{v}_8$	$\mathbf{v}_9$	$\mathbf{v}_{10}$
$\mathbf{v}_1$	0	0	0	$\mathbf{v}_2$	$\mathbf{v}_3$	0	$\mathbf{v}_1$	$2\mathbf{v}_7$	$-2\mathbf{v}_4$	$-2\mathbf{v}_5$
$\mathbf{v}_2$	0	0	0	$-\mathbf{v}_1$	0	$\mathbf{v}_3$	$\mathbf{v}_2$	$2\mathbf{v}_4$	$2\mathbf{v}_7$	$-2\mathbf{v}_6$
$\mathbf{v}_3$	0	0	0	0	$-\mathbf{v}_1$	$-\mathbf{v}_2$	$\mathbf{v}_3$	$2\mathbf{v}_5$	$2\mathbf{v}_6$	$2\mathbf{v}_7$
$\mathbf{v}_4$	$-\mathbf{v}_2$	$\mathbf{v}_1$	0	0	$-\mathbf{v}_6$	$\mathbf{v}_5$	0	$-\mathbf{v}_9$	$\mathbf{v}_8$	0
$\mathbf{v}_5$	$-\mathbf{v}_3$	0	$\mathbf{v}_1$	$\mathbf{v}_6$	0	$-\mathbf{v}_4$	0	$-\mathbf{v}_{10}$	0	$\mathbf{v}_8$
$\mathbf{v}_6$	0	$-\mathbf{v}_3$	$\mathbf{v}_2$	$-\mathbf{v}_5$	$\mathbf{v}_4$	0	0	0	$-\mathbf{v}_{10}$	$\mathbf{v}_9$
$\mathbf{v}_7$	$-\mathbf{v}_1$	$-\mathbf{v}_2$	$-\mathbf{v}_3$	0	0	0	0	$\mathbf{v}_8$	$\mathbf{v}_9$	$\mathbf{v}_{10}$
$\mathbf{v}_8$	$-2\mathbf{v}_7$	$-2\mathbf{v}_4$	$-2\mathbf{v}_5$	$\mathbf{v}_9$	$\mathbf{v}_{10}$	0	$-\mathbf{v}_8$	0	0	0
$\mathbf{v}_9$	$2\mathbf{v}_4$	$-2\mathbf{v}_7$	$-2\mathbf{v}_6$	$-\mathbf{v}_8$	0	$\mathbf{v}_{10}$	$-\mathbf{v}_9$	0	0	0
$\mathbf{v}_{10}$	$2\mathbf{v}_5$	$2\mathbf{v}_6$	$-2\mathbf{v}_7$	0	$-\mathbf{v}_8$	$-\mathbf{v}_9$	$-\mathbf{v}_{10}$	0	0	0

[14]) of any variational functional with (2) as the associated Euler-Lagrange equation. Note that if two functionals lead to the same Euler-Lagrange equation, then they have the same collection of infinitesimal divergence symmetries. This interest is motivated by the fact that, in virtue of the Bessel-Hagen's extension of Noether's theorem, each variational symmetry of a given self-adjoint equation corresponds to a conservation law admitted by the smooth solutions of the equation. Thus, if a vector field $\mathbf{v}$ of form (16) is found to generate a variational symmetry of Eq. (2), then Bessel-Hagen's extension of Noether's theorem implies the existence of a conserved current, which, in the present case, is a couple of differential functions P^α such that

Table 3 Finite transformations. Here, the entries in the "finite transformation" column give the transformed points $\exp(\varepsilon_j \mathbf{v}_j)(x^1, x^2, w)$ and $\chi = (r^2 + w^2)$

Groups	Finite Transformations
$G(\mathbf{v}_1)$	$(x^1 + \varepsilon_1, x^2, w)$
$G(\mathbf{v}_2)$	$(x^1, x^2 + \varepsilon_2, w)$
$G(\mathbf{v}_3)$	$(x^1, x^2, w + \varepsilon_3)$
$G(\mathbf{v}_4)$	$(x^1 \cos\varepsilon_4 - x^2 \sin\varepsilon_4, x^2 \cos\varepsilon_4 + x^1 \sin\varepsilon_4, w)$
$G(\mathbf{v}_5)$	$(x^1 \cos\varepsilon_5 - w \sin\varepsilon_5, x^2, w \cos\varepsilon_5 + x^1 \sin\varepsilon_5)$
$G(\mathbf{v}_6)$	$(x^1, x^2 \cos\varepsilon_6 - w \sin\varepsilon_6, w \cos\varepsilon_6 + x^2 \sin\varepsilon_6)$
$G(\mathbf{v}_7)$	$(e^{\varepsilon_7} x^1, e^{\varepsilon_7} x^2, e^{\varepsilon_7} w)$
$G(\mathbf{v}_8)$	$(\frac{x^1 - \varepsilon_8 (r^2+w^2)}{1-2\varepsilon_8 x^1 + \varepsilon_8^2 (r^2+w^2)}, \frac{x^2}{1-2\varepsilon_8 x^1 + \varepsilon_8^2 (r^2+w^2)}, \frac{w}{1-2\varepsilon_8 x^1 + \varepsilon_8^2 (r^2+w^2)})$
$G(\mathbf{v}_9)$	$(\frac{x^1}{1-2\varepsilon_9 x^2 + \varepsilon_9^2 (r^2+w^2)}, \frac{x^2 - \varepsilon_9 (r^2+w^2)}{1-2\varepsilon_9 x^2 + \varepsilon_9^2 (r^2+w^2)}, \frac{w}{1-2\varepsilon_9 x^2 + \varepsilon_9^2 (r^2+w^2)})$
$G(\mathbf{v}_{10})$	$(\frac{x^1}{1-2\varepsilon_{10} w + \varepsilon_{10}^2 (r^2+w^2)}, \frac{x^2}{1-2\varepsilon_{10} w + \varepsilon_{10}^2 (r^2+w^2)}, \frac{w - \varepsilon_{10} (r^2+w^2)}{1-2\varepsilon_{10} w + \varepsilon_{10}^2 (r^2+w^2)})$

$$D_\alpha P^\alpha = QE(L) \tag{18}$$

where Q is the characteristic of the vector field $\mathbf{v}$. By definition

$$Q = \eta - w_\mu \xi^\mu. \tag{19}$$

The total divergence of the conserved current P^α vanishes on the smooth solutions of Eq. (2) and so we have the conservation law

$$D_\alpha P^\alpha = 0 \tag{20}$$

(18) being its expression in characteristic form, and Q—its characteristic.

To derive the conservation laws of the foregoing type, one can proceed by first determining the variational symmetries of the equation considered on the ground of the invariance criterion

$$\mathrm{pr}^{(2)}\mathbf{v}(L) + (D_\mu \xi^\mu) L = D_\mu B^\mu$$

where B^α are certain differential functions. Then using their characteristics (19) to find, from equality (18), explicit expressions for the corresponding conserved currents P^α.

It is well-known (see [30]), that the Willmore functional (1) is invariant under the conformal transformations of a closed surface $\mathcal{S}$. This follows from the invariance of the functional

$$\int_{\mathcal{S}} (H^2 - K)\, \mathrm{d}A \tag{21}$$

under the group of conformal transformations (see [2, 28]) and the Gauss-Bonnet theorem which states that the area-integral over the Gaussian curvature is a topological invariant.

All vector fields $\mathbf{v}_j,\ j = 1, \ldots, 10$ are variational symmetries of the Willmore equation (2) and hence, ten linearly independent conservation laws

$$D_\alpha P_j^\alpha = 0, \qquad j = 1, \ldots, 10$$

exist that hold on its smooth solutions. The corresponding conserved currents are

$$\begin{aligned} P_j^\alpha &= N_j^\alpha L, \quad j = 1, \ldots, 7 \\ P_8^\alpha &= N_8^\alpha L - Q^{\alpha 1} \\ P_9^\alpha &= N_9^\alpha L - Q^{\alpha 2} \\ P_{10}^\alpha &= N_{10}^\alpha L + \frac{2}{\sqrt{g}} \delta^{\alpha\mu} w_\mu \end{aligned}$$

where

$$N_j^\alpha = \xi_j^\alpha - \frac{1}{2} Q_j D_\mu \frac{\partial}{\partial w_{\alpha\mu}} - \frac{1}{2} Q_j D_\mu \frac{\partial}{\partial w_{\mu\alpha}} + \frac{1}{2} \left(D_\mu Q_j\right) \frac{\partial}{\partial w_{\alpha\mu}} + \frac{1}{2} \left(D_\mu Q_j\right) \frac{\partial}{\partial w_{\mu\alpha}}$$

are the so-called Noether operators (cf. [10]), corresponding to the vector fields $\mathbf{v}_j$ with characteristics $Q_j,\ j = 1, \ldots, 10$, and

$$Q^{\alpha\beta} = -\frac{2}{\sqrt{g}} e^{\alpha\mu} e^{\beta\nu} g_{\mu\nu}.$$

Note that

$$H = D_\alpha \left(\frac{1}{2\sqrt{g}} \delta^{\alpha\mu} w_\mu \right), \quad 4H \delta^{\alpha\mu} w_\mu = D_\mu Q^{\alpha\mu}.$$

4 Lie Group Analysis of the Membrane Shape Equation

One of the principle results in Sect. 3 is that the ten-parameter Lie group of special conformal transformations in $\mathbb{R}^3$ is the largest group of geometric (point) transformations of the involved independent and dependent variables that a generic equation of the form (15) could admit. Using this result one can easily ascertain by inspection that the symmetry group of the membrane shape equation (6) is restricted to the group of motions in $\mathbb{R}^3$ whose basic generators $\mathbf{v}_j$ $(j = 1 \ldots 6)$ and their characteristics Q_j are given in Table 4.

Table 4 Generators and characteristics of the group of motions in $\mathbb{R}^3$

Generators	Characteristics
Translations	
$\mathbf{v}_1 = \frac{\partial}{\partial x^1}$	$Q_1 = -w_1$
$\mathbf{v}_2 = \frac{\partial}{\partial x^2}$	$Q_2 = -w_2$
$\mathbf{v}_3 = \frac{\partial}{\partial w}$	$Q_3 = 1$
Rotations	
$\mathbf{v}_4 = -x^2\frac{\partial}{\partial x^1} + x^1\frac{\partial}{\partial x^2}$	$Q_4 = x^2 w_1 - x^1 w_2$
$\mathbf{v}_5 = -w\frac{\partial}{\partial x^1} + x^1\frac{\partial}{\partial w}$	$Q_5 = x^1 + w w_1$
$\mathbf{v}_6 = -w\frac{\partial}{\partial x^2} + x^2\frac{\partial}{\partial w}$	$Q_6 = x^2 + w w_2$

All vector fields $\mathbf{v}_j$ $(j = 1, \ldots, 6)$ are variational symmetries of the membrane shape equation (6) and hence, in virtue of Noether's theorem, six linearly independent conservation laws

$$D_\alpha P_j^\alpha = Q_j E(L), \qquad \alpha = 1, 2, \qquad j = 1, \ldots, 6$$

exist that hold on the smooth solutions of this equation. The respective conserved currents P_j^α are

$$P_j^\alpha = N_j^\alpha L.$$

5 Concluding Remarks

It should be noted that de Matteis and co-authors [3, 4] also studied the symmetry properties of Eq. (2) and (6), but following a different approach to the problem. Actually, they construct a system of four second and first order nonlinear differential equations on the basis of Eq. (2) or (6) and Gauss-Codazzi-Mainardi equations of the surface and apply Lie group analysis to that system. It is established in [4] that some of the symmetry groups determined therein could be transformed to the first seven symmetries, presented in Table 1. It seems that the other symmetry groups obtained in [3, 4] can be interpreted as generalized symmetries of the Willmore equation (2) in Monge representation, but this matter will be clarified elsewhere. Recalling our comments concerning the invariance of the Willmore functional in subsection 3.2 it has been clear since 1973 that each symmetry generated by the vector fields presented in Table 1 is a variational symmetry of the Willmore equation (2). Thus, what is found in the sequent papers concerning the symmetries of this equation is that the vector fields in Table 1 are the generators of the largest point symmetry group admitted by the Willmore equation.

In closing, it is worth noting a result in [26] that each vector field $\mathbf{v} \in L_{SCT}$ can be mapped by a suitable inner automorphism (adjoint map) of the algebra L_{SCT} to one of the following representatives of conjugacy classes of one-dimensional subalgebras of this algebra

$$\begin{gathered}\langle \mathbf{v}_1\rangle, \quad \langle \mathbf{v}_4\rangle, \quad \langle \mathbf{v}_4 \pm \mathbf{v}_3\rangle, \quad \langle \mathbf{v}_7\rangle, \quad \langle \mathbf{v}_{10}\rangle \\ \langle \mathbf{v}_7 + a_1\mathbf{v}_4\rangle, \quad \langle \mathbf{v}_{10} + a_3\mathbf{v}_4\rangle, \quad \langle \mathbf{v}_7 + a_2\mathbf{v}_4 \pm \mathbf{v}_1\rangle \\ \langle \mathbf{v}_{10} + a_4\mathbf{v}_4 \pm \mathbf{v}_3\rangle, \quad \langle \mathbf{v}_{10} + a_5\mathbf{v}_4 + a_6\mathbf{v}_3 \pm \mathbf{v}_1\rangle\end{gathered} \tag{22}$$

where $a_1, \ldots, a_6$ are real numbers. Thus, these vector fields constitute an optimal system of one-dimensional subalgebras and the essentially different group-invariant solutions depending on one independent variable correspond to the symmetry groups generated by the vector fields (22). In parametric representation de Mateis and Manno proved in [4] that the optimal system of one-dimensional subalgebras consists of one representative and that there do not exist solutions of their equations invariant with respect to two-dimensional subalgebras.

References

1. Bernoulli D.: The 26th letter to Euler. In: Fuss, P.H. (ed.) Correspondence Mathématique et Physique de Quelques Célèbres Géomètres, vol. 2. St.-Pétersbourgh (1843)
2. Chen, B.-Y.: An Invariant of Conformal Mappings. Proc. Amer. Math. Soc. **40**, 563–564 (1973)
3. de Matteis G.: Group analysis of the membrane shape equation. In: Ablowitz, M., Boiti, M., Pempinelli, F., Prinari, B. (eds.) Nonlinear Physics: Theory and Experiment II, pp. 221–226. World Scientific, Singapore (2002). (see also cond-mat/0201044)
4. de Matteis, G., Manno, G.: Lie algebra symmetry analysis of the Helfrich and Willmore surface shape equations. Commun. Pur. Appl. Anal. **13**, 453–481 (2014)
5. Djondjorov, P., Vassilev, V., Mladenov, I.: Analytic description and explicit parametrisation of the equilibrium shapes of elastic rings and tubes under uniform hydrostatic pressure. Int. J. Mech. Sci. **53**, 355–364 (2011)
6. Garay Ö. J.: Riemannian submanifolds shaped by the bending energy and its allies. In: Suh, Y. J. (ed.) Proceedings of the 16th International Workshop on Differential Geometry and the 5th KNUGRG-OCAMI Differential Geometry Workshop, National Institute for Mathematical Sciences and Grassmann Research Group, pp. 55–68 (2012)
7. Ginsparg P. and Moore G., Lectures on 2D Gravity and 2D String Theory, hep-th9304011.
8. Grunau, H-Ch.: Nonlinear questions in clamped plate models. Milan J. Math. **77**, 171–204 (2009)
9. Helfrich W.: Elastic properties of lipid bilayers: theory and possible experiments. Z. Naturforsch **28c**, 693–703 (1973)
10. Ibragimov, N.: Transformation Groups Applied to Mathematical Physics. Nauka, Moscow (1983) (English transl., Riedel, Boston 1985)
11. Kleinert, H.: The membrane properties of condensing strings. Phys. Lett. B **174**, 335–341 (1986)
12. Konopelchenko, B.: On solutions of the shape equation for membranes and strings. Phys. Lett. B **414**, 58–64 (1997)
13. Libai, A., Simmonds, J.: The Nonlinear Theory of Elastic Shells. Cambridge University Press, Cambridge (1998)
14. Olver, P.: Applications of Lie Groups to Differential Equations. Graduate Texts in Mathematics, vol. 107, Second Edition edn. Springer, New York (1993)

15. Ou-Yang, Z.C., Helfrich, W.: Instability and deformation of a spherical vesicle by pressure. Phys. Rev. Lett. **59**, 2486–2488 (1987)
16. Ou-Yang, Z.C., Helfrich, W.: Bending energy of vesicle membranes: general expressions for the first, second, and third variation of the shape energy and applications to spheres and cylinders. Phys. Rev. A **39**, 5280–5288 (1989)
17. Ou-Yang, Z.C., Ji-Xing, L., Yu-Zhang, X.: Geometric Methods in the Elastic Theory of Membranes in Liquid Crystal Phases. World Scientific, Hong Kong (1999)
18. Ou-Yang, Z.C., Tu, Z.C.: Overview of the study of complex shapes of fluid membranes, the Helfrich model and new applications. Int. J. Mod. Phys. B **28**, 1330022 (2014)
19. Ovsiannikov L.: Group Analysis of Differential Equations. Nauka, Moscow (1978) (English transl. Ames, W.F. (ed.) Academic Press, New York (1982))
20. Poisson S.D.: Mèmoire sur les surfaces elastiques. Cl. Sci. Mathem. Phys. Inst. de France **2**, 167–225 (1812)
21. Polyakov, A.: The membrane properties of condensing strings. Phys. Lett. **B103**, 207–210 (1981)
22. Polyakov, A.: Fine structure of strings. Nucl. Phys. B **286**, 406–412 (1986)
23. Thomsen, G.: Grundlagen der Konformen Flächentheorie. Abhandlungen aus dem Mathematischen Seminar der Universität Hamburg **3**, 31–56 (1924)
24. Tu, Z.-C.: Challenges in the theoretical investigations of lipid membrane configurations. Chin. Phys. B **22**, 028701 (2013)
25. Tu Z.C., Ou-Yang Z.C.: Recent Theoretical Advances in Elasticity of Membranes Following Helfrich's Spontaneous Curvature Model. Advances in Colloid and Interface Science, 2014, in press.
26. Vassilev, V., Mladenov, I.: Geometric symmetry groups, conservation laws and group-invariant solutions of the Willmore equation. Proceedings of The Fifth International Conference on Geometry, Integrability and Quantization, SOFTEX, Sofia, pp. 246–265 (2004)
27. Vassilev V., Djondjorov P., Mladenov I.: Cylindrical equilibrium shapes of fluid membranes. J. Phys. A Math. Theor. **41** 435201 (16pp) (2008)
28. White, J.: A global invariant of conformal mappings in space. Proc. Amer. Math. Soc. **38**, 162–164 (1973)
29. Willmore T.: Note on Embedded Surfaces. An. Ş tiinţ. Univ. "Al. I. Cuza" Iaşi Seçt. Ia Mat. **11** 493–496 (1965)
30. Willmore, T.: Total Curvature in Riemannian Geometry. Wiley, New York (1982)

MIX
Papier aus verantwortungsvollen Quellen
Paper from responsible sources
FSC® C105338

If you have any concerns about our products,
you can contact us on
ProductSafety@springernature.com

In case Publisher is established outside the EU,
the EU authorized representative is:
Springer Nature Customer Service Center GmbH
Europaplatz 3, 69115 Heidelberg, Germany

Printed by Libri Plureos GmbH
in Hamburg, Germany